Ziggurat International Conference on Materials Science and Engineering (ZICMSE 2020)

IOP Conference Series: Materials Science and Engineering
Volume 987

London, United Kingdom
5 - 6 October 2020

ISBN: 978-1-7138-3300-0
ISSN: 1757-8981

Printed from e-media with permission by:

Curran Associates, Inc.
57 Morehouse Lane
Red Hook, NY 12571

Some format issues inherent in the e-media version may also appear in this print version.

Institute of Physics
Dirac House, Temple Back
Bristol BS1 6BE UK

Phone: 44 1 17 929 7481
Fax: 44 1 17 920 0979

techtracking@iop.org

Additional copies of this publication are available from:

Curran Associates, Inc.
57 Morehouse Lane
Red Hook, NY 12571 USA
Phone: 845-758-0400
Fax: 845-758-2633
Email: curran@proceedings.com
Web: www.proceedings.com

TABLE OF CONTENTS

CERAMICS, POLYMERS AND COMPOSITE MATERIALS IN CONSTRUCTION AND CIVIL ENGINEERING

MATERIALS FOR ENERGY GENERATION, STORAGE, AND UTILISATION

MATERIALS: SYNTHESIS, PROCESSING, AND DESIGNING

IOP Conf. Series: Materials Science and Engineering **987** (2020) 011001 doi:10.1088/1757-899X/987/1/011001

Preface

Published by IOP Conference Series: Materials Science and Engineering, indexed by Scopus (CiteScore 2019: 0.6) and organised by Ziggurat Publishing in London, the Ziggurat International Conference on Materials Science and Engineering (ZICMSE 2020) was held in London from 5th to 6th of October 2020.

ZICMSE 2020 offers a fantastic opportunity to meet and make new contacts in the field of Materials Science and Engineering. It allows delegates to have issues addressed in Materials Science by recognised global experts who are fully acquainted with the latest developments in the Materials Science field as well as information on new techniques and technologies. The ZICMSE 2020 features world-renowned keynote speakers, plenary speeches, young research forums, technical workshops, and career guidance sessions.

The high-profile keynote speakers of the ZICMSE 2020 are Prof Kenneth Dalgarno (Lead investigator for the EPSRC Centre for Doctoral Training in Additive Manufacture and 3D Printing, Newcastle University, UK), Prof Joydeep Dutta (Chair of Functional Materials, KTH Royal Institute of Technology, Stockholm, Sweden), and Prof Olga E. Glukhova (Chair of Radiotechnology and Electrodynamics, Saratov State University and Professor of I.M.Sechenov First Moscow State Medical University, Russia).

The conference attracted 243 researchers from the United Kingdom, United States, Italy, Lithuania, Oman, Iraq, United Arab Emirates, Egypt, Canada, Jordan, Russia, and China. Manuscripts were double-blind peer-reviewed by experts in the field of materials science and engineering. The acceptance rate of the conference was 59% as only high-quality academic research papers were accepted for publication. Thirty-three oral presentations were delivered over two days without parallel sessions. A total of 6 Medals (Ziggurat Publishing Medals) were awarded to the best three papers and the best three presentations in the conference. Female scientists and engineers are under-represented in global research. Ziggurat Publishing has argued the need for initiatives to increase their opportunities and participation, so we are delighted to announce an awards programme that aims to do both. Ziggurat Publishing's Certificates of Excellence were awarded to female researchers who are able to engage in Science, Technology, Engineering and Mathematics (STEM).

The ZICMSE was planned to take place in London, but due to the current worldwide pandemic COVID-19, the ZICMSE 2020 Organising Committee decided to keep the conference in place but as a virtual event. Organising an event such as this would not be possible without the help of many colleagues. Ziggurat Publishing is grateful to the members of the Organising Committee, the International Scientific Committee, our sponsors and all those colleagues who helped in the review of many full papers. This required meticulous attention to detail and strict adherence to very tight deadlines. However large or small a conference is, the effort required to make the local arrangements work for all is not insignificant. We are also indebted to all the contributors to the conference, particularly the researchers and practitioners.

ZICMSE is an annual conference. For more information about this year conference and the upcoming conferences, please visit the conference website: https://zigguratpublishing.com/index.php/zicmse

IOP Conf. Series: Materials Science and Engineering **987** (2020) 011001 doi:10.1088/1757-899X/987/1/011001

Organisers and Sponsors

https://zigguratpublishing.com

The conference is organised and sponsored by Ziggurat Publishing. Ziggurat Publishing is a scholarly open-access publisher based in London that has the mission to foster open knowledge exchange in all forms and across all disciplines. High-profile academic editors support our diverse, peer-reviewing, and open access journals. We serve scholars from around the world to ensure that the latest research is freely available. Journals published by Ziggurat Publishing are open access where all published articles are available free of charge from the moment of publication. All articles are distributed under the terms and conditions of the Creative Commons Attribution (CC BY) license 4.0. Ziggurat Publishing offers high-quality translation and academic editing and proofreading services in many languages.

IOP Conf. Series: Materials Science and Engineering **987** (2020) 011001 doi:10.1088/1757-899X/987/1/011001

Conference Organising Committee

Conference Chair

Prof Adel Sharif	The University of Surrey, Faculty of Engineering and Physical Sciences, United Kingdom

Organising Committee Members

Prof Adawiya Jumaa Haider	Department of Applied Sciences, University of Technology, Iraq
Prof Ing Abdulkader Kadauw	TU Bergakademie Freiberg, Department of IMKF, Freiberg, Germany
Prof Quanquan Han	Shandong University, China
Assist Prof Dr Hatef Sadeghi	School of Engineering, University of Warwick, United Kingdom
Prof Adnan Shamkhi Jabur	College of Engineering, University of Basrah, Iraq
Prof Dr Abdulraheem A. Ali	College of Materials Engineering, Babylon University, Iraq
Assist Prof Dr Qusiy Al-Galiby	Director of Scientific Affairs Dept, University of Al-Qadisiyah, Iraq

IOP Conf. Series: Materials Science and Engineering **987** (2020) 011002 doi:10.1088/1757-899X/987/1/011002

Peer review declaration

All papers published in this volume of *IOP Conference Series: Materials Science and Engineering* have been peer reviewed through processes administered by the Editors. Reviews were conducted by expert referees to the professional and scientific standards expected of a proceedings journal published by IOP Publishing.

- **Type of peer review: Double-blind**
- **Conference submission management system: OJS**
- **Number of submissions received: 56**
- **Number of submissions sent for review: 56**
- **Number of submissions accepted: 33**
- **Acceptance Rate (Number of Submissions Accepted / Number of Submissions Received X 100): 59%**
- **Average number of reviews per paper: 2**
- **Total number of reviewers involved: 100**
- **Any additional info on review process:**
- **Contact person for queries: Dr Zeyad Al-Shibaany (zeyad@zigguratpublishing.com)**

IOP Conf. Series: Materials Science and Engineering **987** (2020) 012001 doi:10.1088/1757-899X/987/1/012001

Thermophysical properties of [EMIM][BF_4] and [HMIM][PF_6] imidazolium ionic liquids with MWCNTs

Huda Majid Hasen[1]* and Prof. Dr Basma A. Abdulmajeed[2]

[1]Department of Chemistry, College of Science, University of Thi-Qar, Thi-Qar, Iraq
[2]Chemical Engineering Department, University of Baghdad, Baghdad, Iraq

*Huda.maged0907@coeng.uobaghdad.edu.iq

Abstract. In this study, several ionanofluids (INFs) were prepared in order to study their efficiency as a cooling medium at 25 °C. The two-step technique is used to prepare ionanofluid (INF) by dispersing multi-walled carbon nanotubes (MWCNTs) in two concentrations 0.5 and 1 wt% in ionic liquid (IL). Two types of ionic liquids (ILs) were used: hydrophilic represented by 1-ethyl-3-methylimidazolium tetrafluoroborate [EMIM][BF_4] and hydrophobic represented by 1-hexyl-3-methylimidazolium hexafluorophosphate [HMIM][PF_6]. The thermophysical properties of the prepared INFs including thermal conductivity (TC), density and viscosity were measured experimentally. The TC measurement showed an enhancement of about 3% for INF and of 1% MWCNT in [EMIM][BF_4] at a temperature of 298.15 K: the TC was 0.186 W/m.K, the kinematic viscosity was 100 centistokes (cSt), and the density was 1.283 g.cm^{-3}. On the other hand, the TC of 1% MWCNT in [HMIM][PF_6] INF enhanced by 5%. In this case, at a temperature of 298.15 K, the TC was 0.158 W/m, the kinematic viscosity was 1200 cSt, and the density was 1.294 g.cm^{-3}. Furthermore, the stability of the prepared INFs was measured using the zeta potential method after 28 days of preparation. The results show very good dispersion of the nanoparticles in the ILs for all the prepared INFs. The zeta potential was -69.30 mV and -45.34 mV for 0.5% and 1% MWCNT in [EMIM][BF_4], respectively. On the other hand, zeta potential was -51.78 and -46.67 mV for 0.5% and 1% MWCNT in [HMIM][PF_6], respectively. According to the obtained results, the preferable INFs to use as a cooling medium at 25 °C was the INF of 1 wt% MWCNT in [EMIM][BF_4], since it provides better thermophysical properties than the other prepared INFs.

1. Introduction

Water was the first medium used as a heat transfer fluid (HTF) due to its low cost, availability, safety, good thermal stability at the low-temperature range, and its thermophysical properties. However, water has its limitations in high-temperature applications due to its high vapor pressure and associated corrosion problems. Thus, it has been replaced by other liquids such as oils and ethylene glycol. There is an urgent need to replace the conventional cooling liquid in the old heat exchanger units. Researchers have started to look for a new efficient, clean, durable, and environmentally friendly substitute with high flash point characteristics. Several researchers evaluated the use of nanofluid (NF), which is a mixture of different types of nanoparticles in the base fluid. Nanofluids (NFs) are tested and used in various applications, such as solar energy units [1][2], car radiators [3], and heat exchangers (HE) [4]. NFs can be prepared using two techniques: in the one-step technique, nanoparticles are formed inside the base fluid directly producing NF, while the two-step technique includes the previous preparation of nanoparticles followed by dispersing them in the base fluid [5]. A few researchers have made

considerable efforts to prepare a surfactant-free NF of MWCNT in a different base fluid. For example, Rehman et al. used Jatropha seed oil as a base fluid using the two-step technique using an ultrasonic probe. This resulted in producing highly stable NF with TC enhancement of 6.76% [6].
NFs of carbon nanotube (CNT) in water or ethylene glycol represent good candidates for replacing conventional HTF. The disadvantage of these NFs are lower stability, higher vapour pressure, increased pumping power, which causes abrasion and attrition problems, and their high cost [7]. To overcome these limitations, NFs were developed to INF, which are a homogenous mixture of nanoparticles with ionic liquids (molten salts). New thermophysical properties were in the resulting ionanofluids, making them suitable to replace many liquids in different applications, especially in heat exchangers.
A few researchers have tried to prepare hybrid mixture of IL and base fluid and to disperse nanoparticles in this mixture. For example, dispersing Al_2O_3 in a $[C_2mim][CH_3SO_3]:H_2O$ mixture in a concentration range of 1–10 wt% causes a TC enhancement of about 3.9% - 10.2% [8].
The current age of manufacturing tends to use economical low-cost and environmentally friendly methods in many factories. Thus, the use of INFs as a cooling medium in heat exchangers could revolutionise the field of heat transfer. Ionanofluid could be considered as a two-phase system, since it contains the liquid phase represented by the ILs and the solid phase represented by the nanoparticles. INFs combine the advantages of nanoparticles and ILs. Nieto de Castro and colleagues for the first time called this type of fluid 'ionanofluids' [9][10], which have enhanced properties and negligible vapor pressure and can be designed for a specific application.

1.1. Carbon nanotubes
Carbon nanotubes (CNTs) are tiny microscopic tubes made of hexagonal graphene network rolled into cylindrical shells. This hexagonal shape is C_6H_6 molecule rings connected by single C-C and double C=C bonds. CNTs represent a suitable candidate for heat transfer enhancement due to their high stability level, great surface area, high aspect ratio, high thermal and electrical conductivity, good mechanical strength, and their low toxicity, making them environmentally friendly. The thermal conduction in carbon nanotubes is caused by phonons [11].
CNTs can be divided into two types: single-wall nanotubes (SWNTs) and MWNTs, depending on the number of shells forming the tubes. Furthermore, three different types of CNTs can be formed: armchair, zigzag, and chiral, depending on the graphene hexagonal network sheet rolling axis and the radius of the closing and the chiral vector, which discriminate CNTs into these shapes [12][13][14].

1.2. Ionanofluid history
Many researchers have studied in depth various specifications of INFs, such as their preparation, stability evaluation, prediction of thermophysical properties, application in different fields, and their suitability as a heat transfer medium. Castro et al. presented several publications about INFs in the heat transfer field, reaching a moderate thermal conductivity enhancement of about 2-9% [10]. Furthermore, in their work, comparing traditional HTFs and ionic liquids showed that the addition of 1 wt% of MWCNT in ionic liquid gave the produced INF superior thermophysical properties compared to base ILs [15]. Wang et al. found that increasing MWCNT concentration by more than 1 wt% caused a reduction in the TC because of aggregation forming [16]. Murshed et al. estimated that, using INFs of MWCNT, the heat transfer areas of shell and tube heat exchanger dispersed in two IL $[C_4mim][NTf_2]$ and $[C_2mim][EtSO_4]$ as HTFs and found that the heat transfer area decreased by 2.5% by using INF of 1 wt% MWCNTs [17]. Vieira et al. studied the TC of INFs prepared by dispersing MWCNTs in two types $[C_4mim][(CF_3SO_2)_2N]$ and $[C_2mim][EtSO_4]$ and reached an improvement of INFs of about 4-26% [18]. Ferreira et al. found that dispersing MWCNTs in different weight fractions and different ILs ($[C_6mim][BF_4]$, $[C_4mim][CF_3SO_3]$, $[C_4mpyrr][(CF_3SO_2)_2N]$, $[C_4mim][PF_6]$, $[C_6mim][PF_6]$) enhanced the TC 2–9% for the prepared INFs in comparison with the base ILs [19]. Furthermore, in other research, Ferreira et al. found that adding 0.1–0.2 wt% of MWCNT to ILs results in moderate TC enhancements of 0.4–1.4%. Murshed et al. found that increasing MWCNT concentration to 1% in INF led to improved thermophysical properties and an increased heat transfer area [20].

IOP Conf. Series: Materials Science and Engineering **987** (2020) 012001 doi:10.1088/1757-899X/987/1/012001

1.3. The objective
Although many researchers in different fields have studied NFs, INF research still requires further investigations, specially the INF type of MWCNT in IL. This work aims to prepare INFs by dispersing MWCNTs in two concentrations (0.5% and 1%) in two types of IL (the hydrophilic type [EMIM][BF_4] and the hydrophobic type [HMIM][PF_6]). Our objective is to find an INF heat transfer medium that has suitable thermophysical properties (i.e. thermal conductivity, density, and viscosity) and that is highly stable over a long period of operation without using any surfactant.

2. Experimental work
2.1. Materials
MWCNT with CAS number 308068 and a purity of 95% was purchased from the USA from Cheap Tubes Company. Table 1 displays the MWCNT's specifications as received from the company.

Table 1. The MWCNT's specifications [21].

Property	Value	Unit
Outer diameter	<8	Nm
Inner diameter	2-5	Nm
Ash content	<1.5	wt.%
Purity	95	%
Length	10-30	Um
Specific surface area	500	m2/g
Electrical conductivity	>100	S/cm
Bulk density	0.27	g/cm3
True density	~2.1	g/cm3

The ILs used in this work were purchased from Shanghai Cheng Jie Chemical Company in China. These ILs include 1-ethyl-3-methylimidazolium tetrafluoroborate [EMIM][BF_4] and 1-hexyl-3-methylimidazolium hexafluorophosphate [HMIM][PF_6]. Both of them have a purity greater than 99% and specifications as shown in Table 2.

Table 2. The IL's specifications [22].

Property	[EMIM][BF_4]	[HMIM][PF_6]
CAS number	(143314-16-3)	(304680-35-1)
Molecular weight	$C_6H_{11}BF_4N_2$	$C_{10}H_{19}F_6N_2P$
Purity	>99 %	>99 %
Glass Transition Temperature	13 °C	-78.05 °C
Structural formula	[EMIM][BF_4] 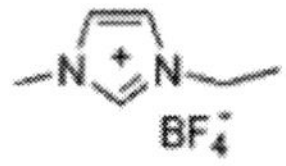	[HMIM][PF_6] 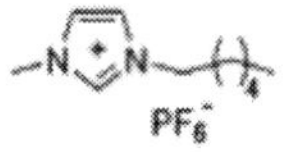

2.2. MWCNT characterisation
MWCNT can be characterised using different methods, including:

1- Raman spectra characterisation
2- Atomic force microscopy (AFM)
3- Transmission electron microscopy (TEM) characterisation
4- X-Ray diffraction (XRD)

2.3. Ionanofluids preparation
INFs are prepared using the two-step technique by directly dispersing MWCNTs separately in the two ionic liquids [EMIM][BF_4] and [HMIM][PF_6] in 0.5 and 1 wt%.
The weight of the nanoparticles and ionic liquids are measured using a highly sensitive electronic balance inside the laboratory hood to avoid air pollution. Following this, the dispersing process starts and includes two steps:
1- Using a magnetic stirrer for ten minutes at room temperature.
2- Sonicating the mixture using an ultrasonic probe homogeniser for 120 minutes.

2.4. Measurement of the thermophysical properties of INFs
Three thermophysical properties of INFs were measured, including the TC, density, and kinematic viscosity.

2.4.1. Thermal conductivity measurement.
The thermal conductivity measurement was carried out using a KD2 Pro apparatus. This device includes different sensors. The KS-1 sensor is a specialised sensor for measuring thermal conductivity/resistivity of liquid samples and insulating materials that have TC < 0.1 W/m·K. The KS-1 sensor must be inserted vertically into the liquid samples to prevent free convection. In addition, the KD2 Pro device must be calibrated using distilled water and glycerine before use.

2.4.2. Density measurement.
The density of the samples was measured using well-calibrated pycnometer with a size of 25 ml corrected to 24.766 ml volume capacity.

2.4.3. Viscosity measurement.
The kinematic viscosity of the samples was measured using capillary tube viscometers with different sizes (C, D, E, F, and G) immersed in a temperature-controlled water bath.

2.5. Zeta potential analysis
The Zeta Plus USA device was used to measure the stability of the prepared INFs. This device requires a very small amount of the sampling and gives accurate results within a short time and is suitable for dark colour samples.

3. Results and discussion
3.1. MWCNT characterisation results
3.1.1. Raman spectra characterisation.
The Raman spectra characterisation test of MWCNT nanoparticles was given by the company that produced it (Figure 1). The figure shows two bands: the disorder band (D-band) and the graphitic band (G-band). The ratio between the intensity of the D-band and the G-band, noted ID/IG, is related to the degree of disorder of the MWCNTs. An increase in ID/IG value corresponds to a higher proportion of sp3 carbon, which is generally attributed to the presence of more structural defects [23].
The G-band in the high-frequency region of the spectrum of 1610 cm^{-1} has an intensity of 5000 a.u. (arb. units), and the D-band of 1360 cm^{-1} has an intensity of 3500 a.u. Therefore, the ratio of ID/IG equals 0.7, which refers to low structure defects [24].

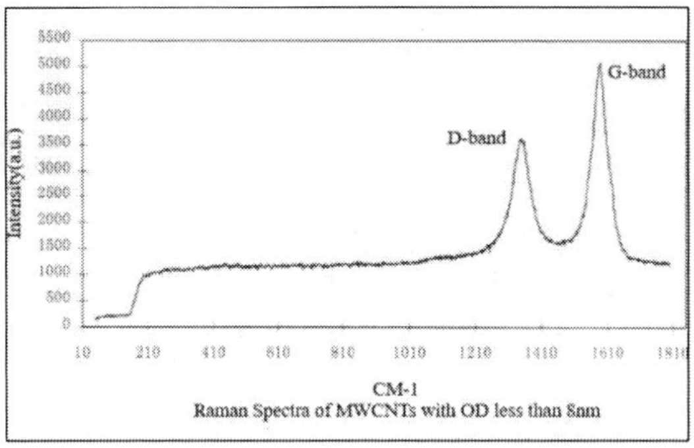

Figure 1. Raman spectra of MWCNT nanoparticles [21].

Figure 2. TEM image of MWCNT nanoparticles [21].

3.1.2. Transmission electron microscopy (TEM).
The TEM image shown in Figure 2 demonstrates the long tubular structure of MWCNTs tangled with each other. This type of structure is useful in increasing the efficiency of heat transfer.

3.1.3. X-ray diffraction (XRD).
The morphology of MWCNT was analysed using the XRD method. Figure 3 shows the XRD result, which displays a three-phase angle at the diffraction 2 theta; two maximum significant peaks at 25.7 and 44 °C refer to the carbon and graphitic structure of MWCNT. The third weak peak appears at 37.5 °C, which could be related to the types of catalyst used during the preparation of MWCNTs. Recent papers give the same result [25].

3.1.4. Atomic force microscopy (AFM).
The surface morphology of MWCNT nanoparticles can be seen using the AFM method. Figures 4 shows an AFM three-dimensional image for MWCNT nanoparticles. The sample size of MWCNTs was 21385*21385 nm. The AFM shows that the maximum apex of particles was 168.48 nm, the average roughness was 42.2 nm, the root mean square was 48.7 nm, the surface skewness was -0.0535, the surface kurtosis was 1.81, and the surface area ratio was 1.68.

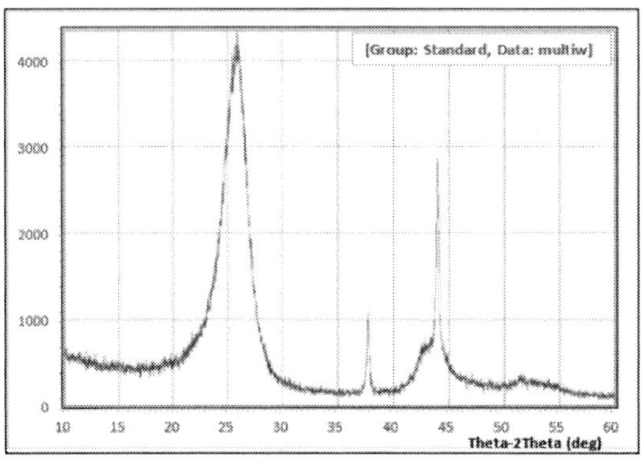

Figure 3. XRD of MWCNT at diffraction angle 2 theta.

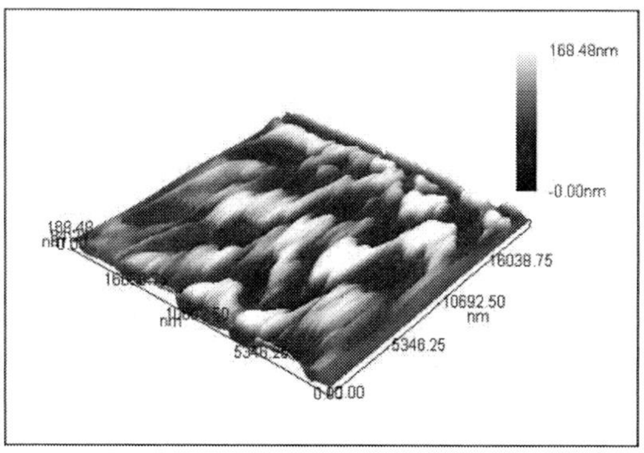

Figure 4. 3-D AFM image of MWCNT.

3.2. Thermophysical properties of ILs and INFs
3.2.1. Thermal conductivity measurement of ILs and INFs.
The TC of the ILs and INFs of concentrations 0.5% and 1% MWCNT are measured in the temperature range 293–330 K and atmospheric pressure. Figure 5 shows the TC of $EMIMBF_4$ and its INFs. It

indicates that the TC increased with increasing MWCNT concentrations. Also, TC increases slightly with temperature, reaching a maximum value of 0.189 w/m.k at 325 K compared with 0.18 w/m.k at 294 K for pure $EMIMBF_4$. This results in a TC improvement of 2.8% (about 3%) for 1% MWCNT INF. Figures 6 shows the TC of $HMIMPF_6$ and its INFs. It shows that the TC increased with increasing MWCNT concentration, with no variation with temperature increase. The maximum TC was 0.158 w/m.k for 1% MWCNT compared with 0.151 w/m.k for pure $HMIMPF_6$. This results in a TC improvement of about 4.6% (about 5%) for 1% MWCNT INF. Castro et al and Ferreira et. al. found nearly the same results of TC with a weak dependence on temperature, as illustrated in the literature [10][19].

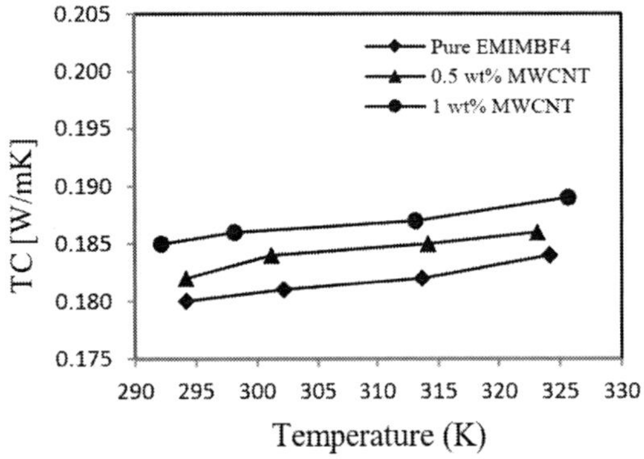

Figure 5. Effect of temperature and MWCNT concentration on the TC of $EMIMBF_4$ and its INFs.

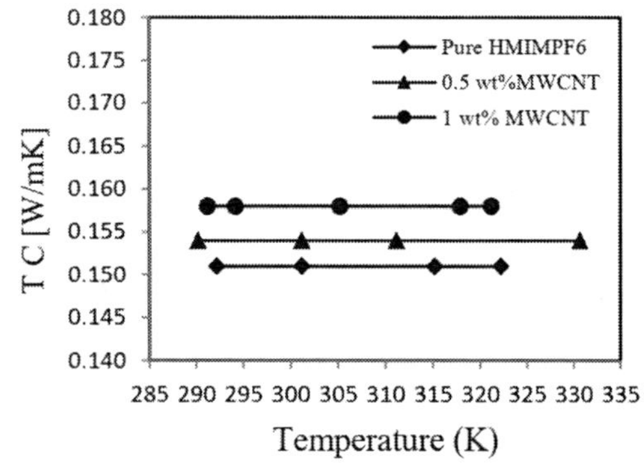

Figure 6. Effect of temperature and MWCNT concentration on the TC of $HMIMPF_6$ and its prepared INFs

3.2.2. *Density measurement of ILs and INFs.*
The density of the two ILs and their prepared INFs at a concentration of 0.5% and 1% were measured in the temperature range 20–80 °C and atmospheric pressure. The density of $EMIMBF_4$ increases with increases in MWCNT percentage by weight. On the other hand, the density of $EMIMBF_4$ and its INFs decreases linearly with temperature, as shown in Figure 7. The same behaviour occurs for the density of $[HMIM][PF_6]$, which increases with increases in MWCNT percentage by weight, as shown in Figure 8.

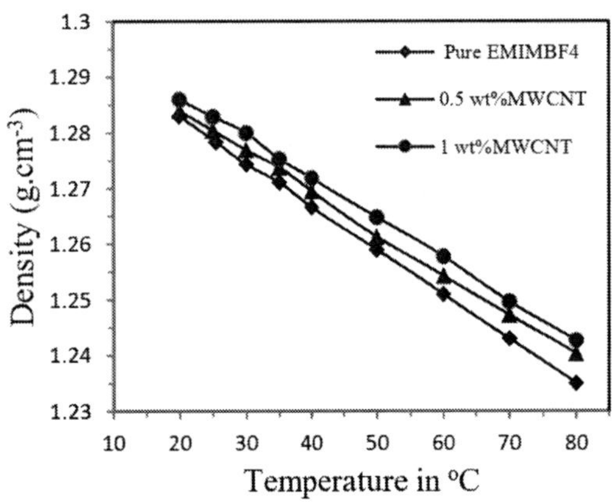

Figure 7. Effect of temperature and MWCNT concentration on the density of $EMIMBF_4$ of $EMIMBF_4$ and its INFs.

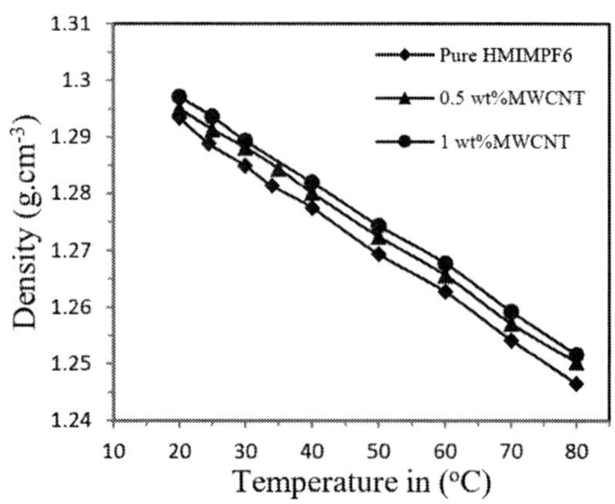

Figure 8. Effect of temperature and MWCNT concentration on the density of $HMIMPF_6$ and its INFs.

3.2.3. *Kinematic viscosity measurement of ILs and INFs.*
The kinematic viscosity of the two ILs and their INFs of the two concentrations of MWCNT nanoparticles were measured in the temperature range 20–90 °C and atmospheric pressure. In general, as shown in Figure 9, the kinematic viscosities of ILs and INFs decrease with temperature. The kinematic viscosity of the $EMIMBF_4$ and its INFs at low concentration decreased almost linearly with increasing temperature. At tge higher concentration of 1% MWCNT in [EMIM][BF_4], kinematic viscosity exhibits a sharp non-linear decline from 122.4 to 72.9 cSt for the temperature range 20–31 °C. Figure 9 also indicates that the kinematic viscosity of the [EMIM][BF_4] and its INFs increases with increases in MWCNT percentage by weight concentration. In Figure 10, the kinematic viscosity of the [HMIM][PF_6] and its INFs exhibits a sharp non-linear decrease with increasing temperature from 20 to 45 °C. This decrease is more pronounced for the high concentration sample of 1% MWCNT in [HMIM][PF_6], which shows a sharp decrease from 2055 to 699 cSt at temperatures range from 20 to 30 °C, respectively.

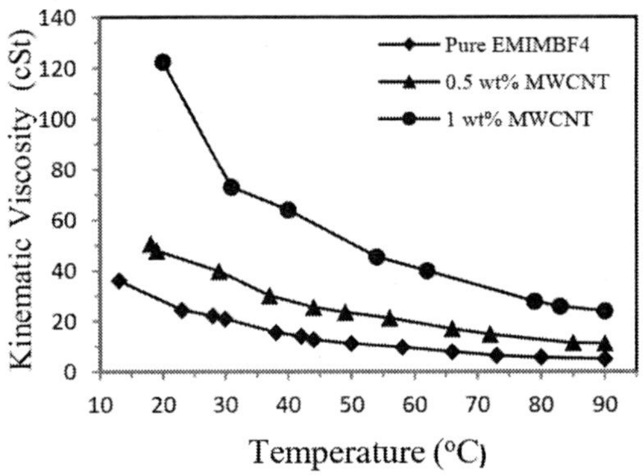

Figure 9. Effect of temperature and MWCNT concentration on the kinematic viscosity of [EMIM][BF_4] and its INFs.

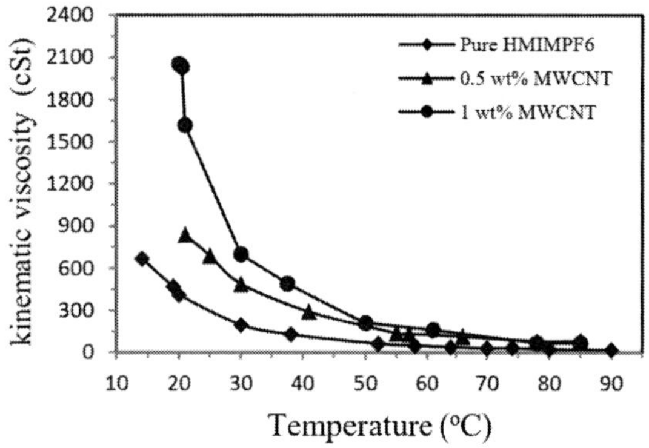

Figure 10. Effect of temperature and MWCNT concentration on the kinematic viscosity of [HMIM][PF_6] and its INFs

3.3. Zeta potential measurement
The stability of the produced INFs was detected using the zeta potential method. This test has been done for the prepared INFs, which includes two samples for $EMIMBF_4$ with 0.5% and 1% MWCNT, and two samples for $HMIMPF_6$ with 0.5% and 1% MWCNT on the 28th day of preparation. The obtained result of the zeta potential of $EMIMBF_4$ with 0.5% MWCNT was -69.30 mV and with 1% MWCNT was -45.34 mV, as shown in Figures 11 and 12, respectively, which evidences the good stability of these INFs.

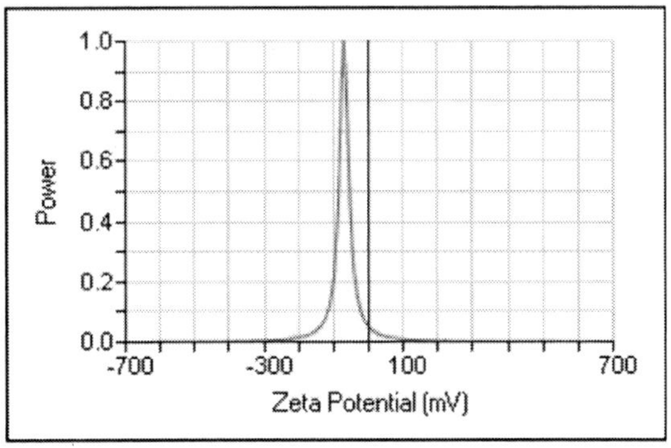

Figure 11. Zeta potential of 0.5% MWCNT in [EMIM][BF_4].

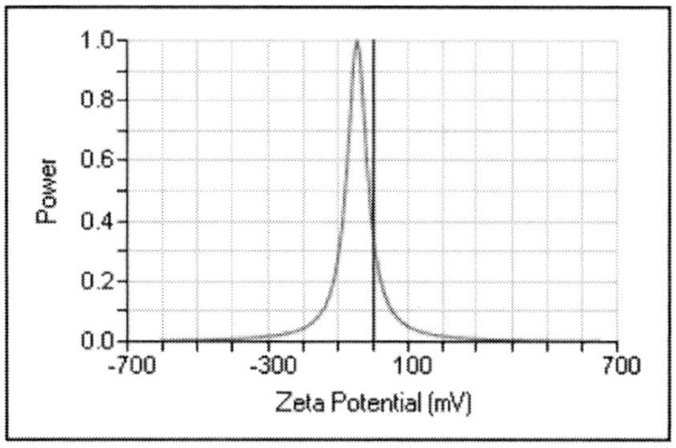

Figure 12. Zeta potential of 1% MWCNT in [EMIM][BF_4].

On the other hand, the zeta potential of 0.5% MWCNT in [HMIM][PF_6] was -51.78 mV and of 1% MWCNT was -46.67 mV, as shown in Figures 13 and 14, respectively, which further supports their good stability.

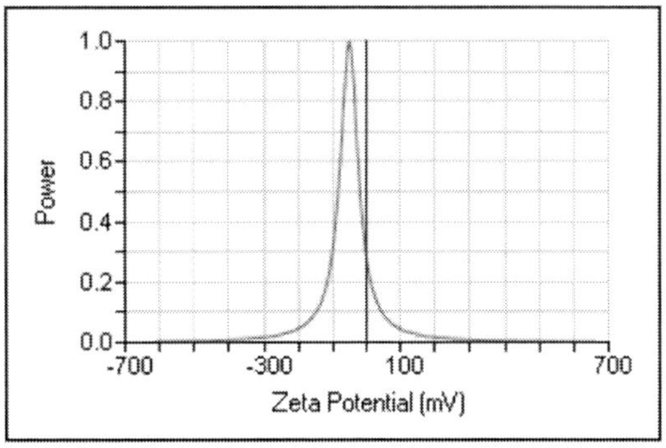

Figure 13. Zeta potential of 0.5% MWCNT in [HMIM][PF_6].

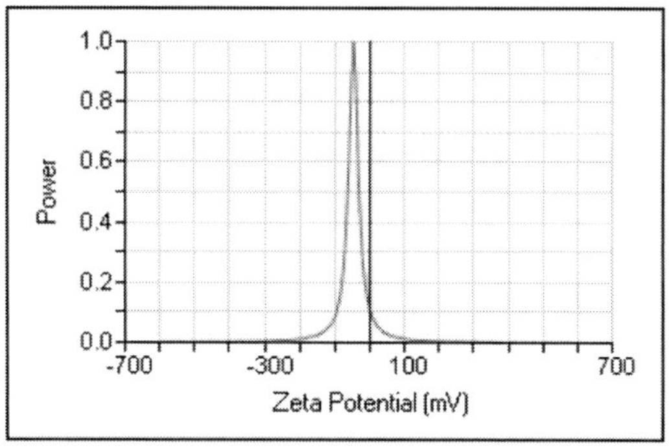

Figure 14. Zeta potential of 1% MWCNT in [HMIM][PF_6].

4. Conclusions

The thermophysical properties of HTF are an essential part for heat transfer applications in different fields. In this research, TC enhancement occurred at about 3–5% for the prepared INFs. A slight increase in the INF density values caused by the addition of MWCNTs. On the other hand, INF viscosity values increased largely with increases in MWCNT concentration in the low-temperature range but decreased to low values at the high-temperature range. This is especially clear for the prepared INFs of [HMIM][PF_6]. The experimental results also demonstrate the impact of different concentrations of nanomaterial upon physical properties. Furthermore, the zeta potential results show very good stability maintained for all the samplings over the tested period beyond the 28 days of preparation. Notably, no surfactant was used in the preparation of these INFs.

In conclusion, the results and data presented in this work are of primary use to heat exchanger designers to provide clear guidance to select heat transfer mediums that are suitable to comply with the heat exchanger design specifications of various exchanger applications, such as electronic and/or solar panel equipment.

5. References

[1] Karami M, Akhavan Bahabadi M A, Delfani S and Ghozatloo A 2014 A new application of carbon nanotubes nanofluid as working fluid of low-temperature direct absorption solar collector *Sol. Energy Mater. Sol. Cells* **121** 114–8

[2] Nerella S, Sudheer N.V.V.S and Bhramara P 2014 Enhancement of Heat Transfer in Solar collectors *International Journal of Innovations in Engineering and Technology* IJIET **3** 120-115

[3] Ganesan T and Kannan P S 2018 Heat-transfer enhancement of nanofluids in a car radiator *Mater Technol.* **52** 507–12

[4] Monjurul Ehsan M, Noor S, Salehin S and Sadrul Islam A K M 2015 *Thermofluid Modeling for Energy Efficiency Applications* (Elsevier Inc.) chapter 4 pp 101-73

[5] Dupont J 2011 From molten salts to ionic liquids: a "nano" journey *Acc. Chem. Res.* **44** 1223–31

[6] Rehman W, Mercian Z M A, Bhat A H, Guan B H, Sulaimon A A, Khan I U, Ullah H, Mukhtar A, Saqib S, Hameed A, Yasir M and Ahmed A 2020 Synthesis and enhancement of thermal conductivity of surfactant free MWCNTs-Jatropha oil-based nanofluid *Mater. Sci. Eng. C* **736**

[7] Tam N T, Phuong N V, Khoi P H, Minh P N, Afrand M, Trinh P V, Thang B H, Zyla G and Estelle P 2020 Carbon nanomaterial-based nanofluids for direct thermal solar absorption *J. Nanomater.* **10**

[8] Chereches E I, Chereches M I, Minea A A, Prado J I and Lugo L 2019 A numerical approach in the assessment of a new class of fluids performance in laminar flow *Mater. Sci. Eng. C* **591**

[9] Oster K, Hardacre C, Jacquemin J. Ribeiro A P C and Elsinawi A 2019 Thermal conductivity enhancement phenomena in ionic liquid-based nanofluids (ionanofluids) *Aust. J. Chem.* **72** 21-33

[10] Nieto De Castro C A, Lourenço M J V, Ribeiro A P C, Langa E, Vieira S I C, Goodrich P and Hardacre C 2010 Thermal properties of ionic liquids and ionanofluids of imidazolium and pyrrolidinium liquids *J. Chem. Eng. Data* **55** 653–61

[11] Dehsheikh H G, Najafi A, Zilabi F and Branch L 2018 Survey of Thermal Conductivity Carbon Nanotubes (TCCN) *J. Transp. Phenom. Nano-Micro Scale* **6** 1–12

[12] Jama M, Singh T, Gamaleldin S M, Koc M, Samara A, Isaifan R J and Atieh M A 2016 Critical review on nanofluids: preparation, characterization, and applications *J. Nanomater.* **2016** 1–22

[13] Choi S U S and Eastman J A 1995 Enhancing thermal conductivity of fluids with nanoparticles *Am. Soc. Mech. Eng., Int. Mech. Eng. Congress & Exposition* (San Francisco) CA.

[14] Zhang M and Li J 2009 Carbon nanotube in different shapes *Mater. Today* **12** (6) 12–8

[15] Nieto de Castro C A, Sohel Murshed S M, Lourenço M , Santos F J V, Manuel L. M, Lopes M M and Franca J M P 2012 Ionanofluids: new heat transfer fluids for green processes development *Green Solvents I Properties and Applications in Chemistry (springer)* ed E (London) chapter 8 pp 233-49

[16] Wang B, Hao J, Li Q and Li H 2014 New insights into thermal conduction mechanisms of multi-walled carbon nanotube / ionic liquid suspensions *Int. J. Therm. Sci.* **83** 89–95

[17] Sohel Murshed S M, Nieto de Castro C, Lourenço M J V, França J M P, Ribeiro A P C, Vieira S I C and Queirós C S 2011 Ionanofluids as novel fluids for advanced heat transfer applications *World Acad. Sci. Eng. Technol International journal of physical and mathematical sciences.* **5** (4) 795–8

[18] França J M P, Vieira S I C, Lourenço M , Sohel Murshed S M and Nieto de Castro C 2013 Thermal conductivity of [C4mim][(CF3SO2) 2N] and [C2mim][EtSO4] and their ionanofluids with carbon nanotubes: experiment and theory *J. Chem. Eng. Data* **58** (2) 467–76

[19] Ferreira A G M, Simões P N, Ferreira A F, Fonseca M A, Oliveira M S A and Trino A S M 2013 Transport and thermal properties of quaternary phosphonium ionic liquids and ionanofluids *J. Chem. Thermodyn.* **64** 80–92

[20] Franca J M P, Lourenco M J V, Sohel Murshed S M, Padua A A H and Nieto de Castro C A 2018 Thermal conductivity of ionic liquids and ionanofluids and their feasibility as heat transfer fluids *Ind. Eng. Chem. Res.* **57** (18) 6516–29

[21] Cheap Tubes Inc (n d) *Multi walled carbon nanotubes products* (Grafton, USA) from https://www.cheaptubes.com/product-categort/multi-walled-carbon-nanotubes

[22] Shanghai Cheng Jie Chemical Co. LTD (n d) Product Center *Imidazole type ionic liquid* (Shanghai Pudong) from http://www.shyfhx.com/work/Default.aspx

[23] Abuilaiwe F A, Laoui T, Al-Harthi M and Atieh MA 2010 Modification and functionalization of multiwalled carbon nanotube (MWCNT) via FISCHER esterification multiwalled carbon nanotube (MWCNT) *Arab J SCI and Eng* **35** 36-48

[24] Janudin N, Chuah A L, Abdullah N, Yasin F M, Saidi N M and Kasim N A M 2017 Comparison and characterization of acid functionalization of multi walled carbon nanotubes using various methods *Mater. Sci. Eng. C* **264** 83–86

[25] Arunkumar T, Ramachandran K, Subramani R R,Viswanathan K and Anish M 2020 Synthesis and characterisation of multi walled carbon nanotubes (MWCNT) *Int. J. Ambient Energy* **41** 452-456

IOP Conf. Series: Materials Science and Engineering **987** (2020) 012002 doi:10.1088/1757-899X/987/1/012002

Cement Concrete Modular Pavement Implementation for Pedestrian and Bicycle Path

A Vaitkus[1], J Gražulytė[1], V Vorobjovas[2], O Šernas[1] and L Žalimienė[1]

[1] Road Research Institute, Vilnius Gediminas Technical University, Linkmenų str. 28, 08217 Vilnius, Lithuania
[2] Department of Roads, Vilnius Gediminas Technical University, Saulėtekio ave. 11, 10223 Vilnius, Lithuania

viktoras.vorobjovas@vgtu.lt

Abstract. Cement concrete modular pavements, also known as precast concrete pavements, designed for pedestrian and bicycle paths is the advanced next-generation technology characterised as high quality, durable, quickly built and easily maintained. Polypropylene fibre reinforced concrete mixture is more frequently used to improve the properties of concrete slabs and to reduce thickness. The thickness of the slab could be reduced to 8–10 cm using the appropriate fibre. This paper focuses on the selection of slab dimensions and verification of fibre reinforced concrete mixture production method when fibres are added to the factory-made concrete mixture (available from the concrete plant) and evaluation of different surface texture of concrete slabs produced by using special texture mats. Seven different combinations of concrete slabs have been developed, constructed and tested in 21 m length pedestrian pathway in Vilnius city. The analysis of mechanical and surface characteristics of these slabs are presented in this paper. The research revealed that the most suitable (pendulum test values ≥ 55) surface texture for the pedestrian pathway is formed by using texture mats.

1. Slab dimensions selection for cement concrete modular pavement

Pedestrian and bicycle paths are an important part of road infrastructure that provides mobility for pedestrians and cyclists in urban areas and nearby. The pavement of pedestrian and bicycle paths should be stable, firm, even, relatively smooth but slip resistant to ensure comfort mobility. Pedestrian and bicycle paths should be incorporated in the surrounded area in order to ensure harmony, express the local community's individuality and improve safety and comfort in urban areas. These issues are achieved using coloured (non-black) asphalt concrete and different laying patterns of interlocking concrete paver blocks. However, these solutions are old-fashioned, and they do not reflect the needs of nowadays society and innovativeness. Previous research has shown [1–2] that cement concrete modular pavements, also known as precast concrete pavements, are suitable for pedestrian and bicycle paths, and successfully address these problems if they are properly designed.

Cement concrete modular pavement consists of prefabricated concrete slabs that are transported to the project site only after the curing period when the desirable concrete strength is achieved and installed on a prepared foundation [3]. Fabrication of slabs in a plant enables the achievement of different pavement texture and style. That is one of the most important factors in the design of pedestrian and bicycles paths in urban areas. In addition to other advantages of slabs fabrication in a plant such as better concrete quality, controlled concrete curing conditions, elimination of materials segregation etc., concrete modular pavements are easily removable, mechanically independent. That is the crucial aspect of getting access to underlying utilities for repair and replacement [1–3].

In accordance with the Lithuanian Roads Technical Regulation KTR 1.01:2008 "Roads", the width of the pavement of general bicycle and pedestrian paths shall be at least 2.5 m (joint) on public roads or nearby. The minimum width of the pedestrian path depends on the number of pedestrians during traffic congestion hours, and the minimum width of the pedestrian path shall be 2.0 m if separate, and the width may be reduced to 1.5 m or 1.0 m in justified cases. The minimum width of the bicycle path shall be 2.5 m, and the width may be reduced to 2.0 m if separate in justified cases. Under the Lithuanian technical regulation for construction STR 2.06.04:2014 "Streets and local roads. General Requirements" on streets and local roads or near to them, the width of the bicycle path shall be at least 1.0 m, if the cycle path is with one lane or at least 2.0 m if the path is with two lanes. The minimum pavement width for common cycling and walking paths shall be 2.3 m. The minimum width of the pavement shall be 1.5 m. Thus, the width of the pedestrian paths should not be less than 1.5 m. The width of bicycles paths with one and two lanes in streets, local roads or along them should be at least 1.0 m and 2.0 m, respectively. Due to that, and ensuring easy transportation, handling and installation it was found that optimal dimensions of concrete slab are 1.0 m in length and 1.5 m in width. The slab dimensions have been selected considering not only predominant loads but also slab extraction from precasting moulds and handling with a vacuum lift.

The objective of this paper is to produce, construct and test concrete modular pavement with different surface texture characteristics and to identify the most suitable one for pedestrian and bicycle paths.

2. Concrete mixture selection for cement concrete modular pavement

Previous research has shown that typically the thickness of concrete slab for roads is 20–30 cm [3] and for the production of these slabs are used C30/37–C40/50 class concrete (compressive strength 27.5–40.0 MPa after 28 days respectively) [4]. However, the weight of the concrete slab of this thickness is too heavy for extraction from precasting moulds by hand and handling with a vacuum lift. Also, the pavement of the pedestrian and bicycle path is subjected to significantly lower loads than pavements of roads. Thus, there is no need to install 20–30 cm thickness concrete slab in the pedestrian and bicycle path. Moreover, the thickness of interlocking concrete paver blocks on pedestrian and bicycle paths is typically 8–10 cm. Consequently, based on selected optimal dimensions of the concrete slab (1.0 m in length and 1.5 m in width) preferable thickness of concrete slab has been calculated that is 5–6 cm. In this point, the question of ensuring a necessary bearing capacity and resistance to cracking during transportation and handling has arisen.

More than 20 years ago, Wafa [5] concluded that using fibres for concrete reinforcement, the thickness of the slabs could be reduced about twice in comparison with plain concrete slabs. Kustermann and co-authors [6] investigated fibre reinforced concrete under high dynamic impact loading. It has been demonstrated that the thickness of the slab could be reduced to 8–10 cm using the appropriate fibre. Fibre reinforced concrete is widely used in civil engineering because fibres are useful in providing greater resistance to cracking of plastic shrinkage and service related cracking. Steel fibres are often used to improve the flexural toughness of concrete while synthetic (polypropylene) fibres are more often used to reduce crack opening due to shrinkage. Buratti and co-authors [7] observed that the performance of steel fibres reinforced concrete is better than macro synthetics reinforced concrete. Jayakumar and co-authors [8] concluded that the flexural strength of propylene fibres reinforced concrete with long and short fibres is higher than in the case of concrete reinforced with short polypropylene and long steel fibres. In addition, it has been concluded that toughness of fibres reinforced concrete is better using long fibres combinations than short fibres and postcrack toughness is better in case of long steel and long polypropylene fibres combinations. Amin and co-authors [9] investigated macro synthetic fibre reinforced concrete. It has been demonstrated that the resistance to stress increased when polypropylene fibres were used during the flexural strength test with increased displacement. Guerini and co-authors [10] observed that stiffness of long fibres (steel and polypropylene) reinforced concrete is better than short fibres reinforced concrete.

Furthermore, it was observed that smoothness of polypropylene reinforced concrete surface is better than steel fibres reinforced concrete, but steel fibres influenced concrete workability less as compared to polypropylene fibres. As the density of polypropylene fibres is about 9 times lower than that of metal fibres, as well as the metal fibres are heavier, there are fewer metal fibres in the volume of the concrete mixture, which makes it easier to mix the concrete mixture.

Based on the literature analysis, it has been selected to reinforce concrete with short and long polypropylene fibres for the production of pedestrian pathway concrete slabs.

Generally, a fibre reinforced concrete mixture is produced at the factory site by adding fibres to a dry mixture of aggregates, binder and additives before adding water. The objective of this paper is to evaluate fibre reinforced concrete mixture production method, when fibres are added to factory-made C30/37–C40/50 class concrete mixture (available from the concrete plant).

3. Experimental research

3.1. Test site and materials

An existing pedestrian pathway located nearby Road Research Institute of Vilnius Gediminas Technical University building in Vilnius (Lithuania) was selected for the installation of concrete modular pavement. The width of this pedestrian pathway is 1.5 m, so it was suitable for the installation of concrete slabs. It was selected a 21 m long straight section. The pavement of selected pedestrian pathway was concrete blocks. So, it was decided to remove these concrete slabs and to install a prefabricated concrete slab after levelling and pre-compacting the existing base course.

Based on literature analysis, it was selected 0.05 m thickness, 1.00 m length, and 1.50 m width concrete slabs. All concrete slabs were produced at the factory using factory-made concrete mixture, which properties are given in table 1. Factory-made concrete mixture was transported to modules pavement production site and reinforced with micro and macro polypropylene fibres in 400 l volume mixing machine. The properties of micro and macro polypropylene fibres are given in table 2. The amount of micro and macro fibres was 0.9 kg/m^3 (0.43% by volume) and 4 kg/m^3 (0.01% by volume) respectively, as it was recommended by the producer. The mixing of fibre reinforced concrete mixture and the addition of fibres were performed according to the instructions of fibre producers. Fibres were added to factory-made concrete mixture. Thus, the processes of design and control of fibre reinforced concrete mixture were skipped in order to test the simplicity and speed of production of such fibre reinforced concrete in the construction site. All slabs were compacted in prepared moulds with needle vibrator.

The surface of 14 concrete slabs was produced by glueing a synthetic special texture mat (see figure 1) into the concrete mould. Eight concrete slabs were made with a rough surface texture using special texture mat (see figure 1 (a)) and two of them were coloured with red pigment. Six concrete slabs were made with less rough surface texture using special texture mat (see figure 1 (b)) and two of them were coloured with red pigment. The surface of two concrete slabs was coloured with red powder. Others concrete slabs were produced typically – smooth texture and grey colour. Overall, there were produced 20 concrete slabs (see table 3).

Table 1. Specification of factory-made concrete mixture used in this study according to standard EN 206

Property	Class
Compressive strength	C35/45
Exposure of corrosion induced by carbonation	XC4
Exposure of corrosion induced by chlorides other than from sea water	XD3
Exposure of freeze/thaw attack	XF4
Chloride content	Cl0.2

IOP Conf. Series: Materials Science and Engineering **987** (2020) 012002 doi:10.1088/1757-899X/987/1/012002

Table 2. Characteristics of micro and macro fibres used in this study

Characteristics	Type or Values	
Fibre Type	Polypropylene	
Description	Micro	Macro
Shape	Smooth	
Length l, mm	13	40
Diameter Ø, mm	0.022	0.7
Aspect Ratio l/ Ø	590	57
Tensile Strength, MPa	380	500
Elastic Modulus, GPa	–	6
Density, kg/m^3	905	922

(a) (b)

Figure 1. Rough (a) and less rough (b) surface texture mats (www.companero.biz)

Table 3. Type of concrete slabs

No.	Quantity	Surface texture and colour
From 1 to 6	6	Rough texture, grey (natural) colour
From 7 to 10	4	Less rough, grey (natural) colour
From 11 to 12	2	Smooth texture, grey (natural) colour
From 13 to 14	2	Smooth texture, coloured using red powder
From 15 to 16	2	Rough texture, red colour pigment
From 17 to 18	2	Less rough texture, red colour pigment
From 19 to 20	2	Smooth texture, grey (natural) colour

Note: The numbering of concrete slabs is in accordance with the installation sequence on the test site.

IOP Conf. Series: Materials Science and Engineering **987** (2020) 012002 doi:10.1088/1757-899X/987/1/012002

(a) (b) (c)

Figure 2. Prefabricated cement concrete slabs (a), their installation using vacuum lift (b) and constructed pedestrian pathway (c)

At factory site produced and from precasting moulds extracted, then transported (after 28 days of concrete curing) to test site concrete slabs were installed using vacuum lift in early fall of 2019 (see figure 2).

3.2. Test methods

During the production of concrete slabs, it was prepared 114 cubic specimens, 76 beam specimens and 76 cylindrical specimens. The compressive, flexure and tensile splitting strength of specimens made of the concrete mixture without and with fibres were determined.

Three cubic specimens of 100 × 100 × 100 mm in size were prepared for each of concrete mixtures. Compression tests were performed after 28 days of concrete curing. Compressive strength was determined by standard EN 12390-3.

Flexural strength was determined by standard EN 12390-5 for each of concrete mixtures. The four-point bending tests were performed on beam specimens sized 100 × 100 × 400 mm (two specimens for each concrete mixture), with a span length of 300 mm. Flexural strength tests were performed after 28 days of concrete curing.

Tensile splitting strength was determined by standard EN 12390-6 for each of concrete mixtures. Tests were performed on cylindrical specimens sized 100 mm in length and diameter after 28 days of concrete curing.

After installation of concrete slabs in the test site, measurements of slip/skid resistance of a cement concrete modular pavement surface were performed by standard EN 13036-4 using pendulum device on the left, axis and right side of each concrete slab. In addition, measurements of surface texture characteristics were done at two randomly selected locations of each concrete slab:

- mean texture depth by standard EN 13036-1 using a volumetric patch technique;
- mean profile depth by standard ISO 13473-1.

4. Results and analysis

4.1. Mechanical characteristics

Results of compressive, flexural and tensile splitting strength for all concrete mixtures are presented in figures 3–5. A box and whisker plot also called a box plot, the analysis was performed, which displays the minimum, first quartile, median, mean, third quartile, and maximum. A level of 95% of confidence was applied.

From the results of compressive strength (see figure 3), it is evident that the usage of fibres for concrete reinforcement reduces the compressive strength. The average compressive strength of plain concrete is 56.13 MPa while of fibre reinforced concrete is 39.45 MPa (reduction is about 30%). The results of compressive strength of plain and fibre reinforced concrete vary in the similar range: 28.4 MPa (from 43.7 MPa to 72.1 MPa) and 41.3 MPa (from 24.1 MPa to 65.4 MPa) respectively.

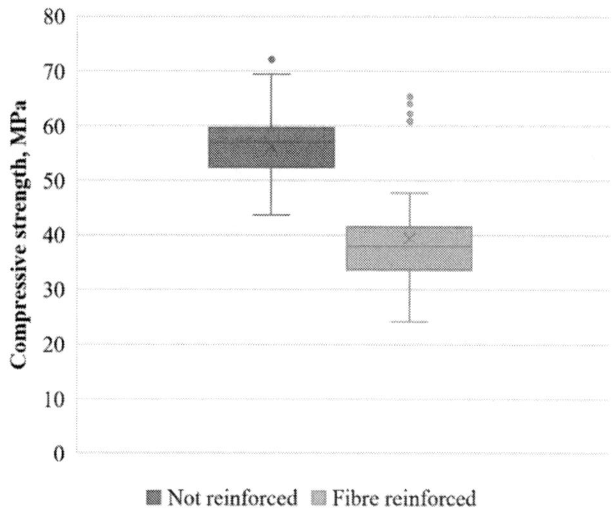

Figure 3. Results of compressive strength

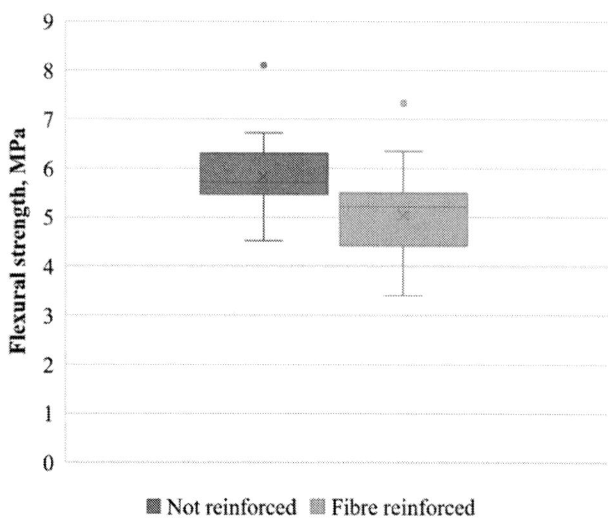

Figure 4. Results of flexural strength

The results of flexural strength are presented in figure 4. The analysis shows that fibres reduce the flexural strength of concrete, similar to that was found in the compression test. The average flexural strength of plain concrete is 5.83 MPa while fibre reinforced concrete revealed the flexural strength of 5.05 MPa (reduction is about 13%). The results of flexural strength of plain and fibre reinforced concrete vary in these ranges: 3.59 MPa (from 4.52 MPa to 8.11 MPa) and 3.92 MPa (from 3.41 MPa to 7.33 MPa) respectively, and both presents outliers above the third quartile.

Tensile splitting strength results (see figure 5) show that the average tensile splitting strength of fibre reinforced concrete (3.03 MPa) is 7% lower than that of plain concrete (3.27 MPa), and this decrease is much lower than that was found during compressive and flexural tests. Also, the results of tensile splitting strength of plain concrete vary in 1.8 times less range than that of fibre reinforced concrete: 1.48 MPa (from 2.42 MPa to 3.90 MPa) and 2.65 MPa (from 2.02 MPa to 4.67 MPa) respectively.

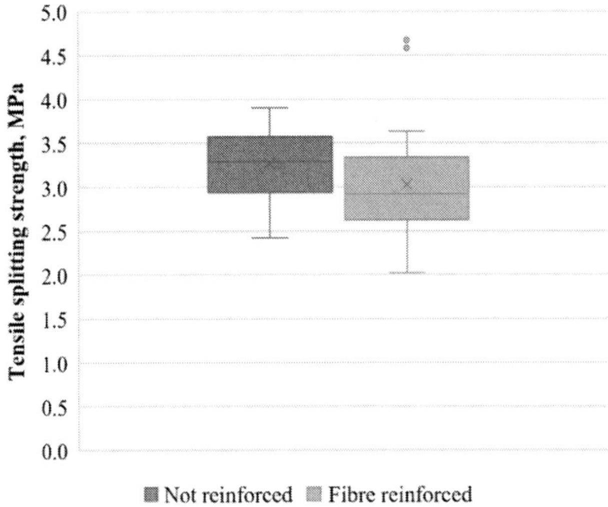

Figure 5. Results of tensile splitting strength

Table 4. Descriptive statics of the mechanical properties (values in MPa)

Mechanical property	Concrete reinforcement	Min	Max	Mean	Standard deviation
Compressive strength	Not reinforced	43.7	72.1	56.13	6.24
	Fibre reinforced	24.1	65.4	39.45	9.83
Flexural strength	Not reinforced	4.52	8.11	5.83	0.68
	Fibre reinforced	3.41	7.33	5.05	0.89
Tensile splitting strength	Not reinforced	2.42	3.90	3.27	0.38
	Fibre reinforced	2.02	4.67	3.03	0.66

Table 4 summarises the results on compression, flexural and tensile splitting tests concrete samples considered in this study. The experimental results showed that mechanical properties, like compressive, flexural and tensile splitting strengths, of fibre reinforced concrete is about 16% lower than that of plain concrete. This fact is proved in the scientific papers [8, 10–12], no significant differences can be observed or mechanical properties, like compressive strength, of fibre reinforced concrete can be even lower than that of plain concrete. In addition, it was observed that after flexural and tensile splitting tests the fibre reinforced concrete specimens split into half (see figure 6), while the fibre reinforced concrete specimens stayed on the fibres after fracture and cracking. So, this indicates sufficient adhesion between fibres and concrete. This property has been confirmed by the research of other scientists [13–15], that there is an appreciable increase in postcrack energy absorption capacity and ductility due to the addition of fibres. Table 4 showed that standard deviations of fibre reinforced concrete mechanical properties is about 54% higher than that of plain concrete. This can be explained by inefficient mixing method of fibres in the concrete mixture. The obtained effect is negative. Therefore, in the production of fibre reinforced concrete mixture, the priority should be given to the addition of fibres before adding water. Also, it is important that the fibres would be dispersed uniformly throughout the mixture. Introduction of the fibres through a wire mesh basket will help even distribution of fibres. Moreover, one of the reasons for poor mixing may be too small fibres content, which was selected according to the fibres manufacturer's recommendations for the minimum quantity. By increasing the minimum amount of fibres specified by the manufacturer, it is likely that better properties of the fibre reinforced concrete mix could be achieved. The latter insight is intended to be researched by the authors in other studies.

Figure 6. A crack in the polypropylene fibre reinforced concrete specimen after the flexural test

After the use of path for 10 months, even with the winter maintenance activities (de-icing with sodium chloride and lightweight snow removal machinery) all slabs does not have any distresses or defects.

4.2. Surface characteristics

Surface characteristics such as mean profile depth, mean texture depth and skid resistance play the main role in pedestrian safety. Figure 7 illustrates the skid resistance of cement concrete modular pavement made of different type of concrete slabs. The analysis of pendulum test values indicates that concrete slabs No. 11, No. 12, No. 13, No. 14, No. 19 and No. 20 with smooth (texture was not formed using texture mats) texture has lower pendulum test values (vary from 38 to 50) than others (vary from 54 to 71). In addition, it was observed that pendulum test values of concrete slabs, which surface was coloured with red powder (concrete slabs No. 13 and No. 14), are lower than 40. And, no significant differences can be observed comparing pendulum test values of concrete slabs, which surface texture was formed using different texture mats (rough and less rough texture).

Results of mean profile depth are presented in figure 8. A mean profile depth of the concrete slabs, which surface was formed using texture mat with rough texture, was about 0.08 mm. The concrete slabs with a smooth texture and concrete slabs, which surface was formed using texture mat with less rough texture, have the lowest mean profile depth (lower than 0.05 mm). It is averagely 0.04 mm, while the mean profile depth of concrete slabs, which surface was coloured with red powder, varies from 0.09 mm to 0.13 mm.

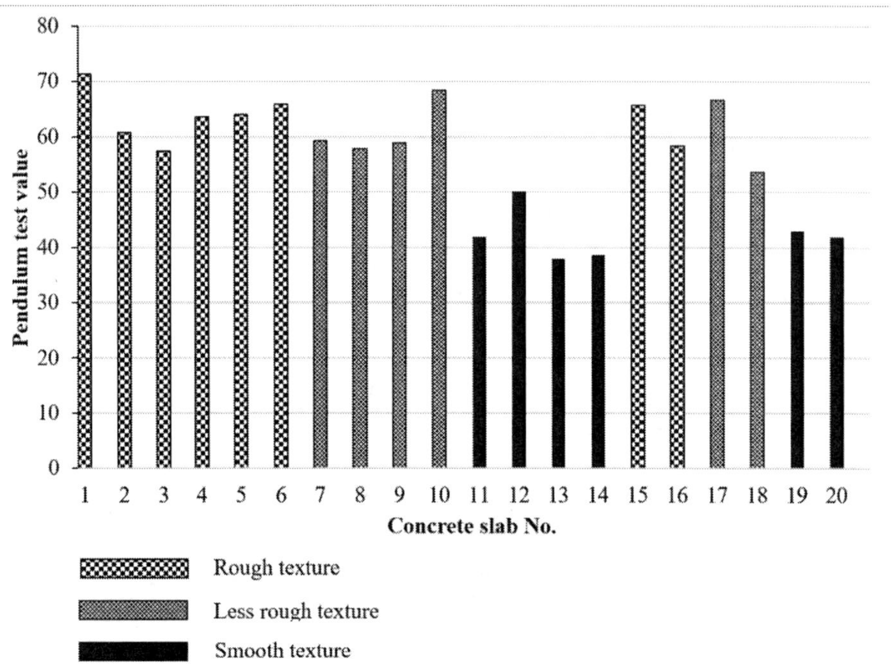

Figure 7. Results of skid resistance

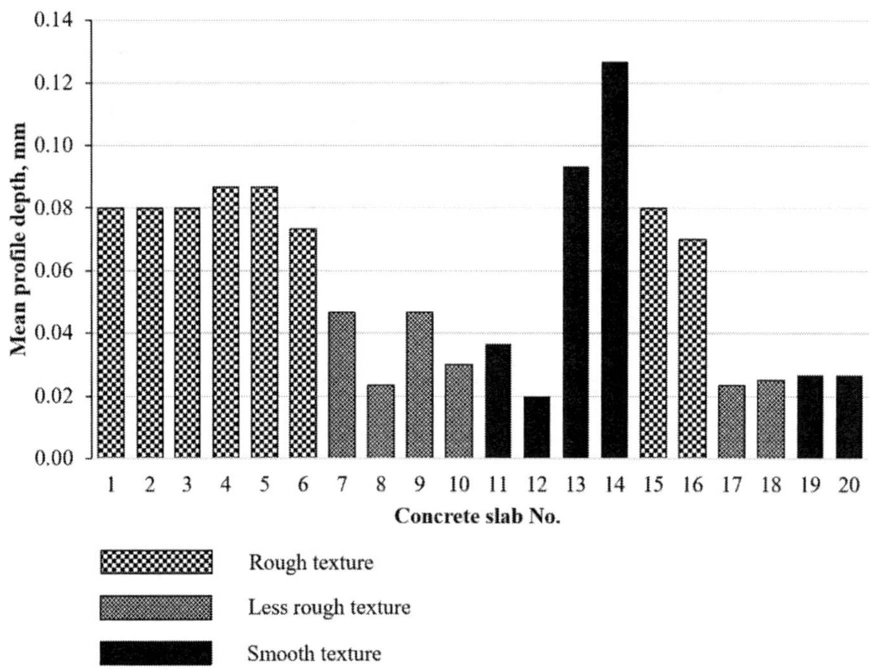

Figure 8. Results of mean profile depth

Results of mean texture depth are presented in figure 9. Mean texture depth of the concrete slabs, which surface was formed using texture mat with rough texture, was about 0.27 mm. The concrete slabs with a smooth texture and concrete slabs, which surface was formed using texture mat with less rough texture, have the lowest mean profile depth. It varies from 0.22 mm to 0.24 mm, while the mean profile depth of concrete slabs, which surface was coloured with red powder, varies from 0.27 mm to 0.30 mm.

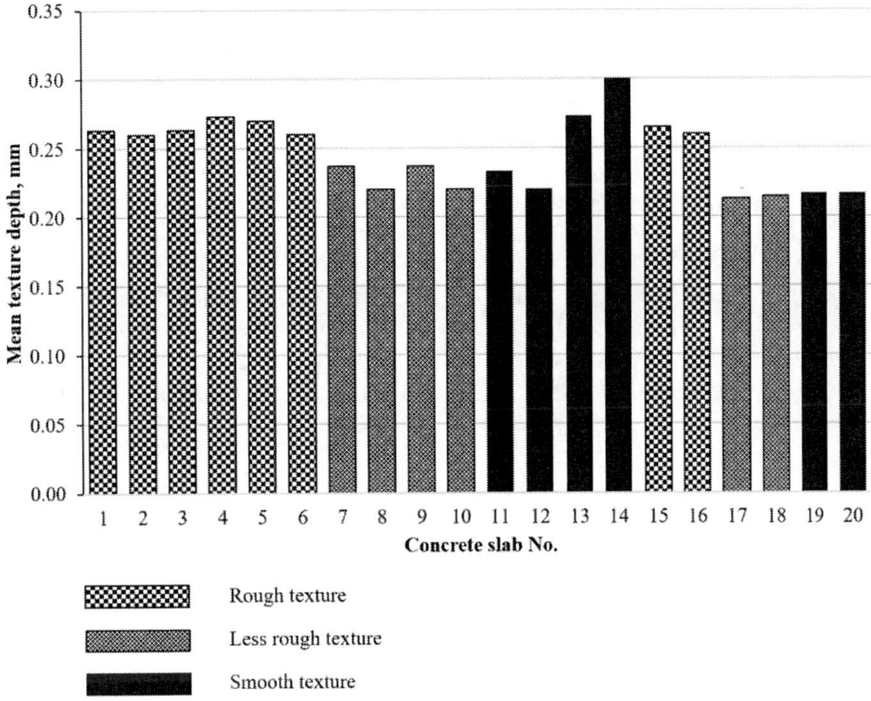

Figure 9. Results of mean texture depth

Mean profile depth, mean texture depth and skid resistance test results shows that the most suitable surface texture for the pedestrian pathway is formed by using texture mat (with rough or less rough surface texture) when sufficient skid resistance is ensured. Also, a rough surface is obtained when the surface was coloured with red powder, but then the skid resistance is insufficient (pendulum test values are less than 40), and values are lower than that of concrete slabs, which were not coloured and surfaces were not formed by using texture mats.

5. Conclusions

This paper presented a study of the design, prefabrication and installation, of thin synthetic fibre reinforced cement concrete slabs with different texture for the pedestrian path. The mechanical and surface characteristics were tested, as well. The results have shown that compressive, flexural and tensile splitting strengths of fibre reinforced concrete is about 16% lower than that of plain concrete. This can be explained by inefficient mixing method when fibres were added not to the aggregate mixture but to the factory-made concrete mixture. The chosen method of addition of fibres to the factory-made concrete mixture is not suitable for improving the strength properties of concrete slabs compared to a standard concrete mix. Measurements of surface characteristics have shown that the most suitable (pendulum test values are not less than 55) surface texture for the pedestrian pathway is formed by using texture mat with rough (skid resistance values vary from 58 to 71) or less rough surface (skid resistance values vary from 54 to 69) texture. Also, a rough surface (mean profile depth is not less than 0.05 mm) was obtained when the surface was coloured with red powder (mean profile depth vary from 0.093 mm to 0.128 mm), but then the skid resistance is insufficient (pendulum test values are less than 40). And, concrete slabs which surface was coloured with red powder has a lower skid resistance than that of concrete slabs, which were not coloured and surfaces were not formed by using texture mats (pendulum test values are greater than 40, but less than 55). The skid resistance of concrete slabs with a smooth texture is not sufficient (lower than 55, and mean profile depth is lower than 0.05). In this case, some texturing actions should be applied, seeking to reach skid resistance limit values. After the use of path for 10 months, even with the winter maintenance activities (de-icing with sodium chloride and lightweight snow removal machinery), all slabs do not have any distresses or defects.

Acknowledgment
This project has received funding from the European Regional Development Fund (project No 01.2.2-LMT-K-718-01-0044) under grant agreement with the Research Council of Lithuania (LMTLT).

References

[1] Gražulytė J, Vaitkus A, Laurinavičius A and Čygas D 2019 Concrete modular pavement type selection based on application area *The 13th Int. Conf. Modern Building Materials, Structures and Techniques* (Vilnius: VGTU Press) pp 318-23.
[2] Vaitkus A, Andriejauskas T, Šernas O, Čygas D and Laurinavičius A 2019 Definition of concrete and composite precast concrete pavements texture *Transport* **34(3)** pp 404–14
[3] Vaitkus A, Gražulytė J, Kleizienė R, Vorobjovas V and Šernas O 2019 Concrete modular pavements - types, issues and challenges *Baltic Journal of Road and Bridge Engineering* **14(1)** pp 80–103
[4] Vaitkus A, Kleizienė R, Vorobjovas V and Čygas D 2019 Mixture strength class and slab dimensions' effect on the precast concrete pavement structural performance *Baltic Journal of Road and Bridge Engineering* **14(3)** pp 443–71
[5] Wafa F F 1990 Properties and applications of fiber reinforced concrete *Engineering Science* **2** pp 49–63
[6] Kustermann A and Keuser M 2004 High strength fibre reinforced concrete (HSFRC) under high dynamic impact loading *The 6th RILEM Symposium on Fibre-reinforced Concretes*, ed M di Prisco, R Felicetti and G A Plizzari (RILEM Publications SARL) pp 1217–26
[7] Buratti N, Mazzotti C and Savoia M 2011 Post-cracking behaviour of steel and macro-synthetic fibre-reinforced concretes *Construction and Building Materials* **25(5)** pp 2713–22
[8] Jayakumar V J, and Anandan S 2014 Composite strain hardening properties of high performance hybrid fibre reinforced concrete *Advances in Civil Engineering* **2014(3)** p 9
[9] Amin A, Foster S J and Gilbert R I 2017 Material characterisation of macro synthetic fibre reinforced concrete *Cement and Concrete Composites* **84** pp 124–33
[10] Guerini V, Conforti A, Plizzari G, and Kawashima S 2018 Influence of steel and macro-synthetic fibers on concrete properties *Fibers* **6(47)** p 14
[11] Caggiano A, Cremona M, Faella C, Lima C and Martinelli E 2012 Fracture behavior of concrete beams reinforced with mixed long/short steel fibers *Construction and Building Materials* **37** pp 832–40
[12] Soulioti D V, Barkoula N M, Paipetis A and Matikas T E 2011 Effects of fibre geometry and volume fraction on the flexural behaviour of steel-fibre reinforced concrete *Strain* **47** pp 535–41
[13] Sivakumar A and Santhanam M 2007 Mechanical properties of high strength concrete reinforced with metallic and non-metallic fibres *Cement and Concrete Composites* **29(8)** pp 603-8
[14] Khaloo A and Afshari M 2005 Flexural behaviour of small steel fibre reinforced concrete slabs *Cement and Concrete Composites* **27** pp 141–9
[15] Pelisser F, Santos N A, Rovere H and Pinto R 2010 Effect of the addition of synthetic fibers to concrete thin slabs on plastic shrinkage cracking *Construction and Building Materials* **24** pp 2171–76

Kevlar-based Composite Material and its Applications in Body Armour: A Short Literature Review

Anand Narayanan Nair[1], Santhosh Sundharesan[1] and Issa Saif Mohammed Al Tubi[2]

[1]Department of Engineering, University of Technology and Applied Sciences-Al Mussanah, Al-Muladha, P.B.191, P.C. 314 Sultanate of Oman
[2]Dean, University of Technology and Applied Sciences-Al Mussanah, Al-Muladha, P.B.191, P.C. 314 Sultanate of Oman
E-mail: anand@act.edu.om

Abstract. Kevlar is a synthetic fibre which has gained worldwide attention due to its numerous industrial applications. It is mostly used in combination with other materials and acts as an effective composite material. Scientifically known as poly(p-phenylene terephthalamide), the Kevlar monomer has 14 carbon atoms, ten hydrogen atoms, two nitrogen atoms and two oxygen atoms. A Kevlar polymer chain can have monomers ranging from five to one million bonded to each other with a molecular formula of $[-CO-C6H4-CO-NH-C6H4-NH-]_n$, where *n* represents the number of monomers. Kevlar has a wide variety of applications ranging from bicycle tyres to aerospace applications. Based on the applications, Kevlar fibres are graded in various ranges. This paper provides insight into major research on Kevlar, especially in defence applications.

Keywords: Kevlar, fibres, armour, composites, polymer-matrix, laminates, bullets, impact tests

1. Introduction

Kevlar has many desirable properties when compared to other synthetic and natural fibres, which gives it a primary position in various load-bearing applications. Its most important property is its high tensile strength to weight ratio, which means that, when compared to a normal material, it has higher tensile strength at a lower weight. Kevlar is stable in most chemical environments, yet it can undergo degradation when exposed to some aqueous acids, bases or sodium hypochlorite. Researchers have proved that it can be used as a substitute for steel, since it is almost five times stronger than steel. Kevlar and polyethylene fibres, when knitted and woven with a composition of 50% each, has shown a higher cut resistance index in composite samples, making it suitable for use in sportswear and domestic applications [1].

Kevlar fabric shows a tensile strength of 3620 MPa and a very low relative density of 1.44, which is ideal for helmet, body armour and ballistic applications, where strength-to-weight ratio is especially important. Armour materials are meant to protect human beings in the defence field from high impact bullets. Currently, several research initiatives are being carried out to reduce the weight of armour using Kevlar-based composite material without compromising the strength of the armour. Synthetic fibres produced through polymerisation are the major constituents of woven Kevlar fabric, which can be used in bulletproof vests [2,3].

Recently, many studies have been done on the development of Kevlar in body armour applications and on optimising the required mechanical properties. Some researchers have focused on optimising the constituent materials in the composites so that the desired properties can be achieved without compromising on weight reduction. As a highly promising area of research, there is still a lot of work to be done related to manufacturing, testing and the environment in which the product is applied. In light of this, this paper provides an overview of some important studies on Kevlar-based composite materials.

2. Types of Kevlar

This section discusses the different varieties of Kevlar materials that are normally available and their usage in various applications.

2.1. Kevlar 29
This grade of Kevlar is normally used in brake linings, armour for vehicles, cables and other industrial applications. It is an ideal candidate in these applications due to its light weight, non-flammability and ability to withstand high temperatures. The breaking strength of this material is 338 N, and it has a tensile modulus of 70.5 GPa. The percentage elongation at breaking point for this Kevlar type is almost in the range of 3–4% [4]. It has also been observed that woven fibre of Kevlar 29, along with Al_2O_3 powder/epoxy acting as a hybrid composite, has a high energy level when impacted with different types of projectiles [5].

2.2 Kevlar 49
This type of fibre is mainly used in aircrafts as reinforcement due to its improved properties of low weight and high tensile strength. The percentage elongation at breaking point for Kevlar 49 has been observed to be in the range of 2–2.5% with a tensile modulus of 112 GPa. This estimate shows that the percentage elongation is 30–40% less and tensile modulus is 30–50% more than Kevlar 29 fibre. The breaking strength of Kevlar 49 is in the range of 264 N. The improved properties are attributed to the perfect orientation of polymer chains in the Kevlar 49 type when compared to the Kevlar 29 type aramids. This is because Kevlar 29 is heat treated at a high temperature under tension to obtain Kevlar 49, which causes less water molecules to pass through [4,6]. The specific heat of both Kevlar 29 and 49 grades is approximately 2010 J/kg K at 100 °C, and the shrinkage of both these fibres was found to be less than 0.1% at 1000 °C [4].

2.3 Kevlar 149
Kevlar 149 has high impact strength when compared to Kevlar 29 and 49, enabling it to be used in applications such as bulletproof vests, tankers and ballistic resistive missiles. However, Kevlar 149 has higher density and toughness when compared to the other types of Kevlar fibres discussed above. The tensile modulus and tensile strength of the Kevlar 149 type was found to be in the range 186 GPa and 3.4 GPa, respectively. The percentage elongation at breaking point was found to be 2% [7]. The comparison of tensile modulus and breaking strength of different grades of Kevlar is shown in Table 1, and a bar chart representing the percentage elongation at breaking point is shown in the Figure 1.

Table 1. Comparison of tensile modulus and breaking strength of Kevlar 29, 49 and 149.

Kevlar grades	Tensile modulus (GPa)	Breaking strength (N)
Kevlar 29	70.5	338
Kevlar 49	112	264
Kevlar 149	186	-

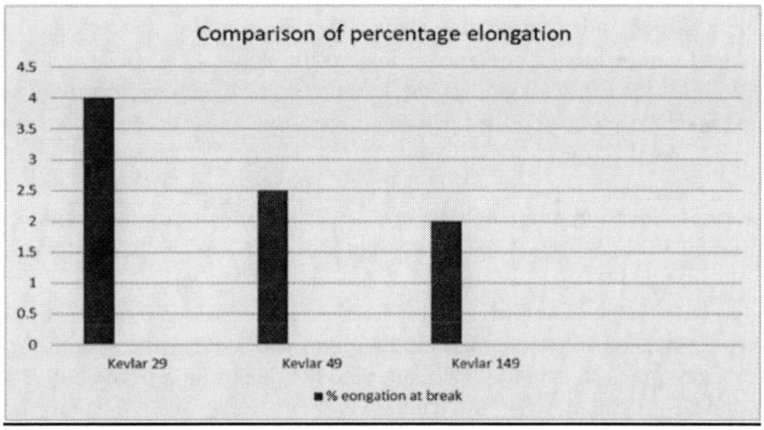

Figure 1. Comparison of percentage elongation of different grades of Kevlar.

3. Kevlar in body armour applications

The use of Kevlar material in body armour has been widely researched since 1970, and some studies have demonstrated excellent results in this area. Previously, steel-based body armour was widely used, but its major drawback was increased weight. In contrast, Kevlar as woven fabric in this application has showcased its capacity to address this concern by minimising the steel material, since it has a high strength-to-weight ratio.

It has been found that there is no direct linear relationship between the type and thickness of Kevlar measured in grams per square metre (GSM), if the combination arranged in layers has the same weight. The capability of different layers of Kevlar material with 160 GSM, 200 GSM and 400 GSM along with ballistic gel was studied, and it was observed that 200 GSM Kevlar is the most effective in stopping a 9 mm projectile [2]. From the modelling and simulation of a 20-layer Kevlar 29 fabric with polyester, it was reported that the composite withstood the energy of a bullet impact with a muzzle velocity of 720 m/s fired from a distance of 10 m [8]. Three combinations of composite plates embedded with Kevlar 29 were fire tested with bullets of 5 mm and 9 mm from a handgun and also from a 7.62 mm bullet from a Kalashnikov gun with bullet speeds of 200 m/s, 360 m/s and 760 m/s, respectively. In this study, the target plate was placed at 5 m from the handguns and 25 m from the Kalashnikov gun, and the kinetic energy absorbed before and after the impact was noted. The tests were also done with an air gap between the plates. The air gap was reported to impose a positive effect on the absorbed energy with the presence of Kevlar, providing high stiffness to the composite plate [9]. Upon being subjected to the strain test, the intra-ply and inter-ply Kevlar/basalt fibres in polypropylene matrix demonstrated that the former failed due to matrix cracks, whereas the latter failed due to delamination and multiple shear. Firing tests were also carried out in this study, and it was observed that the presence of Kevlar alone could not provide resistance to bullet firings. Firing resistance depended on many other factors, such as the grade and thickness of Kevlar fibres and the presence of other backup reinforcement like ceramics. Apart from these, a hybridisation scheme along with firing parameters like the type and speed of the bullet and the angle of firing determined the strength of the composite material [10].

Composite laminates were prepared with epoxy resin and Kevlar 29 woven fabrics with plies ranging from 8, 12, 15 and 19 with different thickness values. With these laminates having different orientations of fibres, the material properties were identified. These laminates were placed at 10 m from the gun point, and the impact of parabellum 9*19 mm bullet was done by firing it at a velocity of 390±10 m/s. The impact and residual velocities were noted for each firing, and five sets of readings with each thickness value of laminates were taken. It was found that the average energy absorption by the laminates increased when the thickness was increased. The energy absorbed by the laminate was equal to the reduction in kinetic energy of the bullet. A very interesting point from the experiment was that the damage on the front side of the plate was less when compared to the back side for all the thickness values of laminates. It was enumerated that, owing to the tensile failure of fibres at the back side, the crack propagation was perpendicular to its orientation. In addition, it was found that the layup of laminates had a significant effect on the energy absorption capacity of plate, with 0/90 layup having maximum value with all wall thicknesses. If the laminate was thin, the front part of the plate mostly failed in compression–shear mode and the back part in tension–shear mode [11].

In a related study, three prototypes were manufactured, and firing tests were done with different types of bullets and guns. The prototypes had a dimension of 100*100 mm, in which prototypes 1 and 2 had eight layers of Kevlar sheets impregnated in epoxy resin and aluminium oxide ceramic materials, respectively. Prototype 3 was manufactured with eight layers of Kevlar sheets stacked with epoxy resin and had no backing material. It was observed that prototype 1 successfully defended the 9 mm bullet fired from a handgun in a range of 4–5 m and also from an MP-5 rifle in a range of 10–11 m. It was also noted that prototype 2 successfully defended the handgun and MP-5 rifle with the same parameters in previous tests. For prototype 3, the material safeguarded the penetration of the bullet through the Kevlar layers. However, it was noted that, when the Kevlar layers were removed, the composite material failed to arrest the movement of the bullet [12]. A higher damping ratio was observed on the Kevlar/epoxy composite when compared to carbon and glass fibre composites due to the fact that Kevlar has higher flexibility in comparison to the other two [13]. Ballistic tests were conducted on Kevlar/fabric and Kevlar/epoxy composites with projectiles launched at different speeds. Kevlar/fabric samples were tested with velocities ranging from 140–260 m/s, and Kevlar/epoxy samples were tested at 60–178 m/s. It was observed that Kevlar fabric can absorb 100% of the projectile's energy when launched at 140 m/s velocity, since the residual velocity was nearly zero, which implies there were no perforations on the material. However, the Kevlar/epoxy composite was able to absorb only 50% of the energy at the same launching speed of the projectile [14].

In a previous study, it was found that Kevlar fabrics treated with 100 nm spherical silica nanoparticles as shear thickening fluids (STF) in different weight percentages contributed to higher energy absorption, which highlights the fact that it can be used in soft body armour applications. The energy absorption for untreated samples at failure was observed at 40% and doubled to 80% when treated with STF [15].

Another piece of research focused on the low velocity impact response of Kevlar and basalt 3D laminates separately and on a hybrid composite laminate that consisted of both Kevlar and basalt, which were manufactured by impregnating polypropylene fibre in between the laminates by using the compression moulding technique. It was shown experimentally that the energy absorption by the hybrid laminate is higher in the range of 180 J when compared to the other two. In this study, the presence of Kevlar fibre in the thickness orientation and its ability to deform in tension was attributed to higher damage extension for hybrid laminates [16].

In a similar study, shear thickening gel (c-STG) combined with Kevlar fabrics offered excellent anti-impact resistance when used in body armour. The low velocity impact test results showed that the maximum centre force of c-STG-Kevlar fabric was in the range of 5700 N, which was only half that of pure Kevlar fabric, and, in the high velocity ballistic test, it significantly absorbed the energy of monolayer c-STG/Kevlar fabric, which amounted to 21% of the impact energy [17].

The ballistic test results of composites made by incorporating basalt and Kevlar fibres with polypropylene arranged symmetrically and non-symmetrically have been compared, and it was proved that symmetric arrangement can withstand the impact of a 9 mm bullet in the range of 365–425 m/s, while non-symmetric arrangement cannot [18].

Short fibres of Kevlar have been used to toughen carbon fibre reinforced polymer laminates, which significantly improved impact strength and failure strain values when tested under low velocity impact. In low velocity impact tests, an indenter having a certain amount of mass is dropped from a height, and the respective velocities are noted with the help of laser attachments, and the impact loads are measured by using piezoelectric load cells [19]. The addition of nano clay and graphene inclusions at 10% weight of the Kevlar/epoxy composite was subjected to a low velocity impact test with a hemispherical drop weight of 15.875 mm diameter and 6.27 kg weight. The results of this study proved that, with an increase in the weight of nano clay, the impact resistance also increases, without any effect on the volume fraction of graphene. The presence of graphene can reduce the damage area of the specimen, UV degradation and water absorption [20]. The characterisation of damaged composite plates has shown that failure occurs by delamination, fibre pull-out and breakage, which depend directly on parameters such as fabric structure, velocity of impact, ply interaction and plate thickness [21]. Hence, in the development of body armour using Kevlar fibres, some important Kevlar composite parameters, test methods and material characterisations are to be considered, as shown in Figure 2.

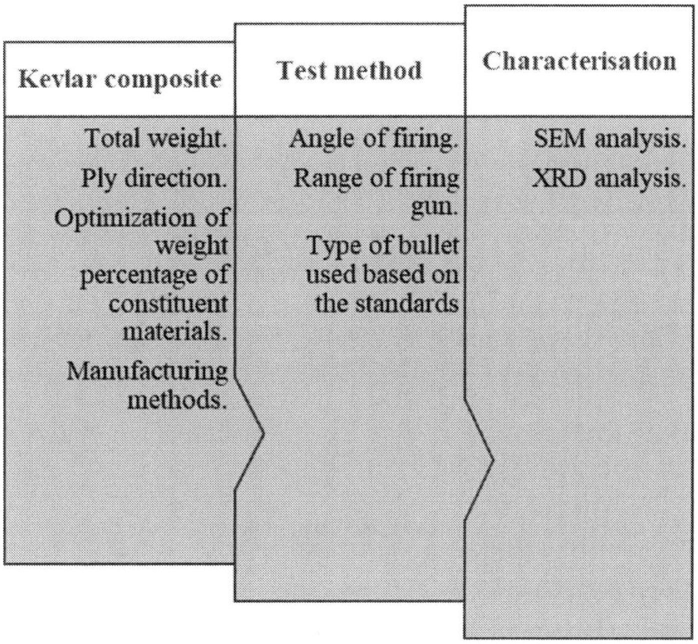

Figure 2. Areas of focus in the development of Kevlar-based body armour.

3.1 Flexibility test

These types of tests are conducted on materials that are supposed to have flexibility properties. In the case of Kevlar body armour, flexibility plays a vital role in absorbing the energy of the bullet after it hits the body. The impact response of Kevlar and rubber composites has been studied, in which a test for flexibility and high velocity impact was performed and compared with finite element simulation methods by using CATIA. Kevlar fabric, having 0.23 mm thickness and 180 GSM, was impregnated with high hardness (HH) rubber and low hardness (LH) rubber as separate specimens by using a hydraulic press. These specimens were made in two sets: one with two layers of Kevlar and the other with four layers, with final thickness of 1 and 2 mm, respectively. The drake tests were conducted by using a weight of 20 g, and the corresponding angle of bend was noted. It was found that, when compared to pure Kevlar fabric, the composite with HH rubber plummeted slightly, which means that neat Kevlar is more flexible. A schematic diagram of the drake flexibility experimental setup is shown in Figure 3. The deformed shape of the specimen after the application of weight is shown with dotted lines, and the angle of deformation is represented as Θ, where the higher the value of Θ, the higher the flexibility of the specimen is [22].

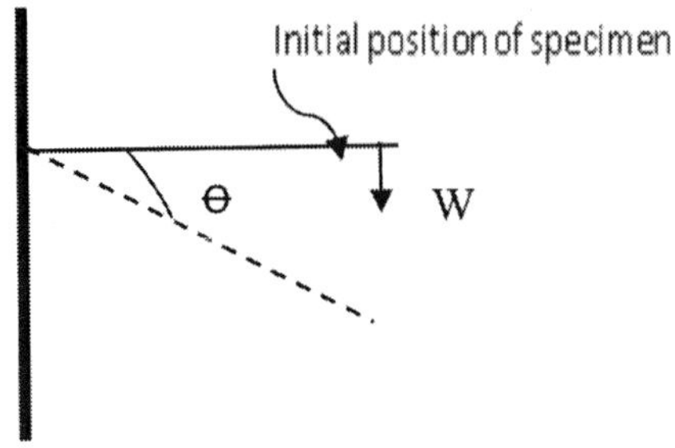

Figure 3. Schematic diagram of the drake test (flexibility test) [22].

3.2 High velocity impact test
The basic theory behind the velocity impact test is finding out the energy absorbed by the target body, which can be found by using the simple kinetic energy equation. Firing tests on the Kevlar/HH rubber composite was conducted with hemispherical bullets fired at a velocity of 150 m/s from a gas gun having a pressurised barrel at 120 bar. The initial velocity is represented by V_i and residual velocity as V_r in m/s. The energy absorbed (ΔE) is calculated by using the following equation (1), where 'm' is the mass of the bullet in kg.

$$\Delta E = 0.5*m*(V_i^2 - V_r^2) \qquad (1)$$

The results of this study prove that the Kevlar/HH rubber composite has better impact performance when compared to the other two specimens in section 3.1. The absorbed energy of the two-layer and four-layer Kevlar/HH rubber composites increased by 53% and 73%, respectively, when compared to neat Kevlar fabric owing to the fact that velocity reduction is directly proportional to the hardness of the rubber [22]. In an experimental study, a hybrid composite was prepared by incorporating epoxy polymer, conch shells and banana fibres in between aluminium plates for effective absorption of impact energy during firing tests. It was observed that the resulting composite material improved tensile and flexural strength, probably due to the presence of banana fibres and conch shells. However, to reveal the effect of each constituent on the bulletproof vest, the scanning electron microscope (SEM) image of bullet marks shows that the presence of banana fibres can cause fire and create a hole in the vest material [23]. A study concentrated on the effect of the stacked basalt and Kevlar hybrid composite in which the high velocity impact test was conducted with the speeds of the projectile at 40, 50 and 60 m/s, where the bullet had a 21.5 mm diameter and 27 g weight. The test plate had four, five and six layers of basalt and Kevlar (both considered), and the plate had a dimension of 13*13 cm. In this study, the number of impacts for failure were noted, and it was found that thickness is directly dependent on the tolerance for impact, i.e. when the thickness increases, the number of impacts required for failure also increases [24]. A multi-layered armour system (MAS) with an epoxy plate and jute fabric at a 30% volume fraction as the intermediate layer was utilised to compare the bullet defence of the armour material with that of Kevlar. The results show that the jute fibre composites are more efficient in absorbing impact energy [25]. In a similar study, improved tensile and flexural strength were observed for a Kevlar/graphene composite under different concentrations of graphene material and different directional orientation of the composites [26].

4. ASTM testing standards used in Kevlar research
The American Society for Testing and Materials (ASTM) provides a streamline for all technical standards that should be followed for testing and developing materials. It is accepted globally by all manufacturers and researchers for ensuring uniformity in standards.

4.1 ASTM D7136/D7136M
This standard specifies the test methods that should be followed in finding the damage resistance of polymer matrix laminates subject to low velocity impacts. The impact tests can be done by dropping a

IOP Conf. Series: Materials Science and Engineering **987** (2020) 012003 doi:10.1088/1757-899X/987/1/012003

mass with a hemispherical indenter, and the height of drop determines the energy of impact. The test specimen, which is normally made in a rectangular shape, will be held in a fixture underneath the weight, and the respective damage resistance properties can be studied. This standard was used in a test where a specimen with a 152.4*101.6 mm cross section was impacted with a 6.27 kg drop weight, the details of which were discussed in section 3 [20].

4.2 ASTM D3039/D3039M-95a

This is the tensile testing standard used for polymer composites in which the force required to break the polymer under the action of tensile load applied by UTM is evaluated. A strain gauge can be used to measure the elongation, and this will assist in plotting the stress–strain diagram to find the tensile modulus. The normal test specimen should have a cross section of 25 mm width and 250 mm length. A test specimen with 10*100 mm dimensions was prepared, and a quarter bridge strain gauge was utilised to measure the strain to figure out the high velocity impact response of Kevlar 29 and Al_2O_3 powder/epoxy composites [5].

4.3 ASTM D7264

The ASTM D7264 standard features a bending test to determine the flexural properties of polymer matrix composites with a rectangular cross section. Under this standard, the test can be conducted on a three-point or four-point loading system. A three-point loading system is simply a supported beam acted upon by a central load. This test can provide data such as maximum flexural stress, flexural strength and maximum strain. The maximum span-to-thickness ratio is specified for this test as 32:1. A specimen with a 100*4.8 mm cross section was tested on a three-point system to identify the damage propagation of a TiNi-Kevlar stitched carbon/epoxy specimen. The displacement rate was given as 1 mm/min, and it was proved that a Kevlar-stitched element requires 20% more energy to deform in comparison to a non-stitched one [27].

5. Conclusion

The use of Kevlar in polymer matrix has a wide scope to open research extremities. The most attractive feature of Kevlar for defence and ballistic applications is its considerable strength at lower weights. It must be noted that only limited research is available utilising Kevlar in ballistic applications, and this paper has covered only the most significant work. The process of hybridisation is mainly utilised by researchers, in which one or more materials in polymer matrix have been developed to understand their behaviour in various applications. The various aspects in the development of Kevlar-based composite material were discussed in this paper with more focus on its body armour applications. Kevlar sheets that are reinforced with varieties of natural fibres in different alignments in a polymer matrix have been studied extensively during recent times. The high specific strength of such fibres positively contributes to the overall mechanical properties of the composite material, and the characterisation of the post-impact bulletproof structure can reveal the vital mechanisms of failure. This can result in optimum arrangements of ply angle and orientation of the natural fibres and Kevlar sheets for better durability. However, there is still need for future research to focus more on the development of Kevlar fibres impregnated between composite plates for the development of body armour. New backup materials need to be explored to study the impact of different types of bullets on the vest material with Kevlar, which provides a network structure for arresting the bullet, thereby safeguarding the user. Hence, the design of experiments is very important in further research, so that the type of bullet for the newly developed material can be clearly defined and applied for specific environments. Furthermore, the optimisation of fibre material in view of how much weight percentage it should have in a polymer matrix governs the strength of the material. The aim of the researcher should be to reduce weight without compromising other desired properties, such as tensile strength, impact strength, energy absorption and fire resistance. Hence, it can be concluded that the applications of Kevlar-based composite structures must be explored extensively, not only in ballistic but also in automobile, aerospace, domestic and medical applications.

6. References

[1] Memon A A, Peerzada M H, Sahito I A, Abbassi S, Jeong S H 2018 Facile fabrication and comparative exploration of high cut resistant woven and knitted composite fabrics using kevlar and polyethylene Fashion and Textiles https://doi.org/10.1186/s40691-017-0122-0

[2] Stopforth R, Adali S 2019 Experimental study of bullet-proofing capabilities of kevlar of different weights and number of layers, with 9 mm projectiles Defence Technology 15(2) 186–192 https://doi.org/10.1016/j.dt.2018.08.006

[3] Kumar N 2016 Bulletproof Vest and Its Improvement – A Review. International Journal of Scientific Development and Research 1(1) 34–39

[4] Aamir M, Yu M 2015 Correlating Structure Property Relationship of Kevlars by Scattering and Birefringence. International Journal of Scientific and ResearchPublications (IJSRP) 5(9) 2250–3153 Retrieved from www.ijsrp.org

[5] Abu Talib A R, Abbud L H, Ali A, Mustapha F 2012 Ballistic impact performance of Kevlar-29 and Al2O3 powder/epoxy targets under high velocity impact Materials and Design, 35, 12–19. https://doi.org/10.1016/j.matdes.2011.08.045

[6] Khusiafan F J 2018 Use of KEVLAR ® 49 in Aircraft Components. Engineering Management Research 7(2) 14 https://doi.org/10.5539/emr.v7n2p14

[7] Manigandan S 2015 Computational Investigation of High Velocity Ballistic Impact Test on Kevlar 149 Applied Mechanics and Materials 766–767 1133–1138 https://doi.org/10.4028/www.scientific.net/amm

[8] Regassa Y 2014 Modeling and Simulation of Bullet Resistant Composite Body Armour Global Journal of Researches in Engineering (A)16(3) 39–44

[9] Ehsan Sabah Al-Ameen, Nathim M Faleh, Salam Abdulrazzaq Hussien 2017 An Invent to Improve Bulletproof Capabilities International Journal of Scientific & Engineering Research 8(10) 1807–1812

[10] Bandaru A K, Chouhan H, Bhatnagar N 2020 High strain rate compression testing of intra-ply and inter-ply hybrid thermoplastic composites reinforced with Kevlar/basalt fibres. Polymer Testing, 84 106407 https://doi.org/10.1016/j.polymertesting.2020.106407

[11] Sikarwar R S, Velmurugan R, Madhu V 2012 Experimental and analytical study of high velocity impact on Kevlar/Epoxy composite plates Central European Journal of Engineering 2(4) 638–649 https://doi.org/10.2478/s13531-012-0029-x R

[12] Nikhil Kumar K, Bharat Kumar C H, Hemanth Kumar K, Noolvi B 2019 Investigation of composite sandwich plates for ballistic armour application Materials Today: Proceedings 27 1738–1742. https://doi.org/10.1016/j.matpr.2020.03.651

[13] Fouad H, Mourad AHI, ALshammari B A, Hassan M K, Abdallah M Y, Hashem M 2020 Fracture toughness, vibration modal analysis and viscoelastic behavior of Kevlar, glass, and carbon fibre/epoxy composites for dental-post applications Journal of the Mechanical Behavior of Biomedical Materials, 101 https://doi.org/10.1016/j.jmbbm.2019.103456

[14] Majzoobi G H, Mohammad zaherib F 2017 Numerical and experimental study of ballistic response of kevlar fabric and kevlar/epoxy composite International Journal of Engineering, Transactions B: Applications 30(5) 791–799 https://doi.org/10.5829/idosi.ije.2017.30.05b.20

[15] Majumdar A, Butola B S, Srivastava A 2013 An analysis of deformation and energy absorption modes of shear thickening fluid treated Kevlar fabrics as soft body armour materials. Materials and Design 51 148–153 https://doi.org/10.1016/j.matdes 2013.04.016

[16] Bandaru A K, Patel S, Sachan Y, Alagirusamy R, Bhatnagar N, Ahmad S 2016 Low velocity impact response of 3D angle-interlock Kevlar/basalt reinforced polypropylene composites Materials and Design 105 323–332 https://doi.org/10.1016/j.matdes.2016.05.075

[17] Zhao C, Wang Y, Cao S, Xuan S, Jiang W, Gong X 2019 Conductive shear thickening gel/Kevlar wearable fabrics: A flexible body armour with mechano-electric coupling ballistic performance Composites Science and Technology 182 https://doi.org/10.1016/j.compscitech.2019.107782

[18] Bandaru A K, Ahmad S, Bhatnagar N 2017 Ballistic performance of hybrid thermoplastic composite armours reinforced with Kevlar and basalt fabrics Composites Part A: Applied Science and Manufacturing 97 151–165 https://doi.org/10.1016/j.compositesa.2016.12.007
[19] Hu Y, Liu W, Shi Y 2019 Low-velocity impact damage research on CFRPs with Kevlar-fibre toughening Composite Structures 216 127–141 https://doi.org/10.1016/j.compstruct.2019.02.051
[20] Rahman A S, Mathur V, Asmatulu R 2018 Effect of nanoclay and graphene inclusions on the low-velocity impact resistance of Kevlar-epoxy laminated composites Composite Structures 87 481–488 https://doi.org/10.1016/j.compstruct.2017.12.054
[21] Naik S, Dandagwhal R D, Loharkar P K 2020 A review on various aspects of Kevlar composites used in ballistic applications Materials Today: Proceedings 21 1366-1374 https://doi.org/10.1016/j.matpr.2020.01.176
[22] Khodadadi A, Liaghat G, Ahmadi H, Bahramian A R, Razmkhah O 2019 Impact response of Kevlar/rubber composite Composites Science and Technology 184 107880 https://doi.org/10.1016/j.compscitech.2019.107880
[23] Arun Kumar G, Ranjith Kumar M, Ramesh Babu A M, Ram Kumar R, Sathish Kumar G, Parameswaran P 2020 Experimental analysis on ballistic performance of newly developed sandwich hybrid natural composites Materials Today: Proceedings 21 41–44 https://doi.org/10.1016/j.matpr.2019.05.357
[24] Khazaie M, Eslami-Farsani R, Saeedi A 2018 Evaluation of repeated high velocity impact on polymer-based composites reinforced with basalt and Kevlar fibres Materials Today Communications 17 76-81 https://doi.org/10.1016/j.mtcomm.2018.08.016
[25] Fernanda Santos da Luza, Edio Pereira Lima Juniora L H L L, Monteiroa S N 1516-1439 Materials Research 18(2) 170–177
[26] Maruthi Gangadhar S, Kotari S, Manikumar M V, Rajasekhar Reddy K Sameeroddin M 2020 Simulation and characterization on effect of graphene in kevlar reinforced laminates Materials Today: Proceedings 27 1696–1702 https://doi.org/10.1016/j.matpr.2020.03.578
[27] Vachon P L, Brailovski V, Terriault P 2013 Impact-induced damage and damage propagation under flexural load in TiNi and Kevlar-stitched carbon/epoxy laminates Composite Structures 100 424–435 https://doi.org/10.1016/j.compstruct.2013.01.011

The Mechanical Properties of Aluminium Metal Matrix Composite (AlMMCs) Reinforced with Ni and SiC_p

Haider Tawfiq Naeem
Chemical Engineering Department, College of Engineering, Al-Muthanna University, 72001 Samawah, Al-Muthanna, Iraq
Corresponding author email: haidertn@mu.edu.iq

Abstract. This investigation studies the effects of nickel (Ni) particles combined with silicon carbide (SiCp) particles on the microstructural and mechanical properties of aluminium(Al)-magnesium(Mg)-manganese(Mn) matrix alloys under different sintering temperatures. The Al-Mg-Mn-SiC_p/Ni_p composite samples were produced via the powder metallurgy route. The sintering process was performed on a green compact of Al-Mg-Mn matrix/SiC_p/ or Ni_p composites at 400 °C and 600 °C for 1.5 hours at a constant heating rate of 50 °C/min. The Al-Mg-Mn-SiC_p/Ni_p samples were evaluated and characterised using a scanning electron microscope and X-ray diffraction. Additionally, the Al-Mg-Mn-SiC_p/Ni_p samples were measured for Vickers hardness and compressive strength. The outcomes show that Al-Mg-Mn-5%SiC_p-6%Ni_p gave the highest Vickers hardness. The results showed that the optimal sintering temperature performed on the Al-Mg-Mn-SiC_p/Ni_p sample was 400 °C compared with 600 °C because the overall dissolution of alloying elements within the Al-matrix led to an enlargement of grains, thus, reducing their mechanical properties.

Keywords: Additive strengthening; Al-Mg-Mn composite; Strength

1. Introduction

Five thousand-series aluminium alloys have acquired prominence in the naval, automotive and aerospace industries due to their specific strength, excellent corrosion resistance, fracture toughness, weldability and thermal and electric conductivities [1, 2]. Many applications need high strength with low density and good formability. A host of techniques, such as heat treatment, thermo-mechanical processing and severe plastic deformation have been used to improve the strength of Al-alloys. In other words, the Al-5000 series is considered a non-heat-treatable alloy, thus, techniques that can improve strength and retain ductility are very limited. Therefore, impediments can be overcome using composite materials to reinforce hard metals or ceramic particles within the matrix of aluminium metal. The technology of aluminium metal matrix composites (AlMMCs) is a powder metallurgy technique used to improve the strength of alloys. AlMMCs are usually reinforced by promising ceramic particles, such as Al_2O_3, SiC, AlN or Y_2O_3 [3–4]. The improvement in mechanical properties of AlMMCs depends greatly on how well the particles are distributed in the metal matrix composite. It also depends on the proper selection of the reinforcing phase and processing techniques [5–6].

Powder metallurgy (PM) is one of the most advantageous techniques in the solid-state process of fabricating AlMMCs because of an isotropic distribution of particles in the matrix and good dimensional accuracy in an economical manner [7–9]. Silicon carbide (SiC) powder is a covalently bonded ceramic material with very stable chemical properties. Other remarkable properties include superior hardness and wear and corrosion resistance, which make it an ideal reinforcing material in metal matrix composites, such as Fe, Al and Mg-based materials [10–11]. Chemical reactions that can occur on the SiC surface during the preparation of SiC metal matrix-composites can worsen the performance of the composite materials. Therefore, metal modification on the SiC surface is often utilised to restrain interface reaction and enhance the interface combination [12–16].

Nickel is a good bonding choice and provides strong reinforcement because of its high strength, stiffness and high-temperature properties. Recently, the production of an Al-based matrix composite has been successfully carried out using powder metallurgy to obtain an Al/Ni-SiC composite with good microstructure and high densification [17–18]. It is also known that the 5000-series aluminium alloys do not respond to different types of heat treatments. Therefore, this study will attempt to change the crystalline and microstructure of the reinforced Al-Mg-Mn (SiCp + Nip) matrix by undergoing the impact of sintering heat treatments.

2. Materials and experimental procedures

Synthesising aluminium with alloy elements using the elemental powder precursors has already been performed (as shown in Table 1). Powder mixtures were used to fabricate four types:
A (Al–5.5%Mg–1%Mn–0.5%Cu–0.4%Zn–0.2%Fe–0.1%Cr)
B (Al–5.5%Mg–1%Mn–0.5%Cu–0.4%Zn–0.2%Fe–0.1% Cr–5%Sic)
C (Al–5.5%Mg–1%Mn–0.5%Cu–0.4%Zn–0.2%Fe–0.1%Cr–5%Sic–3%Ni)
D (Al–5.5%Mg–1%Mn–0.5%Cu–0.4%Zn–0.2%Fe–0.1%Cr–5%Sic–6%Ni)
All compositions are expressed in weight percentage (wt. %). The powder mixture composition analyses for types A, B, C and D were carried out using X-ray fluorescence spectrometry, as listed in Table 1. Al-mixture samples were produced via powder metallurgy processing, which included mixing the elemental powders (Al, Mg, Mn, Cu, etc.) using a turbulent mixer, followed by a cold press by uniaxial-die compaction (hand-operated hydraulic press at 350 MP) to produce the green compact-bulk (Figure 1). Finally, sintering treatment was carried under the two conditions given in Table 2.
To evaluate the microstructural characterisation and phase components of the sintered Al-matrix composite samples, scanning-electron-microscopy (SEM) and X-ray diffraction analysis (XRD) was used. Subsequently, Vickers hardness and compressive strength were tested according to ASTM E92–82.

Table 1: Chemical compositions of the four types of Al-composite samples with and without additives

Standard ident	Solid specimen A	Solid specimen B	Solid specimen C	Solid specimen D
Al	92.36 %	87.415%	84.443%	80.5%
Mg	5.2%	5.28%	4.97%	5.61%
Mn	1.03%	0.901%	0.95%	1.2%
Cu	0.45%	0.501%	0.504%	0.57%
Cr	0.35%	0.297%	0.291%	0.25%
Fe	0.306%	0.301%	0.31%	0.29%
Zn	0.301%	0.295%	0.303%	0.41%
Ni	-	-	3.00%	6.00%
SiC	-	5.00%	5.00%	5.00%

Figure 1: Macro-structure of Al-composite samples under (a) 400 °C and (b) 600 °C sintering conditions

Table 2: Sintering conditions during this study

Sintering conditions	Soaking time and heat rating	Name of compacted Al-composites
400 °C	1.5 h with a heating rate of 40 °C/ min.	A, B, C, D
600 °C		A1, B1, C1, D1

3. Results and discussion

The SEM in Figure 2 shows the microstructure of the sintered Al-Mg-Mn samples. The dark areas denote the alpha aluminium and contain alloying elements within the matrix. Furthermore, some pores were noticed in the matrix because of the sintering operations [20].

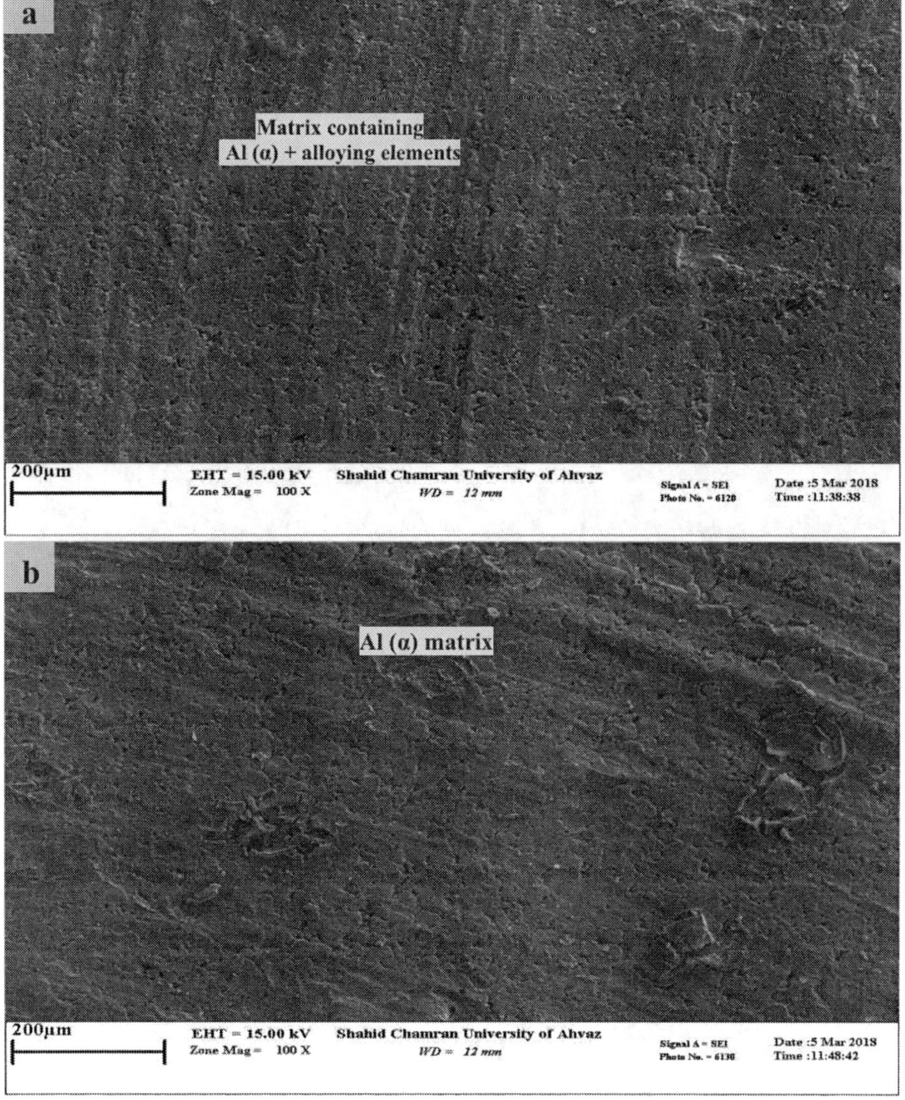

Figure 2: SEM micrographs of the compacted Al-5000 alloys under sintering treatment at (a) 400 °C and (b) 600 °C

The micrograph of the Al-Mg-Mn-5%SiC composite sample under sintering treatment at 400 °C is shown in Figure 3 (a). It indicates the presence of SiC particles that have attached and dispersed throughout the aluminium mixture. The high magnification of SEM (as shown in Figure 3b) noted the good bonding interface between the SiCp within the Al-Mg-Mn matrix.

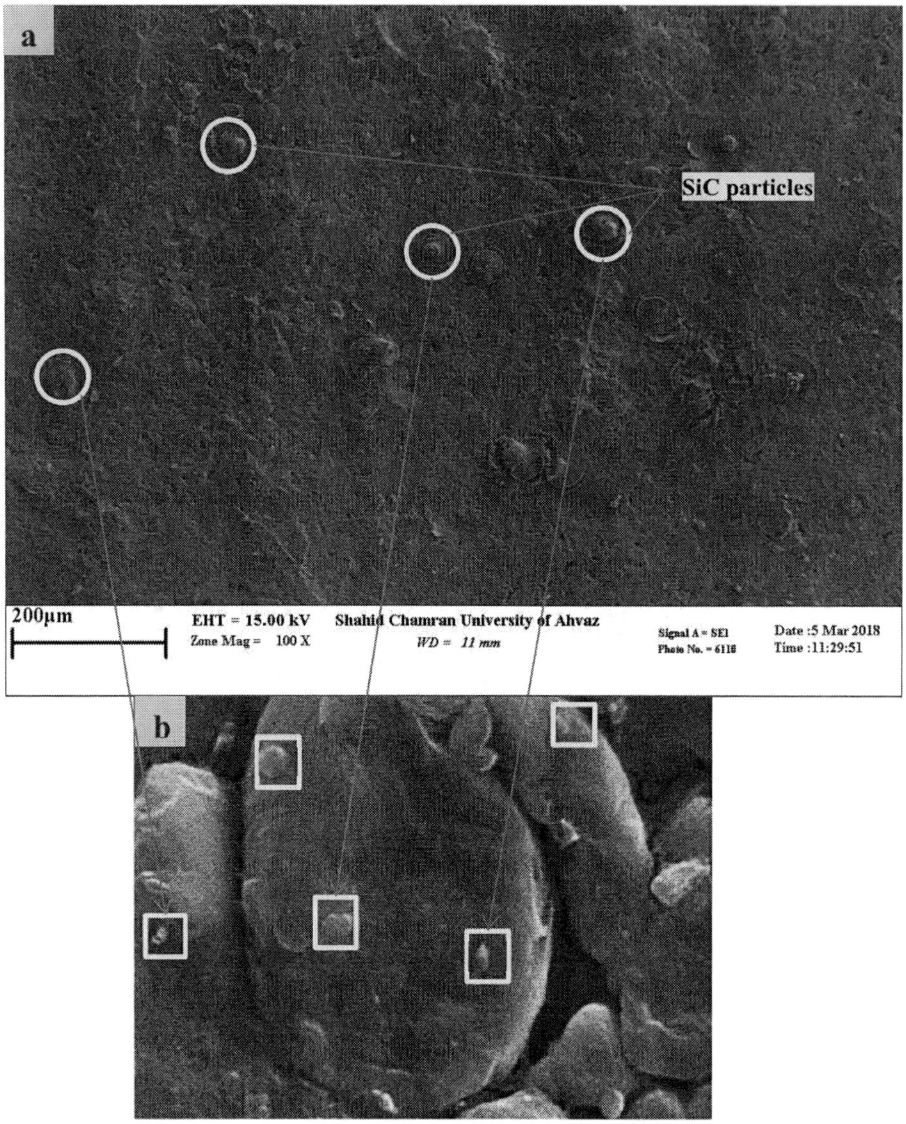

Figure 3: SEM micrographs of the compacted Al-Mg-Mn-5%Sic composite under sintering treatment at (a) 400 °C and (b) high magnification

Figure 4 (a) exhibits the micrograph of the Al-Mg-Mn-5%SiC composite sample under sintering treatment at 600 °C. It shows voids and pores due to micro-cracks throughout the aluminium mixture-reinforced SiCp. These impurities were created due to overheating at 600 °C as reported in [20–21], despite the good distribution and adherence of SiCp within the aluminium-nickel mixture.

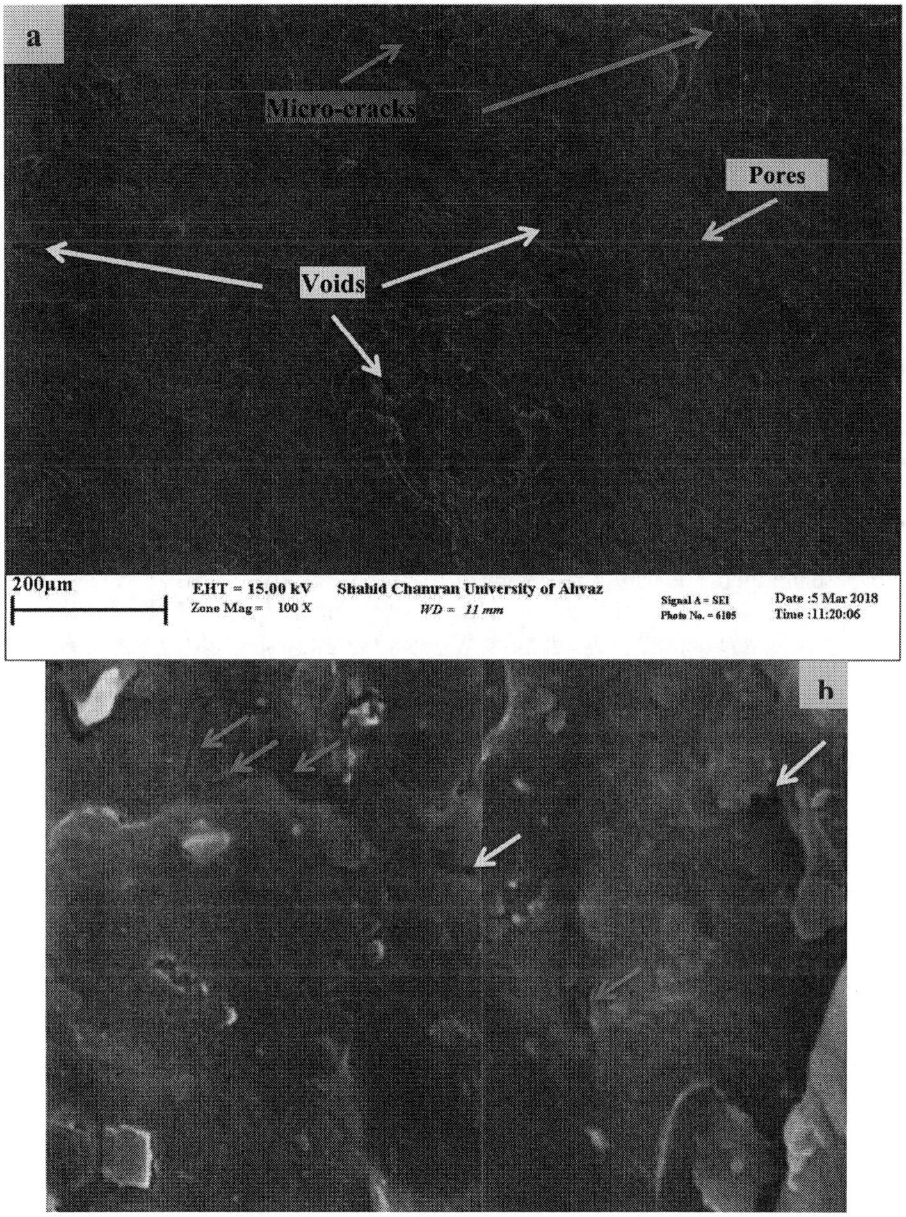

Figure 4: SEM micrographs of the compacted Al-Mg-Mn-Sic-6%Ni composite under sintering treatment at (a) 600 °C and (b) high magnification

Figure 5 presents the XRD spectrum of the aluminium composites (A, B, C, D samples) after sintering treatment at 400°C for 1.5 hours. The X-ray diffraction patterns of mixture A (Figure 5 A) for the Al-alloy represents Al–5.5%Mg–1%Mn–0.5%Cu–0.4%Zn–0.2%Fe–0.1%Cr and exhibited three high peaks corresponding to Al and AlMgMn. The peak intensities decreased after applying the sintering treatment due to the partial dissolution of Mg, Mn, Zn, and Cu in the Al lattice to form a solid solution, the reduction in grain size of the elemental powder induced by the heat and the loss of very small particles to their crystalline structures. The XRD profiles of the aluminium matrix-Sic composite (shown in Figure 5 B) represents **B** (Al–5.5%Mg–1%Mn–0.5%Cu–0.4%Zn–0.2%Fe–0.1% Cr–5%Sic). During this stage, new peaks were observed for SiCp and Al_2MgMn. All constituent elements were identified in the initial powder. In Figure 4 **C** diffraction peaks are strong but peaks containing nickel observed in Al_3Ni_2, AlMgNi and AlMnNi were weak. The peaks of other additive elements were not observed because of the limitation of the filtered X-ray to detect phases with volume fractions less than 2% wt. [22].

After increasing the percentage of nickel up to 6% wt. within Al-composite alloy **D** (Al–5.5%Mg–1%Mn–0.5%Cu–0.4%Zn–0.2%Fe–0.1%Cr–5%Sic–6%Ni), the XRD analytic was presented in Figure 5D. There was a noticeable presence of several peaks of AlNi because of the dissolution of nickel particles within the Al-Mg-Mn matrix alloy. Figure 6 shows the X-ray diffraction peaks in the Al-Mg-Mn matrix-reinforced SiCp-Nickel undergoing sintering treatment; A1 presents AlMg peaks and **B1** displays the XRD results of the Al-Mg-Mn matrix alloy after adding 5% SiCp, which changed the microstructure of the Al-composite. It reveals many peaks (Al, AlMg, AlMn, AlMgMn and SiC). Furthermore, sintering the Al-composite at 600 °C for 1.5 hours led to saturated dissolutions of alloying elements (Mg, Mn, Zn, and Cu) within the aluminium matrix. Moreover, it showed new peaks of alloys **C1** and **D1** (Figure 6 **C1** and **D1**) compared to those in Figure 5 **C** and **D**, respectively.

Vickers hardness values for the Al-matrix composite samples (listed in Table 1) underwent sintering as shown in Figure 7. Sample D has the highest hardness value compared to sample A under the same sintering conditions at 400 °C. A significant increase in hardness for sample D was because of the presence of reinforced 5% wt. SiCp coupled with 5% wt. nickel within the Al-Mg-Mn matrix. Generally, the bonded nickel particles with aluminium created stronger compounds that strengthened the matrix of the alloy (the XRD analytics above provide many compounds of Ni), as evidenced by previous studies [22–23]. That resulted in growth grains are too much and precipitate compounds outside grains boundaries than the weak in hardness property.

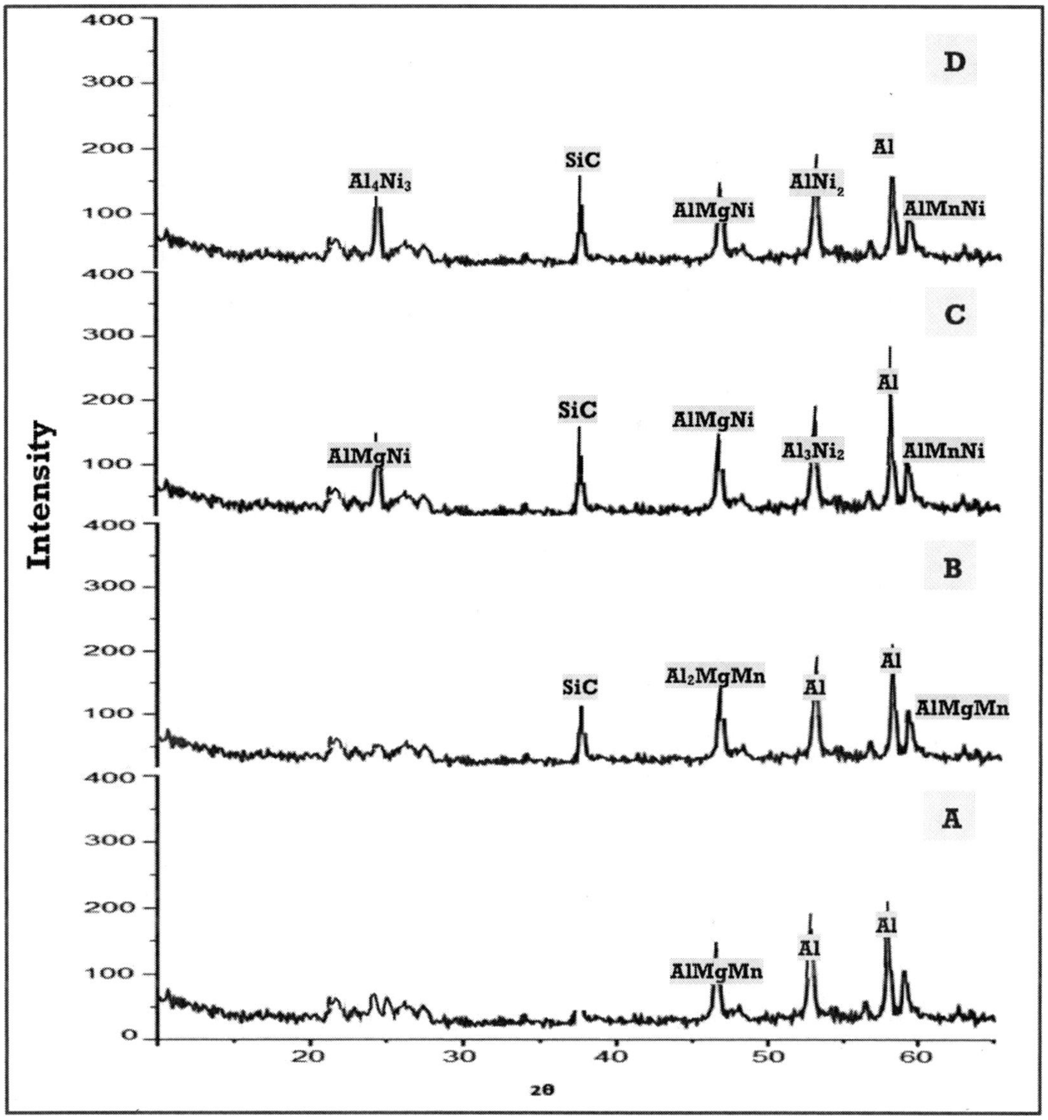

Figure 5: XRD analytic for the A, B, C, D samples undergoing sintering treatment at 400 °C

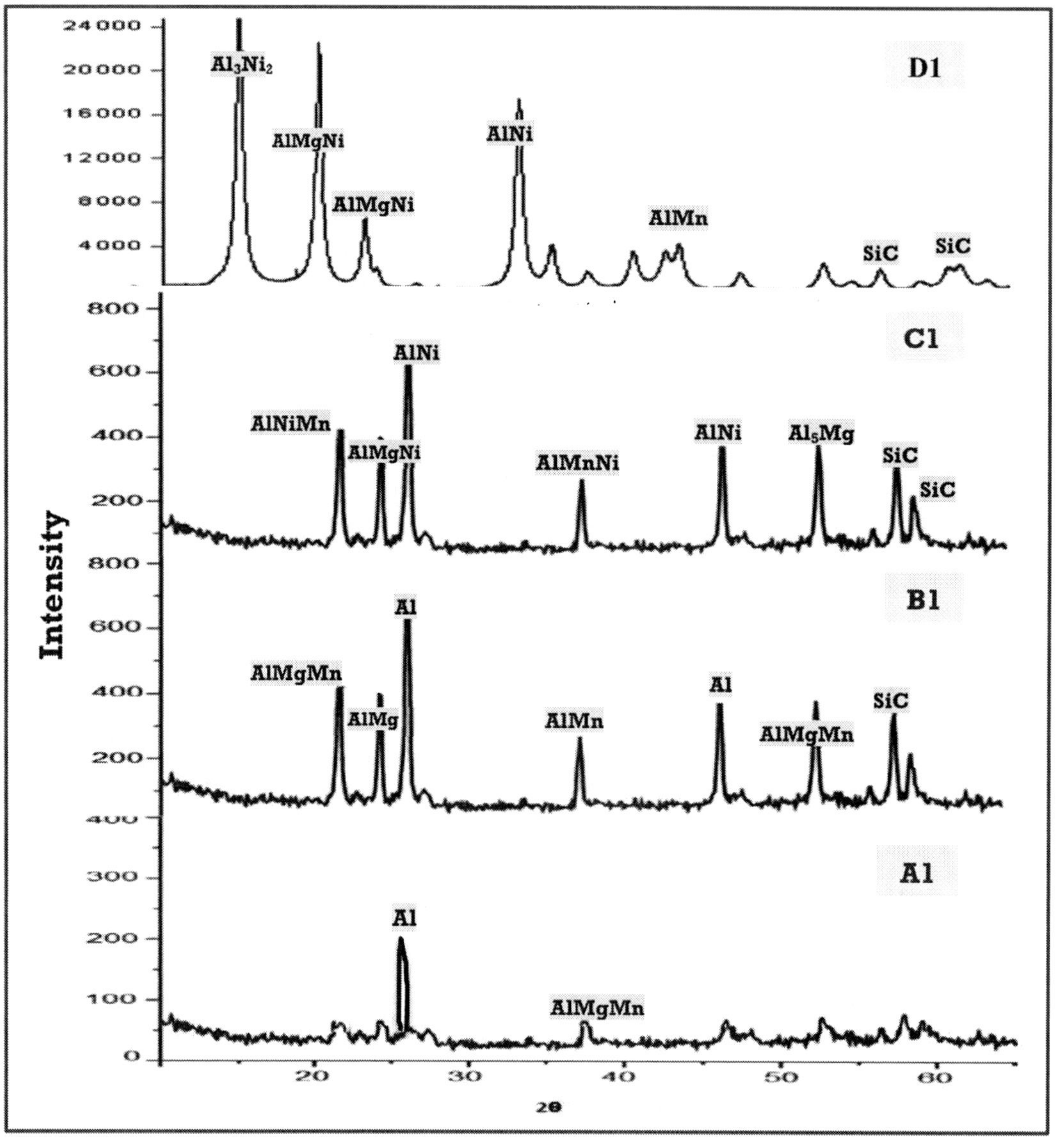

Figure 6: XRD analytic for the A1, B1, C1 and D1 samples undergoing sintering at 600 °C/1.5 h

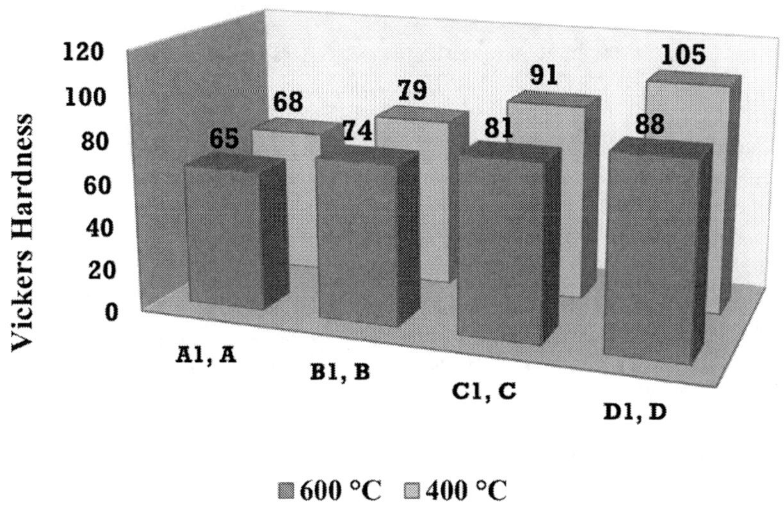

Figure 7: Vickers hardness values of different Al-composites undergoing sintering treatment at 400 °C and 600 °C

4. Conclusion

In this research, powder metallurgy was used to synthesise nickel particles combined with silicon carbide particles within aluminium-magnesium-manganese matrix alloys under sintering conditions of 400 °C and 600 °C for 1.5 hours at a constant heating rate of 50 °C/min. The Al-Mg-Mn-SiCp/Nip samples were evaluated and characterised using a scanning electron microscope and X-ray diffraction. The samples were also measured for Vickers hardness. Microstructural observations showed that after sintering treatment, the compacted Al-Mg-Mn-SiCp/Nip matrix had more precipitation particles with Ni-inter metallics in the sintered Al-matrix. Meanwhile, the Al-PM matrix was free from defects. The results demonstrate that the optimal sintering temperature on the Al-Mg-Mn-SiCp/Nip sample was 400 °C compared with 600 °C because the overall dissolutions of alloying elements within the Al-matrix led to the enlargement of grains, thus, reducing their mechanical properties.

5. References

[1] Hariprasad, T., K. Varatharajan, and S. Ravi. "Wear characteristics of B4C and Al2O3 reinforced with Al 5083 metal matrix based hybrid composite." *Procedia Engineering* 97 (2014): 925-929.
[2] Kumar, CN Shyam, Ranjit Bauri, and Devinder Yadav. "Wear properties of 5083 Al–W surface composite fabricated by friction stir processing." *Tribology International* 101 (2016): 284-290.
[3] Alizadeh, Ali, Alireza Abdollahi, and Hootan Biukani. "Creep behavior and wear resistance of Al 5083 based hybrid composites reinforced with carbon nanotubes (CNTs) and boron carbide (B4C)." *Journal of Alloys and Compounds* 650 (2015): 783-793.
[4] Adebisi, Adetayo Abdulmumin, et al. "Effect of variable particle size reinforcement on mechanical and wear properties of 6061Al–SiCp composite." *Composite Interfaces* 23.6 (2016): 533-547.
[5] Li, Meijuan, et al. "Synthesis and mechanical behavior of nanostructured Al 5083/n-TiB2 metal matrix composites." *Materials Science and Engineering: A* 656 (2016): 241-248.

[6] Nagaral, Madeva, et al. "Tensile behavior of B4C particulate reinforced Al2024 alloy metal matrix composites." *FME Transactions* 45.1 (2017): 93-96.
[7] Jeevan, V., et al. "Fabrication and Characterization of AA6082 ZTA Composites by Powder Metallurgy Process." *Materials Today: Proceedings* 5.1 (2018): 254-260.
[8] Muthuchamy, A., et al. "Microstructural and Electrochemical behaviour of Aluminium Alloy Composites Produced Using Different Sintering Techniques." *Materials Research* 21.3 (2018).
[9] Singh, Lavish K., Alok Bhadauria, and Tapas Laha. "Al-MWCNT nanocomposite synthesized via spark plasma sintering: effect of powder milling and reinforcement addition on sintering kinetics and mechanical properties." *Journal of Materials Research and Technology* (2018).
[10] Bathula, Sivaiah, et al. "Microstructural features and mechanical properties of Al 5083/SiCp metal matrix nanocomposites produced by high energy ball milling and spark plasma sintering." *Materials Science and Engineering: A* 545 (2012): 97-102.
[11] Reddy, A. Prasad, et al. "Silicon carbide reinforced aluminium metal matrix nano composites-a review." *Materials Today: Proceedings* 4.2 (2017): 3959-3971.
[12] Alizadeh, Ali, Alireza Abdollahi, and Hootan Biukani. "Creep behavior and wear resistance of Al 5083 based hybrid composites reinforced with carbon nanotubes (CNTs) and boron carbide (B4C)." *Journal of Alloys and Compounds* 650 (2015): 783-793.
[13] Padmavathi, K. R., and R. Ramakrishnan. "Tribological behaviour of aluminium hybrid metal matrix composite." *Procedia Engineering* 97 (2014): 660-667.
[14] Kretz F, Gacsi Z, Kovacs J, Pieczonka T. The electroless deposition of nickel on SiC particles for aluminum matrix composites. Surf Coat Technol 2004; 575:180–1.
[15] Abolkassem, Shimaa A., et al. "Effect of consolidation techniques on the properties of Al matrix composite reinforced with nano Ni-coated SiC." *Results in Physics* 9 (2018): 1102-1111.
[16] Wu XM, Sha W. Experimental study of the voids in the electroless copper deposits and the direct measurement of the void fraction based on the scanning electron microscopy images. Appl Surf Sci 2009;255: 4259–66.
[17] Luo LM, Yu J, Luo J, Li J. Preparation and characterization of Ni-coated Cr3C2 powder by room temperature ultrasonic-assisted electroless plating. Ceram Int 2010;36: 1989–92.
[18] Fathy A, Elkady O, Abu-Oqail A. Synthesis and characterization of CueZrO2 nanocomposite produced by thermochemical process. J Alloy Comp 2017;719: 411–9.
[19] H. T. Naeem, K. S. Mohammed, R. Khairel, and A. Rahmat,"EFFECTIVENESS OF ALUMINA DISPERSOIDS PARTICLES WITHIN (7XXX SERIES) ALUMINUM ALLOY UNDER THE RETROGRESSION AND REAGING TREATMENTS", Digest Journal of Nanomaterials and Biostructures Vol. 9, No. 1, January - March 2014, p. 295 – 304.
[20] Schaffer, G. B., Hall, B. J., Bonner, S. J., Huo, S. H., & Sercombe, T. B. (2006). The effect of the atmosphere and the role of pore filling on the sintering of aluminium, Acta Materialia, 54(1), 131-138.
[21] Yuan, X. N., Aminossadati, S. M., & Qian, M. (2016). Variation in pore distribution along sample length in sintered 7xxx aluminum alloy. Transactions of Nonferrous Metals Society of China, 26(8), 2019-2028.
[22] Naeem, H. T., Mohammad, K. S., Ahmad, K. R., & Rahmat, A. (2014). Characteristics of Al–Zn–Mg–Cu alloys with nickel additives synthesized via mechanical alloying, cold compaction, and heat treatment. Arabian Journal for Science and Engineering, 39(12), 9039-9048.
[23] Mohammed, Kahtan S., Haider T. Naeem, and Siti Nadira Iskak. "Study of the feasibility of producing Al–Ni intermetallic compounds by mechanical alloying." The Physics of Metals and Metallography 117.8 (2016): 795-804.

IOP Conf. Series: Materials Science and Engineering **987** (2020) 012005 doi:10.1088/1757-899X/987/1/012005

Evaluation of Mechanical Properties of Composite Laminates by Fractal Dimension

Amjed Saleh Mahmood

Electromechanial Engineering Department, College of Engineering, University of Samarra, Samarra, Salah Ad Din, Iraq
dramjed78@gmail.com

Abstract. This study focuses on the evaluation of mechanical properties of composite laminates by fractal dimension (FD). The fibre distribution in a laminate or FD could be changed according to different manufacturing processes, such as hand lay-up, resin infusion and resin transfer moulding. These manufacturing processes are considered in this study. Woven Kevlar / epoxy composite panels were manufactured with a $[0/90]_{2S}$ stacking sequence. The materials were characterised for interlaminar shear strength (ILSS) and flexural strength. The fibre volume fraction (V_f) and FD of the laminates were determined from optical microscopic images of polished cross-sections using ImageJ software (box counting method) with the Fractal and Lacunarity add-in. The results show significant relationships between the mechanical performance of the composite laminates and the FD.

1. Introduction

Fibre reinforced composites are widely used globally in the industry. They are usually used for lightweight structures where they can reduce energy consumption, especially in transport applications, such as space vehicles, aircraft, trains, cars and yachts. The quantification of microstructures of composite laminates requires image analysis and processing. It could be achieved by histograms derived from tessellation (e.g. Voronoi) or as a single number using fractal dimension (FD), which is a non-integer physical dimension. Numerous studies [1–7] have investigated the impact of fibre distribution on the performance of the composite laminates. They found that the reinforcement alignment has a significant impact on the physical properties. These studies also concluded that the waviness or random orientation of fibres can result in a decrease in the mechanical properties due to the low properties of resin. As resin has low elastic modulus, the resin areas show a high strain relative to the fibres. Lomov et al. [8] concluded that the resin areas have minimal strain, especially regarding carbon fibre composites. Sudarisman and Davies [9] studied the influence of several factors, such as holding time, compressive pressure and epoxy concentration, on the performance of laminates. The results of this study confirm weak bonding between individual prepreg layers due to resin areas. A crowded region of the fibres could lead to resin areas and the resulting strain concentration [10]. Despite this, none of the above studies have characterised the microstructure of the composites or studied the fibre distribution effect.

The measurement of the uniformity and characterisation of complex composite microstructures has evolved over the last few decades [11–16]. A study by Picu et al. [17] stated an effective argument between composites and fractal microstructure. FD is a powerful tool for measuring the degree of roughness of highly irregular objects [18,19] and is used to characterise the performance of composite laminates, such as connecting the toughening procedures with rupture surfaces (e.g. [11, 20]) and the development of voids in composite materials [21] and their impact on different mechanical and physical properties (e.g. [17, 18, 22]). The uniformity of fibre distribution, which is influenced by the manufacturing pressure, weave style and the stacking sequence, is an important parameter that has a large impact on the mechanical properties of composites [23, 24].

The objects that are considered in physics or mathematics are generally continuous, linear and smooth. However, nature has shown us that this is not always the case. Ordinary objects are normally rough and irregular [25]. Using FD to characterise irregular objects began scientifically approximately four decades ago [26] to quantify the fracture surfaces of different levels [27] and characterise the level of roughness of irregular objects [28]. Furthermore, fractal geometry deals with highly disordered morphologies [29]. Therefore, FD, which is measured using the box counting method, is widely used in quantifying the mechanical properties in metals, alloys and composites materials [30–34].

The aim of this study is to find out whether it is possible to use the FD to evaluate the performance parameters of composite material laminates by analysing the microstructure images. Through this, it would be possible to link the mechanical properties with microstructure images, which could avoid repeatedly performing mechanical tests when manufacturing new composites.

2. Experimental Methodology

Woven kevalr / epoxy composite panels were manufactured with a [0/90]2S stacking sequence with six layers using different manufacturing processes. The following three manufacturing processes were chosen: hand lay-up, resin transfer moulding (RTM) and infusion. The reinforcement was woven kevlar (310 g / m2). The resin used in this study was SR8100 and SD8824 as the hardener, and the weight mixing ratio was 100 / 22. The RTM procedure was applied with a fixed two mm cavity, and the mould was closed using a vacuum. The vacuum was maintained while the resin cured. Five plates for each manufacturing process were manufactured. The Image J software was used to calculate the fibre distribution for 60 images. Image J has been widely used by researchers [35–38]. The area of the object under consideration that is represented by one pixel is a function of the magnification used when obtaining the images. This study considers images obtained at a magnification that allows imaging of the complete sample thickness in sensible digital image sizes. The polishing process of the samples is illustrated in Table 1. Before using the image analysis, all images should have the same size and background (e.g. black background with a black resin colour) and should set the Image J options to get the same sized grid calibres. All microscopic images have been changed to binary images, as shown in Figure 1. The most efficient cover FD is generated from box counting data over multiple grid positions, as shown in Figure 2, with different grid positions. In each grid position, the minimum box size was one pixel, and the maximum box size was 45 % of the region of interest. The microstructures were quantified when the background and resin pixels were the same colour (black) and the fibre pixels (white) were characterised as the foreground.

Therefore, if we keep using a smaller box size, the count will be higher, and the count approaches infinity when the box size approaches zero. The limit was found to be the slope of the regression line; therefore, the FD was computed by Equation 1 [35–38] for each grid position. The FD of the whole image has been computed by Equation 2 [35–38].

$$FD = \lim \left[\log (F\varepsilon) / \log (\varepsilon)\right]$$
$$FD = \text{slope}\left[\ln (\text{Boxes with Foreground Pixels}) / \ln(\varepsilon)\right] \quad (1)$$
$$\text{Mean FD} = \Sigma\, (FD) / \text{GRIDS} \quad (2)$$

Where $F\varepsilon$ = the foreground pixels
ε = the box size.

IOP Conf. Series: Materials Science and Engineering **987** (2020) 012005 doi:10.1088/1757-899X/987/1/012005

Table 1. The polishing process.

Grinding (SiC Grade)	Force (Newton)	Time (minute)	Rotation Speed (rev / min)
320	25	5	300
400	25	5	300
600	25	5	300
800	25	5	300
1200	25	5	300
Polishing Grade (μm)	**Force (Newton)**	**Time (minute)**	**Rotation Speed (rev / min)**
15	Hand force	5	300
6	Hand force	5	300
1	Hand force	5	300

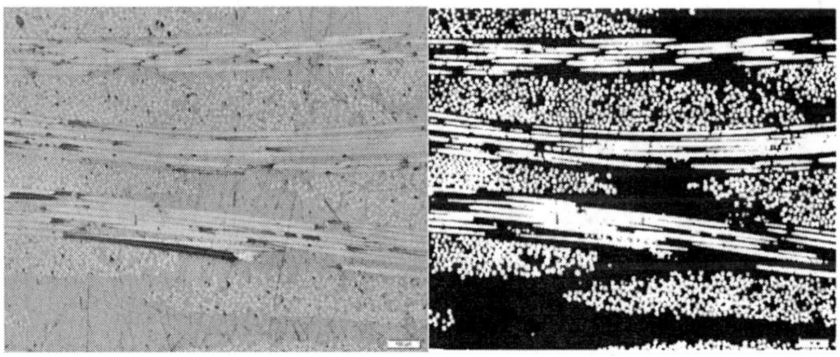

Figure 1. Microscopic image with its binary image.

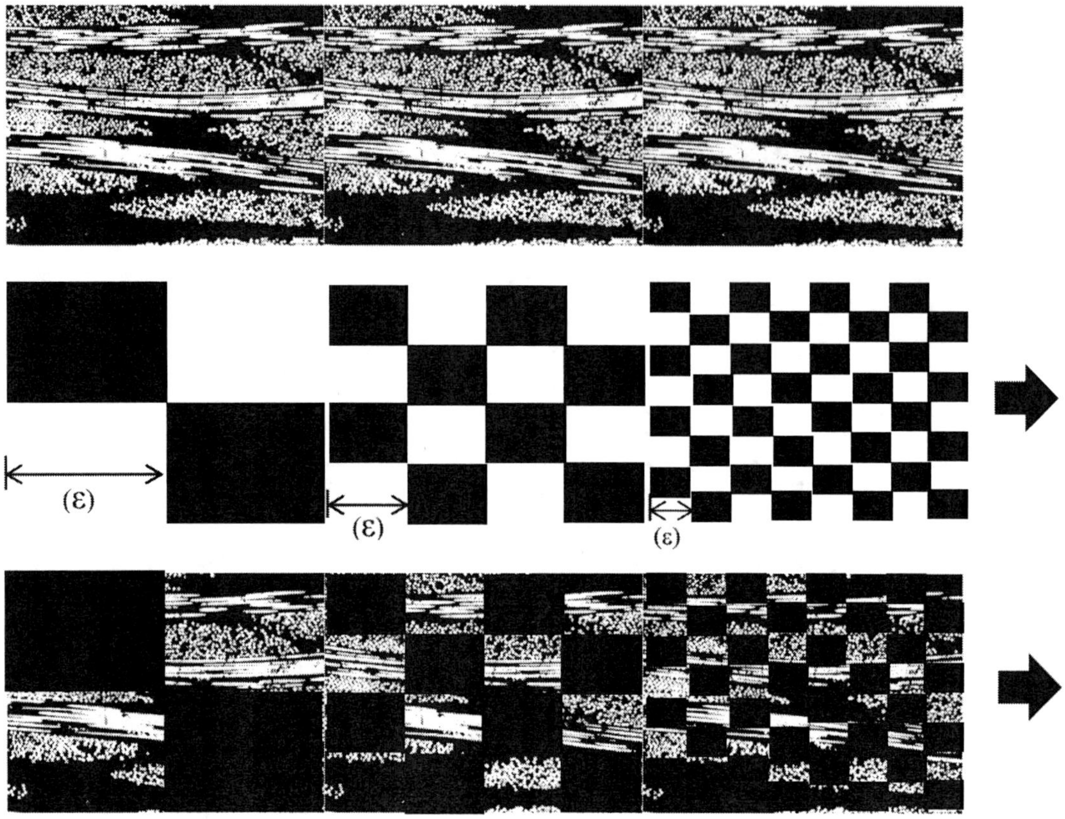

Figure 2. Increasing the box number or decreasing the box size.

3. Mechanical Tests

The interlaminar shear strength (ILSS) was used to measure the bonding force between the fibre and resin and to confirm that the plates are manufactured at an acceptable level. The ILSS test was conducted according to EN ISO 14130: 1998 standards. Three rectangular samples with 10 * 20 * 2 mm (L x W x T) were cut from five different plates for each type of manufacturing process (45 samples in total). The 4- point bending test was conducted according to EN ISO 14125:1998 standards for polymer composites. The Instron Electropuls E300 (model 5527-103, serial number 107190) with a ± five k Newton dynamic load cell and the test speed was one mm per min. The flexural stress σ (Pa) was evaluated using Equation 3.

$$\sigma = \frac{3Fa}{bh^2} \quad (3)$$

Where F is the force (N), a is the distance between the supports (mm) and applied force, h is the thickness of the samples (mm) and b is the width of the samples (mm).
The flexural modulus of elasticity was calculated using Equation 4.

$$E_f = 0.21\,\frac{\Delta F}{\Delta s}\,\frac{L^3}{bh^3} \quad (4)$$

Where E_f in (MPa), L is the span (mm) and s is the beam mid-point deflection (mm). The fibre volume fraction (V_f) for specimens with a different type of manufacturing process was calculated using Equation 5 [38].

$$V_f = n\ AF / \rho_f t \quad (5)$$

Where n is the number of layers, AF is the areal weight of the fabric (kg / m2), ρf is the volume weight of fibre (kg/m3) and t is the laminate thickness (m).

4. Results and Discussion

The microstructures were characterised for composite laminates. The FD and the V_f were calculated. ILSS and flexural tests were conducted. The ILSS, flexural elastic modulus and ultimate flexural stress (UFS) were calculated. Figure (3) illustrates that the FD increases when the manufacturing process is changed from hand lay-up or infusion to RTM. This increase depends on the level of consolidation because more pressure leads to an increase in the V_f of the laminate, which leads to a high ILSS, as illustrated in Figure (4). This means that the FD correlates with the ILSS and V_f. Figure 5 shows that the UFS and flexural elastic modulus is proportional to the level of consolidation due to the increase in FD, V_f and ILSS, as shown in Figures 3 and 4. All the results show that the FD correlates with the mechanical properties for the composite laminates. A high FD means a high V_f and high mechanical performance when we analyse the resin as a background. However, if we analyse the fibre as a background, the issue will be reversed.

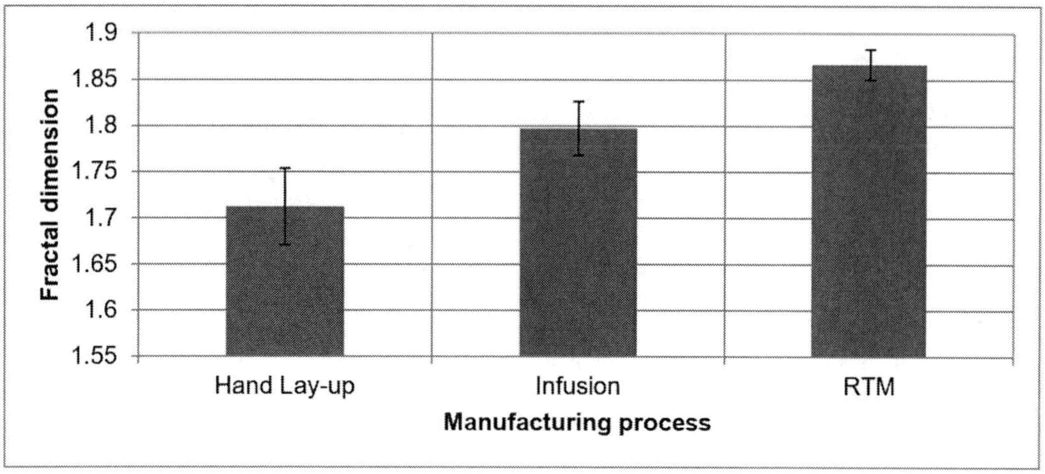

Figure 3. The FD against the type of manufacturing process.

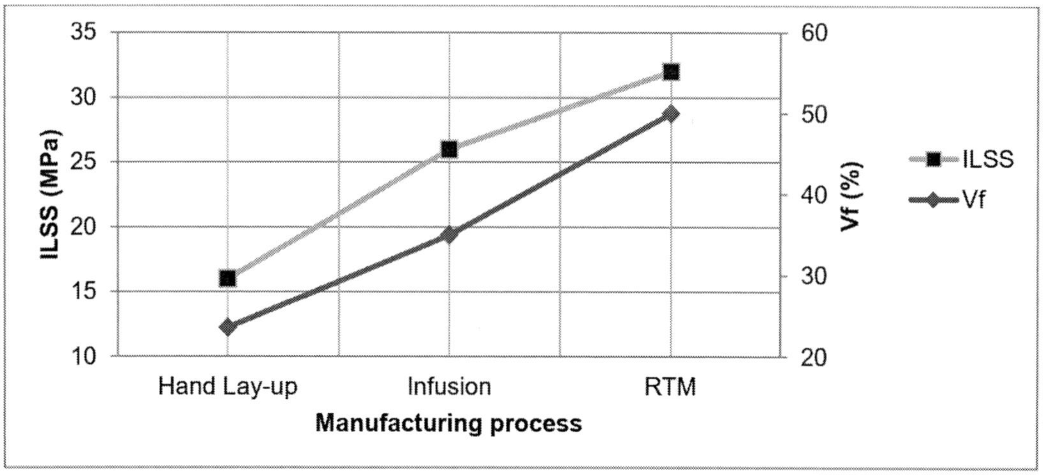

Figure 4. The ILSS and the V_f against the type of manufacturing process.

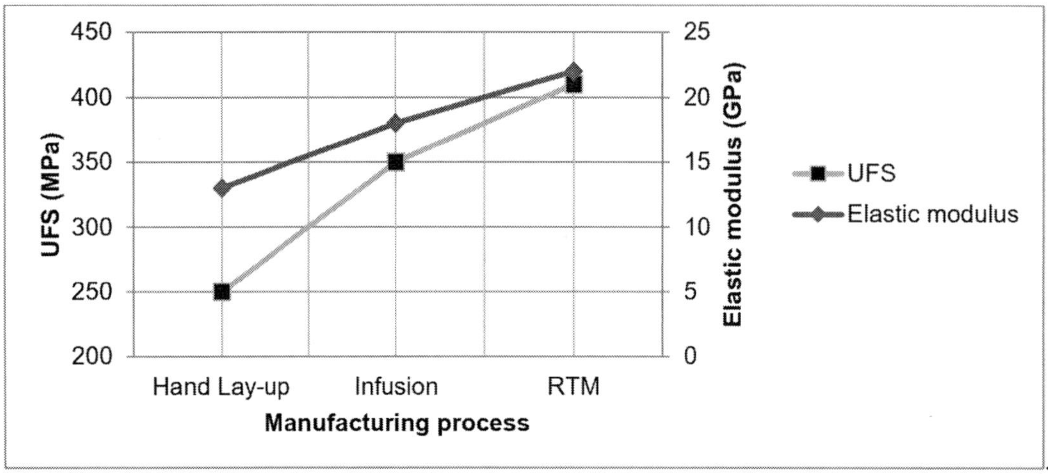

Figure 5. The UFS and elastic modulus against the type of manufacturing process.

5. Conclusion

The study focuses on the evaluation of mechanical properties of composite laminates by FD to replace the mechanical tests that are expensive and time consuming by analysing the microstructure images. The results illustrate that the FD differs according to the method of manufacturing process. The highest value of FD was in the RTM process due to the increase in V_f. FD, V_f, ILSS, UFS and the elastic flexural modulus changed according to the type of the manufacturing process. The results show a significant relationship between the mechanical performance of the composite laminates and the FD. The results concluded that FD could be utilised as a factor to characterise the performance of composite laminates.

References

[1] Basford, D. M., P. R. Griffin, S.M. Grove & J. Summerscales. (1995). "Relationship between mechanical performance and microstructure in composites fabricated with flow-enhancing fabrics." Composites 26(9): 675-679.

[2] Aziz, S. H. and M. P. Ansell (2004). "The effect of alkalization and fibre alignment on the mechanical and thermal properties of kenaf and hemp bast fibre composites: Part 1 – polyester resin matrix." Composites Science and Technology 64(9): 1219-1230.

[3] Gojny, F. H., M. H. G. Wichmann, et al. (2005). "Influence of nano-modification on the mechanical and electrical properties of conventional fibre-reinforced composites." Composites Part A: Applied Science and Manufacturing 36(11): 1525-1535.

[4] Tzetzis, D. and Hogg, P. J. (2006) 'Bondline toughening of vacuum infused composite repairs', Composites Part A: Applied Science and Manufacturing, 37 (9), pp. 1239-1251.

[5] Dhakal, H. N., Z. Y. Zhang & M.O.W. Richardson (2007). "Effect of water absorption on the mechanical properties of hemp fibre reinforced unsaturated polyester composites." Composites Science and Technology 67(7–8): 1674-1683.

[6] Liu, R. M and Liang, D. K. (2010) 'Experimental study of carbon fiber reinforced plastic with embedded optical fibers', Materials and Design, 31 (2), pp. 994-998.

[7] Amjed Saleh Mahmood, M Neil James, John Summerscales. "Process-property-performance relationships in CFRP composites using fractal dimension", IOP Conference Series: Materials Science and Engineering, 2018

[8] Lomov, S. V., Ivanov, D. S., Verpoest, I., Zako, M., Kurashiki, T., Nakai, H., Molimard, J. & Vautrin, A. (2008) 'Full-field strain measurements for validation of meso-FE analysis of textile composites', Composites Part A: Applied Science and Manufacturing, 39 (8), pp. 1218-1231.

[9] Sudarisman & Davies, I. J. (2008) 'The effect of processing parameters on the flexural properties of unidirectional carbon fibre-reinforced polymer (CFRP) composites', Materials Science and Engineering: A, 498 (1-2), pp. 65-68.

[10] Campos, K. A. de, Augusto, J., Pereira, T. A. & Hein, L. R. d. O., '3-D reconstruction by extended depth-of-field in failure analysis – Case study II: Fractal analysis of interlaminar fracture in carbon/epoxy composites'. Engineering Failure Analysis, 2012, 25 pp 271-279.

[11] Worrall, C.M. and Wells, G.M., Fibre distribution in discontinuous fibre reinforced plastics: characterisation and effect on material performance', Proc. 7th European Conference on Composite Materials, London, 1996, pp. 247-252.

[12] Gao, S.-L., Mäder, E. & Zhandarov, S. F., 'Carbon fibers and composites with epoxy resins: Topography, fractography and interphases'. Carbon, 2004, 42 (3). pp 515-529.

[13] Pearce, N. R. L., Guild, F. J. & Summerscales, J., 'An investigation into the effects of fabric architecture on the processing and properties of fibre reinforced composites produced by resin transfer moulding'. Composites Part A: Applied Science and Manufacturing, 1998, 29 (1-2). pp 19-27.

[14] Pearce, N.R.L., Summerscales, J. and Guild, F.J., Improving the resin transfer moulding process for fabric-reinforced composites by modification of the fabric architecture, Composites Part A: Applied Science and Manufacturing, 2000, 31(12), pp 1433-1441.

[15] Summerscales, J., Guild, F.J., Pearce, N.R.L. and Russell, P.M., Voronoi cells, fractal dimensions and fibre composites, Journal of Microscopy-Oxford, 2001, 201(2), pp 153-162.

[16] Summerscales, J., Russell, P.M., Lomov, S., Verpoest, I. and Parnas R.S., The fractal dimension of X-ray tomographic sections of a woven composite, Advanced Composites Letters, 2004, 13(2), pp 113-121.

[17] Mikhaluk, D. S., Truong, T. C., Borovkov, A. I., Lomov, S. V. & Verpoest, I., 'Experimental observations and finite element modelling of damage initiation and evolution in carbon/epoxy non-crimp fabric composites'. Engineering Fracture Mechanics, 2008, 75 (9). pp 2751-2766.

[18] Picu, R.C., Li, Z., Soare, M.A., Sorohan, S., Constantinescu, D.M. and Nutu, E., 'Composites with fractal microstructure: The effect of long range correlations on elastic–plastic and damping behaviour', Mechanics of Materials, 2014, 69 pp 251-261.

[19] Allen, M., Brown, G. J. & Miles, N. J., 'Measurement of boundary fractal dimensions: review of current techniques'. Powder Technology, 1995, 84 (1). pp 1-14.

[20] Pimenta, S. and Pinho, S.T., An analytical modle for the translaminar fractrue toughness of fibre composites with stochastic quasi-fractal fracture surfaces, Journal of the Mechanics and Physics of Solids, 2014, 66 pp 78-102.

[21] Aniszewska, D. and Rybaczuk, M., 'Fractal characteristics of defects evolution in parallel fibre reinforced composite in quasi-static process of fracture'. Theoretical and Applied Fracture Mechanics, 2009, 52 pp 91-95.

[22] Mishnaevsky, L., 'Hierarchical composites: Analysis of damage evolution based on fiber bundle model'. Composites Science and Technology 2011, 71 pp 450-460.

[23] Hale, R. D., 'An experimental investigation into strain distribution in 2D and 3D textile composites'. Composites Science and Technology, 2003, 63 (15). pp 2171-2185.

[24] Mahadik, Y., Brown, K. A. R. & Hallett, S. R., 'Characterisation of 3D woven composite internal architecture and effect of compaction'. Composites Part A: Applied Science and Manufacturing, 2010, 41 (7). pp 872-880.

[25] Pitchumani, R. and B. Ramakrishnan, *A fractal geometry model for evaluating permeabilities of porous preforms used in liquid composite molding.* International journal of heat and mass transfer, **42**(12): p. 2219-2232, 1999.

[26] Campos, K.A.d., et al., *3-D reconstruction by extended depth-of-field in failure analysis – Case study II: Fractal analysis of interlaminar fracture in carbon/epoxy composites.* Engineering Failure Analysis, **25**(0): p. 271-279, 2012.

[27.] Gao, S.-L., E. Mäder, and S.F. Zhandarov, *Carbon fibers and composites with epoxy resins: Topography, fractography and interphases.* Carbon, **42**(3): p. 515-529, 2004.

[28] Allen, M., G.J. Brown, and N.J. Miles, *Measurement of boundary fractal dimensions: review of current techniques.* Powder Technology, **84**(1): p. 1-14, 1995.

[29] Carpinteri, A. and B. Chiaia, *Power scaling laws and dimensional transitions in solid mechanics.* Chaos, Solitons & Fractals, **7**(9): p. 1343-1364, 1996.

[30] Celli, A., et al., *Fractal analysis of cracks in alumina–zirconia composites.* Journal of the European Ceramic Society, **23**(3): p. 469-479, 2003.

[31] Biancolini, M.E., et al., *Fatigue cracks nucleation on steel, acoustic emission and fractal analysis.* International Journal of Fatigue, **28**(12): p. 1820-1825, 2006.

[32] Kuznetsov, P.V., V.E. Panin, and J. Schreiber, *Fractal dimension as a characteristic of deformation stages of austenite stainless steel under tensile load.* Theoretical and Applied Fracture Mechanics, **35**(2): p. 171-177, 2001.

[33] Venkatesh, B., D.L. Chen, and S.D. Bhole, *Three-dimensional fractal analysis of fracture surfaces in a titanium alloy for biomedical applications.* Scripta materialia, **59**(4): p. 391-394, 2008.

[34] Yuan, C.Q., et al., *The use of the fractal description to characterize engineering surfaces and wear particles.* Wear, **255**(1–6): p. 315-326, 2003.

[35] Foroutan-Pour, K., Dutilleul, P. & Smith, D., 'Advances in the implementation of the box-counting method of fractal dimension estimation'. Applied Mathematics and Computation, 1999, 105 (2). pp 195-210.

[36] Pitchumani, R. & Ramakrishnan, B., 'A fractal geometry model for evaluating permeabilities of porous preforms used in liquid composite molding'. International journal of heat and mass transfer, 1999, 42 (12). pp 2219-2232.

[37] Li, J., Du, Q. & Sun, C., 'An improved box-counting method for image fractal dimension estimation'. Pattern Recognition, 2009, 42 (11). pp 2460-2469.

[38] Amjed Saleh Mahmood, Processing-performance relationships for fibre-reinforced composites'. PhD thesis 2016, University of Plymouth.

Rheological and Mechanical Investigation of PET, LDPE and HDPE Wastes Blended for Re-Use in Bio-Pipes

Ruqaya Khairy Jabir[1], Nizar Jawad Hadi[2], and Ali Abd-alameer AL-Zubiedy[3]
[1,2,3]Polymer and Petrochemical Industries Department,
College of Material Engineering/ University of Babylon, Iraq
ruqayaalmusawi965@gmail.com[1], nizarjawadhadi@yahoo.com[2],
mat.ali.alzubiedy@uobabylon.edu.iq[3]

Abstract. This work focuses on the re-use of biopolymer wastes to produce the pipes and reduce the impact of these materials on the environment. The ratios of 10, 20, 30 and 40wt% of recycled polyethylene terephthalate (RPET) were added to the reference blend, which consists of recycled low-density polyethylene (RLDPE) and recycled high-density polyethylene (RHDPE). Rheological and mechanical tests were performed on these blends. The blend of RLDPE and RHDPE was already successful in the manufacturing of pipes. The capillary rheometer was used to check the shear viscosity and shear stress behaviour with the shear rate increasing for different blends. The density, tensile strength, elastic modulus and impact strength were also tested for all blends. The results showed that the shear viscosity decreases and the shear stress increases with the shear rate increasing for all blends. In general, the addition of RPET to the reference blend decreases the viscosity at each shear rate. The blends of ten wt% and 20wt% are more compatible with the reference blend, while the blends of 30wt% and 40wt% exhibit a clear deviation after a shear rate of 300. The density indicates an increase with the increase of RPET to the reference blend. The tensile strength increases and impact strength decreases with the increase of RPET up to 30%. After that, the tensile strength decreases and impact strength increases up to the 40wt% RPET. The results showed that the rheological test can be used to predict the mechanical behaviour. Additionally, there was a good agreement observed between the rheological and mechanical tests. The ten wt% and 20wt% blends were more suitable for this task.
Keywords. Polymers Wastes, Rheological Behaviour, Mechanical Tests, Bio Pipes.

1. Introduction

Plastics are synthetic polymers with a high resistance to acids, bases and chemicals. Plastics are resistant to microorganisms; therefore, they are not biodegradable. One plastic that is extensively used in certain industries, such as textile and packing, is polyethylene terephthalate (PET) [1]. PET is a linear polyester [2].PET's important applications in packaging include the manufacturing of soft drink bottles and other beverages and fibre production.

PET has many important features, including high compressive and tensile strength, good chemical resistance, high impact resistance, transparency, fracture resistance, gas penetration resistance, heat resistance, good electrical properties, low weight and a lower price compared to other packaging materials, such as glass and metals. PET has a long life due to its resistance to biodegradation, which leads to the accumulation of large amounts of PET waste. Due to the development of new technologies, it is

impossible to avoid the use of plastics. Therefore, plastic materials should be recycled [1]. Recycling is the main process to reduce waste material [3]. The following two important techniques are used to recycle waste PET: chemical recycling and mechanical recycling. Mechanical recycling of PET waste typically includes the elimination of contamination by separating, washing, drying and melting [1]. Mechanical recycling is the most desirable approach because of its low cost and high reliability. In general, mechanical recycling keeps the molecular structure of the polymer molecule intact [4]. A mechanical recycling method that involves blending the plastics is one of the solutions for recycling plastics [5]. However, exposure to several rounds of reprocessing recycling may change the molecular structure and molecular weight of PET, which changes the rheological and mechanical properties of PET. Therefore, a mixture of PET with other polymers, in particular polyolefin, may produce a significant equilibrium in the mechanical properties and processability [1].
High-density polyethylene (HDPE) and PET have been widely used in packaging applications and constitute a large portion of post-consumer wastes[5]. HDPE is mostly linear and much more crystalline [6]. It is one of the largest used commodities, thermoplastic, for industrial and household applications. Its mechanical properties make it an ideal material for moulding products, and it is a 100% recyclable material [7]. Low-density polyethylene(LDPE) is a highly branched form [8] and a highly flexible material because it contains numerous side chains that increase the distance between the main C-C chains [9]. LDPE contains short-chain propyl- and butyl branches, as well as a few long-chain branches. Table 1illustrates the structures of these recycled polymers [8].

Table1.Structures of recycled polymers.

Names	Structure
Polyethylene (PE)	$-[CH_2-CH_2]_n-$
Polyethylene terephthalate (PET)	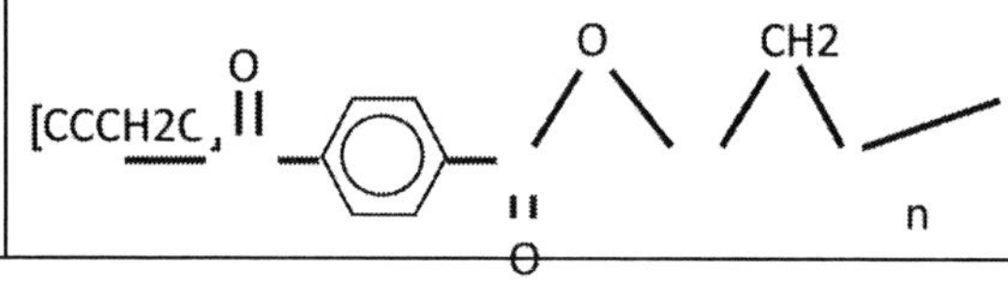

Rheology is a branch of material science that deals with the deformation and stream of matter under stress [10].In academic studies, rheology science is a significant field, and it affects the plastics industry, sprays, detergents, medical research, food and others [11]. Rheological and physical properties are used as an indication as to whether the recycling process is going in the right direction [12]. The elastic modulus is a measure of the stiffness of the material [13].
A capillary rheometer is commonly used in both industries and academia to determine the rheological behaviour of polymer melts at high shear rate levels before evaluating their processability on a full industrial scale [14]. The capillary rheometer is the most frequently used method for determining the flow properties of polymer melts. The fluid property results provided in the capillary rheometer depend on the device's design, with the same die giving different results when used in different device models [15].

2. Materials and Methods

Recycled low-density polyethylene(RLDPE) from agricultural covers waste and recycled high-density polyethylene(RHDPE) from fruits boxes and chairs waste were provided from the Al Ghadeer factory in Karbala in Iraq as pellets. The RLDPE and RHDPE blend is a reference blend that was previously used to produce pipes, as seen in Figure 1.The recycled polyethylene terephthalate (RPET)that was used in this study was collected from water bottle waste from the Iraqi market. After crushing, washing, drying and converting the wastes into flakes, the properties of these recycled polymers are shown inTable 2.

Figure1. The pipes manufactured from the RLDPE / RHDPE blend.

The outer and inner diameters are 31.9 mm and 21.7 mm, respectively.

Table 2. Properties of the used RPET,RLDPE and RHDPE.

Properties	RPET	RLDPE(Branch)	RHDPE(Linear)
MFR (g / 10min)	0.72	9.60	5.34
Density (g/cm3)	1.1584	0.8696	0.9066
Melting point (°C)	250	115	137

3.RPET

The waste from the old plastic's water was collected from multiple sources, the dirt and stickers were washed away and it was dried. Then, the old plastics were cut, which is usually done to reduce the initial volume. The process is conducted using scissors to cut it into longitudinal slices and then small pieces. Then, the cutting machine is used, which contains a series of rotary blades that are driven by an electric motor. The small pieces of PET are placed in the slicing machine via the hopper located above the rotor blade. The slicing product is a pile of plastic flakes, which are rough and irregularly shaped, and they can then be treated. Every 50 old plastic pieces weigh 320 g.

4. Sample Preparation

The polymer blend of RLDPE, RHDPE and RPET are extruded using a co-rotating twin-screw extruder model (SIJ-30A). The diameter of the screw is 30 mm, and the screw speed iszero–320 rpm. The polymer blends are prepared in the following procedures: RLDPE with 62,55,49,43 and 37 wt% , RHDPE with 38,35,31,27 and 23 wt% and RPET with ten,20,30 and 40wt%. These polymers were mixed at room temperature according to the ratios displayed in Table 3to produce sheets for testing.

Table 3. Blends ratios that were used in this work.

Sample NO.	Constituent		
	RPET (wt%)	RLDPE (wt%)	RHDPE (wt%)
1	0	62	38
2	10	55	35
3	20	49	31
4	30	43	27
5	40	37	23

The mixed materials were placed in a co-rotating twin-screw extruder at a constant temperature (210°C) and constant speed (30 rpm) until the resulting sheets were ready to be used in mechanical tests.

5. Characterisation

5.1. Rheological Measurement

The rheological behaviours of different (RLDPE/RHDPE/RPET) blends were determined using a capillary rheometer (SR20) (INSTRON CEAST, Country of Origin Italy) that was equipped with a capillary length of 20 mm and a diameter of one mm. The sample weights were approximately ten g with the shear rate in the range of 100–500 s^{-1} at a temperature of 250°C. The samples were added to the capillary, heated to a specific temperature, held for ten minutes, and then tested at different stresses on their rheological parts. All rheological measurements were performed in rheological parameters and were collected directly from the supplier's computer software (TRIOS, TA Instruments, New Castle, DE, USA).

5.2. Density Test

According to the ASTM D-792, density is measured at room temperature via a digital accuracy of ± 0.0001 g/cm3 using Matsu Haku High Precision DENSITY TESTER GP-12OS. This method can be used with a sheet, rod, tube, moulded articles and powders. The specimen was weighed in both the air and distilled water at 23°C using a sinker and wire to hold the specimen completely submerged as required.

5.3. Mechanical Characterisation

5.3.1. Stiffness of the Pipe

Pipe stiffness is an inherent strength that is measured according to the standard ASTM D2665 test method for the determination of external loading characteristics of plastic pipes. The pipe is deflected at five per cent to the internal diameter, and the load is recorded.

5.3.2. Tensile Test
Instron tester type 5556 Universal Testing Machine (WDW/5E) in accordance with ASTM D638-IV was used to test the tension properties at a crosshead speed of five mm / min. The elastic modulus was also calculated depending on the data from the tensile test.

5.3.3. Impact Strength
The Charpy notched impact samples, according to the ASTM D-256,with dimensions of $65\times13\times3.2$ Pmm^3 were tested by a cast pendulum impact tester. A blunt notch with a notch tip radius of 0.25 mm was introduced to the impact specimens with a CEAST notch opener.

6. Results and Discussion

6.1. Rheological Properties

6.1.1. Viscosity Curve
The shear viscosity behaviour curve shows significant indications about the structure changes of different blends, as shown in Figure 2. Most polymeric materials in the viscosity curve show shear-thinning behaviour, and the polymers used in this research are among them. The shear behaviour indicates changes that occur in the structure and chains of the polymer during the extrusion process. These changes are related to the rheological behaviour during the extrusion process, as well as to the final specifications of the polymer hardened after extrusion. Therefore, it is important to identify the shear viscosity behaviour of the blends and study the changes that take place.

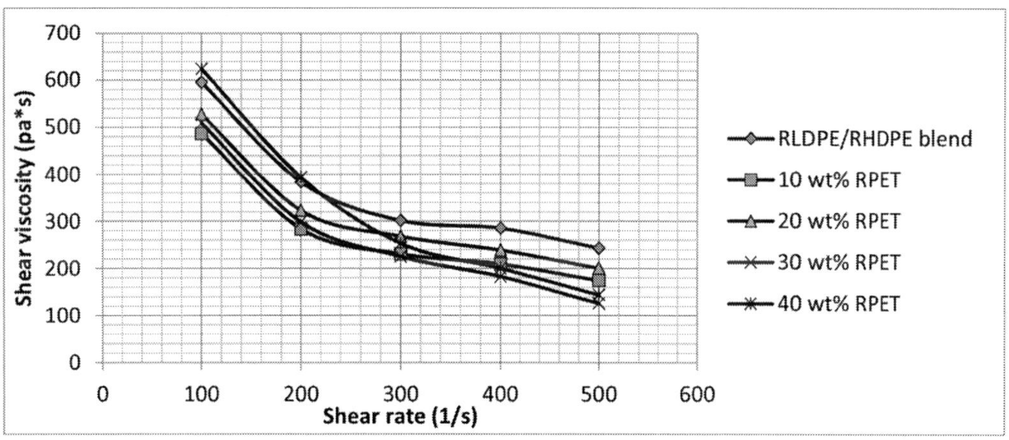

Figure 2. Shear viscosity behaviour vs. shear rate of different LDPE/HDPE/PET recycled blends.

In Figure 2, the shear viscosity decreases with the increase of the shear rate of the first blend(RLDPE and RHDPE). This blend is to prove success in the manufacturing of the pipe. The viscosity decreases rapidly from 100 to 200 shear rates, and then gradually slopes to 300 shear rates. Finally, the viscosity becomes more stable with the shear rate of up to 500. After adding the RPET to the first blend by ten wt%, the same style of shear-thinning behaviour is observed with an almost equal decrease along the curve by 100pa.s. When adding 20wt% of RPET to the first blend, the resulting viscosity curve is less than the viscosity of the first blend by approximately 60pa.s. However, when adding 30wt% of RPET to the first blend, the viscosity behaviour is similar to the previous three blends up to the shear rate of 300. After that, the viscosity deviates from the usual behaviour. Finally, when adding 40wt% of the RPET, the shear viscosity decreases rapidly until the 300 shear rate. Then, the speed continues to decrease with less intensity but more values compared to the rest of the ratios.

According to the foregoing, the closest blend to the first blend with the rheological behaviour is the blend with a ratio of 20wt%, followed by the blend with a ratio of ten wt%. The last *blend with a ratio of* 30wt% and 40wt% is unstable and differs to the behaviour of the first blend [16].

6.1.2. Flow Curve

Figure 3shows that the shear stress gradually increases with the shear rate increase of the first blend until the shear rate is 300. Then, it rises faster and moves towards stability.

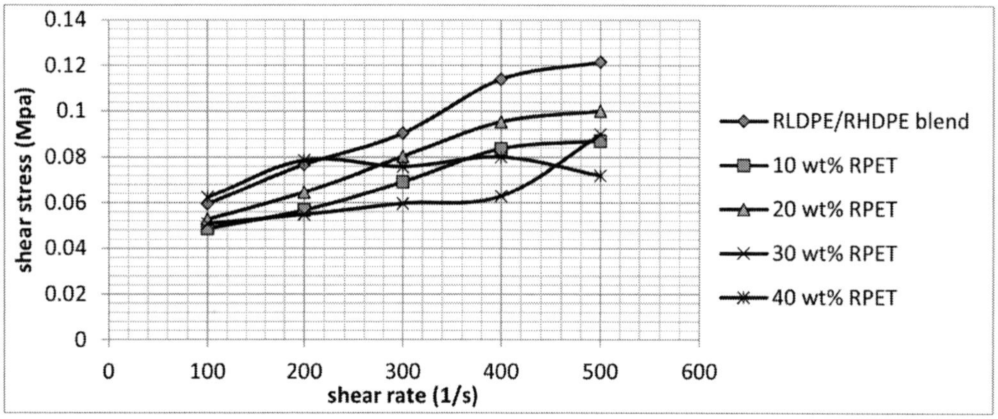

Figure 3. Shear stress behaviour vs shear rate of different LDPE/HDPE/PET recycled blends.

When adding ten and 20wt% RPET to the first blend, the shear stress of these two blends behaves harmoniously and similar to the first blend. Note that the increase in stress is gradual from 100 to 400 shear rates, and then the trend leans towards stability for both blends. Shear stress increases gradually in the 30wt% RPET and 40wt% RPET blends until the shear rate is 400. This oscillation and diffraction that occurs from the natural pathway reflects the change in the composition of the polymers during movement and is consistent with the behaviour of the shear viscosity of the same blends in Figure 2. In general, the ten wt% and 20wt% blends are both closer to the first blend until the shear rate is 300. Therefore, they are more suitable for the aim of this work.

6.2. Density Test

Figure 4 indicates an increase in the density of all mixtures by adding different percentages of RPET because its density is greater than the other two materials. The density values shown in this figure are reasonable based on the densities of the three substances in Table 2, as the highest density is 1.07(g/cm^3).

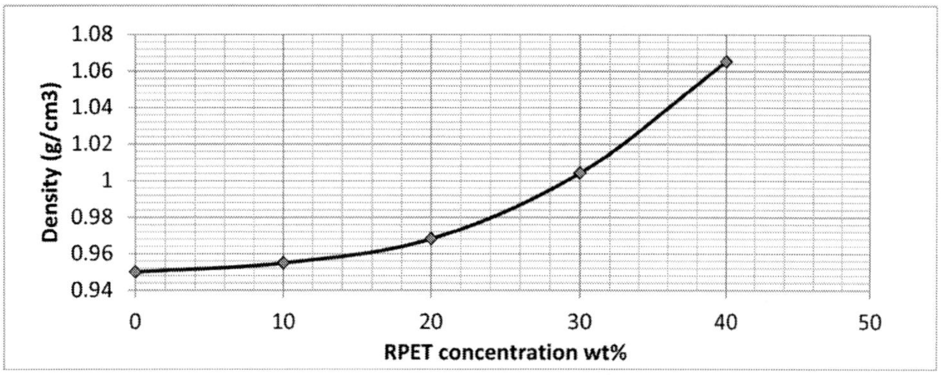

Figure 4. Density of the RPET / RLDPE/ RHDPE blends as a function of the RPET concentration.

The crystallinity in polymeric materials is directly proportional to the density and inversely proportional to the branching of its chains. This means that an increase in the percentage of crystallisation reduces the proportion of branching. This change in the composition of the mixtures directly affects the remaining mechanical, physical and thermal properties. Therefore, it is possible to predict what might happen with the two characteristics of tension and impact resistance. When they were examined, the first increased and the second decreased. From the foregoing, density testing can be used to either limit the most acceptable options or obtain initial indications about products [17].

6.3. Mechanical Properties

6.3.1. Stiffness of the Pipe

Pipe stiffness is a characteristic strength of the pipe. Pipe stiffness refers to the resistance to deflection. The greater the pipe's wall thickness, the greater the stiffness, and therefore, its ability to resist external bending loads. The pipe manufactured from the RLDPE / RHDPE blend was tested, and the applied load and the amount of vertical deflection were 0.084 kN and 1.085 mm, respectively. The stiffness of the pipe ring was 5109.7 MPa and can resist cracks and return to its original position without permanent deformation. The stiffness of the manufactured pipes from the RLDPE / RHDPE blend was used as a reference to check the mechanical properties of the suggested blends. There is a direct relation between the tensile strength, Young's modulus and stiffness. Increasing the tensile strength and Young's modulus increases the stiffness, which is important for during pipe manufacturing. In general, the ten wt% and 20wt% blends are more acceptable regarding this property.

6.3.2. Tensile Strength

Figure 5 shows that the tensile strength increases in all mixtures with an increase in the threshold ratio of 30wt%, after which it tends to decrease. Its highest value is 35 MPa. This increase is due to the increase in the percentage of crystallisation and density and the decrease in the proportion of the branching chains in the compounds.

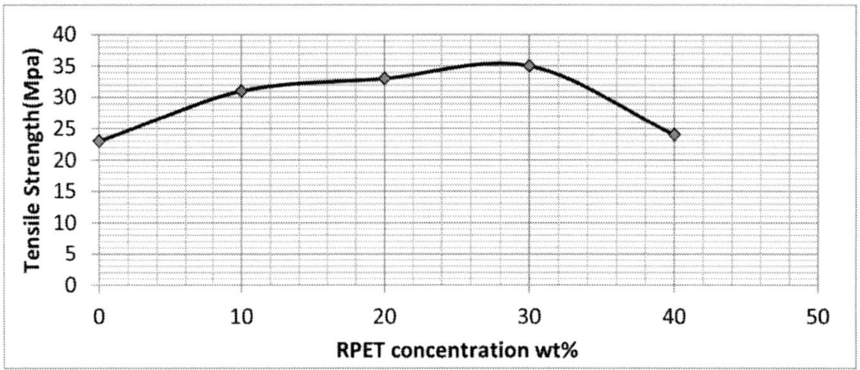

Figure5. Tensile strength of the RPET / RLDPE/ RHDPE blends as a function of the RPET concentration.
Therefore, these results are identical to the results of the density test, which is consistent with the previous analysis in Figure 4. The decrease in tensile strength at 40wt% is due to a change in the crystallisation of the chains' branching ratio and the non-homogeneous distribution in this blend. The tested samples for tensile strength were produced at a certain shear rate. However, the viscosity and chains branching decrease and crystallinity increases with the RPET additions in each shear rate. The melting points of each polymer in Table 2 are another indication of the crystallinity content, as well as the structured content in the blends later [17,18].

6.3.3. Elastic Modulus (E)
In Figure 6, the modulus of the elasticity (tensile modulus) increases in all mixtures when increasing the RPET content, as it reaches its highest value of 100 MPa at 30wt% of the RPET content. This increase is attributed to an increase in the percentage of crystallinity, tensile strength, density and a decrease in the percentage of branching chains in the compounds. The elastic modulus decreased to70 MPa for 40wt% RPET, which is higher than the value of the elasticity of the mixture before adding RPET.

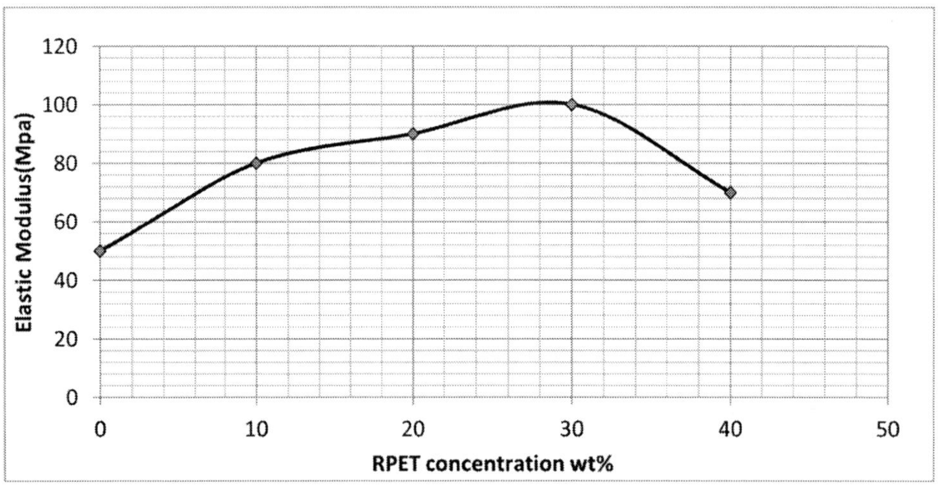

Figure 6. Elastic modulus of the RPET / RLDPE/ RHDPE blends as a function of the RPET concentrations.

The elastic modulus is a measure of the stiffness of the material. Therefore, increasing the elastic modulus increases the stiffness of the material [13].

IOP Conf. Series: Materials Science and Engineering **987** (2020) 012006 doi:10.1088/1757-899X/987/1/012006

6.3.4. Impact Strength
One of the major mechanical tests where the material is subjected to a fast, dynamic load is the impact test. The impact test was conducted at room temperature on the samples using the Charpy method. Depending on the energy required for a sample fracture, the impact resistance was calculated for the polymer blend. Figure 7 shows the reference blend's impact strength as a function of the RPET content.

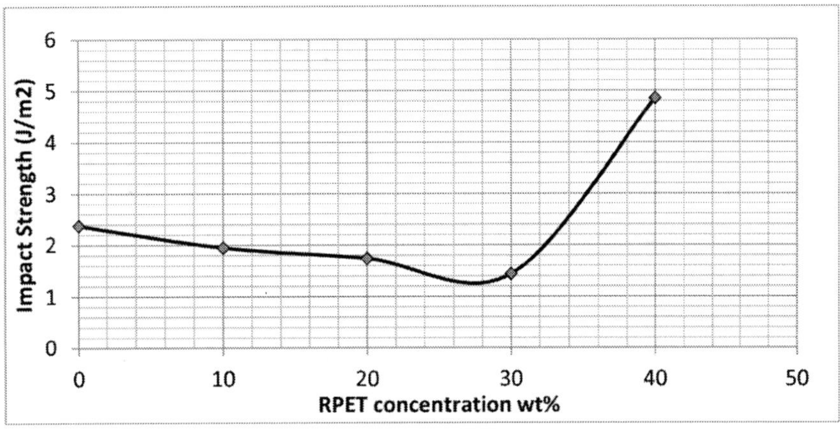

Figure7. The impact strength of the RPET / RLDPE/ RHDPE blends as a function of the RPET concentrations.

First, the impact strength decreases gradually and slightly at ten and 20wt% RPET, and it reaches its minimum value at 30wt% RPET. Then, it rapidly increases and reaches a maximum value of 40wt% RPET. This result matched the tensile, density and viscosity tests. The impact strength decreases with the viscosity, the and chains branching decreases at each shear rate.

From the above results, and according to the rheological and mechanical properties, it may be possible to manufacture similar pipes with better specifications. Ten wt% and 20wt% blends are more suitable for this task [19].

7. Conclusions

In the present work, PET waste (bottles and cans) was recycled and mixed with RLDPE and RHDPE in different ratios. The RLDPE and RHDPE blend that was produced was a successful pipe. The effect of the RPET content on the rheological and mechanical properties of the RLDPE / RHDPE blend was studied. The polymer blends exhibited shear-thinning behaviour. The shear viscosity decreased and shear stress increased as the shear rate increased for all blends. In general, the viscosity decreased as RPET was added at each shear rate. The rheological test can be used as a good indication of the mechanical properties of thermoplastic polymers, which reduces the cost, time and effort to trying to find the optimum results. There was an increase in the density and tensile strength and a decrease in the impact strength with the RPET added at a certain shear rate. The increase in the tensile strength and Young's modulus indicates an increase in the stiffness, which is required for pipe manufacturing. The tensile strength and impact strength showed a good compatibility. Additionally, there was a good match between the rheological and mechanical behaviour.
In general, adding RPET to the reference blend decreases the viscosity at each shear rate. The decrease in the chains branching and density and the increase in the crystallinity is associated with the decrease in viscosity.

The increase of RPET to the reference blend up to 20wt% produced the same shear viscosity and shear stress behaviour with a lower value up to the shear rate 500. The blends of 30 and 40wt% RPET were compatible with the reference blend up to the 300 shear rate. After that, deviation occurred. According to the rheological and mechanical properties, similar pipes could be manufactured with better specifications. Ten wt% and 20wt% blends are more suitable for this task.

8. References

[1] Taghavi S K, Shahrajabian H and Hosseini H M 2018 Detailed comparison of compatibilizers MAPE and SEBS-g-MA on the mechanical/thermal properties, and morphology in ternary blend of recycled PET/HDPE/MAPE and recycled PET/HDPE/SEBS-g-MA *J. of Elastomers & Plastics* **50**(1) 13-35.

[2] Ahani M, Khatibzadeh M and Mohseni M 2016 Preparation and characterization of poly (ethylene terephthalate)/hyper branched polymer nano composites by nelt blending *Nano composites* **2**(1) 29-36.

[3] Hadi N J, Saad N A and Saadi Z A 2014 Study the relation between mechanical and thermal behavior of PP- runner waste blends *J. of Int. Academic Research for Multidisciplinary* **1** 12.

[4] Hamad K, Kaseem M and Deri F 2013 Recycling of waste from polymer materials: an overview of the recent E\works *Polymer degradation and Stability* **98**(12) 2801-2812.

[5] Baker W E, Scott C E, Hu G H and Akkapeddi M K 2001 *Reactive Polymer Blending* vol 1 (Munich: Hanser).

[6] Ravve A 2013 *Principles of Polymer Chemistry* (Springer Science & Business Media).

[7] Shao Y F, Xi Q F , Bernd L and Yiu W M 2009 Processing and impact deformation behavior of high density polyethylene (HDPE)–calcium carbonate nano composites *J. of Macromol Materials Eng.* **2**(14) 35-70.

[8] Subramanian M N 2017 *Polymer Blends and Composites: Chemistry and Technology* (John Wiley & Sons).

[9] Jnr A K L, Yunana D, Kamsouloum P, Webster M, Wilson D C and Cheeseman C 2018 Recycling waste plastics in developing countries: use of low-density polyethylene water sachets to form plastic bonded sand blocks *Waste Manag.* **80** 112-118.

[10] Hadi N J, Saad N A and Alkhfagy D 2016 Rheological behavior of waste polypropylene reinforced with zinc oxide nano particles *WIT Transactions on Eng. Sci.* **105** 201-211.

[11] Mustafa A M K 2017 Experimental and modeling of thermoplastic polymer rheology in capillary rheometer Published MSc University of Babylon Iraq.

[12] Hadi N J, Saad N A and Saadi Z A 2013 Study the rheological behavior of PP-runner waste blends towards using for medical syringes production *Int. J. of Eng. and Technol.* **3**(10).

[13] Balani K, Verma V, Agarwal A and Narayan R 2015 *Physical, Thermal, and Mechanical Properties of Polymers* (Bio Surfaces: A Materials Science and Engineering Perspective) pp 329-344.

[14] Ansari M, Zisis T, Hatzikiriakos S G and Mitsoulis E 2012 Capillary flow of low-density polyethylene *Polymer Eng. & Sci.* **52**(3) 649-662.

[15] Sombatsompop N and Dangtungee R 2001 Flow visualization and extrudate swell of natural rubber in a capillary rheometer: effect of die/barrel system *J. of Applied Polymer Sci.* **82**(10) 2525-2533.

[16] Ogah A O and Afiukwa J N 2012 The effect of linear low density polyethylene (LLDPE) on the mechanical properties of high density polyethylene (HDPE) film blends *Int. J. Eng. Manag. Sci.* **3**(2) 85-90.

[17] Mancini S D and Zanin M 1999 Recyclability of PET from virgin resin *Materials Research* **2**(1) 33-38.

[18] Jarukumjorn K and Chareunkvun S 2007 Compatibilization of recycled high density polyethylene (HDPE)/polyethylene terephthalate (PET) blends *Suranaree J. Sci. Technol* **14**(1).

[19] Ronkay F 2011 Impact of fiber reinforcement on polymer blend properties *Society of Plastics Engineers* DOI 10.

The Rheological Properties of LDPE Nanocomposites per a Capillary Rheometer: An Experimental and Numerical Study

Zahra B Latef[1], Nizar J Hadi[2] and N A Saad[3]

[1,2,3]Polymer and Petrochemical Industries Department, College of Material Engineering/ University of Babylon (Iraq)

zahrabasemlatif@gmail.com[1], nizarjawadhadi@yahoo.com[2], Jasim_910@yahoo.com[3]

Abstract. Rheological properties are very significant in determining processing performance. The increase of only a few percent of nanoparticles by weight will result in substantial improvement in rheological properties. This work focuses on studies of shear viscosity, shear stress and pressure at a shear rate range of (3-1500) s^{-1} and temperature of 170 °C. The flow properties of low-density polyethylene melt and nanocomposite experimentally studied in a single-bore capillary rheometer with a length-to-diameter ratio of (10:1) and numerically using POLYFLOW-Ansys version 15.0 software based on the power-law model. The experimental results show that the viscosity of nanocomposites decreased while the shear stress increased with the shear rate increase. The pressure drop decreases with the shear rate increases for all additional ratios. The experimental and numerical studies of polymer nanocomposites in a capillary die, the shear viscosity, shear stress and pressure behaviour numerically reveal good agreement with the corresponding experimental results.

Keywords: Capillary Rheometer, Viscosity, Shear Stress, Pressure, LDPE nanocomposite, ANSYS / POLYFLOW

1. Introduction

Numerical simulation, an extremely efficient technique, can well be predicted by a complicated phenomenon. However, most recent researchers are still aiming to solve simplified numerical examples, including the hypothesis of material parameters and characteristics of the flow domain. It may be appropriate for numerical algorithms to be established. However, from an industrial point of view, it is essential to regulate entire flow patterns by using a more accurate model for practical flow problems [1]. The analysis of (shear viscosity, shear stress and pressure) in polymer melts is of great importance in many polymer processing operations and has been the subject of interest, both in experiments and in modelling. Recent developments in numerical computation of generalised models, integral models and differential models to solve viscoelastic constitutive equations have enabled simulated predictions of both the entry and exit flow of polymer melts. Therefore, the importance of numerical simulation in predicting the stress distribution and the extrudate swell has grown with experimental data [2].

In the last few years, POLYFLOW has been developed into a sophisticated numerical package for steady polymer flow, dependant on finite element numerical techniques [3,4]. There are three POLFLOW fluid flow analysis systems available in Workbench: The Fluid Flow (POLFLOW) system provides the full simulation capabilities of POLFLOW; the Fluid Flow-Blow Moulding (POLFLOW) system provides only the application-specific capabilities of POLFLOW that are suited to blow-moulding simulations; and the Fluid Flow-Extrusion POLFLOW) system provides only the application-specific capabilities of POLFLOW that are suited to extrusion simulations [4].

Numerical studies of the extrusion process and moulding are conducted in different numerical and software work areas. For example, the ANSYS program is a collection of programs that contain many specialisations. The ANSYS POLYFLOW-based finite element method is designed primarily to simulate applications where flexible viscous and viscous flows play important roles. The flows may be dependent on isothermal or non-isothermal, two- or three-dimensional, stable or time elements. A software instrument equation, governing the conduct of this element that solves them all, creates a comprehensive construction of how the system acts completely. These outcomes are then presented in tabulated or graphic forms. This type of analysis is typically used for the design and optimisation of a system that would otherwise be too complex to analyse by hand. A system that might fit into this class is one that is difficult owing to its geometry, scale or principal equation [5].

POLYFLOW is primarily used in polymer and rubber processing, food rheology, glass furnaces and many other rheological applications to solve flow problems [6]. POLYFLOW's finite element numerical technique includes originally solving the equations for a Newtonian fluid [10]. Then, a parameter of evolution is implemented which gradually transfers the constitutive Newtonian equation to that of the viscoelastic type (generalised model) selected. Using POLYFLOW software, numerical simulations were carried out. POLYFLOW is a commercial finite (version 15.0) for which Crochet and his team created an element package [3, 4].

2. MATERIALS USED

The products used by Subic Company (Iran) under the trade name (HP7022) are low-density polyethylene pellets widely used in the extrusion process; LDPE properties include [(MFR=1.92g/10min at 190°C and 2.16 kg), density=0.91g / cm^3, Mw=79,200g.mol^{-1}]. LDPE melts are known to be very sensitive to temperature and pressure compared to LLDPE and HDPE. Zinc oxide (ZnO) nanoparticles were supplied by (Shijiazhuang Sun power Technology Co., Ltd, China) and had a density of 5.6 g/cm^3 and a diameter mean of d=30–40 nm, and silver (Ag) nanoparticles were supplied by (Intelligent Materials, NANOSHEL Pvt., Ltd, India) and had a density of 10.5 g/cm^3 with particle size about 20nm.

3. Measurement of Rheological Properties by Capillary Rheometer

A capillary rheometer is a device intended to assess shear viscosity and other characteristics of rheology (= flow) [7]. Capillary rheometers as shown in Figure 1 are piston die systems that are intended to assess the viscosity of polymer melts as a result of temperature and deformation rates. They can test fundamental polymers, compounds and multiple composites with small strengthening particles or fibres, metal injection moulding feedstock and similar materials [8]. The capillary is the small cylinder channel through which the materials flow during the experiment. It is defined by a length and diameter, and the ratio of the two is named L/D. It is machined by the die, and sometimes the L/ D is immediately referred to the die [9]. The temperature, sample, shape and preheating parameters refer to the indication given for the basic rheological test.

IOP Conf. Series: Materials Science and Engineering **987** (2020) 012007 doi:10.1088/1757-899X/987/1/012007

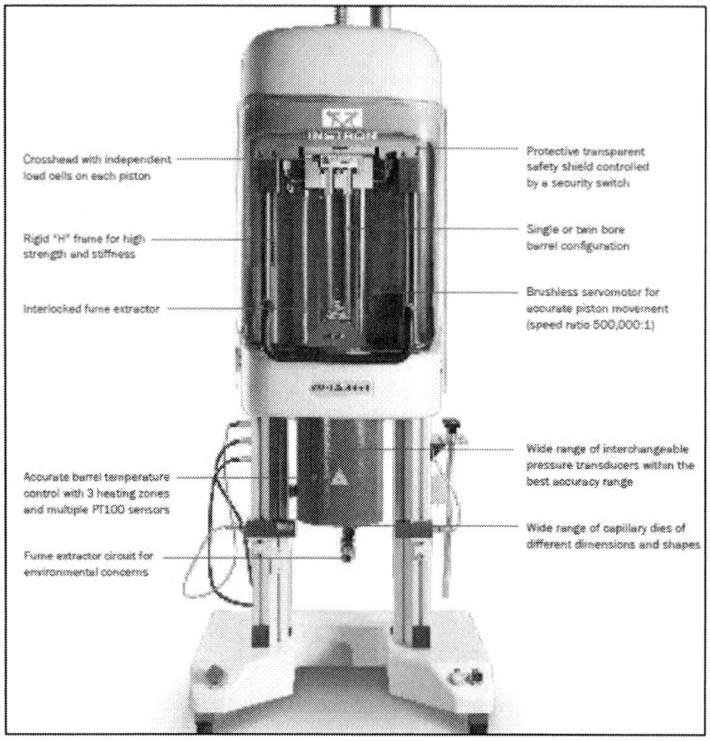

Figure 1. Capillary Rheometer.

4. Main Assumptions

The capillary die flow model can be simulated with the following main assumptions:
1. There is no slip between the molten polymer and the inner surface of the die
2. A 2D model is used for capillary simulation
3. The melted polymer is incompressible
4. A fully developed, steady, isothermal and laminar flow
5. The viscosity is dependent on the shear rate

The power law model was used to check the non-Newtonian viscosity between experimental and numerical according to the following formula:

$$\eta = k\,(\gamma)^{n-1} \quad (1)$$

The finite element numerical technique used by POLYFLOW involves solving the equation's non-Newtonian fluid. The ANSYS program/ POLYFLOW based on the finite element technique was used to simulate the flow in a capillary rheometer [10, 11]. The geometry of the model was created using the ANSYS program (POLYFLOW, 2D planer). First, the researchers determined (inlet) fluid entry area, (wall) reservoir and die, the free surface as shown in Figure 2.A. Second, the researchers determined the contact area between the end of the die and the free surface through Figure 2.B. Third, the researchers applied the mesh to the finite element method to divide the model into several elements and nodes, as shown in Figure 2.D regarding sizing of mesh (fine) (5569 nodes and 5304 elements).

IOP Conf. Series: Materials Science and Engineering **987** (2020) 012007 doi:10.1088/1757-899X/987/1/012007

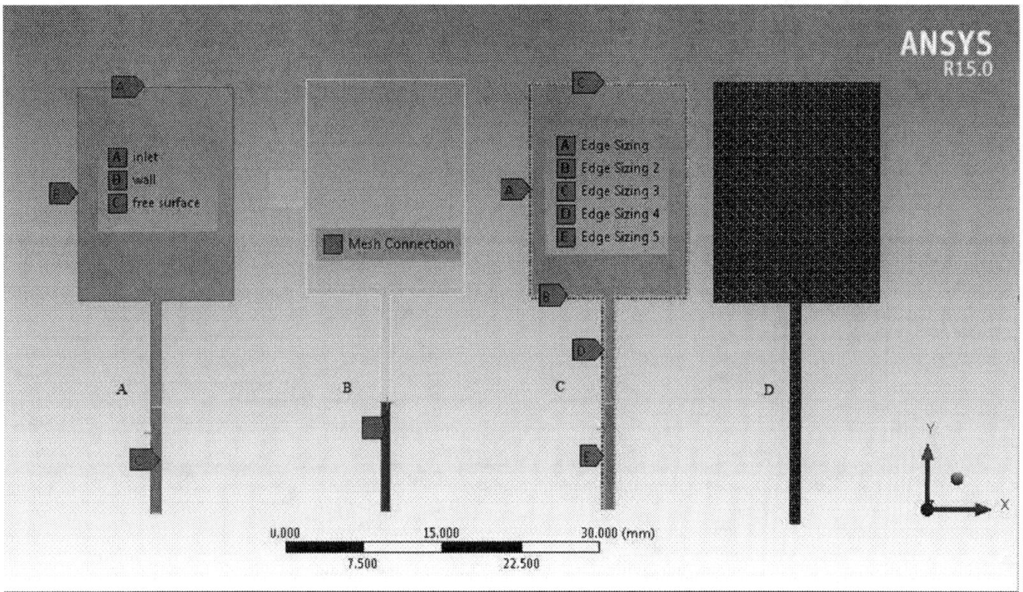

Figure 2. Meshing of A- Inlet, wall and Free Surface B- Mesh connection C- Sizing D- Mesh of model.

5. RESULTS AND DISCUSSION

5.1 Experimental Result

5.1.1 Shear Viscosity

The shear viscosity, shear stress, and pressure were measured over the shear rate range through the one die at a temperature of 170°C. The numerical results exhibited good agreement with the experimental results; therefore, comparisons for shear viscosity, shear stress and pressure were performed quantitatively. The difference between the numerical and the experimental behaviour occurred due to the input data, which is produced from experimental results and calculation. These data are applied in the simulation program as an average value for a certain boundary condition. Figure 3 shows the comparison between experimental and numerical shear viscosity of nanocomposites at different concentrations and additives of (ZnO and Ag) in a capillary die for L/D =10 and 170°C. The experimental shear viscosity indicates lower values compared with its numerical test at a high shear rate. It was clear to understand that the viscosity of the LDPE melt was more sensitive to temperature due to the structure of the chain. The increasing temperature was attributed to an increase in intermolecular interactions, which affected the amount of free volume available between molecular chains of the polymer [12]. Therefore, as the viscosity of the LDPE melts decreased, the flow became easier through the capillary die; then, the pressure drop decreased [13].

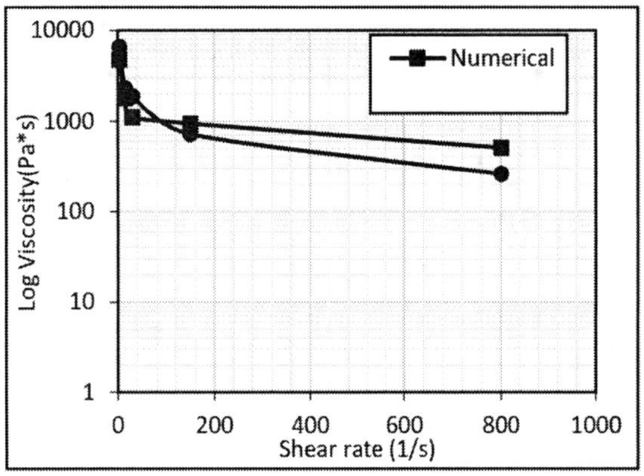

a

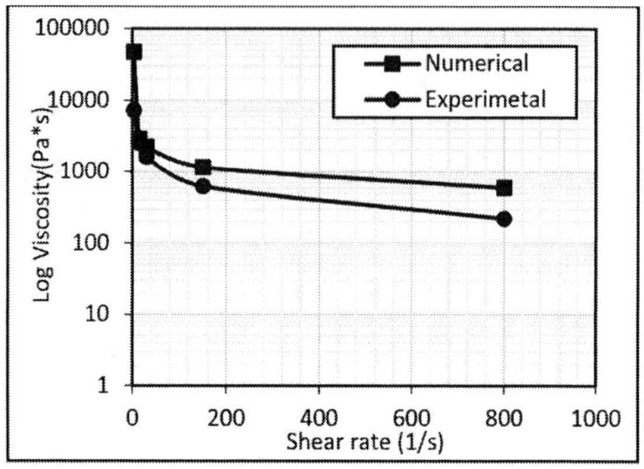

b

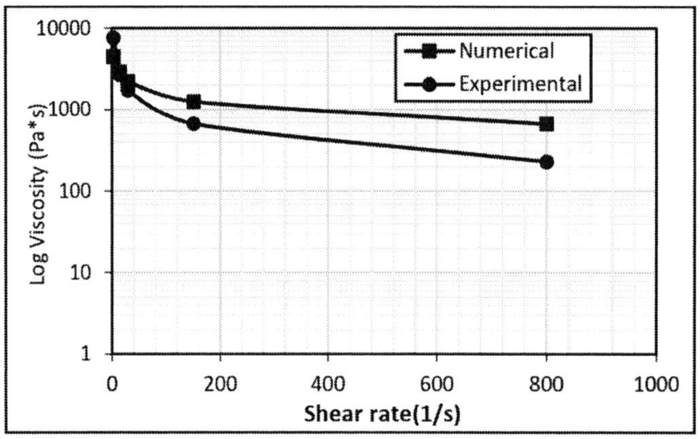

c

Figure 3. Experimental and numerical viscosity curve of LDPE melt and nanocomposites in a capillary die at a) 1% ZnO b) 3% ZnO c) 5%ZnO.

The difference between these two cases converged gradually toward a high shear rate due to the stable input experimental data there. Figure 4 presents the effect of the addition of Ag nanoparticles on the general viscosity behaviour.

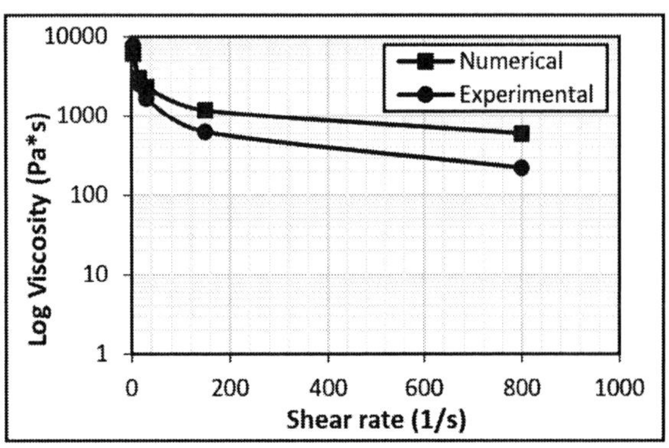

a

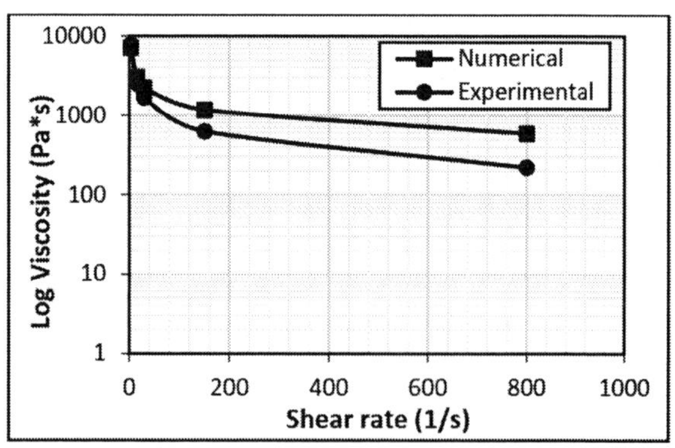

b

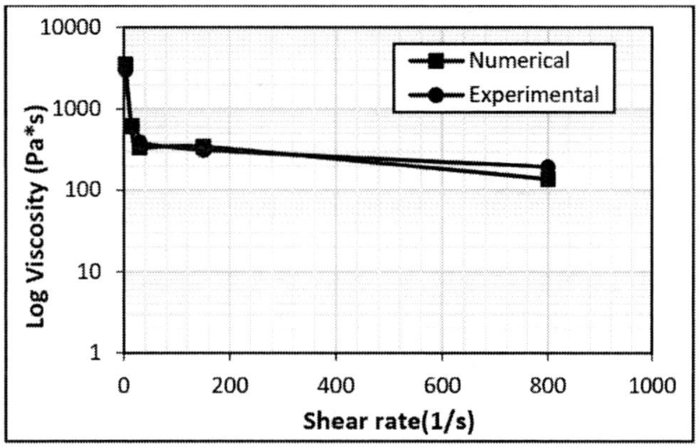

c

Figure 4. Experimental and numerical shear viscosity curve of nanocomposites in a capillary die at (a) 1% Ag (b) 3% Ag (c) 5% Ag.

5.1.2 Shear Stress

The shear stresses of nanoparticle systems were studied for a capillary rheometer with a diameter of 1 mm at 170°C as functions of the shear rate. It can be seen that nanocomposites showed pseudoplastic conduct as shear stress improved in a nonlinear way with shear rate. Figure 5 illustrates that shear stress increases for all additional concentrations' experimental and numerical tests. The numerical behaviour again indicates higher values compared with the experimental results, due to the average input data selected from the real test.

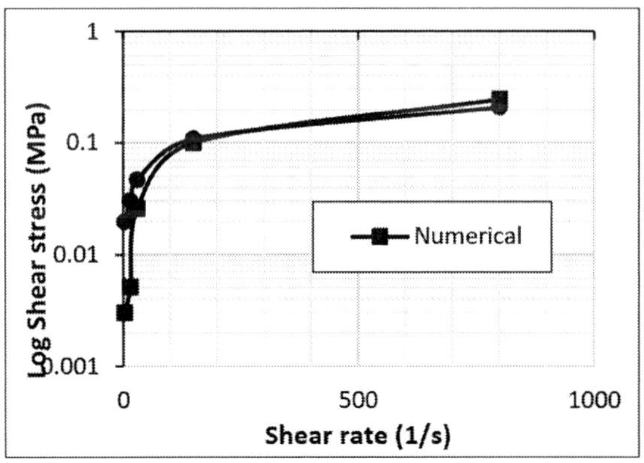

a

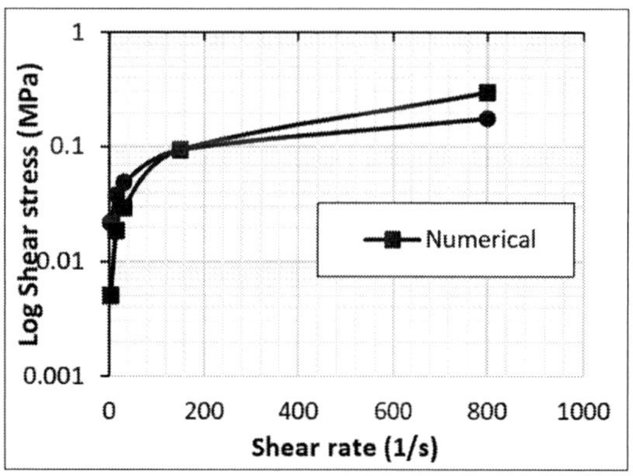

b

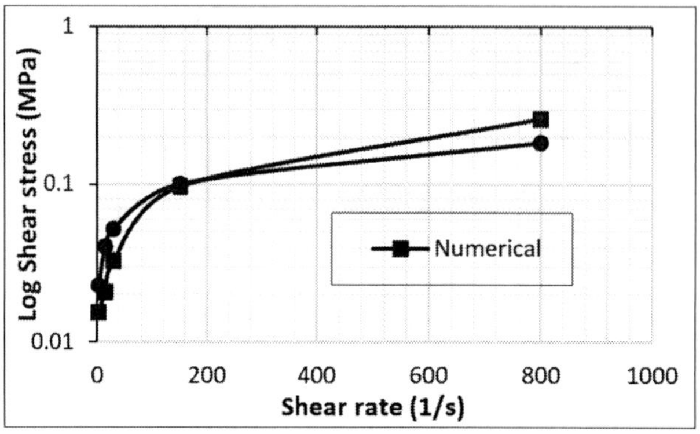

c

Figure 5. Experimental and numerical shear stress curve of nanocomposites in a capillary die at temperature 170°C a) 1% ZnO b) 3% ZnO c) 5% ZnO.

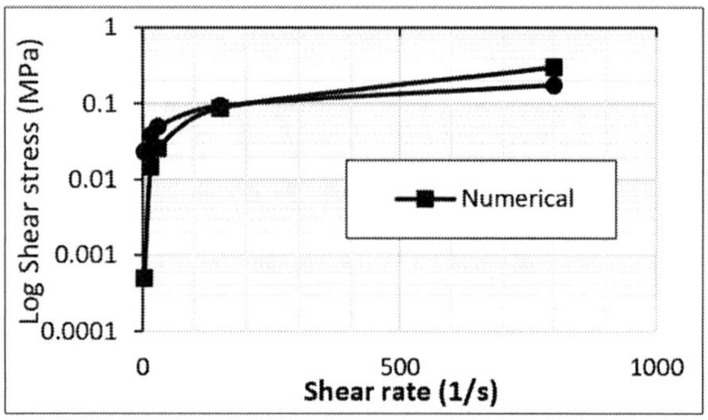

a

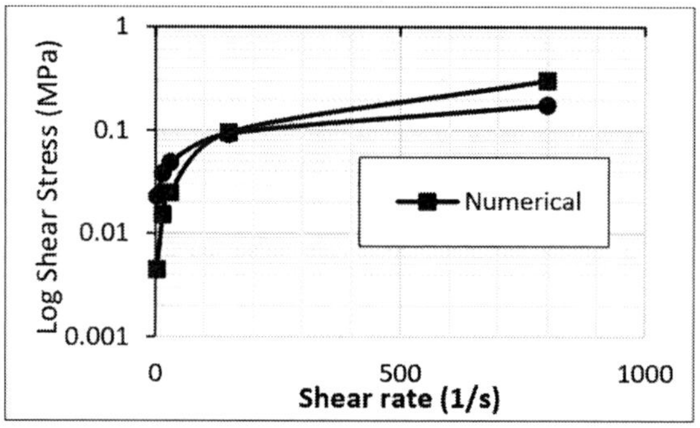

b

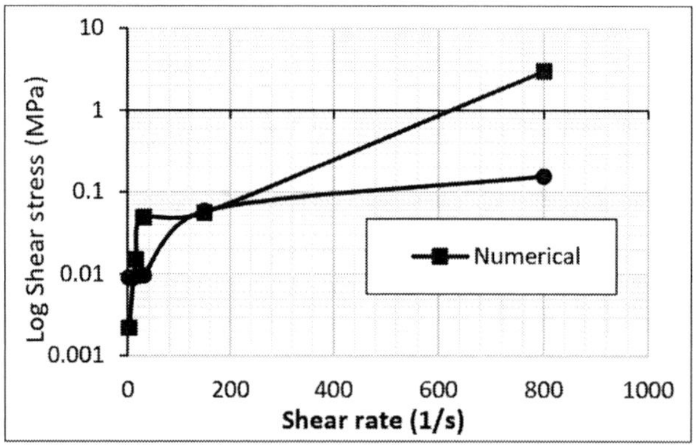

c

Figure 6. Experimental and numerical shear stress curve of nanocomposites in a capillary die at temperature 170°C (a) 1% Ag (b) 3% Ag (c) 5% Ag.

5.1.3 Pressure Drop

The pressure drop of LDPE nanocomposite increased with shear rate increases. Figures 7a, 7b and 7c show the test and simulation pressure comparisons. Under the current shear rate, the influence of shear rate on pressure is evident, and the results of the experiments are lower than those obtained through simulated results.

There are two reasons why simulated outcomes may differ: on the one hand, there is the difference between the mathematical model and the practical processing conditions, such as the die-length-to-diameter ratio, the non-slip hypothesis whose influence is ignored. On the other hand, there is the inherent defect of the constitutive equation, which cannot correctly describe polymer rheological characteristics. Quantitatively predicting field variables directly through experimental measurement is still difficult. However, this can be achieved both qualitatively and quantitatively through numerical simulation technology, thereby revealing the mechanism of material forming.

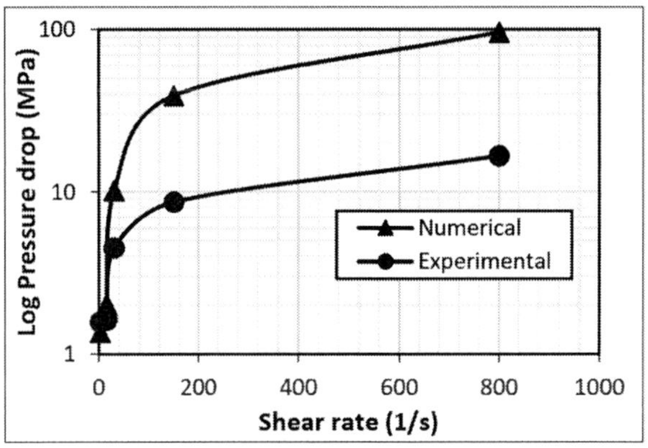

a

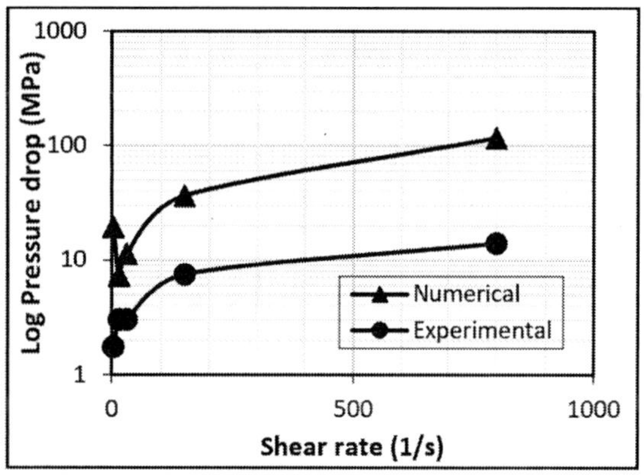

b

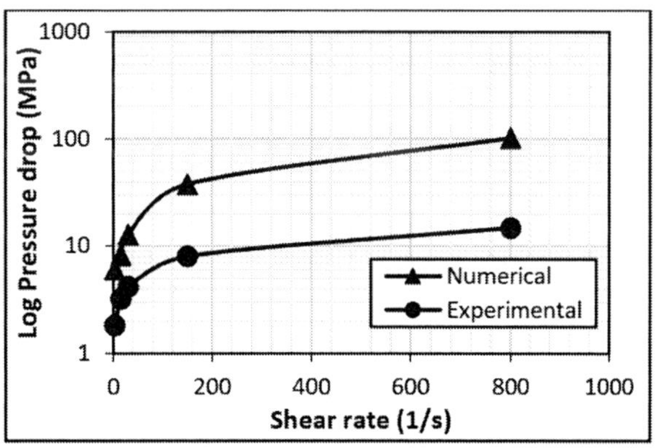

c

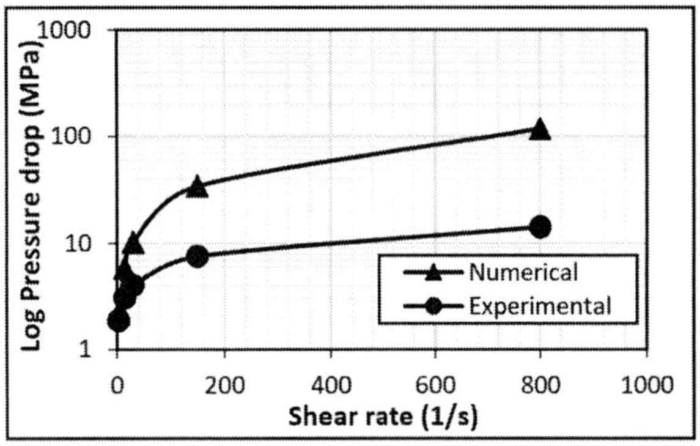

d

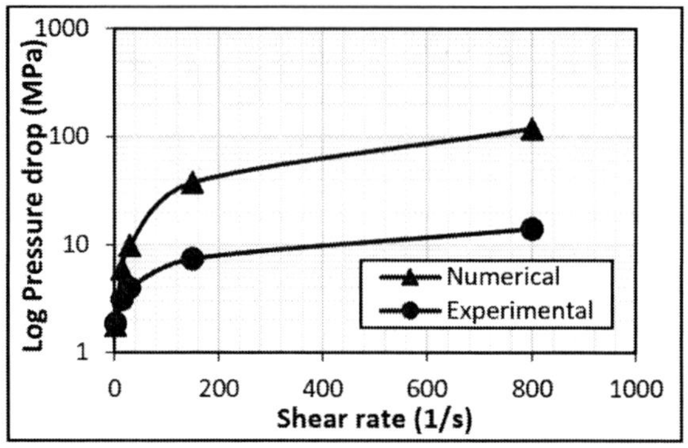

e

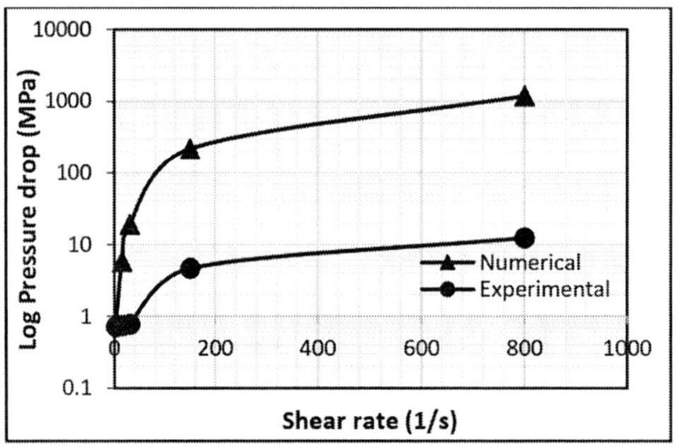

f

Figure 7. Experimental and numerical pressure curve of nanocomposites in a capillary die at temperature 170°C, (a) 1% ZnO (b) 3% ZnO (c) 5% ZnO d) 1% Ag e) 3% Ag f) 5% Ag.

6. NUMERICAL RESULT

6.1 Shear Viscosity

Figures 8 and 9 visualise the 2D steady viscosity, shear rate contours of LDPE melt and nanocomposites in a capillary rheometer at 170°C, L/D=10, and shear rate (3-800) s^{-1}. At the centre of the capillary die, the viscosity reaches a maximum value, gradually reducing towards the die wall; at the same time, shear rate indicates opposite behaviour at the top of the capillary die: the pressure reaches a maximum value. These contours were compatible with the quantitative viscosity curve.

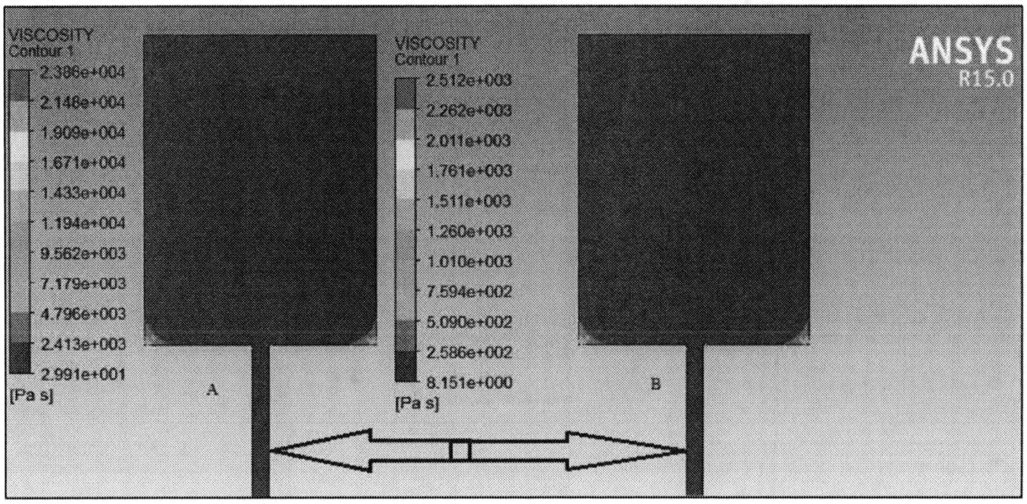

a

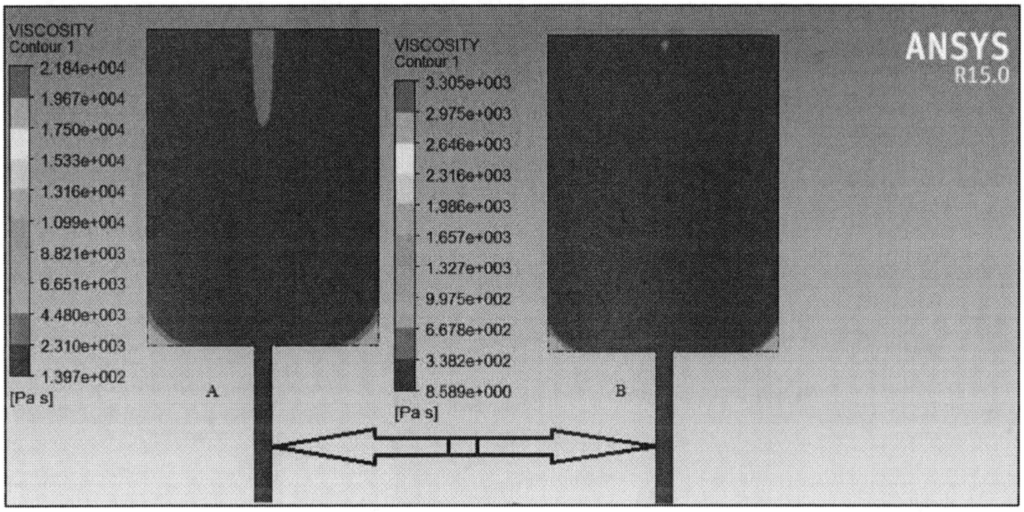

b

Figure 8. Simulation of LDPE nanocomposite in a capillary rheometer of L/D =10, shear rate A) 3 s^{-1} B) 800 s^{-1} a) 1% ZnO b) 5% ZnO.

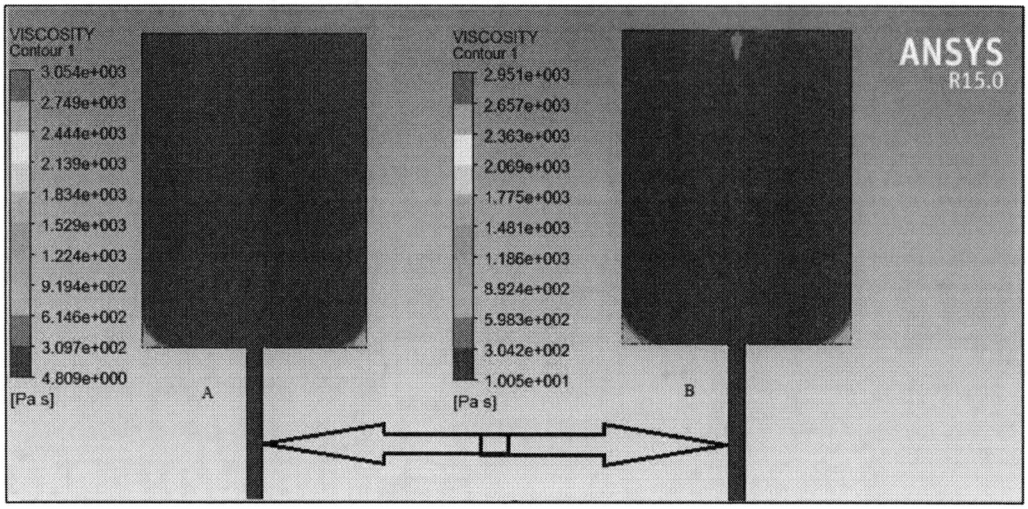

a

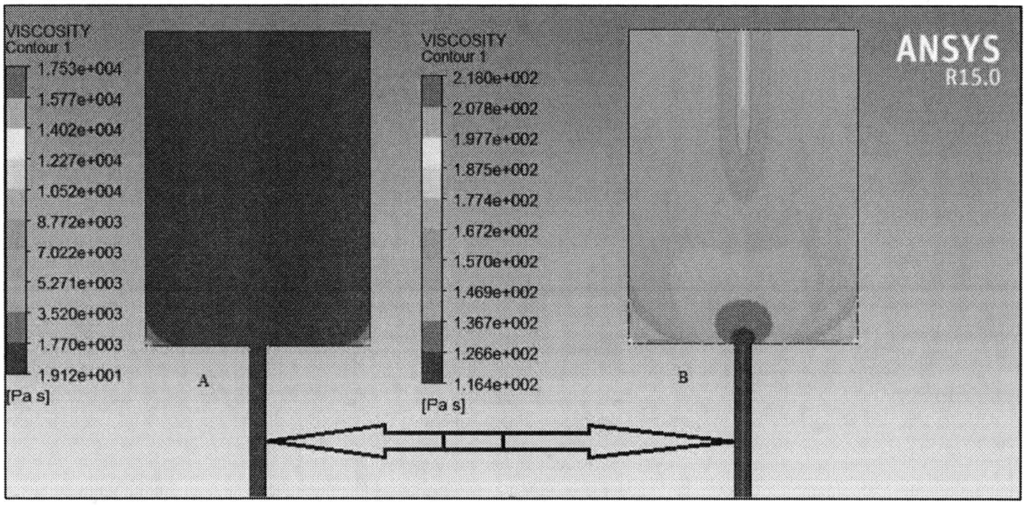

b

Figure 9. Simulation of LDPE nanocomposite in a capillary rheometer of L/D =10, shear rate A) 3 s^{-1} B) 800 s^{-1} a) 1% Ag b) 5% Ag.

6.2 Pressure and Shear Rate Relation

According to Figures 10 and 11, which represent the numerical result of pressure with shear rate, it is very clear that the behaviour demonstrated with pressure value increasing with the increase of the shear rate (3-800 s^{-1}) is good agreement with the experimental results of pressure and shear rate. Figure 10 reveals that the mean of numerical simulation and the power law model are very active in predicting the changes in the pressure with the increase of shear rate. This creates a very accurate means by which to predict the value of the increasing value in the pressure, and it is very active in predicting the effect of the different parameters which are affected by the pressure, such as a nano-effect. The results of the model agree with researchers such as Mustafa [14] as well as with experimental results.

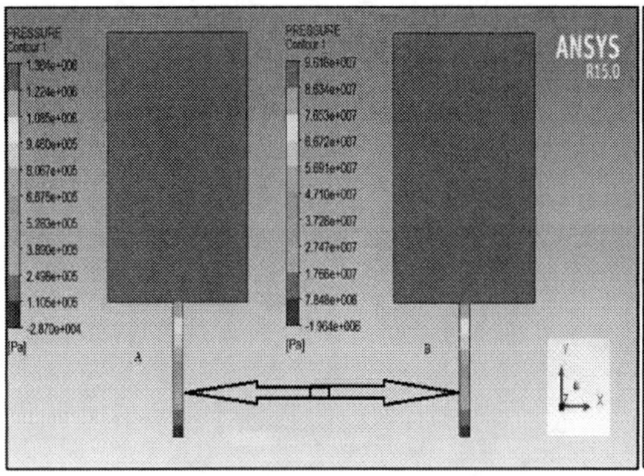

a

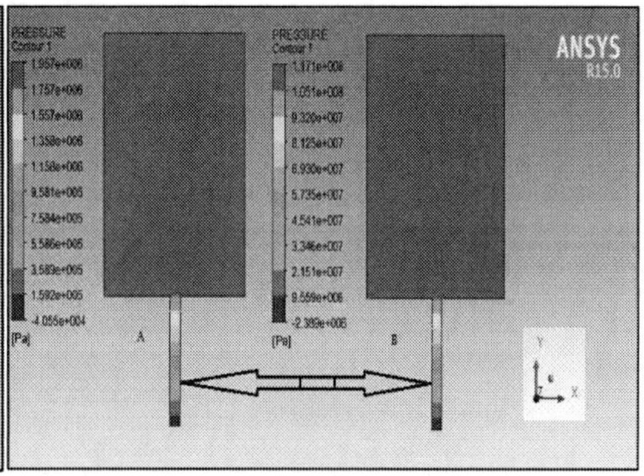

b

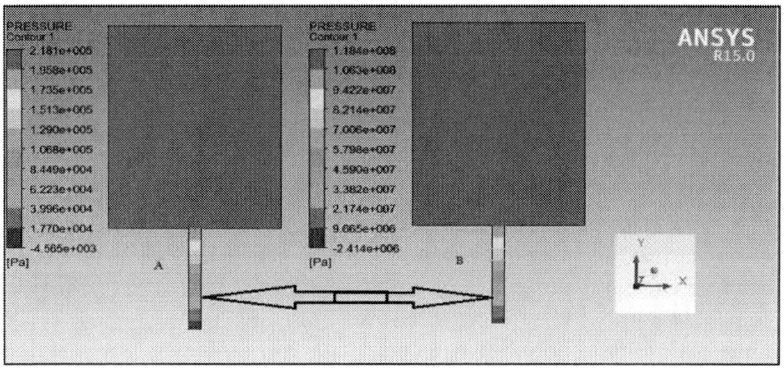

c

Figure 10. Simulation of LDPE nanocomposite in a capillary rheometer of L/D =10 and 170°C at shear rate A) 3 s^{-1} B) 800 s^{-1} a) 1% ZnO b) 3% ZnO c) 5% ZnO.

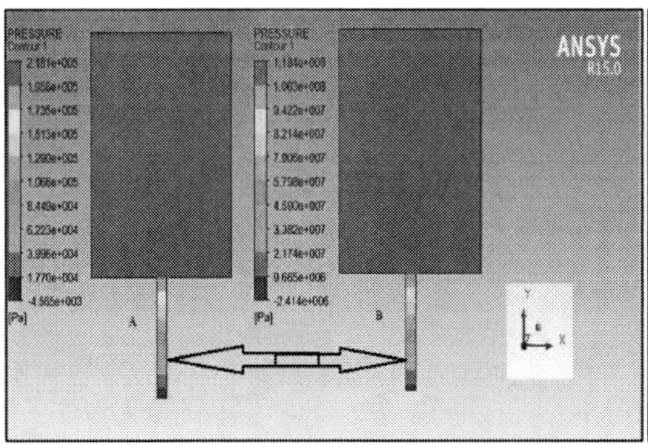

a

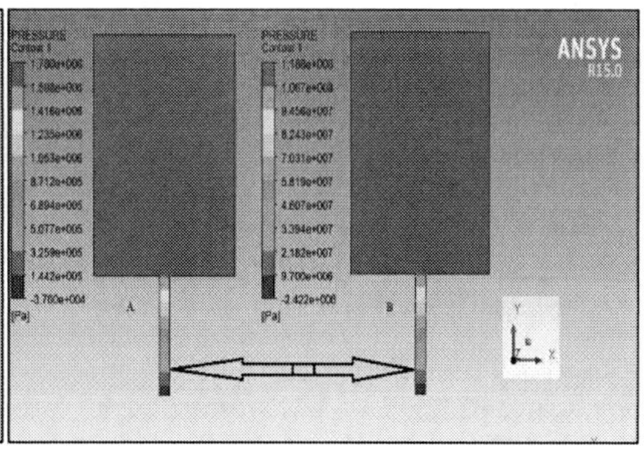

b

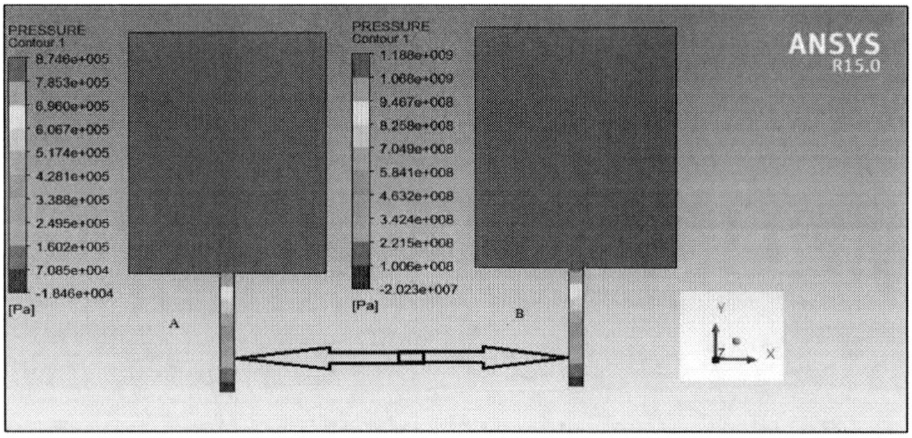

c

Figure 11. Simulation of LDPE nanocomposite in a capillary rheometer of L/D =10 and 170°C at shear rate A) 3 s^{-1} B) 800 s^{-1} (a) 1% Ag (b) 3% Ag (c) 5% Ag.

7. CONCLUSIONS

As the shear viscosity, shear stress and pressure were tested experimentally and numerically, this study concluded that, as the shear rate increases, the shear viscosity decreases, while the pressure and shear stress increase. The numerical simulation of the 2D non-Newtonian flow based on the power law model showed good agreement regarding the shear viscosity, both quantitatively and qualitatively. The experimental and numerical study of polymer nanocomposites as non-Newtonian melt indicates a good agreement between each other.

REFERENCES

[1] Sunarso A, Yamamoto T and Mori N 2007 Numerical analysis of wall slip effects on flow of Newtonian and non-Newtonian fluids in macro and micro contraction channels *J. of fluids eng.* **129**(1) 23-30.

[2] Crochet M J 1990 POLYFLOW theoretical background *VKI Computational F. Dynamics for Industrial Flows.*

[3] Hertel D, Valette R and Münstedt H 2008 Three-dimensional entrance flow of a low-density polyethylene (LDPE) and a linear low-density polyethylene (LLDPE) into a slit die *J. of Non-Newtonian Fluid Mechanics* **153**(2-3) 82-94.

[4] Hertel D and Münstedt H 2008 Dependence of the secondary flow of a low-density polyethylene on processing parameters as investigated by laser-Doppler velocimetry *J. of Non-Newtonian Fluid Mechanics* **153**(2-3) 73-81.

[5] Pepliński K and Mozer A 2011 Design of extrusion die for plastic profile using ansys polyflow software *J. of Polish CIMAC* **6**(3) 221-226.

[6] 'Software/Ansys/17.0/ help,' 2017. https://www.sharcnet.com.

[7] Ansari M, Hatzikiriakos S and Mitsoulis E 2012 Capillary flow of low-density polyethylene *Poly. Eng. and Science* **52**(3) 649-662.

[8] Bernstein B, Kearsley E and Zapas L 1963 A study of stress relaxation with finite strain *Transactions of the Society of Rheology* **7**(1) 391-410.

[9] Kažys R and Rekuvienė R 2011 Viscosity and density measurement methods for polymer melts *Ultragarsas" Ultrasound"* **66**(4) 20-25.

[10] Yang B and Lee L 1987 Effect of die temperature on the flow of polymer melts. Part II: Extrudate swell *Poly. Eng. Sci.* vol. **27**, no. 14 pp 1088–1094.

[11] Hertel D and Valette R 2008 Three-dimensional entrance flow of a low-density polyethylene (LDPE) and a linear low-density polyethylene (LLDPE) into a slit die *J. Nonewton. Fluid Mech.* vol. **153**, no. 2 pp 82–94.

[12] Bettencourt A and Almeida A 2012 Poly (methyl methacrylate) particulate carriers in drug delivery *J. of microencapsulation* **29**(4) 353-367.

[13] Qaiser A, Qayyum A and Rafiq R 2009 Rheological properties of ABS at low shear rates: effects of phase heterogeneity *Malaysian Poly. Journ* **4**, 29-36.

[14] Mustafa A 2017 Experimental and modeling of thermoplastic polymer rheology in capillary rheometer *Int. J. of Mech. Eng. and Technol (IJMET)* vol. **9** pp 193–203

IOP Conf. Series: Materials Science and Engineering **987** (2020) 012008 doi:10.1088/1757-899X/987/1/012008

Preparation of Sustainable Thermal Insulators from Waste Materials

Iman A. Mashkoor [1], Haider M. Mohammad [2] and Prof. Dr Saadoon F. Dakhil [1]

[1]Basra Technical College, Southern Technical University, Iraq

[2]College of Engineering, University of Basrah, Iraq

imanmashkoor@stu.edu.iq, haider.mohammed@uobasrah.edu.iq, s.albahadili@stu.edu.iq

Abstract. This work aims to produce an experimental and theoretical analysis of thermal insulator specifications for buildings with sustainable requirements. In the experimental work, three categories of thermal insulators were prepared from composite materials, and each category had ten models. These composites included the addition of two types of waste (sawdust and tyre waste) as fillers for two types of matrices (liquid polyurethane and polyurethane foam) to obtain composite materials for thermal insulation samples. The prepared samples were subjected to tests to show their thermal properties, such as thermal conductivity and specific heat capacity as well as undergoing a hardness test. The theoretical analysis included the discovery of empirical equations for thermal properties as functions of two variables (temperature and mass ratio) and hardness as a function of one variable (mass ratio). A genetic algorithm optimisation technique was used to find the optimum mass ratio of the composite that produced the required insulation specifications. The results showed that thermal conductivity decreased when the sawdust mass ratio and the rubber waste mass ratio increased but remained under the thermal insulation range. Furthermore, the prepared insulator samples showed an improvement in thermal storage and the hardness of tyre waste (liquid polyurethane composites) and sawdust (polyurethane foam composites). Finally, optimum results were obtained using the optimisation technique.

1. Introduction

The first definition of sustainable development was created by the World Commission on Environment and Development (WCED) in 1987: 'Sustainable development is the development that meets the needs of the present without compromising the ability of future generations to meet their own needs' [1]. The 2005 World Summit showed that sustainable development is based on three fundamental pillars (social, economic and environmental) [1, 2].

Recently, a new area of sustainable development, known as sustainability science, has emerged [3]. Now, sustainability is an area for solving a problem that exists and is directed towards creating an area that helps make relevant decisions in solving this problem [4].

'The building industry is at a crossroads and the question is where do we go from here? The "green" train has left' [5].

In recent years, many researchers have looked at solutions for social mobilisation that are closely related to technological development, using terms such as zero emissions, zero energy housing, energy efficiency, etc. [5]. The biggest energy consumers are buildings, and to reduce consumption there has been an increase in the manufacture of thermal insulators, especially in hotter and colder areas around the world. However, energy sources are constantly depleted, so waste is used as an alternative energy source for sustainable development [6]. The construction sector consumes around 40% of the world's energy and this necessitates the use of sustainable architecture standards. Proper application of building insulation is the best measure of conserving energy in buildings [7].

Thermal insulators can be defined as substances that reduce the transfer of heat between surfaces of a specific object. The thermal insulation principle is based on the replacement of a short heat flow pathway, which has low thermal resistance with long-lasting and high thermal resistance [8].

Many researchers have presented studies on the manufacture of thermal insulators using different methods. Khunthon et al. (2007) tested the manufacture of insulating sheets from waste tissue paper and corn husk and used polystyrene foam as a packaging surface to enhance performance. The study concluded that thermal conductivity was less corrected with an increased amount of corn husk in the mixture. Alternatively, the addition of waste tissue paper increased the strength of the insulating plates. The addition of polystyrene foam has greatly improved mechanical properties [9]. Panyakaew and Fotios (2008) used agricultural waste in Thailand as a thermal insulator for walls and ceilings. After adding certain materials, they chose the best insulators after testing thermal conductivity. The materials used in the test were rice straw, coconut husks, sugar cane waste, palm oil leaves, corn stalks and Durian husks. The first three waste materials recorded the highest values of thermal insulation [10]. Monika et al. (2011) discussed the effect of adding natural fibres to enhance the value of thermal conductivity, using bamboo fibres for their study. The results showed a decrease in the values of conductivity when the number of natural fibres was increased. The results were compared with the addition of synthetic fibres, which were less expensive and less harmful to the environment and improved mechanical properties [11].

Lakrafli et al. (2013) used wood and leather waste to conduct tests of their thermal properties. These waste materials were used as fillers for hollow samples or as separating material for cement slabs. The samples significantly reduced thermal conductivity when dry. As the moisture content increased, the thermal conductivity of the materials increased. Furthermore, the effects of weight, volume and humidity on thermal conductivity and the probability of conductivity increased with increasing moisture content [12].

Binici et al. (2014) used waste materials to manufacture thermal insulators for buildings. Rectangular samples were used measuring 30 * 40 * 2.5 cm. Two types of epoxy were used as the bonding material. The waste materials used were cotton and the roots and stems of sunflowers. Samples were tested for mechanical properties and thermal conductivity, and the obtained results met Turkish Standard 601 TS 805 EN. The study solved two industrial problems: first, useful building materials were produced and second, waste was disposed of in a manner that was less harmful to the environment. Results also showed that the gain in electric power was up by 8% with respect to ISO operating conditions, while thermal efficiency increased by 1.5% [13].

Binici and Kekili (2015) investigated the use of vegetable and animal fibres for manufacturing thermal insulation for buildings in Turkey to reduce energy consumption and heating costs in the winter and cooling in the summer. Several types of plant and animal fibres were used and plaster was used as a bond. The results showed that most of the thermal coefficients of the samples matched Turkish standards. Similarly, these materials could be used to produce local thermal insulation more economically and provide a solution for the disposal of agricultural waste sustainably [14].

El Wazna et al. (2018) manufactured thermal insulators with new materials that were non-woven fabric waste and conducted thermal and physical tests to determine their efficiency in terms of thermal conductivity and performance. Low values of thermal conductivity and good thermal performance demonstrated the feasibility of producing new competitive low-cost insulation materials that met sustainability criteria [15].

IOP Conf. Series: Materials Science and Engineering **987** (2020) 012008 doi:10.1088/1757-899X/987/1/012008

This work investigates the manufacture of thermal insulators for buildings from composite materials where waste materials (sawdust and tyre waste) will be added as a filling to matrices. They will be tested to find several properties, including thermal and mechanical properties. The optimum thermal insulator will then be extracted depending on a genetic algorithm optimisation technique that achieves sustainability requirements.

2. Materials and Method

2.1 Sawdust

Sawdust is a type of waste from wood processing (wood dust). Particles of different sizes are formed either from drilling, cutting, grinding or milling using wood grinding tools or a saw [12]. Samples were collected from a carpentry workshop in Zubayr (Basrah). They were screened to the required size of 500 μm, as shown in Figure 1.

2.2 Tyre waste

Tens of millions of tyres are abandoned in the Middle East every year. Disposal of tyre waste is a difficult task because they have a long, unbreakable life. The traditional way of managing tyre waste is to store, dump or landfill illegally, all of which are short-term solutions. The tyres provide a fertile ground for breeding mosquitoes, insects and snakes. Sudden or accidental fires in tyre dumps have also released toxic gases for months [16].

One solution to this problem is to recycle these materials. In this research, these waste products were used as raw materials in the manufacture of environmentally friendly thermal insulation. Tyres were treated in the welding workshop at the Faculty of Engineering in Basrah to obtain the tyre powder, and it was screened at 500 μm or less, as shown in Figure 2.

Figure 1 Sawdust

Figure 2 Rubber waste

2.3 Polymer (PU)

Polyurethane (PUR and PU) is a polymer composed of organic units joined by urethane links. Most polyurethanes are thermosetting polymers that do not melt when heated, but thermoplastic polyurethanes are also available [17]. Polyurethane is produced by the reaction of isocyanates containing two or more groups of isocyanates for each molecule (R- (N = C = O) n) with polyol containing the average of two or more hydroxyl groups in the molecule (R '- (OH) n) in the presence of a stimulant or tonic using ultraviolet light [18]. The polyurethane used in this work was manufactured by Henkel Polybit Industries Ltd and is a two-component solvent-free epoxy resin-based primer and sealer coat for polyurethane-based coatings and toppings [19].

3. Preparation of Composites

The samples consisted of sawdust (liquid polyurethane), tyre waste (liquid polyurethane) and sawdust (polyurethane foam). The samples were prepared in the metal laboratory in the Department of Materials Engineering, College of Engineering, University of Basra. Models were made in normal weather conditions of pressure and temperature. The composites were 35 mm in diameter and 5 mm thickness depending on subsequent tests. At first, the waste materials had different mass ratios as different percentages were added to the polymer (5, 10, 15, 20, 25, 30, 35, 40, 45 and 50%); then they were hardened. The samples were mixed until they were homogenised, and then they were poured into moulds and left to cross for 24 hours. They were extracted and placed in an oven at 80 °C for one hour to complete the crucifixion [20]. Except for the polyurethane foam (sawdust) the samples did not need a post-curing process. Figure 3 shows a sample of sawdust (liquid polyurethane).

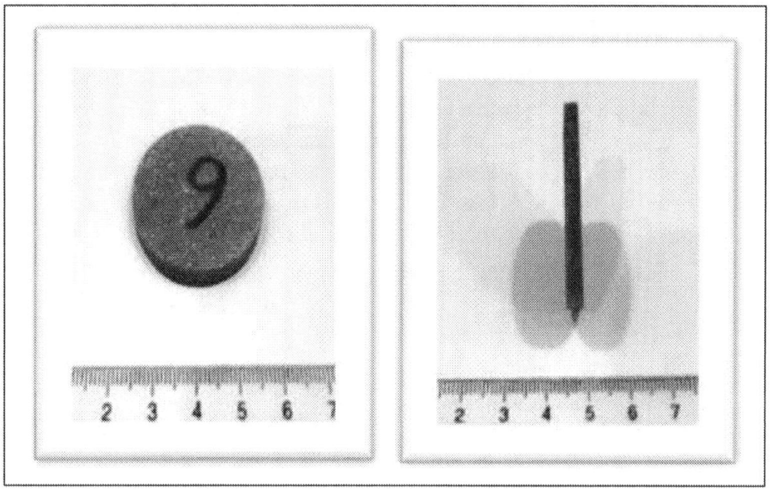

Figure 3 Sample of sawdust (liquid polyurethane) representing its diameter and thickness

4. Experiment

To evaluate insulation performance under field conditions, dimensioning the heating and cooling systems correctly is necessary, as is assessing the functionality of the design. This can be done by conducting the following tests.

4.1 Thermal conductivity measurement

In this work, a Lee's-disk method was used to evaluate the thermal conductivity of bad conductor materials. Measurement was performed at a steady state in the metal laboratory of the Engineering College Material's Department according to BS 4745.

Thermal conductivity is equal to the heat flux through the material layer's unit thickness when the temperature of the material on opposite sides of the layer differs by 1 °K [8].

Fourier's law of heat conduction is expressed as:

$$\boldsymbol{k} = \frac{Q.\Delta x}{A.\Delta T} \quad (1)$$

where $\boldsymbol{k}$ is the thermal conductivity coefficient of the material of dimensions w/m.K, Q is the heat flow at steady state, A is the cross-section area in mm^2, x is the sample thickness in mm and $\Delta T = T_h - T_c$ is the temperature gradient at the thickness of the sample.

Figure 4 shows the structure of Lee's device and the temperature gradient through the sample. The heat lost from the sides of the sample was neglected and, on this assumption, thermal conductivity was calculated [21, 22].

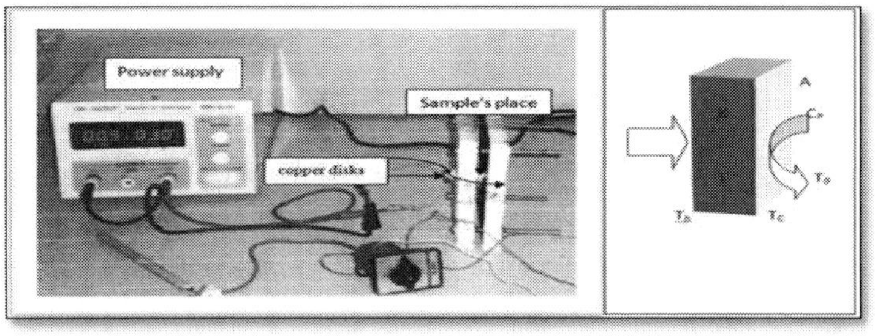

Figure 4 Lee tested with graded temperatures during the sample

A Lee's-disk device consists of two brass discs and an electric heater source to cause a heat gradient through the sample; the sample is placed between the two brass discs. The test continues until thermal equilibrium is reached, where T_c is constant for more than 10 min (the steady-state of heat transfer).
To determine the heat transfer, Q, heat entering the brass plate (2), equals the rate of heat loss due to cooling (at the equilibrium). The heat loss can be determined by measuring the cooling rate of the equilibrium temperature T_C of the brass plate (2) placed with the insulator (sample) as shown in Figure 5. If the disc cools at a rate of $\frac{dT}{dt}$, then the rate of heat loss is given by:

$$\boldsymbol{Q = mc_p \frac{dT}{dt}} \quad (2)$$

where M is the mass of the plate and c_p is the heat capacity of brass $(0.38)\frac{kJ}{kg.K}$. The cooling curve of the brass plate determines the slope $\left[\frac{dT}{dt} = \frac{\Delta T}{\Delta t}\right]$ at temperature T_C [22, 23].

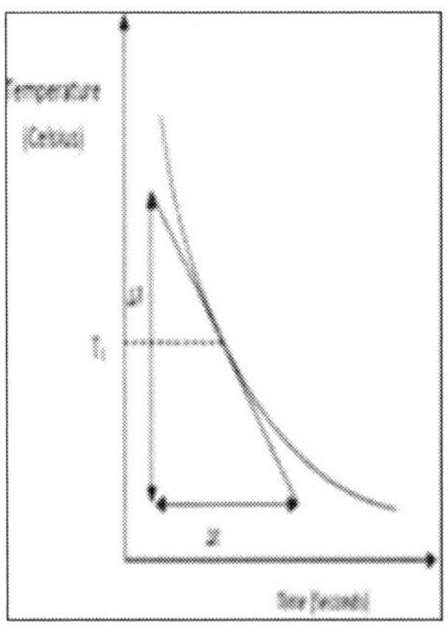

Figure 5 The temperature–time relation for the calculation of Q [23]

4.2 Specific heat capacity (C_P)
Specific heat capacity is known as the amount of heat energy emitted or soaked up under the cooling or heating of the material sample with a unit mass over a temperature change of 1 °K [24].
The specific heat capacity of inorganic building materials (concrete, break, natural stone) is varied within 0.75–0.92 kJ/ (kg °C); for wood, it is 0.7 kJ/ (kg °C).

The specific heat capacity of composite materials can vary if temperature variations are accompanied by phase changes [25]. A high-quality insulator has a higher specific heat capacity because it takes time to absorb more heat before it truly heats up (temperature rising) to transfer the heat [25].

4.2.1 Specific heat capacity measurement

This test is easy to implement as the tools used for measuring are simple and inexpensive. Its reliability for composite materials is better for samples weighing more than 5 mg to ensure the excellent mixing of samples [26]. Specific thermal capacity (expressed as $\mathbf{C_P}$; the pressure of solids is often constant unless stated) is related to the heat index, as illustrated by the following equation:

$$\mathbf{Q = m\, C_P\, \Delta T} \quad \mathbf{(3)}$$

where Q represents the amount of heat, m represents the mass of the substance and ΔT represents the difference of temperature [26]. The temperature difference will cause a continuation of heat exchange between the hot and cold body until it becomes stable, as shown by the law of conservation of energy in the following equation:

$$\mathbf{m_h\, C_{Ph}\, \Delta T_h = m_c\, C_{Pc}\, \Delta T_c} + Q' \quad (4)$$

$\mathbf{m_h}$, $\mathbf{C_{Ph}}$ and $\mathbf{\Delta T_h}$ represent the mass, specific heat capacity and temperature difference of a high-temperature object, while $\mathbf{m_c}$, $\mathbf{C_{Pc}}$ and $\mathbf{\Delta T_c}$ represent the mass, specific heat capacity and temperature difference of a low-temperature object.

Equation (4) contains an energy loss (Q′) that can be cured in two ways. The first is to use a double insulation barrel system and the second is by using a calliper made of a low thermal conductivity material so that it can be neglected. In the test, the second method was used. The equation will be:

$$\mathbf{m_h\, C_{Ph}\, \Delta T_h = m_c\, C_{Pc}\, \Delta T_c} \quad \mathbf{(5)}$$

The hot side represents water and the cold side represents the specimen [26]. Instruments used in the test consisted of simple tools that are calorific and sensors of temperature, as shown in Figure 6.

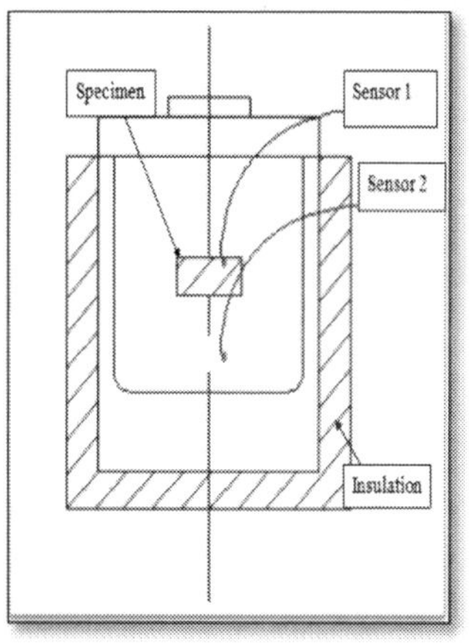

Figure 6 Diagram of heat exchange test

The procedures of the test included:

1- The water was heated to a certain temperature and quickly placed in the calliper.
2- The sample (at temperature -20 °C) was placed in the calliper.
3- The temperature difference was recorded as required.
4- The specified heat capacity was calculated from equation (5) [26].

4.3 Hardness

The hardness test is the most widely used test for evaluating the mechanical properties of materials. The purpose of the hardness test is to determine the suitability of a material for a given application, conformance to a specification, standard, or particular treatment to which the material has been subjected (heat treatment, thermal process) [27].

The hardness of the samples (wood polymer and rubber polymer) was measured by placing a load of 60 kg on each sample to measure the resistance shown by the body. The hardness of the foam and wood samples was tested using a Shore D scale.

5. Theoretical Analysis

The theoretical work includes empirical analysis and optimisation techniques. Empirical analysis is an evidence-based approach to the study and interpretation of information. The empirical approach relies on real-world data, metrics and results rather than theories and concepts.

In addition to the analytical data of the thermal properties (thermal conductivity, specific heat capacity) and hardness test data, the equations of the composite materials were used to explain their behaviour as composite material behaves very differently from the original materials depending on the method of mixing (mass ratio, volumetric ratio).

Finally, the optimisation technique was used to find the optimum result that achieves sustainability requirements.

5.1 Empirical analysis

Depending on the experimental approach, the data on the practical side was explained. The relationship between two or more variables was found due to their importance in engineering applications. Several points were made.

5.1.1 Empirical equation for thermal conductivity k. This was based on the 10-point experimental data for each composite. The equation for each compound was discovered based on two independent variables (temperature and mass ratio).

5.1.1.1 Sawdust (liquid polyurethane) composite. With 10 experimental points and two independent variables (temperature and mass fraction), the empirical equation was found using the LAB Fit program. The empirical equation can be expressed as:

$$k = T_1/(A+(B*M^2\%)) \quad (6)$$

where k represents the thermal conductivity in W/m.°C, T_1 represents the temperature of the hot surface in °C and A&B are the equation's constants. The values of the regression correlation coefficient are as follows:

- R=0.998865
- R^2=0.997

5.1.1.2 Tyre waste (liquid polyurethane) composite. LAB Fit was used to discover the empirical equation. The constant of the equation and the regression correlation coefficient is as follows:

$$k = T1/(A-(B*M^2))$$

- R=0.999536
- R^2=0.999

5.1.1.3 Sawdust (polyurethane foam) composite. The empirical equation, the constant of the equation and the regression correlation coefficient are as follows:

$$k = A*((T_1/M)^B) \quad (7)$$

- R=0.987273
- R^2=0.974

5.1.2 Empirical equation for specific heat capacity

The experimental equations for specific heat capacity were discovered by the same method. The following are the regression results for each category.

5.1.2.1 Sawdust (liquid polyurethane) composite. The model of this category is linear and depends on two variables: the temperature and mass ratio of wood waste. The equation was discovered and its parameters added to the regression correlation coefficient:

$$C_p = A*T+B*M^2 \quad (8)$$

- R=0.994816
- R^2=0.989

5.1.2.2 Tyre waste (liquid polyurethane) composite. The model of this category was smaller and its result is below:

$$C_p = A*M+B*T^2$$

- R=0.981291
- R^2=0.962

5.1.2.3 Sawdust (polyurethane foam). The equation of this model was exponential and the following is the result of this category:

$$C_p = A*M^{(B*M)} \quad (9)$$

- R=0.997142
- R^2=0.994

5.1.3 Empirical equation for hardness

The experimental results of the hardness tests were based on one variable, mass ratio, and 10 points were tested for each category. The equations derived from the empirical test and the equation constants in addition to the convergence coefficients are as follows:

5.1.3.1 Sawdust (liquid polyurethane) composite

$$H = -A*M-B \quad (10)$$

- R=0.993225
- R^2=0.986

5.1.3.2 Tyre waste (liquid polyurethane) composite

$$H = -A*M^3+B*M^2-C*M+D \quad (11)$$

- R=0.993897
- R^2=0.987

5.1.3.3 Sawdust (polyurethane foam) composite

$$H = A*M^2-B*M+C \quad (12)$$

- R=0.988438
- R^2=0.97

5.2 Genetic algorithm optimisation technique

Optimisation represents the optimum result under given circumstances and enables engineers to make important decisions (technology and managerial) in the construction, maintenance and design of any engineering system. It is used to estimate the maximum or minimum value of a condition function [28]. Optimisation techniques can be divided into two categories: traditional and non-traditional. In response to solving complex optimisation problems, traditional, or modern, methods have been developed [28]. In this work, a genetic algorithm (GA) was adopted to find the optimal values of the thermal insulators manufactured using the properties of thermal conductivity and hardness. GAs are extremely suitable for solving problems as they can find the best global solution with high probability [28]. GA is a method for solving optimisation problems based on natural selection: the process that drives biological evolution [29]. It acts as a random search system for a set of extracted solutions. At each step, the GA selects individuals at random to be parents and uses them to produce children for the next generation. Step by step, the generations reach an optimal solution. A GA uses three types of operators at each step to create the next generation from the present population [30]: representation, crossover and mutation. Solving a problem and finding optimal solutions via a GA goes through five stages:

- **Initial population**
- **Fitness function**
- **Selection**
- **Crossover**
- **Mutation**

The initial population represents a set of individuals who are selected to solve a problem. Fitness function represents the environment that the solution lives in. The selection operator represents the best fitness function of individuals and copies them. Crossover matches individuals from the mating pool as pairs. Mutation represents a change in the gene of a single chromosome [28, 30]. Figure 7 represents a simple flowchart of the working principle of a genetic algorithm [30].

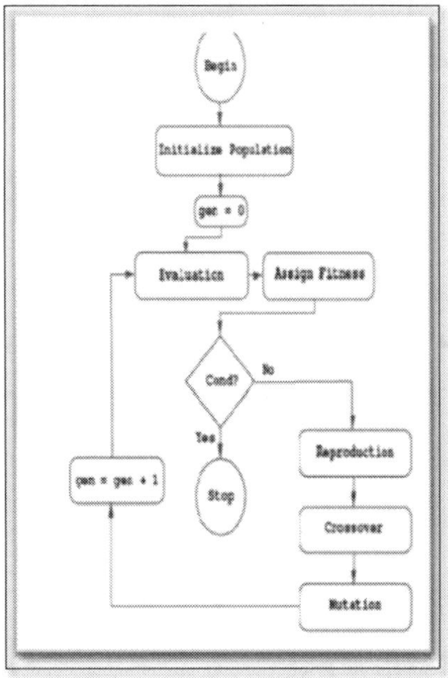

Figure 7 Simple flowchart of the working principle of a genetic algorithm [30]

The purpose of this study is to find the optimum values of two objective functions (thermal conductivity and hardness) for the thermal insulation of buildings based on design variables that include temperature and mass fraction.

We needed to improve more than one objective, so the multiple objective optimisation (MOO) method was used. Although MOO is tough, it presents realistic issues whenever the objectives are usually conflicting, which prevents the synchronic improvement of every objective [31]. The GA is considered as one of the most fashionable heuristic approaches for the resolution of multi-objective improvement issues because the multi-objective genetic algorithm (MOGA) does not need the user to prioritise, scale or weight objectives [31].

There are three steps in solving MOO problems using a GA [31–33].

1. *Fitness function assignment and selection*

In MOO, the fitness assignment is not straightforward; therefore, many approaches have been proposed to solve this issue, and Pareto-ranking is the most popular [31–33]. In this approach, the population is ranked according to a dominance rule, and then each solution is assigned a fitness value based on its rank in the population, not its actual objective function value [32]. MOO is different from single-objective optimisation (SOO); in SOO, there is just one optimum solution while in MOO, there is a set of optimal solutions called the 'Pareto optimal set' [31]. Furthermore, the set of solutions is known as non-inferior or non-dominated [34].

2. *Diversity*

Diverse solutions are intended for distribution on the border, which gives a clear idea of the decision-making about the true trade-off of the MOO problem [31].

3. *Elitism*

Elitism means always promoting the best individual to the next generation. This improves the GAs performance. There are several possible elitist solutions; therefore, two strategies have been adopted to overcome this problem: (i) maintaining elitist solutions in the population; (ii) storing elitist solutions in an external secondary list and reintroducing them to the population [32].

5.2.1 Multi-objective optimisation of this study using GA

The extraction of the optimum values of thermal insulators manufactured in this study depends on two objective functions in terms of their design variables.

One of the most important thermal properties was chosen as the first objective: thermal conductivity. The second objective is hardness, which is one of the important mechanical properties of thermal insulators.

The design variables are the temperature and mass fraction of waste material. The optimisation process was performed using MOO (gamultiobj) in MATLAB R2014, which is software that optimises the two objectives simultaneously.

The options in MOO (gamultiobj) were specified as follows:

1. Population size = 100
2. Population initial range = [25 5; 50 50]
3. Lower bound = [25; 5]; Upper bound = [600; 50], which were considered as objective constraints
4. Creation function = Feasible population
5. Selection function = Tournament; tournament size = 10
6. Crossover function = Two-point crossover; crossover fraction = 0.8
7. Mutation function = Adaptive feasible
8. Pareto fraction = 0.7
9. Plot function = Pareto front

6. Results and Discussion

6.1 Thermal conductivity

6.1.1 The effects of temperature on the thermal conductivity of the composites

The effects of temperature on thermal conductivity are shown in Figures 8, 9 and 10. For all composites, increasing the temperature causes an increase in thermal conductivity and this can be explained as follows: the connection of heat in any matter from the hot point to the cold point depends on the movement of electrons, phonons, or both. The movement of electrons and phonons is linked to kinetic energy. However, the thermal conductivity values of the three categories remain within the thermal isolation range. The highest values were recorded at 45 °C (0.035332, 0.179394, 0.049224 W/m.°C) for each of the composites (sawdust [liquid polyurethane]; tyre waste [liquid polyurethane]; sawdust [polyurethane foam]), respectively.

6.1.2 The effects of waste addition on the thermal conductivity of the composites

The thermal conductivity of the composites was reduced after adding the sawdust; as shown in Figure 8, it reduced from 0.019629 W/m.°C to 0.008389W/m.°C at 25 °C. The addition of sawdust, whose thermal conductivity is less than the polymer, in addition to the formation of small air chambers, reduced the contact points in the polymer matrix, thus, dispersing the electrons, which led to reduced thermal conductivity [12, 25].

For the R-PU thermal insulator in Figure 9, thermal conductivity increased with an increasing mass fraction of rubber particles. This is attributed to the thermal conductivity of waste rubber, which is larger than the thermal conductivity of PU. Where the waste rubber contained the compound (black carbon), the thermal conductivity of the composite increased. When the percentage of waste rubber increased, particles joined together to form filler conductive chains, which greatly contributed to the thermal conductivities of the composite, but the values of thermal conductivity remained within the range of thermal insulations [16].

The foam structure contained many small air pockets making it well insulated. The addition of sawdust led to the dispersion of the particles between the air chambers and the foam material, which led to a greater dispersion of electrons as well as reducing the vibrations, which, in turn, reduced the thermal conductivity of the compound (as shown in Figure 10) [24].

Figure 11 shows the comparison of thermal conductivity between the three compounds. Sawdust (liquid polyurethane) is the lowest and tyre waste (liquid polyurethane) is the highest.

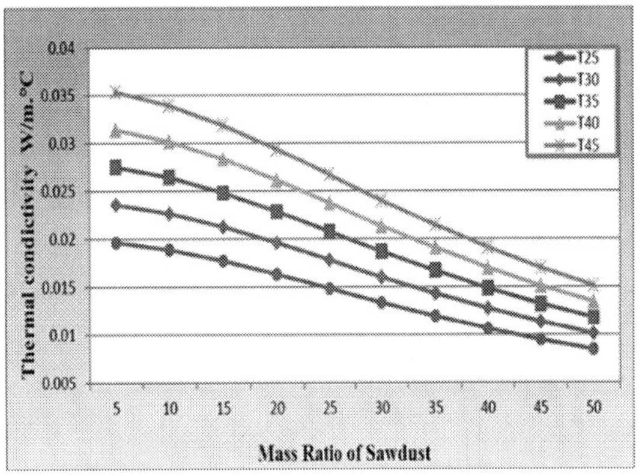

Figure 8 The effect of mass ratio and temperature on the thermal conductivity of sawdust (liquid polyurethane)

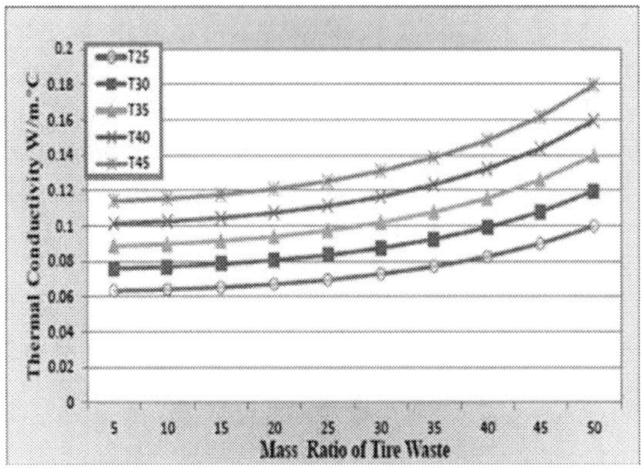

Figure 9 The effect of mass ratio and temperature on the thermal conductivity of tyre waste (liquid polyurethane)

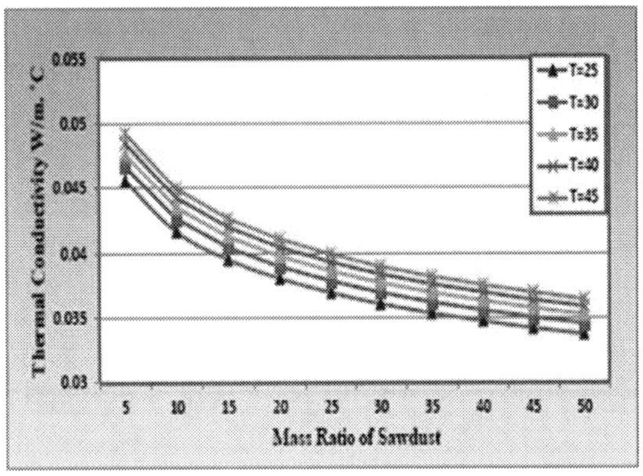

Figure 10 The effect of mass ratio and temperature on the thermal conductivity of sawdust (polyurethane foam)

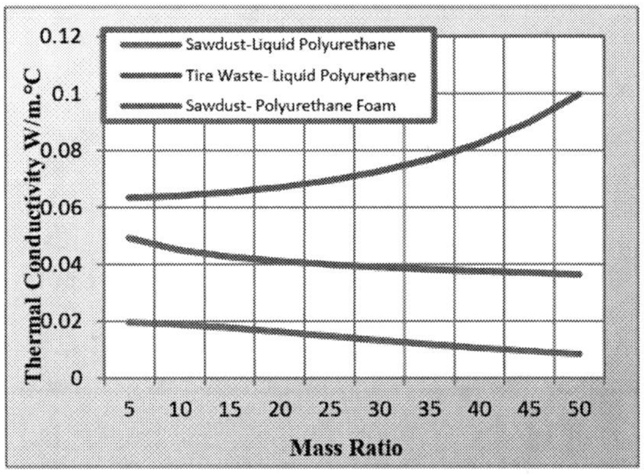

Figure 11 Comparison between thermal insulators at T=25 °C

6.2 Specific heat capacity (C_P)

6.2.1 The effects of waste addition on the specific heat capacity of the composites

Figures 12, 13 and 14 present the variations in specific heat capacity (C_P) with the loading of sawdust and tyre waste on the polymer matrix and the loading of sawdust on the foam matrix, respectively. For Figures 12 and 13, the increase in content increases specific heat capacity because the specific heat capacity of the sawdust (2.3 kJ/kg.°C) and waste rubber (2.01 kJ/kg.°C) is greater than the specific heat capacity of the polymer (1.800 kJ/kg.°C); the specific heat capacity of the sawdust is less than the foam (2.32 kJ/kg.°C). From previous measurements, the loading of sawdust and tyre waste in the polymer matrix resulted in the composites improving their specific heat capacity. Furthermore, sawdust (polyurethane foam) had the least specific heat capacity of all the composites. This approach is fully consistent with the equation below, which represents the weighted average of the specific heat capacity of the components [24].

$C_P = C_{P1}M_1 + C_{P2}M_2$

6.2.2 The effects of temperature on the specific heat capacity of the composites

Figures 12 and 13 display the effects of temperature on the specific heat capacity of the composites. When the temperature increases, the specific heat capacity of the composites also increases. This approach demonstrates that kinetic energy is kept within the material through the vibrations of the lattice, the rotations of molecules and the molecules themselves. Every movement or rotation is alleged to be a degree of freedom. Heat is required to extend the temperature of a material as a result of the K.E. per degree of freedom; this must increase because the temperature will increase. The more degrees of freedom, the greater the specific heat capacity [24].
Figure 14 shows the effect of temperature on the specific heat capacity of the compound made of foam and wood. The figure shows that an increase in temperature leads to a decrease in specific heat capacity because the foam contains air pockets. This reduces the holes in the foam and, thus, leads to the formation of more compact material. As a result, this reduces the degree of freedom, leading to a decrease in the capacity of specific heat [34].

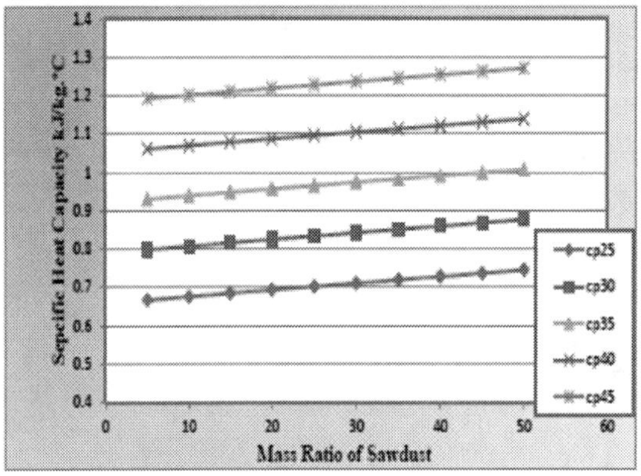

Figure 12 The effect of mass ratio and temperature on the specific heat capacity of sawdust (liquid polyurethane)

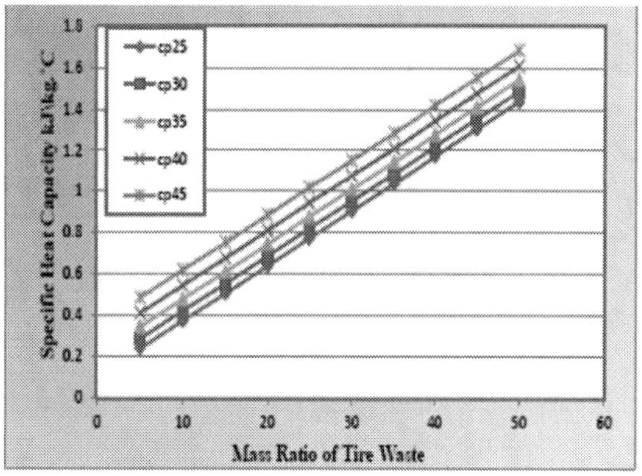

Figure 13 The effect of mass ratio and temperature on the specific heat capacity of tyre waste (liquid polyurethane)

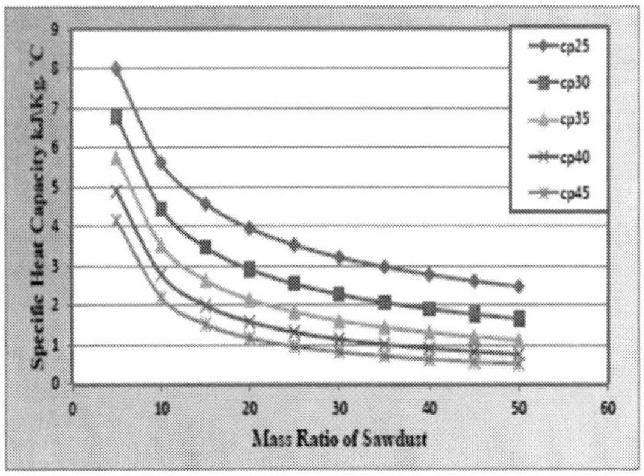

Figure 14 The effect of mass ratio and temperature on the specific heat capacity of sawdust (polyurethane foam)

6.3 Hardness
Figure 15 explains the variation of the hardness of the sawdust (liquid polyurethane) composite with a mass fraction of sawdust. The figure reveals a drop in hardness from 96.45 at 5% to 18.75 at 50%.

This behaviour is because the particles of sawdust may cause porosity, which weakens the composite [24].

Figure 16 shows the variation of hardness in the tyre waste (liquid polyurethane) composite with a mass fraction of rubber waste. First, notice a decrease in hardness until the rubber ratio reaches 20%, and then the hardness of the composite increases with the increasing ratio of waste rubber. The reason for this drop is initially due to the combination of particles in the waste rubber. Additionally, flipping the material into the air leads to the introduction of air and the formation of voids, which reduces thermal insulation hardness. The increase in the percentage of waste rubber, characterised by its dominance, leads to a more uniform distribution and collection of rubber particles, which leads to an increase in hardness; a 20% addition of tyre waste resulted in 63.78, whereas a 50% addition results in the insulator hardening to 122.52 [36].

Figure 17 shows how the sawdust load was affected by the hardness of the polyurethane foam. The increase in the composite hardness can be attributed to the structure of the foam matrix, which is generally softer because it contains bubbles or interstitial spaces. As content increases, regular dispersion begins with the penetration of fine particles from the sawdust between the polymer matrix molecules, reducing interstitial distances and increasing bond strength, thus, increasing the hardness of the composite [35].

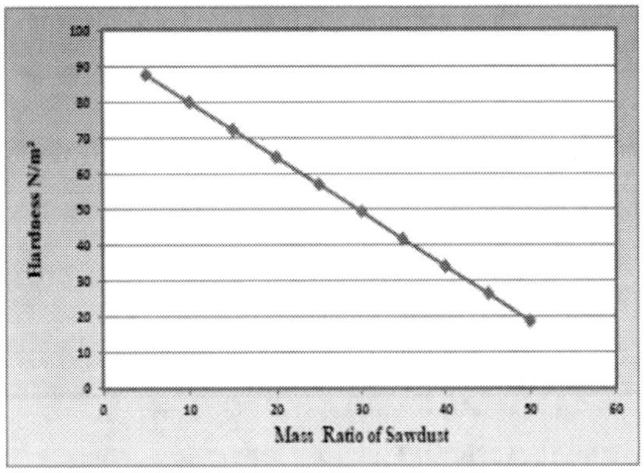

Figure 15 The effect of mass fraction of sawdust (liquid polyurethane) on hardness

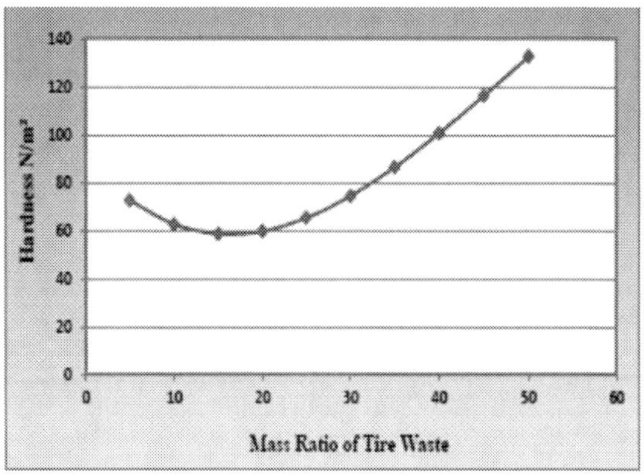

Figure 16 The effect of mass fraction of waste tyre (liquid polyurethane) on hardness

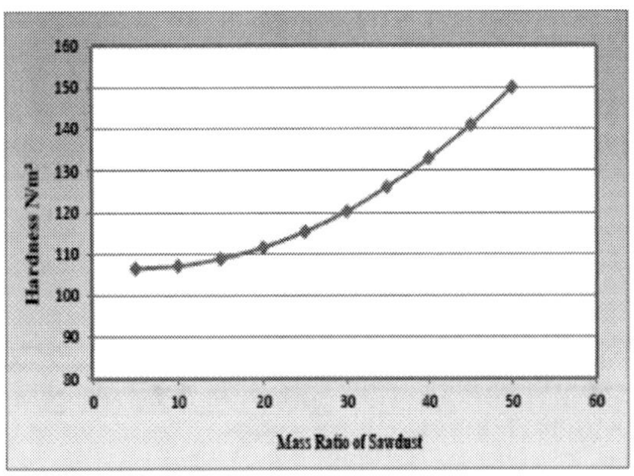

Figure 17 The effect of mass fraction of sawdust (polyurethane foam) on hardness

IOP Conf. Series: Materials Science and Engineering **987** (2020) 012008 doi:10.1088/1757-899X/987/1/012008

6.4 The result of optimisation

Figures 18, 19 and 20 show the results of the optimisation technique (Pareto optimal set) on thermal insulation composites. Table 1 below contains the optimum values of thermal conductivity, hardness and their variables.

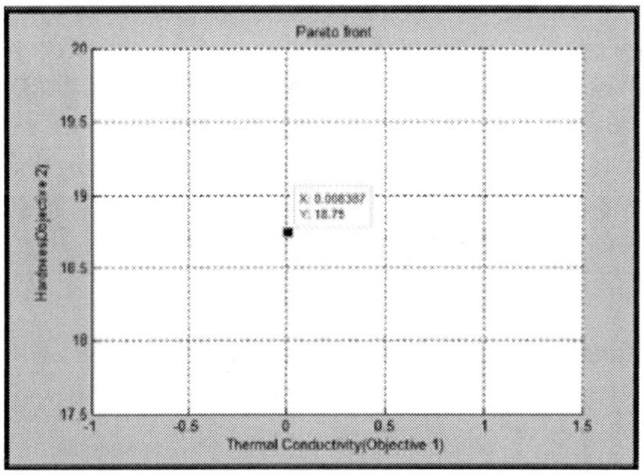

Figure 18 Pareto front set (Sawdust: liquid polyurethane)

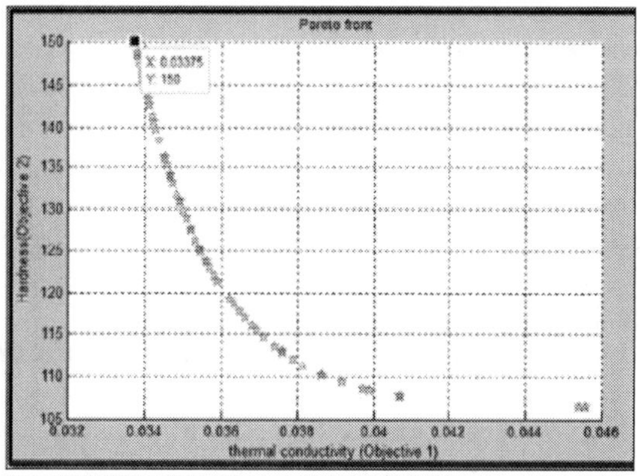

Figure 19 Pareto front set (Sawdust: polyurethane foam)

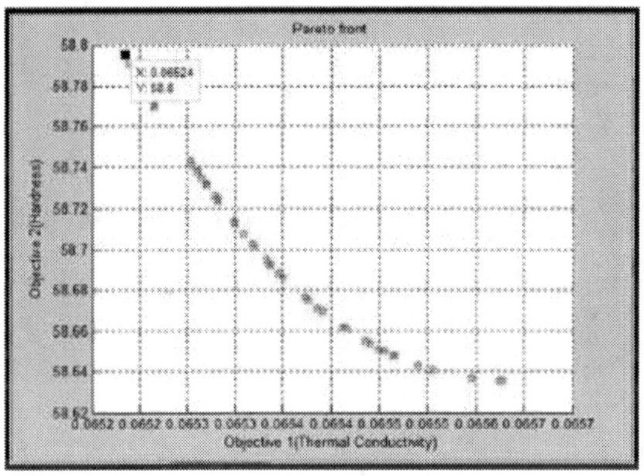

Figure 20 Pareto front set (Tyre waste: liquid polyurethane)

Table1: Optimum values of thermal conductivity, hardness, and their variables

Composites	k W/m^2.°C	H	T °C	Mass Fr of additives %
Sawdust: liquid polyurethane	0.008387	18.75	25.	50
Sawdust: polyurethane foam	0.033745	149.9965	25	49.99815
Tyre waste: liquid polyurethane	0.063313	72.645	25	5

7. Conclusion

Filler-loading sawdust decreased the thermal conductivity of thermal insulator composites. Filler-loading tyre waste increased the thermal conductivity of thermal insulator composites, but it remained within the range of thermal insulation. Thermal conductivity for all composites increased when the temperature increased. Specific heat capacity increased for the sawdust (liquid polyurethane) and tyre waste (liquid polyurethane) composites as the temperature increased. The mechanical properties of tyre waste (liquid polyurethane) and sawdust (polyurethane foam) improved with an increase in filler loading. The use of wood and tyre waste achieves the standard of sustainability in the thermal insulator industry because it reduces the cost of manufacturing thermal insulators. However, the use of these thermal insulators leads to a reduction in the energy consumption of the building, as well as ridding the environment of proper waste. Using the optimisation method to extract optimal values contributes to reducing the cost of thermal insulation manufacturing, which contributes to standards of sustainability.

References

[1] Giovannoni E and Fabietti G 2013 What is sustainability? A review of the concept and its Applications *Spr .Cham.* 21-40.
[2] Molamohamadi Z and Ismail N 2013 Developing a New Scheme for Sustainable Manufacturing *Int. J. Mat. M. M.* **1(1)** 1-5.
[3] Kates R W 2010 Readings in sustainability science and technology *Center for International Development at Harvard University.*
[4] Clark W C and Dickson N M 2003 Sustainability science: the emerging research program *Proceedings of the national academy of sciences* **100(14)** 8059-8061.
[5] Thorsell T 2012 Advances in Thermal Insulation: Vacuum Insulation Panels and Thermal Efficiency to Reduce Energy Usage in Buildings. PhD Thesis. *KTH Royal Institute of Technology.*
[6] Vasilache M, Pruteanu M and Avram C 2011 Use of waste materials for thermal insulation in buildings *Environmental Eng. and Management J* **9** 1275-1280.
[7] Akadiri P O, Chinyio E A and Olomolaiye P O 2012 Design of a sustainable building: A conceptual framework for implementing sustainability in the building sector *Buildings* **2(2)** 126-152.
[8] Al-Homoud M S 2005 Performance characteristics and practical applications of common building thermal insulation materials *Building and environment* **40(3)** 353-366.
[9] Lertsutthiwong P, Khunthon S, Siralertmukul K, Noomun K and Chandrkrachang S 2008 New insulating particleboards prepared from mixture of solid wastes from tissue paper manufacturing and corn peel *Bioresource technology* **99(11)** 4841-4845.
[10] Panyakaew S and Fotios S 2008 321: Agricultural Waste Materials as Thermal Insulation for Dwellings in Thailand: Preliminary Results. *In 25 Conference on Passive and Low Energy Architecture Dublin.*
[11] Mounika M, Ramaniah K, Prasad A R, Rao K M and Reddy K H C 2012 Thermal conductivity characterization of bamboo fiber reinforced polyester composite *J. Mater. Environ. Sci* **3(6)** 1109-1116.
[12] Lakrafli H, Tahiri S, Albizane A, Bouhria M and El Otmani M E 2013 Experimental study of thermal conductivity of leather and carpentry wastes *Construction and Building Materials* **48** 566-574.
[13] Binici H, Eken M Kara M and Dolaz M 2013 An environment-friendly thermal insulation material from sunflower stalk, textile waste and stubble fibers *ICRERA* 833-846 .
[14] Binici H and Kekili M 2015 The Production of Thermal Insulation by Waste Animal and Vegetable Fibres *J Material Sci Eng* **4(186)** 2169-0022.
[15] Wazna M E, Gounni A, Bouari A E, Alami M E and Cherkaoui O 2019 Development, characterization and thermal performance of insulating nonwoven fabrics made from textile waste. *Journal of Industrial Textiles* **48(7)** 1167-1183.

[16] Aliabdo A A, Elmoaty A E M A and Abd Elbaset M M 2015 Utilization of waste Tire in non-structural applications *Construction and Building Materials* **91** 195-207.
[17] Pandya J, Shah S and Dave S 2017 Feasibility of Utilization of Industrial Polyurethane (PU) Tire Waste in Geopolymer Concrete *Kalpa Publications in Civil Engineering* **1** 285-290.
[18] Rad A S and Ardjmand M 2008 Studying on the mechanism and raw materials used to manufacturing polyurethane *Transportation* **1** 23-8.
[19] henkelpolybit@henkelpolybit.com.
[20] Saleh S A 1999 Synthesis and Evaluation Properties of Epoxy – Resolic IPN'S
[21] Nelkon M Parker P 1988 Advanced level physics *3rd e. London: Heinman*.336–8.
[22] Lees' Conductivity Apparatus (Electrical Method) LL44-590I.S. 1122/7302.
[23] Ganeshkumar A G, Shylin H, Jose.S and Silambarasan R 2017 Review on Thermal Behaviour of Synthetic Rubber Composite *International Journal of ChemTech Research* **10 14** 93-100.
[24] Chung D D 2010 Composite materials: science and applications *Springer Science and Business Media.*
[25] Marchenko A V 2008 Thermo-mechanical properties of materials *Cold Regions Science and Marine Technology*.
[26] Ma B, Zhou X Y, Liu J, You Z, Wei K and Huang X F 2016 Determination of specific heat capacity on composite shape-stabilized phase change materials and asphalt mixtures by heat exchange system *Materials* **9(5)** 389.
[27] Chandler H 1999 Introduction to hardness testing *Hardness testing. USA: ASM International* 1-13.
[28] Rao S S 2009 Engineering optimization: theory and practice. *John Wiley & Sons.*
[29] Hristakeva M and Shrestha D 2004 Solving the 0-1 knapsack problem with genetic algorithms *In Midwest instruction and computing symposium.*
[30] Onwubolu G C and Babu B V 2013 New optimization techniques in engineering *Springer*. **141.**
[31] Long Q, Wu C, Huang T and Wang X 2015 A genetic algorithm for unconstrained multi-objective optimization. *Swarm and Evolutionary Computation* **22** 1-14.
[32] Konak A, Coit D W and Smith A E 2006 Multi-objective optimization using genetic algorithms: A tutorial. *Reliability Engineering and System Safety* **91(9)** 992-1007
[33] Zitzler E Evolutionary 1999 algorithms for multiobjective optimization: Methods and applications *Ithaca: Shaker* **63.**
[34] Man K F, Tang K S and Kwong S 1996 Genetic algorithms: concepts and applications. *IEEE transactions on Industrial Electronics* **43(5)** 519-534.
[35] Shi D, Shi L, Zhang J Y and Cheng J 2012 Preparation and properties of a novel nonflammable thermal insulation material *In Advanced Materials Research* **450** 1504-1512.
[36] Kader M A, Abdel-wehab S M, Helal M A and Hassan H H 2012 Evaluation of thermal insulation and mechanical properties of waste rubber/natural rubber composite *HBRC Journal* **8(1)** 69-74.

IOP Conf. Series: Materials Science and Engineering **987** (2020) 012009 doi:10.1088/1757-899X/987/1/012009

Manufacturing of High-Load Engine Mounts From Rubber Composites (Natural Rubber and Carbon Black) with Novel Properties

Ahmed Abdulameer Subeh[1], Mohammed H. Al-maamori[2], Ahmed Fadhil Hamzah[3]

[1]Karbala Technical Institute, Al-Furat Al-Awsat Technical University, Karbala, Iraq.
[2,3]College of Materials Engineering, University of Babylon, Babylon, Iraq.

[1]ahmedeng6217@yahoo.com, [1]ahmedeng@ikr.atu.edu.iq,
[1]ahmed.abdulameersubeh@uobabylon.edu.iq
[2]mat.mohammed.hamzah@uobabylon.edu.iq,
[3]mat.ahmed.fadhil@uobabylon.edu.iq

ABSTRACT This paper explains the successful and simple process of generating a rubber mount with novel properties by incorporating carbon black particles type ISAF 220 at different ratios ([46, 56, 66, 76 , 86] [pphr]). Rubber mounting was made using 100 % natural rubber and carbon powder that was applied to the master batch ratio (60 g). To achieve greater homogeneity, rubber and carbon were combined for several minutes. The rubber mount has been subjected to several tests at Babylon tires factory and Babylon University (tensile, elongation, modulus, compression, hardness, rebound resilience, specific gravity and fatigue tests). The best results were obtained from the rubber mount, as it had the ability to resist fatigue and could be used for a long period of time. It could also increase the number of cycles to 900 rpm. The rubber mount that was reinforced with carbon had a better resistance to the mechanical properties (tensile, hardness, fatigue and resilience). The results of the tensile strength, hardness, fatigue and resilience tests showed that additional quantities of carbon at 66 pphr increased the values by up to 18.95 Mpa, 64 Mpa, 600 Mpa and 55.5 Mpa, respectively.

1. INTRODUCTION

Engine mounts are constructed from rubber and steel. The metal part attaches to the motor on one side and the frame on the other. In the meantime, the rubber has enough stability to ensure that the shaking engine does not cause the vehicle to rock. The number of engine mounts varies between the truck and car [1].

The mount's movements come from the following two distinct points: 1. the engine noises to be extracted are usually in the range of 30Hz–200 Hz, with amplitudes that are typically less than 0.3 mm; 2. the frame, which is influenced by disturbances of the ground surface through the suspension mechanism. Usually, these frequencies are in the one Hz to 30 Hz range and have an amplitude greater than 0.3 mm. The optimal dynamic stiffness for mounting an engine is shown in the figure below.

IOP Conf. Series: Materials Science and Engineering **987** (2020) 012009 doi:10.1088/1757-899X/987/1/012009

High damping for shock excitation is required for low frequencies to prevent engine bounce and to provide driving stability [1].
Natural rubber is produced from plants and is known as a chemical compound of polymerA with large molecules of the same kind, which are formed from many smaller molecules. Some polymers spontaneously occur, and others are manufactured in laboratories and factories.
A polymer is a chemical compound that is composed of many smaller molecules of the same kind with large molecules.
Rubber's properties include high strength and the potential to stretch several times without cracking. Natural rubber compounds are highly versatile, strong and resistant to many corrosive substances [2].
Carbon black is a substance that is formed by the incomplete combustion of heavy petroleum products, such as Fluid catalytic cracking tar, coal tar or cracking tar with ethylene. Carbon black is a type of paracristalline carbon with a high ratio of surface-area-to-volume, which is lower than activated carbon. It is dissimilar to soot due to its significantly higher surface-area-to-volume ratio and significantly lower polycyclic aromatic hydrocarbon (PAH) content (negligible and non-bioavailable). However, carbon black is commonly used for diesel oxidation studies as a model compound for diesel soot. [3] Additionally, it is used mostly as a protective filler for tires and other rubber goods, as well as a dye additive in fibres, paints and inks [4]. Among the previous studies in this field are the following:
In Al-Kawas's [5] study, the natural rubber type NR (SMR5) and neoprene rubber (CR) with different loading levels of carbon black (35, 45, 55, 60, and 65 pphr) were used separately.
The tensile, compressive, shear, stiffness and ultraviolet light effect tests were planned to obtain the best properties with carbon black reinforced NR (60 pphr) and carbon black reinforced CR (70 pphr).

2. Materials and Method

2.1 Used Materials

The material that was used in this research was natural rubber (100 pphr), and the materials that were used to prepare the rubber recipe were the activator (zinc oxide), sulfur, (CBS), anox (HB), (IPPD), antiozone (WAX A 111), activator (stearic acid), aromatic oil and carbon black. These were imported by the Diwaniyah tire company.

2.2 Preparation Method

Mixing was done in a two-roll mill. The kneading and mixing operations were conducted on this machine as per ASTM D 15. The shape of the mixing machine and the manufactured samples, both mechanical or physical, are as follows:

Figure 1. Mixing machine.

Figure 2. Vulcanisation test.

Figure 3. Tensile, hardness and resilience specimen.

Figure 4. Compression specimen.

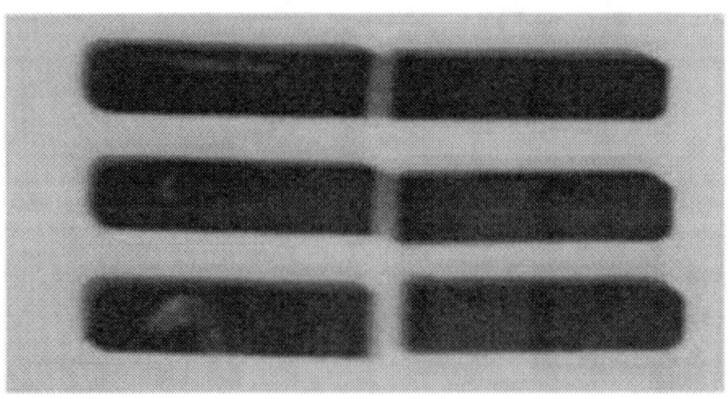

Figure 5. Flexibility specimen.

3. Results and Discussion

3.1 Mechanical Properties

Figure (6) shows high tensile strength due to the wide surface area and black carbon winding with a high degree of carbon black in the rubber recipes.

Strong interlacing systems improve the mechanical properties of the new compound as they come into contact with rubber (NR) and make the composite material resistant to tensile strength and drag. This resistance persists with the rise in carbon black but at higher loading ratios (56 pphr and above). The tensile resistance starts to decrease due to the increased fragility of the rubber material, which is compatible with the findings of other researchers [5,6].

Figure (7) shows the rate of elongation declining as the percentage of carbon black rises. This is because raising the carbon black reduces the amount of rubber base material covering the carbon black minutes, which improves the fragility and hardening of the rubber compound. This decreases the elongation when cutting. This finding follows the findings of other researchers [7,5].

The elastic modulus seen in Figure 8 increases as carbon black rises for the same reasons as above to increase tensile resistance. This finding correlates with the findings of other researchers [7,5].

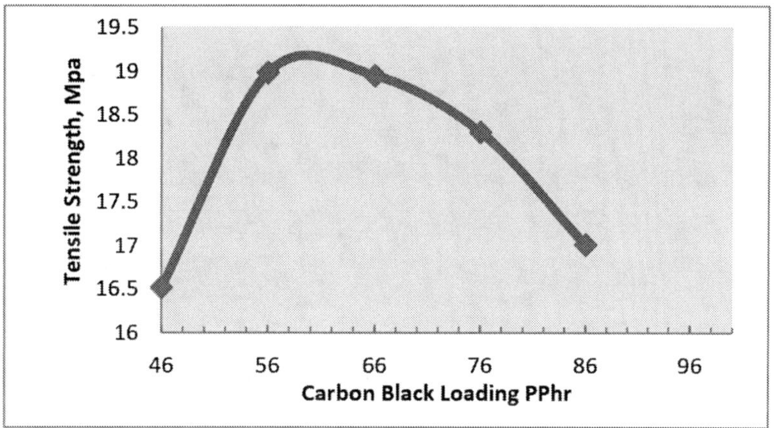

Figure 6. Effect of CB (pphr) on the tensile strength of NR matrix composite.

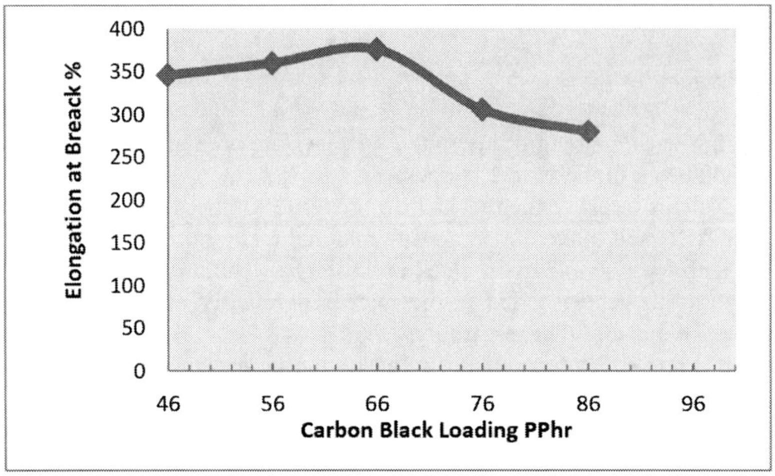

Figure 7. Effect of CB (pphr) on the elongation of the NR matrix composite.

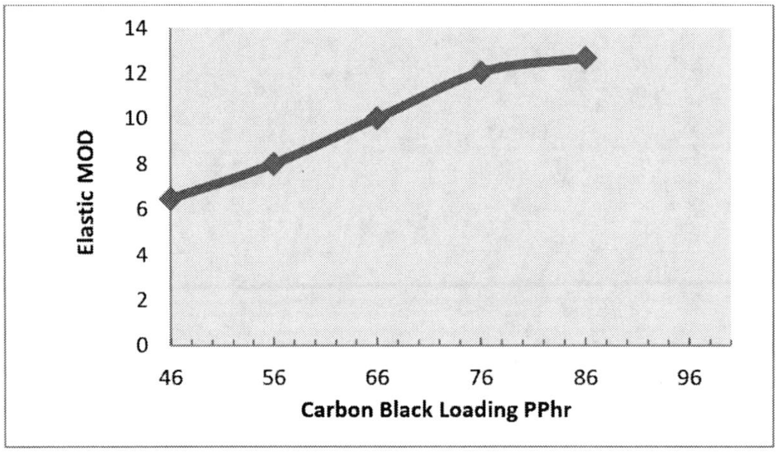

Figure 8. Effect of CB (pphr) on the elastic modulus of the NR matrix composite.

When applying the carbon black form N-220, which has a granular size of 350 microns to the natural rubber with different loading ratios (pphr [46, 56, 66, 76, 86]), there is an increase in the hardness values. This is because the carbon black granules have an active and winding surface, which increases the tangles or bonds between them and the rubber chains when they come into contact with the rubber. This increases the resistance to the external body penetration of the compound due to the increased surface tension, as seen in Figure (9) [8,6,5].

IOP Conf. Series: Materials Science and Engineering **987** (2020) 012009 doi:10.1088/1757-899X/987/1/012009

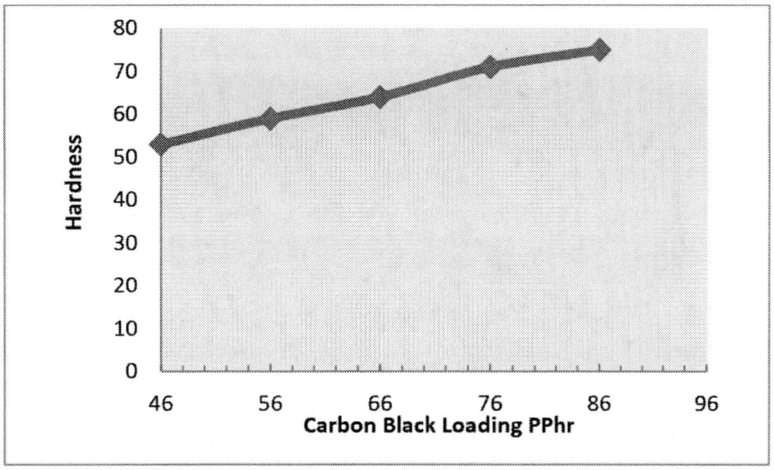

Figure 9. Effect of CB (pphr) on the hardness of the NR matrix composite.

Figure 10 shows a decline in the rebound values. This is due to the increased strength of the compound rubber material with the rise in the percentage of carbon black, which gives the rubber material a greater ability to absorb kinetic energy and transform it into thermal energy and less propensity to return it. This finding correlates with previous research [5].

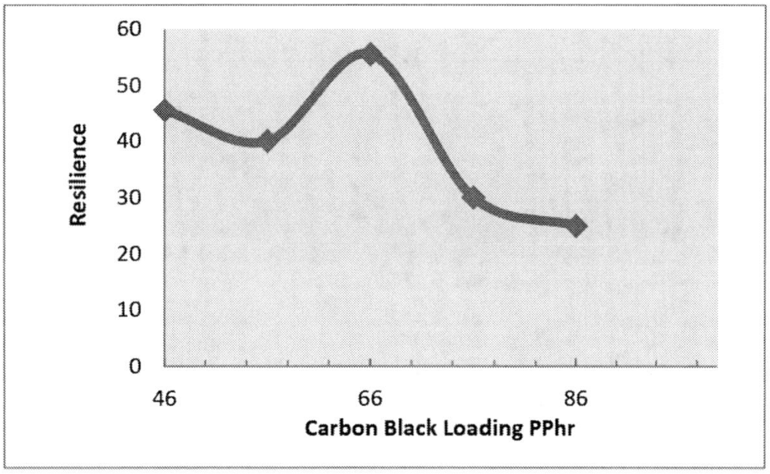

Figure 10. Effect of CB (pphr) on the resilience of the NR matrix composite.

The compression range increases with the increasing carbon black, which is attributed to the rise in the entanglement density (the same reasons for the increase in the tensile strength, as described above), the decrease of the movement of rubber strings and the increase of the rigidity of the compound rubber. These entanglements limit the portability of the processing of compound rubber [9,10].

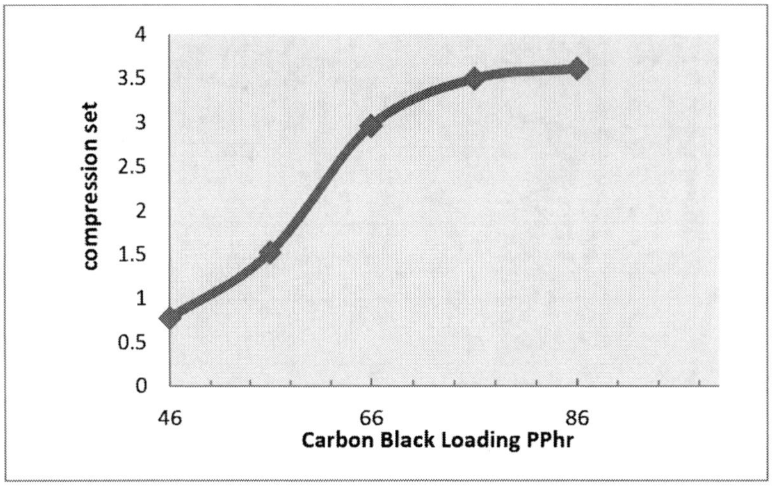

Figure 11. Effect of CB (pphr) on the compression set of the NR matrix composite.

This property represents the materials that are resistant to crack growth. In small proportions, carbon, as shown in Figure 12, has a low resistance to crack growth. A higher number of particles increases fatigue resistance, and the introduction of fillers possessing uniform geometries effectively reduce the crack propagation rates. This is due to the structures of the filler network, especially at high tearing energies [11].

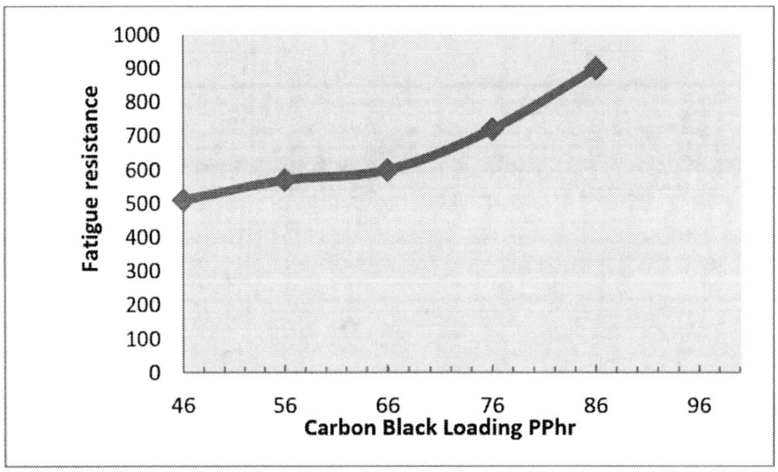

Figure 12. Effect of CB (pphr) on the fatigue resistance of the NR matrix composite.

3.2. Physical Properties

Figure 8 indicates an improvement in the viscosity values, which is an indicator of the elasticity of the rubber component and its ability to produce the rubber compound content with an increase in carbon black due to an increase in the interconnections and a decrease in the freedom of mobility of the rubber chains. This refers to the stability between the carbon granules and rubber chains.

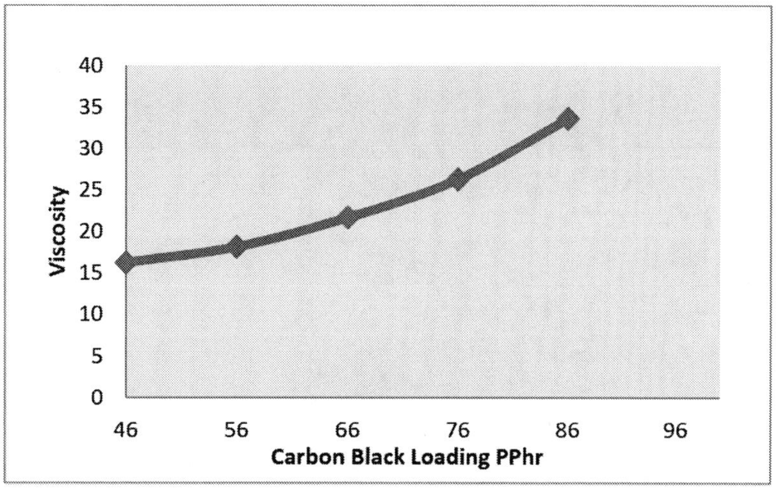

Figure 13. Effect of CB (pphr) on the viscosity of the NR composite.

Figure 14 indicates a steady rise in torque with an increase in carbon black for the same reasons as the increase in viscosity. These findings are compatible with previous research [5].

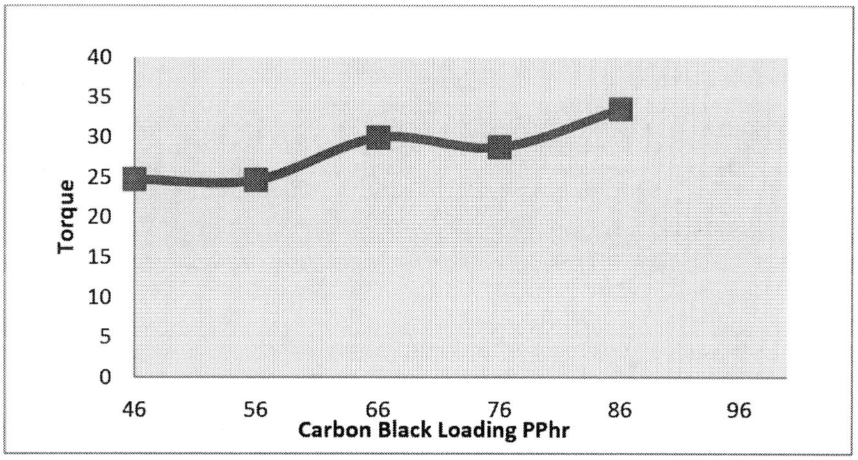

Figure 14. Effect of CB (pphr) on the max. torque of the NR composite.

Figure 15 shows a decrease in the combustion time with an increase in carbon black. The increase in the surface area of the carbon particulate matter is due to meandering in the carbon surface. This means that carbon black acts to provide the required heat to complete the vulcanisation process of the internal structures of the rubber formula due to its contact with the interconnections.
The cure time in Fig. 16 declines as the amount of carbon black in the rubber mixture increases as carbon black joins the networking connections and thereby contributes to early maturation [5].

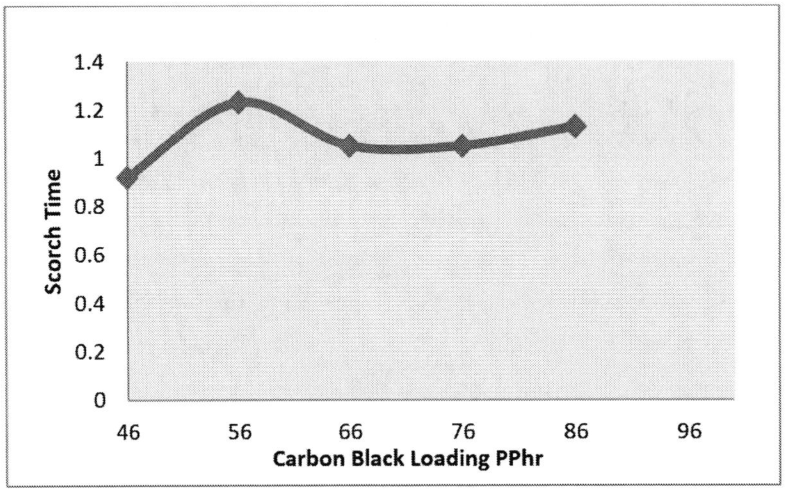

Figure 15. Effect of CB (pphr) on the scorch time of the NR composite.

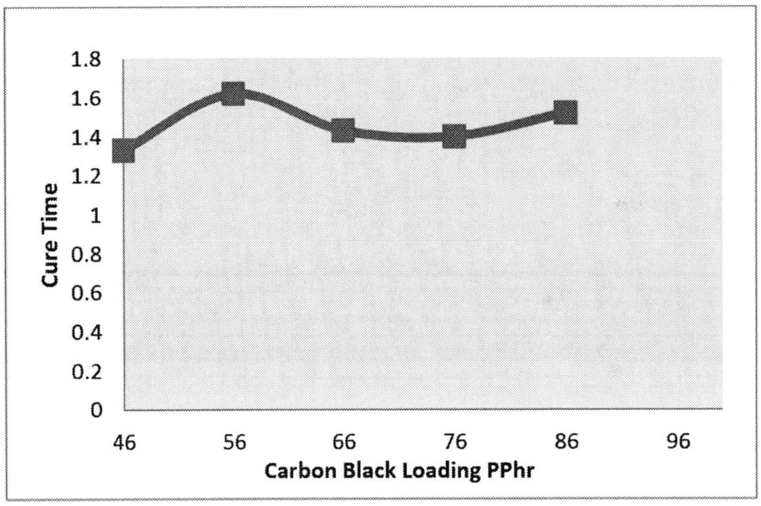

Figure 16. Effect of CB (pphr) on the cure time of the NR composite.

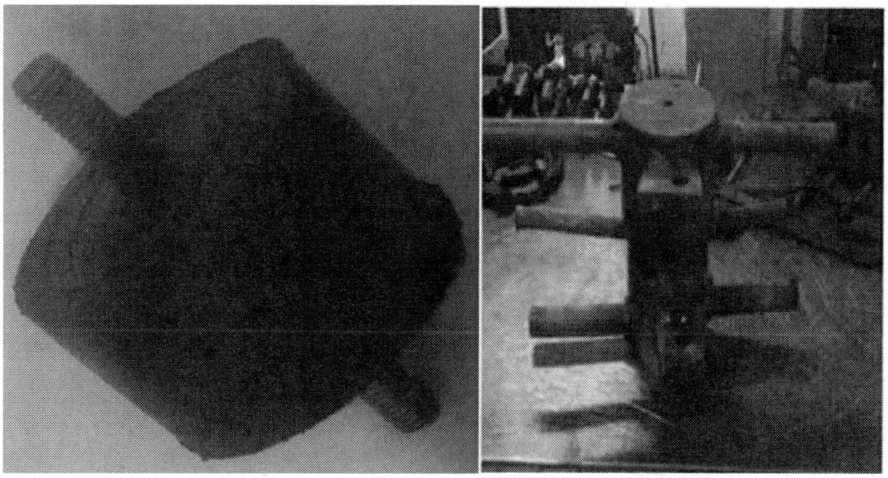

Figure 17. The rubber mount product, as manufactured.

4. CONCLUSIONS

The problems of the mount, which are hardness, fatigue and resilience, were improved by adding fillers where it can be operated at a high load at 900 cycles per minute without any defects. The addition of carbon to the rubber improved the mechanical and physical properties, especially tensile strength, elongation, hardness, fatigue and resilience. However, adding large quantities of fillers led to aggregates of material, which weakened the properties. The best percentage of fillers for the best result is 66 pphr.

References

[1] Waleed F. Faris, Fazidah Saad 2010 Dynamic characterisation of elastomeric engine mounts of different geometry used in automotive industry: simulation approach, *International Journal of Vehicle Noise and Vibration*, 2010 Vol.**6** No.**1**, pp.1 – 39.

[2] Vijayaram, T. R. 2009 A technical review on rubber. *Int. J. Des. Manuf. Tech.* 3:25–36.

[3] C. Arnal, M.U. Alzueta, A. Millera and R. Bilbao 2011 Experimental and kinetic study of the interaction of a commercial soot toward no at high temperature, Aragón Institute of Engineering Research (I3A), University of Zaragoza. Campus Río Ebro.

[4] Ceresana 2013 Market Study: Carbon Black.

[5] Al-Kawaz, Ammar 2008 Preparation of rubber composite as bearing pad and study some of the physicals and mechanical properties for bridges pad applications , *M. D. Thesis*, University of Babylon, Iraq.

[6] Al-Hatammi, Mujed 1999 Improve the properties of Part in contact with the ground for Babylon tire, *M. D. Thesis*, University of Babylon, Iraq.

[7] Al - Maamori, M.H 2005 Mechanical & Physical Properties of Rubber Composite For Engine Mounting *PH. D. Thesis*, University of Technology, Baghdad, Iraq.

[8] Muhannad Kh. D. 2007 Design and Industrialization Recipes of Cloth Drawing Rollers From Rubber composite material *M. D. Thesis*, University of Babylon, Iraq.

[9] Schuur MVD, Gaymans RJ. 2005 Influence of chemical crosslinks on the elastic behaviour of segmented block copolymers. Polymer;**46:**6862–8.

[10] A. Mostafa, et al 2009 J. of Materials and Design ,**30**, 1561–1568 .

[11] Guo, H., Jerrams, S., Xu, Z. et al 2020 Enhanced fatigue and durability of carbon black/natural rubber composites reinforced with graphene oxide and carbon nanotubes. Engineering Fracture Mechanics, 223,106764. doi:10.1016/j.engfracmech.2019.106764.

Measurement of Alpha Activity in Several Types of Iraqi Ceramic and Cement in the Iraqi Markets

Eman Ibrahim Awad[1], Shymaa Al-Rubaye[2], Azhar Lateef Jebur[3],
Abbas Fadhel Hadi[4], Nagwa Ibrahim Awad[5]

[1,3] Department of Environmental Health, Applied Medical Sciences College,Kerbala University, Iraq.
[2] Department of Physics, College of education for Pure Sciences, University of Babylon, Iraq.
[4] Department of Physics, College of education Pure Sciences, Kerbala University,Iraq.
[5] Basic Sciences Branch, Faculty of Dentistry, Kerbala University, Iraq.
E-mail: pure.shyma@uobabylon.edu.iq

Abstract. In this study fifteen samples of ceramic and cement collected from Iraqi markets have been determined for the radium efficacy and radon exhalation rates using the CN-85 plastic track detectors. For the estimation of the effective radium content and exhale rate, these samples comprised 15 different Iraqi materials. The effective radium content values ranged from 2.108 to 21.689 Bq / kg to an average of 9.399 Bq / kg and standard devolution of 6.950. The mass exhalation levels for radon ranged from 0.106 to 1.092 mBq / kg.day, with an average value of 0.473 mBq / kg.day and standard deviation of 0.312 Bq / kg. The exhalation rate for the radon surfaces ranged from 1.889 to 19.44 mBq / kg.day. The mean radon exhalation levels, which are an effective radium element, were found to fall below the world average of 57600 Bq / m2 in the samples of ceramic and concrete materials and Soil: 35 Bq / kg, In addition, the study's showed that the effective radium content in this study and the associated radon exhalation do not pose a human health risk when they are used in daily life in Iraqi markets.

1. Introduction

The world is radiated and contains normal levels of radiation, which plays a significant role in our lives. There is radiation in the soil and the human body, as well as in the air that we breathe in [1]. Most exposure to radiation comes from terrestrial and extra-terrestrial natural sources. Extra-terrestrial radiation is produced in the universe and enters the atmosphere as primary cosmic rays. Terrestrial radiation is produced in the Earth's crust from small quantities of radioactive nuclides, including soils and rocks, that expose the organisms and hydrosphere. These types of radiation are also emitted through food chains or inhalations from nuclear weapons, which are transferred to people and deposited into their tissues. Additionally, people are always exposed to natural ionising radiation [2]. Exposure to elevated levels of radon in the population over an extended period contributes primarily to several pathological effects, including respiratory and pulmonary cancers [3]. Knowledge of the radioactivity that is present in building materials allows the use of such materials to evaluate all possible radiological risks to humanity. Therefore, the radioactivity of common construction materials is important considering the above evidence. From the perspective of radiation safety, the concentration values and levels of exhalation rates for building materials are very important due to the high attenuation power, as the alpha-emitting radioactive substance can harm the human body's normal tissues. Their chemical and radioactive toxicity may damage the normal tissues of different organisms [4]. Most hazards are due to low inhalation. The risk of radon is due to its radioactive progeny, which uses its physical properties to spread in or attach to the lungs and deposit energy into

the tissues. This produces greater ionisation densities than beta or gamma rays. Radon's energy is also deposited in the lungs. The health effects of inhaling radon-decay products include lung cancer, skin cancer and children diseases [5]. Radionuclide activity levels are important for assessing population exposure in construction materials and their components, as most people spend 80 % of their time inside buildings [6]. Radon gas and dijas are inhaled, which expose the pulmonary tissue to nuclear radiation that can raise the risk of pulmonary cancer. Radon gas is one of the key causes of skin cancer and is likely to cause epithelial cell damage because of the accumulation of radon on the skin. In some people who are exposed to radon, kidney-related diseases have also been found. The reason for this is that after radon has been transmitted from the pulmonary to the kidney, a higher dose is absorbed in comparison to other organisms [7].
In nearly all soil, rock and water, uranium is naturally present. The dominant isotope, uranium238, comprises a wide list of declined components, including radium-226 and radon-222. In the natural series of uranium, radone is the decay product of radium. The distance, which is determined by several variables, including the diffusion rate, effective soil permeability and own half-life, travels freely from the radon's source as inert gas. Radon can be identified as a normal alpha-emitter by an alpha-sensitive detector [8]. Alpha emitters must be studied in ceramic and cement materials regarding the role of causers of different diseases, especially cancer. The aim of this work is to determine the radon levels from various ceramics and cement materials, as well as to calculate the effective radium content and exhalation rate in Iraqi building materials.

2. Materials and Methods

Passive technologies of solid-state detectors used the 'sealed-can technique' for the assessment of radium content and exhalation rate of radon [9,12].

2.1. *Collection and Preparation of Samples*

Fifteen samples from various markets in Iraq were collected from ceramic and cement. This material is widely used in all Iraqi cities for the packaging and coating of buildings, both inside and out. In addition, we collected seven ceramic samples and eight cement samples with different origins in Iraq and abroad. Moreover, these samples were dried, milled crushed and sieved using a two-mm mesh, and 85 grams per sample were placed in a cylindrical plastic container in a diffusion chamber. The samples also faced a CN-85 detector track, as shown in Fig.1. The procedure involved putting approximately 85 grams of ceramic and cement samples into an emanation container, which was then closed to balance the radium with the radon for three weeks.

The CN-85 detectors were placed at the closed end of the plastic cup (diameter 7.8 cm and length 9.5 cm). The sampling distance was six cm, and the ceramics and cement samples were kept in plastic canisters at a volume of 167.2 cm3, as shown in Fig. 1.

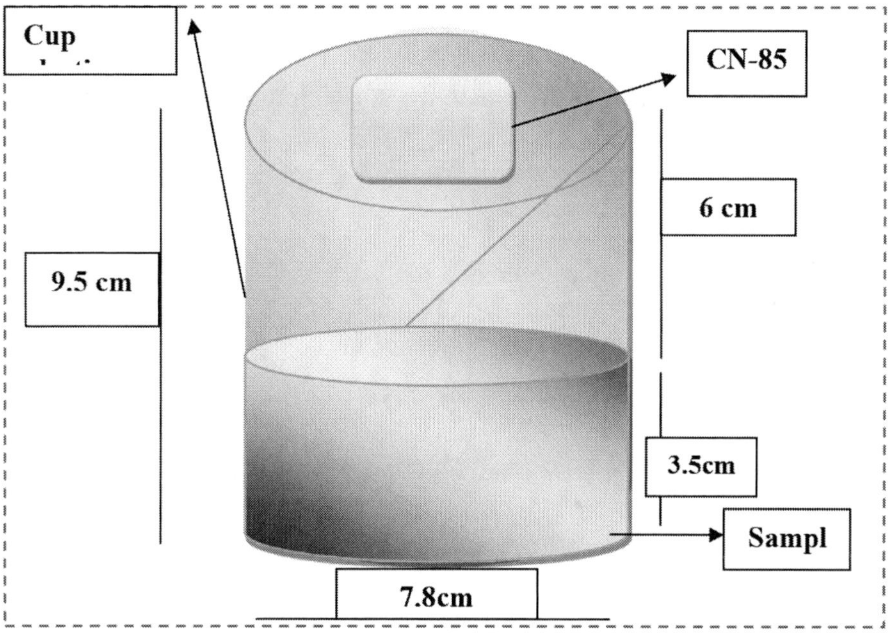

Figure 1: Experimental set-up for the radon exhalation rate measurement of the ceramic and cement samples.

2.2. *Etching and Scanning Process*

The detector reports traces of α-particles that form radon gas, as formed by the α-decay of the sample radium. The detectors had been exposed for approximately 60 days. The detectors were removed and etched in a 6.25N NaOH solution at 60 °C and kept in a constant bath for one hour after exposure to the tracks. The detectors were then washed and dried, and the α-tracks were counted using a 400x magnification (40x objective and 10x eye piece) optical-mission microscope (kruss-mbl 2000).

3. Theoretical Considerations

The densities of the tracks were calculated using the equation below [13]:

$$\rho = \frac{\sum_i N_i}{nA} \quad (1)$$

Where A is the view field area, Ni is the total number of tracks and n = the total number of view fields. Therefore, by using the conversion factor (K), the track density observed was transformed into radium concentration [14].

$$\rho = K\, C_{Ra}\, T_e \quad (2)$$

where K is the plastic detector sensitivity factor, CN-85 is the 0,0092 sensor cm-2h^{-1}/(Bq m-3) [15] and Te refers to the effective exposure time [16].

$$T_e = [T - \lambda_{Rn}^{-1}(1 - e^{-\lambda_{Rn}T})] \quad (3)$$

Given that 226Ra's half-life is approximately 1600 years and 222Rne's is 3.82 days, it should be assumed that the radium-radon members of the decline series will have an effective balance (approximately 98%) in three weeks. After a radioactive balance has been established, the radon alpha analysis can be used to determine the stable level of radium activity. After closure according to relation, the activity concentration of radon starts to increase with time [17].

$$C_{Rn} = C_{Ra}(1 - e^{-\lambda_{Rn}T}) \quad (4)$$

Where CRa is a good radius element of the sample and Rn is a constant decrease of 222Rn, since a plastic track detector measures the time-integrated value of the above expression, i.e. the total number of alpha disintegrations in the unit volume of the can during exposure time T with a K sensitivity.

3.1. Effective Radium Content

The effective radium content was measured using the following formula [18, 20]:

$$C_{Ra}(Bq.\text{kg}^{-1}) = \left(\frac{\rho}{KT_e}\right)\left(\frac{hA}{M}\right) \quad (5)$$

Where M is the mass of building sample materials in kg, A is the cross-sectional region of the dust in m2 and h is the height between the detectors and the top surface of the sample (0.03 m).

3.2. Radon Exhalation Rates

With the expression, the sample surface (area) exhalation rate can be calculated for the release of radon [21, 22].

$$E_A = \frac{C_{Rn} \ V\lambda}{A[T + \lambda^{-1}(e^{-\lambda T} - 1)]} \quad (mBq/m^2.day) \qquad (6)$$

Where EA is the mBq / m.day area-exhalation radius, CRn is the built-in radon exposure, V is the cup's efficient volume in m3, T is the hourly exposure, CRn is the 222Rn (h-1) decline constant and A is the m2 cup area. The sample exhalation rate for radon release can be calculated using the following expression [21, 22]:

$$E_M = \frac{C_{Rn} V\lambda}{M[T + \lambda^{-1}(e^{-\lambda T} - 1)]} \quad (mBq/kg.day) \qquad (7)$$

Where EM is the mass exhalation of radon in terms of mBq / kg.day, and M is the mass of the sample in terms of kg.

4. Result and Discussion

Fifteen different samples were analysed in this work using the ceramic and cement technique. The effective radium content, surface exhalation and mass exhalation rates are indicated in Table 1. This is because the CN-85, as a component to other solid-state detectors for nuclear plotting, was measured using the CN-85 detector as a passive radon dosimeter. In Iraq's ceramic and cement samples, the effective radium content was distributed, while the relative comparison of mass variation and surface exhale radon with radium content was found, the radon exhalation rates match the estimated values of the radium content. The exhalation value of the mass and surface radon in Table 1 is below what many scientists have recorded [10, 23–25]. In the current sample, the radon exhalation levels are slightly lower than the 57 600 mBq / m2 h world average (0.016 Bq / m). Therefore, residents are safe from health risks [26]. The values reported in Table 1 for radium content are smaller than those shown by scientists [27, 28]. In the present study, the effective radium values for the ceramic and cement samples were below the recommended action level of 370 Bq / kg [29]. In addition, there was less than 35 Bq / kg in the soil, which is also the average global value [30]. Therefore, the result shows that the radium health risks are safe for this material. We noticed from the data that in the samples, the effective radium content differs with a mean value of 9.399 bq / kg from 2.108 to 21.689. The mass exhalation rates also varied with an average value of 0.473 / mBq / kg.day from 0,106 to 1,092 mBq / kg.day, a standardised exhalation rate from 1 889 to 19.44 / m2.day to 1 8,524 mBq / m2.day and a standardised exhalation level of 5,5522. The highest values were calculated in Iranian cement in the cement sample -1 value and the lowest in Iraqi-resistant cement. However, the highest values were in the Spanish ceramic -1 value and the least were in the Chinese ceramics -2 value. This means that samples of ceramic and cement material from the Iraqi market cannot produce dangerous indoor radon levels when used as materials.

Table 1: Collected from Iraqi market efficient radium content and exhale rate of radon of ceramic and cement materials.

No.	**Sample**	**$\rho \times 10^4$ (Trac / cm^2)**	**CRa Bq / kg**	**Em mBq / kg.day**	**ES mBq / m^2.day**
1	Cement ordinary Iranians	3.680	10.317	.519	9.247
2	Iranian Cement -1	4.832	13.547	.682	12.142
3	Iranian Cement -2	2.208	6.190	.311	5.548
4	Cement ordinary Iraqis -1	2.408	6.751	.340	6.051
5	Cement ordinary Iraqis -2	2.912	8.164	.411	7.317
6	Cement-resistant Iraqis	1.272	3.566	.179	3.196
7	Cement Lebanese	3.360	9.420	.474	8.443
8	Spaniard Ceramic -1	6.720	18.840	.949	16.887
9	Spaniard Ceramic -2	7.736	21.689	1.092	19.440
10	Iranian ceramics	1.336	3.746	.188	3.357
11	Ceramic Syrian	4.000	11.214	.565	10.051
12	Egyptian ceramics -1	1.576	4.418	.222	3.960
13	Egyptian ceramics -2	1.096	3.073	.154	2.754
14	Chinese Ceramics -1	6.400	17.943	.904	16.083
15	Chinese Ceramics -2	.752	2.108	.106	1.889
Maximum		7.736	21.689	1.092	19.440
Minimum		.752	2.108	.106	1.889
Mean		3.352	9.399	.473	8.424
Standard Deviation		2.209	6.195	.312	5.552

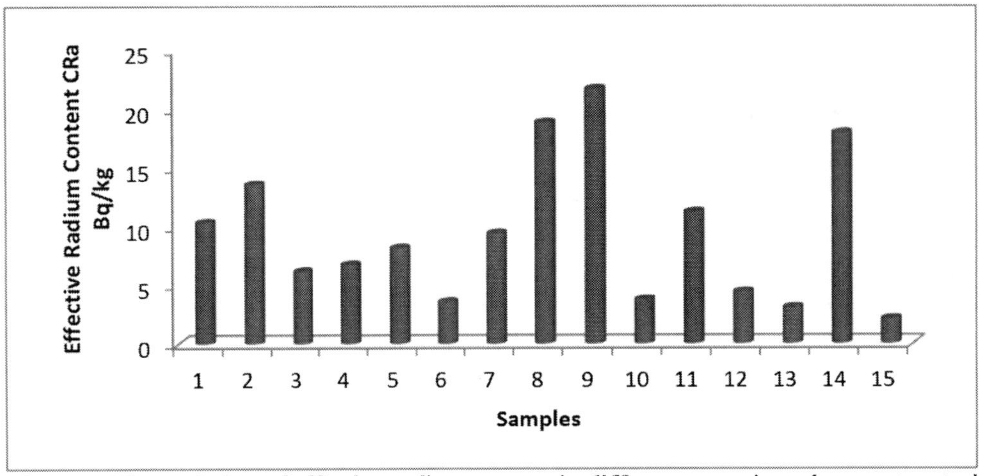

Figure 2: The variation of effective radium content in different ceramic and cement materials.

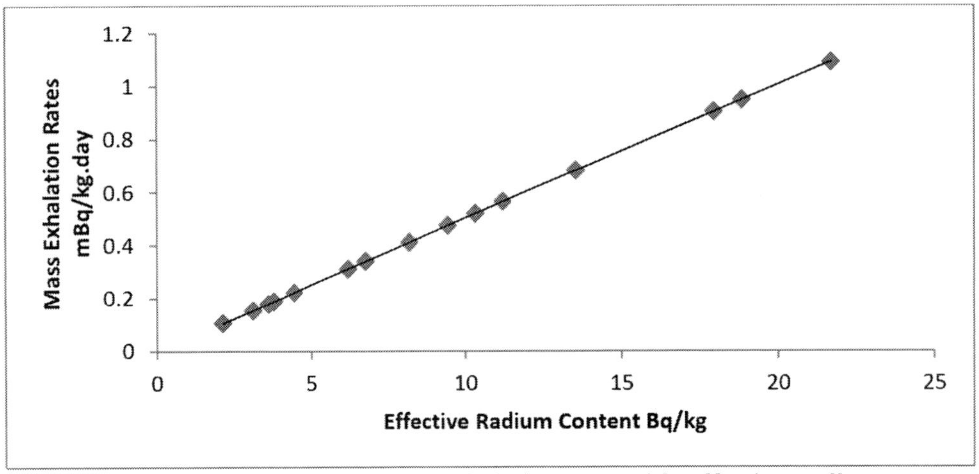

Figure 3: The variation of the mass exhalation rate with effective radium content in ceramic and cement.

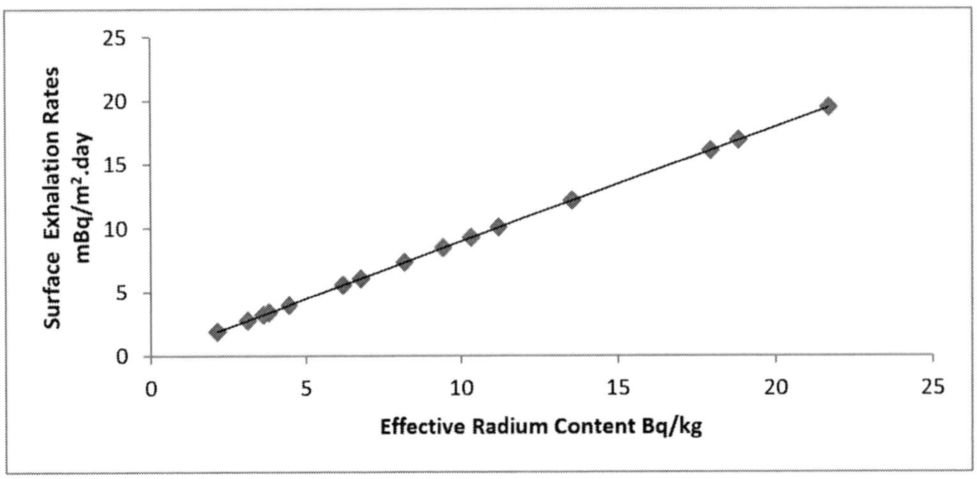

Figure 4: Surface exhalation rate change with effective radium in ceramic and cement.

5. Conclusion

Effective rates of the radium content of radon were successfully measured using CN-85 plastic track detectors and the sealed-can technology. The distribution of radium in ceramic and cement samples is heterogeneous, as radium content varies from sample to sample in the Iraqi markets. The average overall radium value for the ceramic and cement samples from this study was 9.399 Bq / kg, whereas the global average radium content value for the Earth's crust is 40 Bq / kg. These samples showed that the exhalation of radon is far less than the world average of 57 600 mBq / m2 h (0.016 Bq / m2 s). Therefore, these materials are not a health risk because they can be used for construction purposes. In the ceramics and cement samples, we found that the effective radium content was low and did not pose a health hazard. In summary, this study found that material samples are safe for use in building materials. They will not produce hazardous levels of alpha radiation in dwellings. The radium content of the ceramic and cement samples correlated with the radium exhalation rate.

References

[1] E. M. Lee, G. Menezes, and E. C. Finch, "Natural radioactivity in building materials in the Republic of Ireland," *Health Phys.*, vol. 86, no. 4, pp. 378–383, 2004.

[2] Namibi, K. S. V., Bapat, V. N., David, M., Sundram, V. K., Sunta, C. M., Soman, S. D., Internal Report: Health Physics Division, B.A.R.C, Bombay,1986.

[3] BEIRVI, Report of the Committee on the Biological effects of Ionizing Radiation), Natl. Res. Council. Natl. Acad. Press, Washington, DC, 1999.

[4] Fleisch, R. L., Price, P. B., Walker, R.M., "Nuclear Tracks in Solids: Principle and application, "University of California Press, Berkelely, pp. 210-215, 1975.

[5] Kumar, S., Chander, S., Yaday, G. S., Shama, A. P.,. Some environmental effect studies on the response of (CR-39) plastic track detector, "*Nuclear Tracks*, vol. 12, pp. 129–132, 1986.

[6] K. Badhan, R. Mehra, R. G. Sonkawade, and S. Singh, "Use of Gamma-Ray spectrometry for assessment of natural radioactive dose in some samples of building materials," *Asian J. Chem.*, vol. 21, no. 10, pp. 207–211, 2009.

[7] Henshaw, D. L., Eatough, J. P., Richardson, R. B., "Radon as a causative factor in induction of myeloid leukemia and other cancers. Lancet, vol. 335, pp. 1008- 1012, 1990.

[8] V. E. Archer, J. D. Gillam, and J. K. Wagoner, "Respiratory Disease Mortality Among Uranium Miners," *Ann. N. Y. Acad. Sci.*, vol. 271, no. 1, pp. 280–293, 1976.

[9] J. Chen, N. M. Rahman, and I. A. Atiya, "Radon exhalation from building materials for decorative use," *J. Environ. Radioact.*, vol. 101, no. 4, pp. 317–322, 2010.

[10] A. J. Khan, R. Prasad, and R. K. Tyagi, "Short Communication Measurement of Radon Exhalation Rate From Some Building Materials*," *Int. J. Radiat. Appl. Instrum*, vol. 20, no. 4, pp. 609–610, 1992.

[11] A. K. Mahur, R. Kumar, D. Sengupta, and R. Prasad, "Estimation of radon exhalation rate, natural radioactivity and radiation doses in fly ash samples from Durgapur thermal power plant, West Bengal, India," *J. Environ. Radioact.*, vol. 99, no. 8, pp. 1289–1293, 2008, doi: 10.1016/j.jenvrad.2008.03.010.

[12] R. G. Sonkawade, K. Kant, S. Muralithar, R. Kumar, and R. C. Ramola, "Natural radioactivity in common building construction and radiation shielding materials," *Atmos. Environ.*, vol. 42, no. 9, pp. 2254–2259, 2008, doi: 10.1016/j.atmosenv.2007.11.037.

[13] N. Ahmad, M. S. Jaafar, S. A. Khan, T. Nasir, S. Ahmad, and M. Rahim, "Measurement of radon exhalation rate, radium activity and annual effective dose from bricks and cement samples collected from Dera Ismail Khan," *Am. J. Appl. Sci.*, vol. 11, no. 2, pp. 240–247, 2013.

[14] M. Shakir Khan, A. H. Naqvi, A. Azam, and D. S. Srivastava, "Radium and radon exhalation studies of soil," *Iran. J. Radiat. Res.*, vol. 8, no. 4, pp. 207–210, 2011.

[15] Tufail, M. , Khan, M.A., Ahmed, N., Khan, M.A., Zafer, M.S., "Measuremen of radon concentration in some cities of Pakistan, "*Radiat. Prot. Dosim.*vol.40, no. 1, pp.39-44, 1992.

[16] Azam, A., Naqvi, AH., Srivastava, DS., "Radium content and radon exhalation measurement using LR-115 type II plastic track detectors: "*Nucl Geophys*, vol. 9, no. 6, pp.653–657, 1995.

[17] M. S. Khan, D. S. Srivastava, and A. Azam, "Study of radium content and radon exhalation rates in soil samples of northern India," *Environ. Earth Sci.*, vol. 67, no. 5, pp. 1363–1371, 2012.

[18] Ali, K., Alsaedi, B., Almayahi, A., Alasadi, A. H.,"Cement Rn and Ra concentration measurements in selected samples from different companies, "*Asian Journal of Natural and Applied Sciences (AJSC)*, vol. 2, no.4, pp. 95-100, 2013.

[19] M. S. Khan, M. Zubair, D. Verma, A. H. Naqvi, A. Azam, and M. K. Bhardwaj, "The study of indoor radon in the urban dwellings using plastic track detectors," *Environ. Earth Sci.*, vol. 63, no. 2, pp. 279–282, 2011.

[20] Mahur, A. K., Khan, M. S., Naqvi, A. H., Prasad, R., Azam, A., "Measurement of effective radium content of sand samples collected from Chhatrapur beach, Orissa, India using track etch technique. Radiat Meas, 2008.

[21] Sonkawade, R. G., Kant, K., Muralithar, S., Kumar, R., Ramola, R. C., "Natural Radioactivity In Common Buildingconstruction And Radiation Shieldingmaterials, "*Atmospheric Environment*, vol. 42, p.p 2254-2259,2008.

[22] Amrani, D. and Cherouati, D. E., "Radon exhalation rate in building materials using plastic track detectors. "*Journal of Radio analytical and Nuclear Chemistry*, vol. 242, pp. 269-271, 1999.

[23] Ismail, A. M. O., Abumurad, K. M., Kullab, M. K., Al-Batanina, B. A., "Measurement of Rn-222 concentrations in Building materials used in Jordan. Second Arab conference on the peaceful uses of atomic energy, Cairo5-9 Nov, pp. 1119-1127, 1994.

[24] Laith, A. Najam., Nada, F. Tawfiq., Rana, Hesham. Mahmood., Radon Concentration in Some Building Materials in Iraq Using CR-39 Track Detector. "*International Journal of Physics*, Vol. 1, No. 3, pp. 73-76, 2013.

[25] Zikovsky, L., "Determination of radon exhalation rates from Canadian building materials with an internal proportional counter. "*Nucl. Tracks Radiat. Meas.*, Vol. 3 pp. 525-527, 1992.

[26] UNSCEAR 2000 report sources and effects of ionization radiation, United Nations Scientific Committee on the Effects of Atomic Radiation, Report to the General Assembly with scientific annexes Volume1United Nations New York, 2000.

[27] S. Singh, A. Kumar, and B. Singh, "Radon level in dwellings and its correlation with uranium and radium content in some areas of Himachal Pradesh, India," *Environ. Int.*, vol. 28, no. 1–2, pp. 97–101, 2002.

[28] D. K. Sharma, A. Kumar, M. Kumar, and S. Singh, "Study of uranium, radium and radon exhalation rate in soil samples from some areas of Kangra district, Himachal Pradesh, India using solid-state nuclear track detectors," *Radiat. Meas.*, vol. 36, no. 1-6 SPEC., pp. 363–366, 2003.

[29] OECD, Exposure to radiation from the natural radioactivity in building materials. Report by a Group of experts of the OECD, Nuclear energy agency, Paris, France, 1979.

[30] UNSCEAR 2000 report sources and effects of ionization radiation, United Nations Scientific Committee on the Effects of Atomic Radiation, Report to the General Assembly with scientific annexes Volume1United Nations New York, 2000.

IOP Conf. Series: Materials Science and Engineering **987** (2020) 012011 doi:10.1088/1757-899X/987/1/012011

Characterisation of Fe Doped Layered Cathode Material as Nano-Rechargeable Batteries

Rusul A.Rsool Abaas[1], Adawiya J. Haider[2*], Faten Abdul Jabbar[3], Mustafa Mohammed Hathal[4]

[1,3,4] Ministry of Industry & Mineral, Baghdad, Iraq.
[2]Applied Sciences Department, Laser Science and Technology Branch, University of Technology, Baghdad-Iraq

*corresponding author:100081@uotechnology.edu.iq

Abstract. This research focuses on the layered doped lithium conductive metal oxides with the universal form Li MO_2 (M = Co, Ni, Ag or Fe doped solutions) adjustment of cathode electrodes for lithium-ion cells (LIBs). Preparation in a low temperature combustion reaction promotes its electrochemical efficiency. The modification starts with the preparation of the LiCo0.5Ni0.45Fe0.5O2 (LCNFO) powder with citric acid, which is used as fuel, along with deionized water (DW). As binding materials, nitrate of ions (Li+) and ions (Fe+) were added to this mixture to create a uniformed mixture. The required amount of citric acid and de-ionised water was added to this unified mixture to create a powder. This powder was cautiously heated to 150 C^0–200C^0 initially, and finally annealed at a range from 600 C^0 to 1000 C^0 for six hours.
The resulting product was submitted to a thermal analysis (TGA), which is a useful tool to explore a wide array of polymer properties. This research demonstrates the different applications of TA methods in the field of polymer-nanomaterial research and displays several examples of implementation of the differential scanning calorimeter (DSC). TGA was deployed to investigate the performance of thermal characteristics, which vary with the temperature, for the illustrations and structural characterisation, X-ray Diffraction (XRD) and analysis. A meticulous annealed temperature will output pure, single phase LCNFO / NP material with electrochemical behaviour. The powder was studied, and the dimensions of the cathode material were designed approximately for the XRD data using the Scherer's equation. Furthermore, we used the Fourier transform for the IR (FT-IR) and surface morphology EDS and SEM analysis. The results focused on modifying the $LiCoO_2$ of the cathode materials by adding materials that are similar to Ni and Fe because these materials are documented as being their own state and not contaminated with the hexagonal geometry arrangement and R three m (gap assembly). We noticed that doped Fe has material that will perform better according to the capacity, power, rate capability stability and low cost, with increased storage capacity due to the nano-size effects.

1. Introduction

Energy embodies a significant function in our life. Industries have pollution, global warming and finite ordinary energy supplies and resources, which make electricity, along with the use of green energy, a global precedence [1]. Nanomaterials are imperative due to their unique mechanical, electrical and elective properties that are measured in a run of one–100 nm. Batteries are gadgets that change the vitality that is discharged by unconstrained chemical responses to power. They have a few advantages can store and recover electrical vitality [2]. There is an increasing importance to ensure that electrical components (energy storage units) fit the requirements, such as the upcoming plug in hybrid electric vehicles (PHEVs), as renewable resources are rare [3,4]. A contemporary community progressively controls additional electricity. The component's progress enables it to store minor energy directly or indirectly, which could be a definite subject for the 21st century [5].

LIBs are electrochemical vitality capacity gadgets that can realise the vitality capacity / discharge using charge / release LIB cells. LIBs are composed of the following three primary components: the anode, cathode and electrolyte. Li-ion battery innovation provides the ideal capacity vitality. For all these applications, to attain commercial victory, the batteries must have a long lifecycle and calendar life, sufficient energy / power, be tolerant of damaging conditions, not be harmful and operate securely. The cathode material is an exceptionally critical part of the supporting particle [3, 5].

The Li-ion cell's structure is well-organised as it supplies portable devices using superior authority necessities. The contemporary cell phones for technology-based civilisation depends on transportable electronics with slim energy resources, and investigations into slim layer cells have increased to accomplish this [6].

It is our responsibility to report the difficulties regarding synthesis, electrochemical stability, performance and security matters of these three phases of lattices from the perspective of Li+ ion motion. The aim of this study is to discover the consequences of new investigations into Li, Li ion and Li-free slim-layer cells, along with crystalline $LiCoO_2$ electrodes, and to summarise and characterise the remarkable characteristics of the nano- and microcrystalline layers in the FeLiO system. The LCNFO / NP's electrode material has more power than $LiCoO_2$ [7,8]. **Firstly**,there are additional motives for regulating the stoichiometry of electrode materials. Substitution and decreasing electrode sources enhance the sources' electro-chemical behaviour in association with the exact power and interfacial density of energy (Li et al. 2002; Park et al. 2001; Veluchamy et al. 2001) [8]. **Secondly**, it may well steady the source and enhance cycling recital (Fey et al. 2003) . **Thirdly**, it might reduce the cost of manufacturing by reducing the use of some luxurious metals, such as cobalt, and using less luxurious metals, such as as Mn or Fe (9). **Finally,** it may decrease the material's toxicity and promote a healthier and safer environment. This work describes an improved preparation of LCNFO / NP using a solid reaction in the air and considers the electrochemical characteristics of the resulting powder and the consequences of the annealed on the crystallisation of the manufactured LCNFO / NP. As a result of the disadvantages of LiCoO2, $LiNiO_2$ and $LiMnO_2$, in the current research, a new cathode material LCNFO / NP was synthesised using the combustion technique using citric acid as energy and D.W as the binding material [10,11].

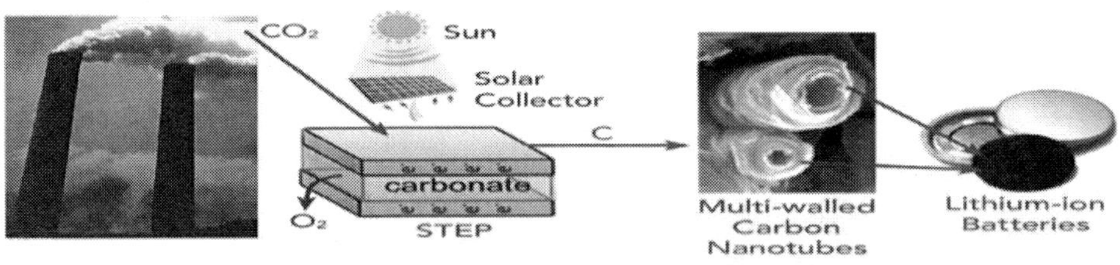

Figure 1: How to reduce Co_2 using LCNFO nanoparticles of a combustion reaction.

2. Experiment
2.1 Powder

Figure 2 displays a chart of the experimental trial for the preparation of nano-fine particles [9–11]. Burning the synthesis of LCNFO in the form of powder, which was made from lithium nitrate {LiNO3 (7.064 g)} nitrate crystal, cobalt (II)- nitrate {$Co(NO_3)_2.6H_2O$ (14.909 g)}, nickel (II-nitrate {Ni $(NO_3)_2$. $6H_2O$] (13.407g)} nitrate crystal, ferric nitrate { $Fe(NO_3)_3$.$9H_2O$ $Fe(NO_3)_3.9H_2O$ (2.069 g)} and citric acid anhydrous C6H8O7 (15.158 g), they were resulted from HARRIS chemicals cooperation in England with decontamination 99%. It was reduced to nano-size using a low preparation temperature of 150 ◦C under atmospheric pressure until firing produced a fine powder material. This produced eight to ten gm of pure cathode. Then, we used a thermal analysis to obtain the optimum annealing temp and divided it into three batches at 600, 800 and 1000 degrees centigrade [12,13].

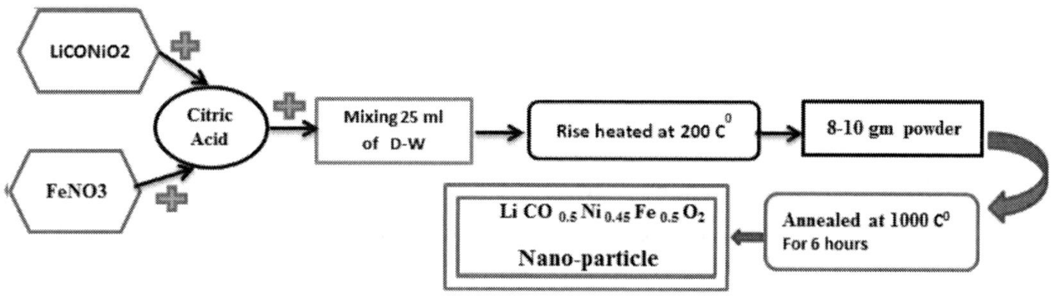

Figure 2. The schism state of the reaction of combustion.

2.2 Thermal Analysis (TGA).
The purpose of this test was to calculate the optimum temperature that eliminated the impurities and phase configuration with the resulting crystallisation. TGA was created using a thermal analyser (STA PT-1000 Model) at a heating rate of 25 co / min through ambient ambiance to conclude the maximum temperature for state configuration and / or the overall solidification of the powder illustration.

2.3 X-Ray Diffraction for the Powder
The powder of LCNFO was defined using the experiments of the X-ray. The measurement was executed and correlated with the American Society of Testing Materials cards by applying Philips PW 1840 X-ray diffract meter of Å from Cu-K. The elements with impurities that were found in the powder targets were analysed by applying the X-ray fluorescence source with the model OLYMPUS \ DELTA.

2.4 FT-IR Spectrophotometer
The optical feature of the powder that was examined by transmission by FTIR was absorption spectroscopy. The measurable wave number area relied on the sensor that was used; the main detectors were TGS (PLATINUM-ATR) 400–7800 cm $^{-1}$ and the semiconductor-type sensor ALPHA. The low sensing limit is approximately 650 cm.

2.5 EDS and SEM of LCNFO Cathode Powder

We examined the morphology, focusing on the particle size distribution of the material. We also had topographic considerations and conducted a microstructure examination. The voltage went up to 30000 v (Ministry of Science and Technology Production Engineering and Metallurgy). The image shows a 3D tomography component. Tescan Vega III, Czech Republic, (USA).

IOP Conf. Series: Materials Science and Engineering **987** (2020) 012011 doi:10.1088/1757-899X/987/1/012011

3. Results and Discussion

In Figure 3, the TGA curve shows three obvious parts of weight loss by m1by m3. The first part (m1-m2) was due to water loss. The second division (m2-m3) was because of the disintegration of the convinced remaining hydrocarbons that did not react.
The third part (m^3) referred to the disintegration of $FeNiO_3$ and the creation of the uncontaminated phase of the layered LCNFO compound and the de-lithia ion of the layered LCNFO by the Li vaporisation. There were weight losses, corresponding masses and molar weights for every part [13,14].

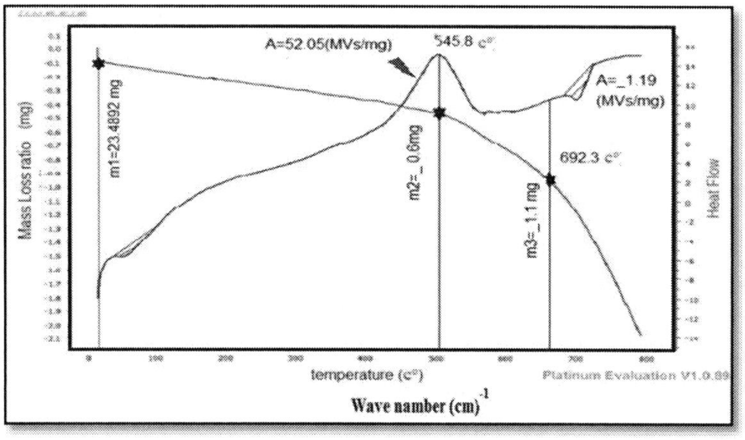

Figure 3: TGA (mass loss and heat flow curves) of the thermally analysed powder.

As a result, the TGA/DSC thermal profile of the LCNFO powder should be studied for the recognition of a convenient annealing temperature that will not have consequences in the creation of the metal / conductivity metal oxide contaminates in the annealed module. For this pattern, the temperature should be high enough to produce pure LCNFO material by-transitory a laser temperature of creation of the metal / conductivity metal oxides. Basically, for well electrochemical facial appearance of the (LCNFO) material, the selected annealing temperature should be more than 973 K, and the samples should be placed in a fully heated furnace of an appropriate temperature. The elevated temperature will enforce the Fe ions into the LiCONiO2 lattice to produce stoichiometric LCNFO material. The selected annealing temperature will produce pure, single phase LCNFO material, as well as electrochemical behaviour. [9,15]. The mass losses of every part are shown in Table 1.

Table 1. Mass losses and structure of the material at TGA curve parts.

Part	T (OC)	Mass (mg)	Mass loss (mg)	Structure
M1	25	23.4892	0.0	1LiCO0.5 Ni0.45Fe0.5O2 + H2O + Co2 + FeNiO3
M2	545.8	22.8892	0.6	LiCO0.5 Ni0.45Fe0.5O2 (from XRD patterns)
M3	692.3	21.7892	1.1	LiCO0.5 Ni0.45Fe0.5O2

3.1 Fourier Transforms IR (FT-IR)

An analysis of the Fourier infrared is necessary to gain spectral visual knowledge about the composite materials, as it indicates the bonds' strength or length in a particular spectral region using the vibrations.

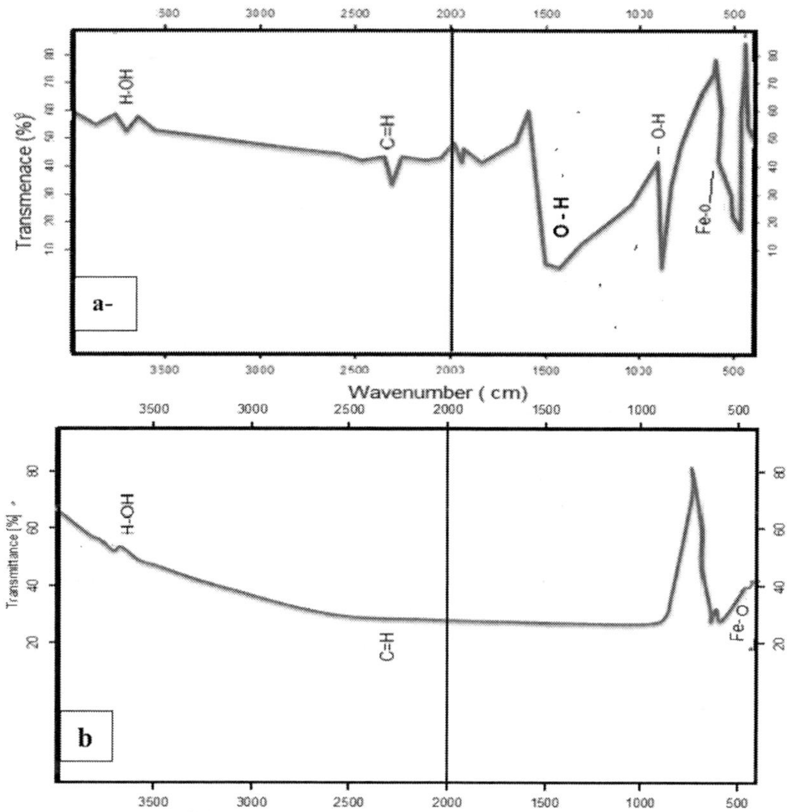

Figure 4. FTIR pattern of the LCNFO / NP (a) at RT °C at 1000 °C and (b) after annealing.

Figures 4a and b display the FTIR spectra of the prepared powder at various temperatures subject to the ambient. The IR is inactive in the area around 2000–4000 Cm^{-1}, which could be due to temperature exposure, as when it is increased, the mass of the particles increases. This is illustrated in Figure 4a. Three absorption areas at 1300–1409.50 Cm^{-1} peak at a broad weak intensity because of bond (C–H), and the area around 1804.20–2500 Cm^{-1} peaks at broad medium. The intense strength comes from bond (C=H). Finally, area 3000–3737 Cm^{-1} IR is inactive because there is no dipolar bond (C=C) or off bond (H-OH). Figure 4b displays the peaks at 421.77–570.62 Cm^{-1}, which is equivalent to the iron-oxide bond Fe–O in the tetrahedral sites. When the group at 916 Cm^{-1} corresponds to the stretching vibration mode, it should correspond with the M–O absorption band. The bigger absorption peak that is located at 3500 Cm^{-1} could refer to the widening of the -OH/HO- assemblies of the arcade water molecules with the hydrogen–hydroxyl bounded groups in both CO and $Fe(OH)_2$. The peaks at approximately 2000–2374.96 Cm^{-1} are due to C=H stretching and the C–O bending vibrations.

The shape of the peak at the broadband at 400 and 570 Cm^{-1} is attributed to the stretching vibrations of Fe^{3+}-O^{2-}, which are considered in all the ferrite modules. The broad at 3653 Cm^{-1} could be due to the vibration state of the chemically bonded hydroxyl groups. The peaks at around 3833.81–3500 Cm^{-1} are attributed to H–OH stretching [12,13].

3.2 -X-Ray Diffraction (XRD)

Figure 5 reviews the annealing procedure, which decreases lattice deformations and damage. However, this technique can create the grains' adhesion. This prejudiced the amplified mean particle dimensions for nanoparticles and the grain dimension from 37.59 nm–484 nm, along with the temperature increase. The determined range of the lattice stricture of the **LCNFO** / NP modules were in accordance with the reference statistics tag (system 98-016-2293) [12]. Model grains are sub-micron dimensional particles with a sphere shape, which are sufficient for improved electrochemical effectiveness. As the heating duration is raised, the diffraction peaks become sharper and superior due to the raise in the crystallinity of the resulting product.

LCNFO powders that were gained through these wet-chemistry procedures showed the precise XRD pattern and the hexagonal doublets (006) / (102) and (018) / (110), along with an obvious splitting. This signalled a higher degree of crystallinity, good hexagonal characteristics and a better layered facial appearance. The XRD consequences proved the configuration of the pure state [14,15].

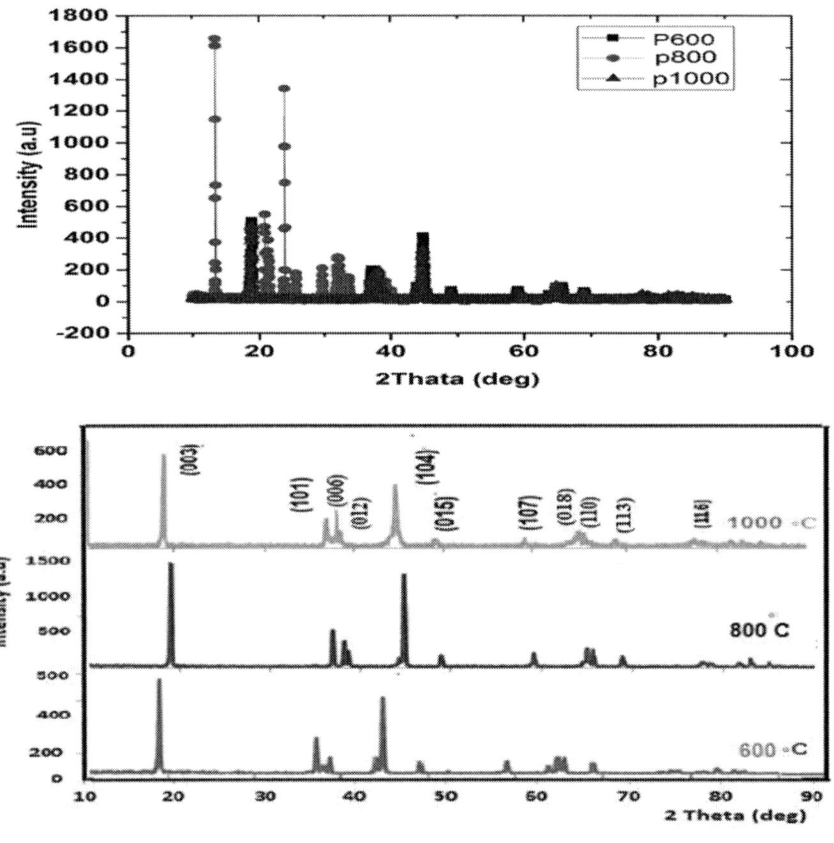

Figure 5 XRD patterns of LCNFO with various annealing temperatures of 600 ◦C, 800 ◦C and 1000 ◦C for a, b and c, respectively.

3.3 Energy Dispersive Spectroscopy of LCNFO Cathode Active Material

Figure 6 reviews the EDS of the LCNFO cathode active material. The EDS results before and after the annealing state that the weight percentages of the cobalt, nickel and iron were 37.80, 31.44 and 3.76, respectively, without changes after the annealing process. These results were in accordance with the compositional structure of the conducted material, as determined from the XRD results [9–16].

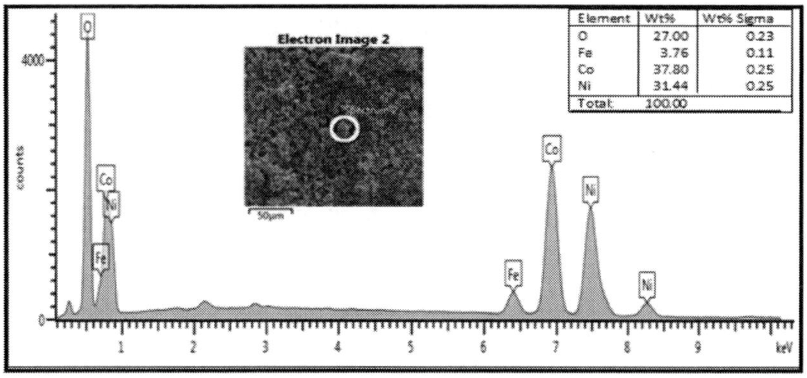

Figure 6 EDS of the LCNFO cathode (active).

3.4 Scanning Electron Microscopy of the LCNFO POWDER

The morphological alterations of the arranged LCNFO powder at various annealing temperatures have been tested by SEM, and the consequences are displayed in Fig. 7a. Previously, annealing had a spherical and springy shape with a length of the small dimension nanoparticles being less than 37 nm. As the temperature rises, the size becomes larger and the accumulation becomes significant, as shown in the SEM image in Fig. 7b. This was in accordance with the results from the XRD measurements [17,18,19].

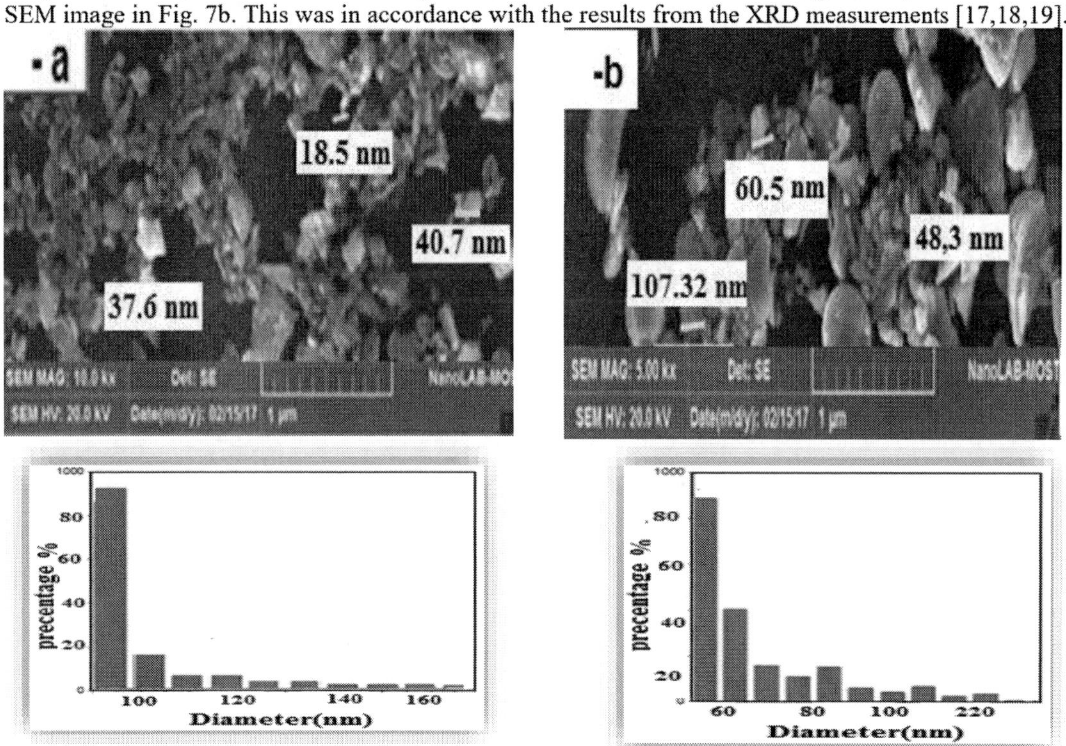

Figure 7 (a, b) Upper SEM images of LCNFO powders that were prepared through combustion and annealed at various temperatures: (a) before annealing, and (b) 1000 ◦C. The lower images display the statistical distribution of the LCNFO powder.

This growth of the uneven, average LCNFO dimensions could be associated with the straight exposure of the annealing temperature on the powder structure. As the temperature rises, the superior energy is utilised by the substance, and there is a possibility to form great particles due to the increased ion distribution and grain creation [20, 21,22, 23].

Conclusion

$LiCo_{0.5}Ni_{0.45}Fe_{0.5}O_2$ layered nanoparticles are defined as developed cathodes with simple transition-metal layered dioxides, in which at least one lithium ion per transition-metal ion could be reversed by extracting / inserting while keeping the materials' cost and toxicity low, and they were successfully synthesised by a firing reaction at low temperature. This was proven through characterisation using an X-ray and EDS with SEM. These materials are important for use in future lithium-ion batteries. Fe doped with LiNiCoO2 makes the movement of the lithium- ions easy due to layered structures, which progress the performance of the electrochemical cells. The TGA curve demonstrates the decomposition of the modules in the single stage. The topographical SEM images show that LCNFO may illustrate encouraging electro-chemical features as intercalation, and the de-intercalation of Li^+ is thought to increase the capacity of the cathode hosts by focusing on nano-sized powders and amorphous materials.

Acknowledgments
Authors sincerely acknowledge to the support of the Nanotechnology team in Ministry of Industry & Mineral, and the Nanotechnology team in State Company for Electrical Industries/Baghdad/Iraq.

References

[1] Julien C, Mauger A, Vijh A, Zaghib K. 2016 *book Springer, Cham (pp. 29-68)*...
[2] Razzak, A. A., Al-Gebori, A. M., & Haider, M. J. 2020 *Journal,an International 14* (1), 19-29.
[3] Godse, P., Sakhare, R., Pawar, S., Chougule, M., Sen, S., Joshi, P., & Patil, V. P. 2011 *Journal of Surface Engineered Materials and Advanced Technology*, *1*(02), 35.
[4] Ensminger, J. 2016 *Conservation, and Environmental Remediation*, **389**
[5] Sathiyamoorthi R., Subramania A., Gangadharan R. and Vasudevan November 2005 *studies for lithium secondary cells*, **65**(5). 973-980.
[6] Leo, C. J., Rao, G. S., & Chowdari, B. V. R. 2002.. *Journal of Materials Chemistry 12*(6), 1848-1853.
[7] Keun S. , Tae W., Ki Lee3 H., J.Shim , 17 November 2008, *Jeonbuk Korea.* , **573**-701,
[8] Aykut, Y., Pourdeyhimi, B. and Khan, S.A., 2013. *Journal of Physics and Chemistry of Solids, 74* (11), pp.1538-1545.
[9] Haider, A. J., Al-Rsool, R. A., & Haider, M. J. 2018. *Journal Plasmonics 13*(5), 1649-1657.
[10] Fadhil, F. A., Sultan, F. I., Haider, A. J., & Rsool, R. A.2019. *In AIP Conference Proceedings* **2190** (1), p. 020056). *AIP Publishing LLC.*
[11] He, D. Z., Guo, Q. X., Yin, H., Li, J. T., & Gong, Z. L.2017.*Int. Journal Electrochem. Sci*, *12*, 455.
[12]Choi, J. W. 2006 *Crystal chemistry, chemical stability, and electrochemical properties of layered oxide cathodes of lithium ion batteries (Doctoral dissertation) 76.*
[13] Philippe B, Mahmoud A, Ledeuil JB, Sougrati MT, Edström K, Dedryvère R, Gonbeau D, Lippens PE. 2014 $MnSn_2$ electrodes for Li-ion batteries: Mechanisms at the nano scale and electrode/electrolyte interface. *Electrochimica Acta*. **123** 72.
[14] Taton T 2000 *Science Journal*, **289**(5485), 1757–1760.
[15] Yousif, Ali A., Adawiya J. Haidar, and Nadir F. Habubi. 2012 ,*Journal of nanoelectronic and materials* **5**. 47-55.
[16] Kamarulzaman, N.; Jaafar, M. H.2012 *Stoichiometry and Materials Scienc, When Number Matters: InTech: Rijeka, Croatia.*
[17] Aruna, S. T.; Mukasyan, A. S. Curr. Opin. 2008 , *Journal Solid State Mater.* Sci. **12**, 44(50.).
[18] Patil, K. C.; Aruna, S. T.;2002 *Mimani, T.Curr. Opi. Solid State Mater.*, **6**, 507{512).

[19] Balakrishnan, P., & Veluchamy, P. 2015 *journl Int. J. Chem. Tech. Res*, ***8***(1), 271-276..
[20] Jaafar, M. H., Kamarulzaman, N., Elong, K., Mohamed, N. S., & Rusdi, R. 2013 *International Journal of Electrochemical Science*, ***8***, 2254-2261. .
[21] Elong, K., Kamarulzaman, N., Rusdi, R., Badar, N. and Jaafar, M.H., 2013. *International Scholarly Research Notices*, ***2013.***
[22]Al-Douri, Y., Haider, A. J., Reshak, A. H., Bouhemadou, A., & Ameri, M. 2016. *Optik*, ***127*** (20), 10102-10107.
[23]Haider, M. J., Hadi, A. A., Saimon, J. A., & Haider, A. J. 2020.. *Journal of Nanoparticle Research*, ***22*** (7), 1-9.
[24]AL-Saedi, Safaa I., Adawiya J. Haider, Asama N. Naje, and Nathalie Bassil 2020 *Improvement of Li-ion batteries energy storage by graphene additive. Energy Reports*(**6**) 64-71.

The effect of adding graphene nanoparticles to the thermal performance of $LiCo_{0.525}Ni_{0.475}O_2$ cathode materials in Li-ion batteries

Safaa I. Al-Saedi[1], Asama N. Naje[2]*, Adawiya J. Haider[3]

[1,2] Department of Physics, College of Science, University of Baghdad, Iraq
[3] Department of Applied Science, University of Technology, Iraq
*Corresponding authors: naje.as75@gmail.com
100081@uotechnology.edu.iq
safaa.ied.h@gmail.com

Abstract. In this work, the effect of adding graphene to the performance of layered $LiCo_{0.525}Ni_{0.475}O_2$ (LCNO) cathode materials was investigated. Cathode materials $LiCo_{0.525}Ni_{0.475}O_2$ (LCNO) were prepared using a self-propagating combustion reaction. Graphene (G) was added to the prepared compound (LCNO) in different weight percentages to obtain LCNO/G (LCNOG) to improve cathode performance. The powder was divided into four parts weighing 10 g each, and the parts were mixed with percentages weighing 0, 0.5, 2, 4 and 6% of graphene with 200 ml of ethanol; they were then stirred several times at room temperature to dry completely The effect of graphene weight percentages on LCNOG mass loss was studied using thermo-gravimetric analysis (TGA). It was concluded that all of the added percentages of graphene behaved the same in all temperatures, except for 6% added graphene, where it demonstrated inverse behaviour at a high temperature. To determine the optimum annealing temperature range, TGA was identified at 650, 750 and 850 °C for 12 hours in the air. The X-ray diffraction (XRD) patterns of the LCNO and LCNOG showed the crystallisation of all powders in the space group R-3m and hexagonal axes for the crystal system. Energy dispersive spectrometry (EDS), field emission scanning electron microscopy (FE-SEM), a vibrating sample magnetometer (VSM) and atomic force microscopy (AFM) were also used. The LCNOG surface morphology consisted of regular coverage of nano-crystalline grains and random orientation.

Keywords: Lithium-ion batteries, graphene, nanostructures, thermal properties

Introduction

Li-ion batteries (LIBs) are considered the most prevalent energy storage device nowadays; therefore, they are widely used in various areas, such as portable electronics and electric vehicles [1, 2] because comprehensive polarisation occurs with greater current densities [3]. The researchers' efforts focused on replacing all of the fossil fuels with other renewable energy resources [4]. The LIBs efficiency and energy density is based on the electrode's physical and chemical properties. In general, the concentration of electrons (or holes) inside semiconductor materials is small, therefore, the interactions of magnetic exchange are super interactions [4–6].

Battery safety is studied in terms of temperature resistance, which is the temperature needed for the thermal runaway process to start. One of the problems of thermal runaway is that the exothermic reactions are chained, which can lead to igniting one after the other [7]. One of the ways to prevent reactions from ending in an explosion or fire is using components with more thermal stability.

IOP Conf. Series: Materials Science and Engineering **987** (2020) 012012 doi:10.1088/1757-899X/987/1/012012

Recently, the researchers became interested in graphene for its unique properties, such as its large surface area compared with mass ratio, very high conductivity and high chemical stability. Therefore, they used graphene to develop energy storage sources [8], improve the performance of solar cells, supercapacitors, fuel cells and Li-air batteries [9]; they also used graphene as an additive material to improve the performance of Li-ion batteries [10].

Graphene could be considered an ideal material to create a composite for use in LIB electrodes. Graphene is a sheet material that consists of sp2-hybridised carbon atoms that are organised in a hexagonal crystal lattice 2-D structure, which means that every atom can be estimated as a surface atom [11]. As graphene is a single layer with a thickness of one atom of carbon, lithium ions can be placed between any two layers to make Li2C6, which enables the fabrication of excellent LIB electrodes compared with ordinary carbon electrodes [12].

There are two main methods of graphene fabrication. The first is the top-down method, and the second is the bottom-up method. In the top-down method, graphene sheets are created by disconnecting, or breaking, the graphite stacked in many single layers using the van der Waals forces that connect carbon layers [13]. This method presents some challenges as defects in the graphene sheets' surfaces and the detached sheets mean that the production method has a low yield [13]. In the second approach (bottom-up), graphene is created using carbon molecules as constructing blocks [14]. However, this method cannot create a graphene sheet with a wide surface area. The most common technique for creating graphene is to make it from graphite using the mechanical cleavage technique, which produces low defect graphene sheets of high quality [15]; nevertheless, this method can be improved.

If improved Li-Ion electrodes lead to an increase in the electrode's ability to store lithium ions, this increases the capacity of the battery. Therefore, researchers believe that batteries with electrodes containing graphene will last longer than conventional batteries [16, 17].

The researchers are focused on energy; as LIBs are one of the few systems with rechargeable energy, they are used widely in different fields [1, 18]. Global energy objectives aim to replace fossil fuels with renewable sources. Therefore, the researchers focused on using alternative and improved cathode materials instead of prepared materials as they have some disadvantages, such as a high cost, high toxicity and weak operating voltage.

The main objective of this work is to improve the performance of the cathode material so that it is less toxic and has a high operating voltage; this was done by adding graphene (G) with different weighted percentages to the $LiCo_{0.525}Ni_{0.475}O_2$ powder.

Experimental work

Lithium nitrate ($LiNO_3$), cobalt nitrate ($Co(NO_3)_2{\cdot}6H_2O$), nickel nitrate ($Ni(NO_3)_2{\cdot}6H_2O$) and citric acid $C_6H_6O_7$ with 99.99% purity, were bought from Harris Chemicals Corporation in England. The weights were set up at 15.274 g of $Co(NO_3)_2{\cdot}6H_2O$, 13.81 g of $Ni(NO_3)_2{\cdot}6H_2O$ and 6.8942 g of $LiNO_3$.

All of the nitrates were dissolved in deionised water then placed on a magnetic stirring device for 15 min at 25 °C. The mixture was put on a hotplate for 30 min at 100 °C [19–22] then placed in a furnace at 400 °C for 15 min to burn the reaction materials.

A combustion reaction is a chemical reaction affected by factors, such as reaction materials and their impurities; combustion reaction occurs in two steps. The first step is heating the mixture to evaporate the water to make the mixture more viscous [23, 24]. Combustion begins when the water concentrate progressively decreases by evaporation until it reaches a critical point of the threshold temperature value, which is the amount of thermal energy needed to start the combustion reaction process. The second step begins at the moment the reaction is formed (nitric acid and hydroxides), which is due to the dissolution of the nitrate minerals in the water; the intermediate compounds are formed when citric acid is added to the mixture.

Graphene (G) was added to the resulting powder at different ratios of sample weight (0%, 0.5 %, 2%, 4% and 6%), and they were labelled S_0, S_1, S_2, S_3 and S_4, respectively.

Thermal analysis was used to choose the optimum temperature, which led to the removal of impurities that gave rise to phase forms along the crystallisation.

For thermo-gravimetric analysis (TGA), STA-PT1000 from Lenses, Germany was used for powder thermal analysis. The measurement factors were set up with a heating rate of 10 °C/min with the atmospheric pressure staying constant.

X-ray diffraction (XRD-6000) was performed using 2 θ, operating at 40 kV and 30 mA X-ray Cu Kα (1.540600 Å) was used to analyse the LCNOG mixture with an FE-SEM. MIRA3 TESCAN was used to investigate the topographies of the LCNOG mixture. To discuss and understand the distribution of particle size, an AFM (Angstrom USA, Spm3000 system) was used. A VSM (Model VSMF 7407) was used to analyse magnetic properties.

Results and discussion

Figure 1 shows the curve of TGA pointing to a continuous mass decrease concurrent with temperature increase. Three points located on the plot support the analysis.

The percentages of mass loss at the first point (100 °C) were 1.99%, 1.67%, 1.68%, 1.27% and 1.62% for the samples S_0, S_1, S_2, S_3 and S_4, respectively; an endothermic peak was the reason for residual water evaporation from the LCNOG mixture. The second point was located at around 541 °C with an exothermic peak, and all the unreacted materials in the LCNOG mixture combusted at the percentages of mass loss 11.59%, 11.11%, 11.51%, 10.59% and 12.37% for samples S_0, S_1, S_2, S_3 and S_4, respectively. The LCNOG pure phase formed at 700 °C, which represents the third point on the curve, which is in agreement with [25–32]. The mass loss for samples S_0, S_1, S_2, S_3 and S_4 were 16.39%, 14.80%, 14.52%, 13.12% and 16.95%, respectively. Figure 2 shows the comparison between the percentage of mass loss for the samples at temperatures 100 °C, 541 °C and 700 °C.

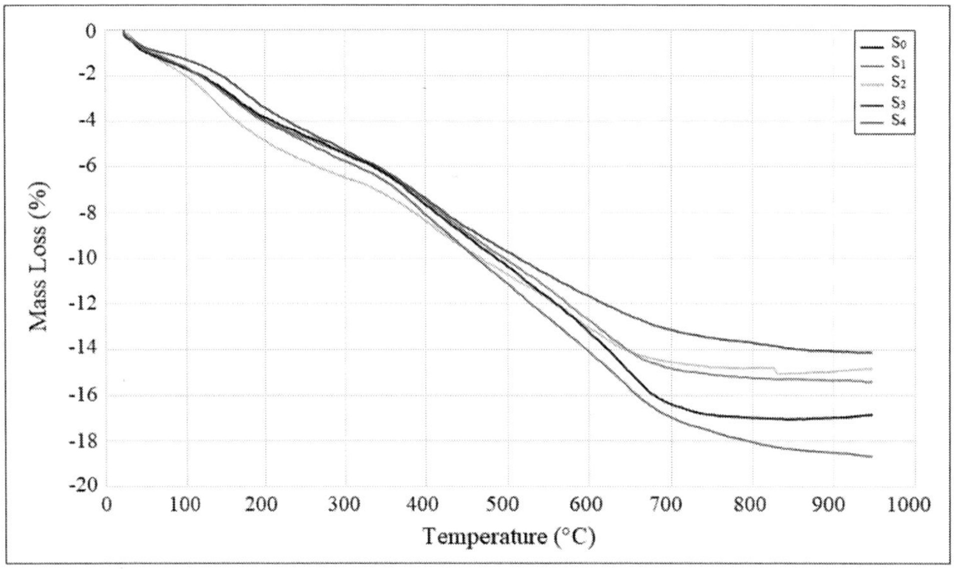

Figure 1 Results of the TGA for samples S_0, S_1, S_2, S_3 and S_4

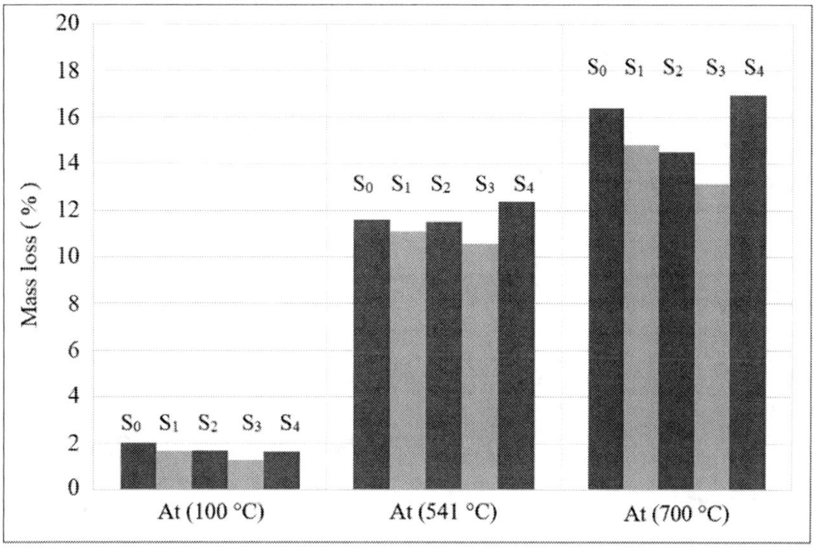

Figure 2 The percentage of mass loss for samples S_0, S_1, S_2, S_3 and S_4 at temperatures 100, 541 and 700 °C

We noted that at 100 °C, adding graphene caused a reduced mass loss from the LCNOG mixture across all proportions, while in the second and third regions at temperatures of 541 and 700 °C, this behaviour was valid for all additions, except for the 6% addition where it caused an increase in mass loss. This was because the graphene sheets acted as a filler, reinforcing particles that attracted the mixture compound, whereas increasing the mass loss at a high-temperature occurred because the reinforcing effect of graphene reduces the mixture components in segmental mobility. This last enhancement that improved the LCNOG mixture was observed when the graphene ratio increased, and this result was accepted in [23].

At the 850 °C annealing temperature and above, lithium oxide decomposed [10]. When long annealing hours are excluded, a structurally better compound may be formed at high temperatures.

Depending on the results of TGA, the temperatures chosen were 650, 750 and 850 °C as the annealing temperatures.

This work used EDS analysis to investigate the components that made up the mixture, and the results are revealed in Figure 3. Table 1 shows the weight percentages of Co, Ni, O and C for sample S_3 at annealing temperatures of 650, 750 and 850 °C. The results agree with the results in [22, 25–27].

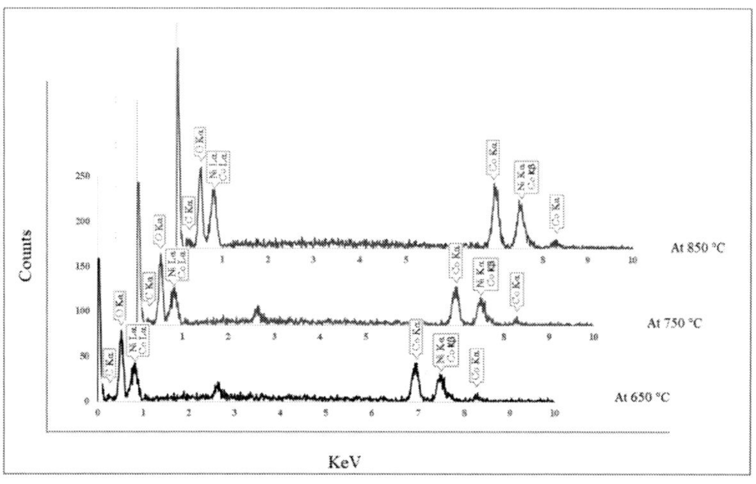

Figure 3 Energy dispersive spectrum of S_3 at annealing temperatures 650, 750 and 850 °C

Table 1 The percentages of powder contents

	Percentage weight at 650 °C	Percentage weight at 750 °C	Percentage weight at 850 °C
Carbon (C)	5.30	5.42	4.29
Oxygen (O)	29.97	40.03	22.98
Cobalt (Co)	35.58	29.19	39.53
Nickel (Ni)	29.15	25.35	33.21

There is no peak appearance for Li in Figure 3 because the X-ray of Li K is ultra-soft (about 55 eV), so detecting it is a great challenge; the emitted signal intensity is very weak because the generated X-ray is so small and the absorption probability high [28].
Figure 4 shows the X-ray pattern results before annealing for S_0, S_1, S_2, S_3 and S_4 powders.
The materials have a hexagonal α- $NaFeO_2$ structure with the R-3m space group, which is also the same structure as LiCoO2. The diffraction patterns indicate that all the materials are pure and single-phase, and no impurity peaks were observed.
Figure 4 demonstrates that the predominant phase for $LiCo_{0.525}Ni_{0.475}O_2$ powders is polycrystalline and they are very clear (2 θ = 44.823) with Miller indices (105).
The X-ray spectra of LCNOG before annealing show peaks located at around 2 θ = 18.841, 37.110, 38.129, 38.767, 44.823, 49.063, 65.781 and 78.647 that was attributed to LiCoO2 (003), (101), LiNiO2 (012), (104), (015) and $LiCoO_2$ (113) and (116), respectively, with an attributed orientation plane. This agrees with the result obtained by Kalyani et al. [29].

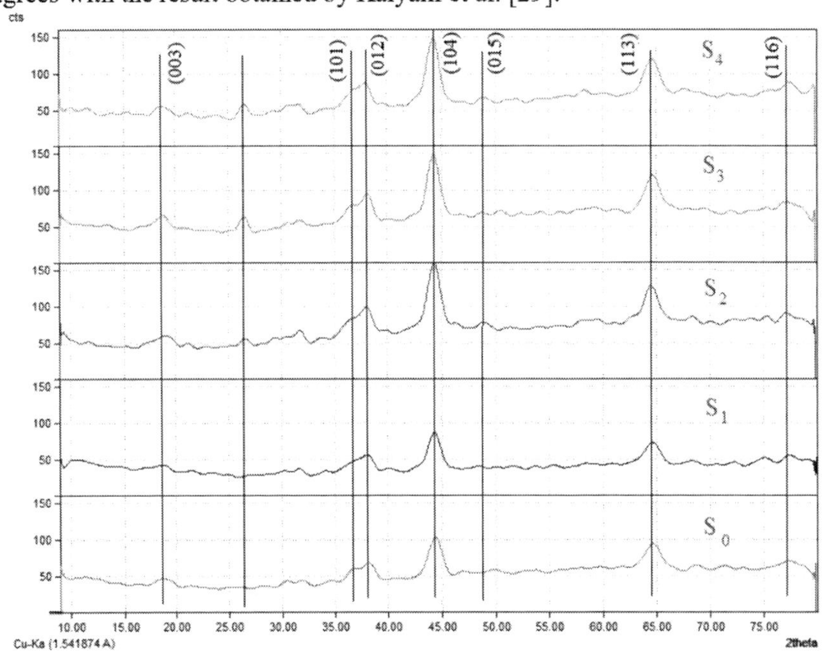

Figure 4 X-ray patterns before annealing for S_0, S_1, S_2, S_3 and S_4 powders

The sharp diffraction peaks in the patterns above indicate the appearance of nano-size crystallite [24, 30–31]. The peak of graphene was detected at 2 θ ~ 26.6° corresponding to a d-spacing of 0.335 nm, which is close to graphite d-spacing. Moreover, there was no shift in the diffraction angle positions of LCNO peaks with added graphene, which might refer to the high mobility of graphene.
The XRD patterns for the S_0, S_1, S_2, S_3 and S_4 samples at annealing temperatures of 650, 750 and 850 °C are shown in Figures 5 a, b and c, respectively.

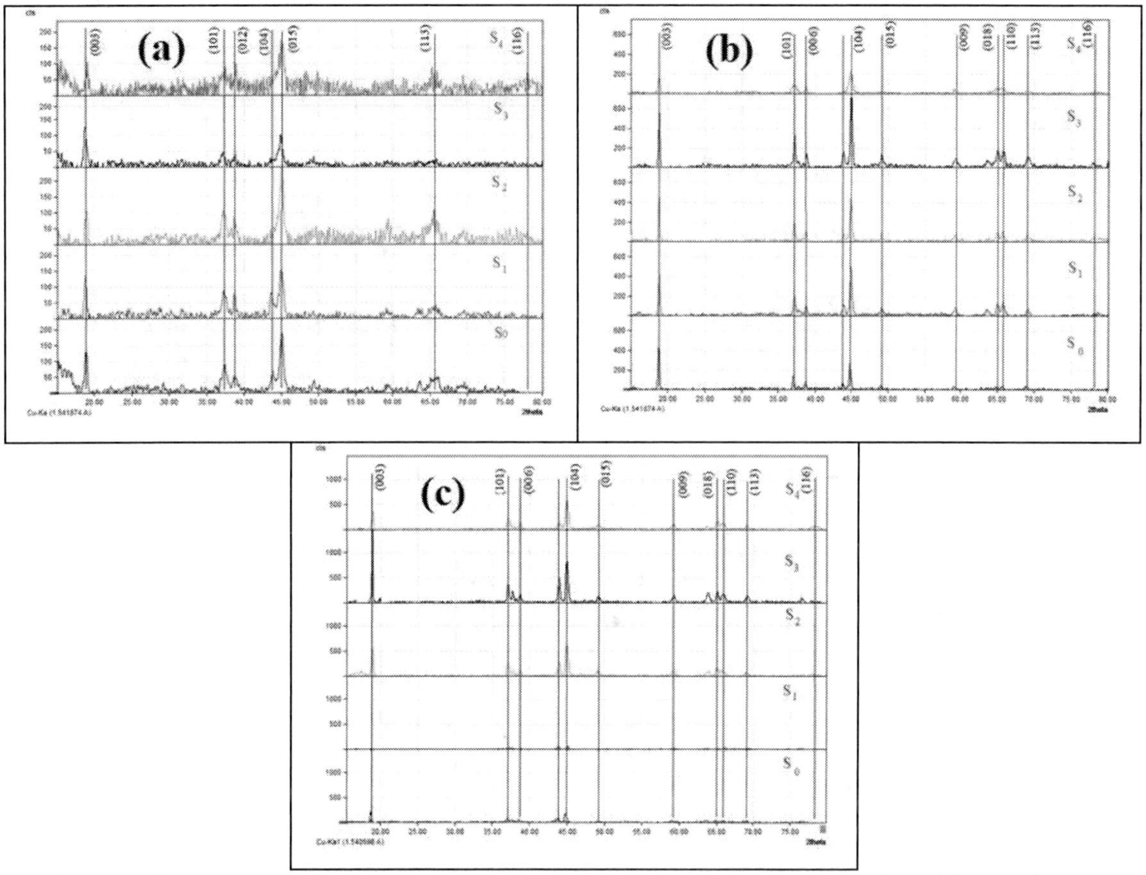

Figure 5 X-ray patterns for LCNOG with G = 0, 0.5, 2, 4 and 6%; S_0, S_1, S_2, S_3 and S_4 powders at annealing temperatures (a=650 °C, b=750 °C, c=850 °C)

The results of XRD patterns for all the annealing temperatures (640, 750 and 850 °C) show high-intensity peaks that indicate good crystallinity. Furthermore, when the annealing temperature increases, the number of peaks increase. This means that the crystalline size increases when the annealing temperature increases because the annealing process reduces lattice defects and increases retraction. This process can also lead to the adhesion of smaller grains so that the average grain size is larger [32]. The annealing peak of graphene was not seen very clearly. This may be attributed to the low-doped concentration of graphene and the high-intensity peaks of LCNO after annealing.

This is a similarity between all peaks and a slight variation in their positions.

The annealing procedure minimises the defects and strains of lattice, but it also causes the coalescence of grains that leads to an increase in the average grain size. The particles of samples are submicron-sized spherical grains, so they can be considered suitable for use in electrochemical applications.

The XRD patterns of LCNOG powders grown by wet-chemistry methods are well-defined with a fine splitting. This refers to a high degree of crystallinity, fine hexagonal ordering and larger characteristics. The XRD results emphasise the composition of the pure phase [33, 34]. However, the characteristics of LCONG powders' crystalline structures confirmed that the material produced maybe be used as a cathode component with electrochemical properties efficiently compared with others [35, 36].

The 3D topographical structure was tested using AFM for sample S4 at 650, 750 and 850 °C annealing temperatures, as shown in Figure 6; the area scanned was approximately 3 µm × 3 µm.

The surface roughness of sample S_3 was around 0.24 nm at annealing temperature 640 °C and up to 0.39 nm for 850 °C. The results showed that the particle sizes were average for sample S_4 at annealing temperatures of 650, 750 and 850 °C, and were around 71.4, 80.3 and 101.5 nm, respectively. The increase in average particle size and their roughness occurred because of the increase in annealing temperature.

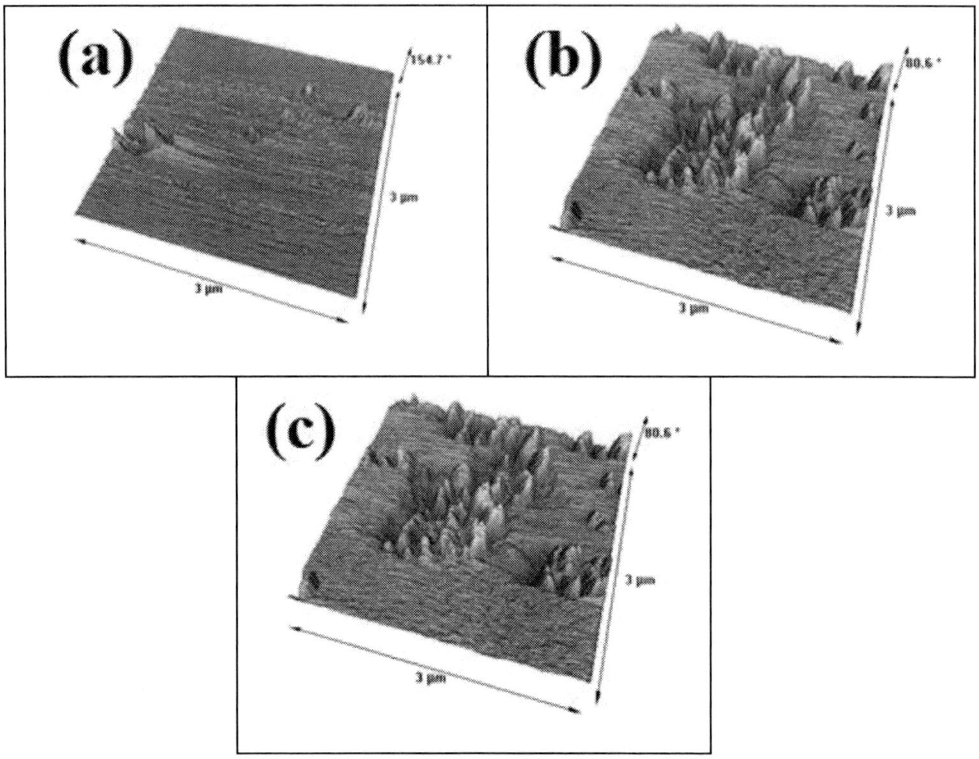

Figure 6 3D reconstructed images of AFM micrographs of sample S_4 at different annealing temperatures (a) 650°C (b) 750 °C (c) 850 °C

The FE-SEM results showed that when the annealing temperature rises, the interstellar distances between particles decrease and become smaller and more regular; at the same time, the particle size increases, as shown in Figure 7. The results agree with [27]. The electrochemical properties of the material in this work (LCNOG) may be considered promising but may be improved by working to reduce the distances between particles and diffusion.

In our previous study [10], the hysteresis loop measurement of the LCNOG material at annealing temperatures of 650, 750 and 850 °C was studied. The results showed that the LCNOG material had paramagnetic properties and gained a magnetisation factor of 0.8 emu/g in a ±15kOe external field [10].

Both the residual magnetisation (Mr) and saturation magnetisation (Ms) decreased when the particle size increased; however, when the heating and annealing temperature increased, the coercivity force (Hc) decreased.

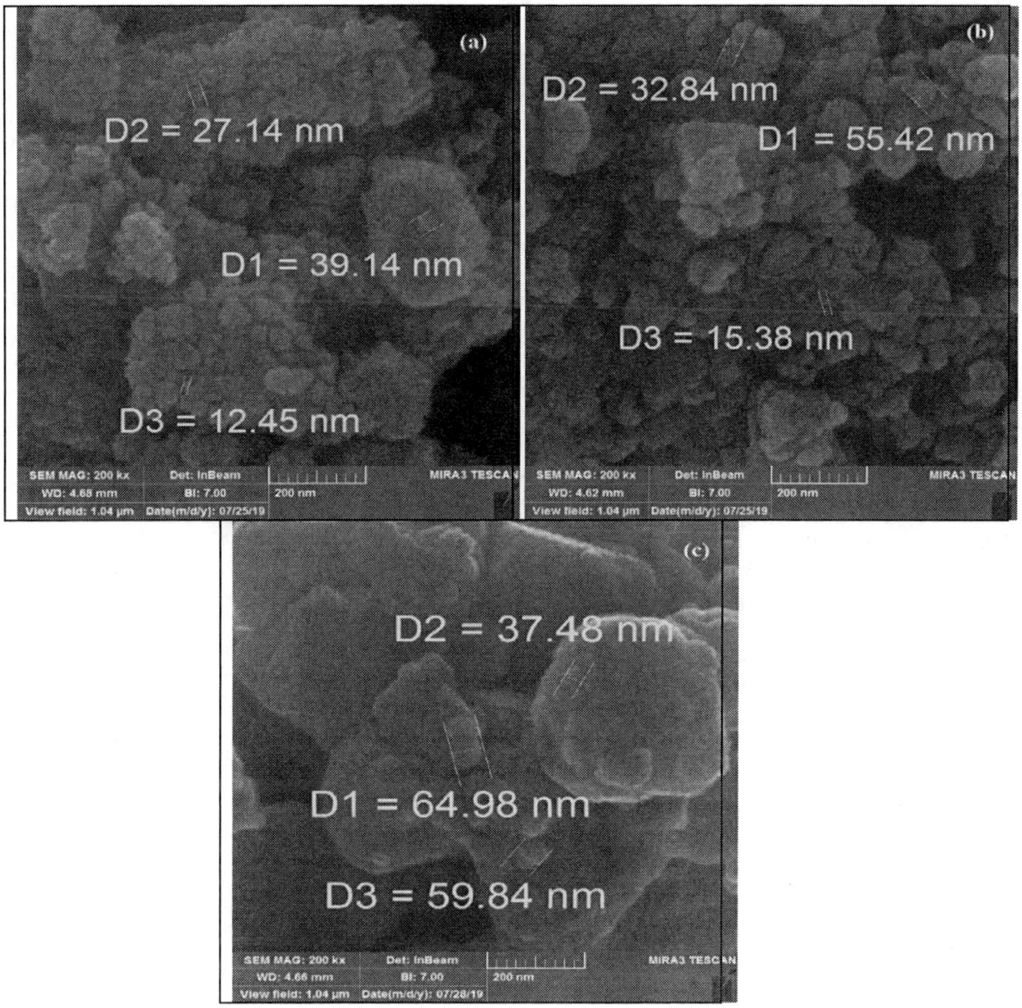

Figure 7 FE-SEM images of S_3 at different annealing temperatures (a) 650 °C (b) 750 °C (c) 850 °C

Conclusion

This paper presented the thermal properties of LCNO powders after adding graphene at different percentages; this fabricated and characterised the cathode material (LCNO and LCNOG). The addition of graphene to LCNO reduces mass loss because its sheets act as a filler, reinforcing particles that attract the mixture compound. Adding 6% of graphene increased the mass loss at a high temperature because the reinforcing effect of graphene reduces the components' segmental mobility. The most suitable addition of graphene was at 4%, which had high thermal stability and low mass loss at the annealing temperature of 750 °C.

References

[1] Al-Tabbakh AA, Kamarulzaman N, Al-Zubaidi AB. 2015 Synthesis and properties of a spinel cathode material for lithium ion battery with flat potential plateau. *Turkish Journal of Physics*. **39** 187.

[2] Philippe B, Mahmoud A, Ledeuil JB, Sougrati MT, Edström K, Dedryvère R, Gonbeau D, Lippens PE. 2014 $MnSn_2$ electrodes for Li-ion batteries: Mechanisms at the nano scale and electrode/electrolyte interface. *Electrochimica Acta*. **123** 72.

[3] Haider AJ, Al-Rsool RA, Haider MJ. 2018 Morphological and Structural Properties of Cathode Compound Material for Lithium-Ion Battery. *Plasmonics*. **13** 1649.

[4] Aykut Y, Pourdeyhimi B, Khan SA. 2013 Synthesis and characterization of silver/lithium cobalt oxide ($Ag/LiCoO_2$) nanofibers via sol–gel electrospinning. *Journal of Physics and Chemistry of Solids*. **74** 1538.

[5] Balakrishnan P and Veluchamy P. 2015 Synthesis and characterization of $CoFe_2O_4$ magnetic nanoparticles using sol-gel method. *Int. J. ChemTech Res*. **8** 271.

[6] Haider AJ, Al-Tabbakh AA, Al-Zubaidi AB, Rsool RA. 2018 Preparation and characterization of $LiCo_{0.5}Ni_{0.45}Ag_{0.05}O_2$ cathode material for lithium–ion battery. Journal of Materials Science: *Materials in Electronics*. **29** 13277.

[7] Lisbona D, Snee T. A 2011 review of hazards associated with primary lithium and lithium-ion batteries. *Process Safety and Environmental Protection*. **89** 434.

[8] Bruce PG, Scrosati B, Tarascon JM. 2008 Nanomaterials for rechargeable lithium batteries. *Angewandte Chemie International Edition*. **47** 2930.

[9] Jafta CJ, Ozoemena KI, Mathe MK, Roos WD. 2012 Synthesis, characterisation and electrochemical intercalation kinetics of nanostructured aluminium-doped $Li[Li_{0.2}Mn_{0.54}Ni_{0.13}Co_{0.13}]O_2$ cathode material for lithium ion battery. *Electrochimica Acta*. **15** 411.

[10] AL-Saedi SI, Haider AJ, Naje AN, Bassil N. 2020 Improvement of Li-ion batteries energy storage by graphene additive. *Energy Reports*. **6** 64.

[11] Eason R, editor. 2007 *Pulsed laser deposition of thin films: applications-led growth of functional materials*. (New Jersey: John Wiley & Sons.)

[12] Waller GH, Lai SY, Rainwater BH, Liu M. 2014 Hydrothermal synthesis of $LiMn_2O_4$ onto carbon fiber paper current collector for binder free lithium-ion battery positive electrodes. *Journal of Power Sources*. **251** 411.

[13] Wang H, Tan TA, Yang P, Lai MO, Lu L. 2011 High-rate performances of the Ru-doped spinel $LiNi_{0.5}Mn_{1.5}O_4$: effects of doping and particle size. *The Journal of Physical Chemistry C*. **115** 6102.

[14] Warner JH, Schaffel F, Rummeli M, Bachmatiuk A. 2012 *Graphene: Fundamentals and emergent applications*. (Waltham: Elsevier)

[15] Goodenough JB, Kim Y. 2010 Challenges for rechargeable Li batteries. *Chemistry of materials*. **22** 587.

[16] Fu J, Bai Y, Liu C, Yu H, Mo Y. 2009 Physical characteristic study of $LiCoO_2$ prepared by molten salt synthesis method in 550–800° C. *Materials Chemistry and Physics*. **115** 105.

[17] Julien C, Mauger A, Vijh A, Zaghib K. 2016 *Lithium batteries*. (Cham: Springer)

[18] Julien C, Letranchant C, Rangan S, Lemal M, Ziolkiewicz S, Castro-García S, El-Farh L, Benkaddour M. 2000 Layered $LiNi_{0.5}Co_{0.5}O_2$ cathode materials grown by soft-chemistry via various solution methods. *Materials Science and Engineering: B*. **76** 145.

[19] Bolotin KI, Sikes KJ, Jiang Z, Klima M, Fudenberg G, Hone J, Kim P, Stormer HL. 2008 Ultrahigh electron mobility in suspended graphene. *Solid state communications* **146** 351.

[20] Kalyani P, Kalaiselvi N. 2005 Various aspects of $LiNiO_2$ chemistry: A review. *Science and technology of advanced materials*. **6** 689.

[21] Kim SK, Jeong WT, Lee HK, Shim J. 2008 Characteristics of $LiNi_{1/3}Co_{1/3}Mn_{1/3}O_2$ cathode powder prepared by different method in lithium rechargeable batteries. *Int. J. Electrochem. Sci*. **3** 1504.

[22] Grande L, Chundi VT, Wei D, Bower C, Andrew P, Ryhaenen T. 2012 Graphene for energy harvesting/storage devices and printed electronics. *Particuology*. **10** 1.

[23] Valentini L, Cardinali M, Bon SB, Bagnis D, Verdejo R, Lopez-Manchado MA, Kenny JM. 2010 Use of butylamine modified graphene sheets in polymer solar cells. *Journal of Materials Chemistry*. **20** 995.

[24] Novoselov KS, Geim AK, Morozov SV, Jiang D, Zhang Y, Dubonos SV, Grigorieva IV, Firsov AA. 2004 Electric field effect in atomically thin carbon films. *Science*. **306** 666.

[25] Hovington P, Timoshevskii V, Burgess S, Demers H, Statham P, Gauvin R, Zaghib K. 2016 Can we detect Li KX-ray in lithium compounds using energy dispersive spectroscopy?. *Scanning* **38** 571.

[26] Wang Q, Deng S, Wang H, Xie M, Liu J, Yan H. 2013 Hydrothermal synthesis of hierarchical $LiFePO_4$ microspheres for lithium ion battery. *Journal of alloys and compounds*. **553** 69.
[27] Edwards RS, Coleman KS. 2013 Graphene synthesis: relationship to applications. *Nanoscale*. **5** 38.
[28] Ruffo R, Wessells C, Huggins RA, Cui Y. 2009 Electrochemical behavior of $LiCoO_2$ as aqueous lithium-ion battery electrodes. *Electrochemistry communications*. **11** 247.
[29] Sathiya M, Prakash AS, Ramesha K, Shukla AK. 2009 Nitrate-melt synthesized HT-$LiCoO_2$ as a superior cathode-material for lithium-ion batteries. *Materials*. **2** 857.
[30] Paek SM, Yoo E, Honma I. 2009 Enhanced cyclic performance and lithium storage capacity of SnO_2/graphene nanoporous electrodes with three-dimensionally delaminated flexible structure. *Nano letters*. **9** 72.
[31] Park S, Ruoff RS. 2009 Chemical methods for the production of graphene. *Nature nanotechnology*.**4** 217.
[32] Tan KS, Reddy MV, Rao GS, Chowdari BV. 2005 High-performance $LiCoO_2$ by molten salt ($LiNO_3$: LiCl) synthesis for Li-ion batteries. *Journal of Power Sources*. **147** 241.
[33] Tarascon JM. 2010 Key challenges in future Li-battery research. Philosophical Transactions of the Royal Society. *Mathematical, Physical and Engineering Sciences*. **368** 3227.
[34] Godse P, Sakhare R, Pawar S, Chougule M, Sen S, Joshi P, Patil VP. 2011 Effect of annealing on structural, morphological, electrical and optical studies of nickel oxide thin films. *Journal of Surface Engineered Materials and Advanced Technology*. **1** 35.
[35] Zhang B, Wang ZX, Guo HJ. 2007 Effect of annealing treatment on electrochemical property of $LiNi_{0.5}Mn_{1.5}O_4$ spinel. *Transactions of Nonferrous Metals Society of China*. **17** 287.
[36] Zhang PP, Zhu KY, Su LQ, Xiao R. 2013 Preparation and properties of graphene/polyamide 6 composites by melt compounding. *Advanced Materials Research*. **621** 31.

High Uniformity Distribution of Fe@Au Preparation by a Micro-Emulsion Method

Maha A. Al-Kinani[1], Adawiya J. Haider[2]* and Sharafaldin Al-musawi[3]

[1, 2]Applied Science Department/ Laser Branch, University of Technology, Baghdad, Iraq
[3]Colleg of Biotechnology, Al-Qasim green University, Iraq
*Corresponding author: adawiyajumaa96@gmail.com

Abstract. In this study, the characterisation and composition of Fe@Au NPs were discussed. Au-coated Fe core-shell is a type of nanoparticle that contains a magnetic Fe NPs core with a fine layer of Au NPs synthesised by the micro-emulsion method. These Fe@Au NPs are characterised by Atomic Force Microscope (AFM), Scanning Electron Microscopy (SEM), High-Resolution Transmission Electron Microscopy (TEM) and UV-Visible Spectrophotometer. The best result and the smallest nanosize was obtained at a temperature of 38°C and HAuCl4.3H2O Volume 200mg. The size of Fe@Au NPs was 63.65 nm, the Fe NP's core diameter was 32.47nm and the Au NP's shell thickness was approximately 31.18 nm. Optimum results mention that temperature and the HAuCl4.3H2O volume play a great role in the formation of small, monodisperse Fe@Au NPs for future use as a drug delivery means to treat different types of cancer.

1. Introduction

Core-shell nanoparticles are specially defined as particles that have a core and a shell and include dimensions in the nanometer value [1]. Nanomaterials are defined as manufactured or natural materials containing particles with one or more external dimensions with sizes ranging from (1–100) nm [2]. In the latest studies, metal-based nanotechnology has widely been advanced because of the material's superior distinctive physicochemical characteristics; for example, their catalytic, magnetic, antimicrobial, electrical, optical and other characteristics are entirely dissimilar from bulk materials of the same chemical composition [3-6]. Au-coated magnetic nanoparticles Fe@Au, a type of nano-particles, have attracted much interest due to their profitable characteristics, such as their super magneticity, inertness, ease of location in the human body, non-toxicity, a magnetic core that remains despite oxidation, catalytic surface and simplified bio-conjugation capability [7]. A lot of effort has been synthetic in preparing the nano-materials to join Fe and Au [8, 9]. The fabrication of magnetic nanoparticles has attracted attention because of prospective applications like medical imaging, targeted delivery [10], cancer treatment, bio-separation [11] and electrochemical sensors [12]. A shell addendum can promote dispersibility and stability of particles core, thereby adjusting or improving their properties [13]. Au was a favoured coating material with respect to its famous optical property, biological reaction, and chemical function. Au can transfer magnetic particles with the required biomedical or chemical properties [14]. The Fe NPs were first transport for an aqueous solution, then covered with an Au shell by a soft reduction of hydrogen tetrachloroaurate hydrate ($HAuCl_4 \cdot 3H_2O$) [15]. A different strategy was advanced by Coworkers and Zhong [16, 17]. There are many manufacturing methods of core-shell, such as pulsed laser ablation [18,19], reverse microemulsion method and nanoprecipitation method. The Fe@Au NPs were synthesised by a micro-emulsion method. The microemulsion process is a flexible formulation mechanism which enables the manipulation of particle properties like size, morphology, geometry, homogeneity and surface area [20]. Microemulsion is a thermodynamically steady dispersion in which there are initially two liquids. The main feature of using microemulsion is that the dimensions producing micelles can be monitored by changing the structure of the microemulsion and the temperature. The surfactant steady droplets supply a micro-environment for the production of nano-particles by forbidding excess aggregation of particles and changing their contents. Also, nanoparticles produced in a microemulsion often show good surface features, so nano-catalysts obtained from microemulsion show better selectivity and activity than those produced by other methods [21]. This study shows Fe@Au NPs prepared and characterised by micro-emulsion methods, studying the uniformity of Fe@Au NPs colloidal for future use as a drug delivery method to treat different types of cancer.

2. Materials and Methods

Fe NPs were produced under nitrogen ambiance by thermal decomposition of iron pentacarbonyl Fe(CO)5; see Figure 1 [22]. In addition, (2.3 g) of oleic acid (OA) was stirred in (21 mL) of octyl ether, the solution was heated to (40 °C), and (0.3 mL) of Fe(CO)5 was injected in a 1:3 molar ratio to the OA. The reaction mixture became orange-yellow. After (30 min) of refluxing, the mixture became black. The mixture was stirred for (2 h) under reflux and then cooled to room temperature. The particles were precipitated by the addition of ethanol and subsequent centrifugation. The precipitate was washed with ethanol.

Figure 1. Fe NPs prepared by thermal decomposition.

Fe@Au NPs were prepared by a micro-emulsion method. First, Fe NPs 10.3 g were dispersed in a 200 ml distilled water solution in a plane base beaker using a magnetic stirrer for 20 minutes. Then, 100 mg-200 mg of HAuCl4·3H2O was dissolved in 5 ml of distilled water using a vortex mixer for five minutes, and then slowly dropped on Fe NPs at a different temperature to admit the adsorption of (Au+3) to the (Fe) surface as shown in Figure 2.

IOP Conf. Series: Materials Science and Engineering **987** (2020) 012013 doi:10.1088/1757-899X/987/1/012013

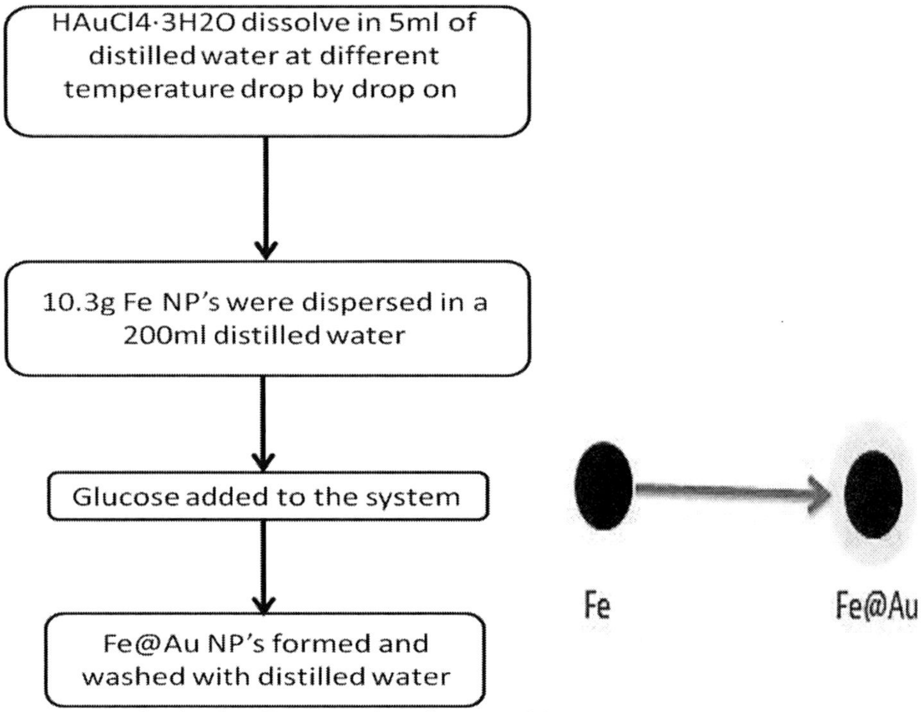

Figure 2. Synthesis of Fe@Au NPs by micro-emulsion method.

Subsequently, 5 mg of glucose was dissolved in 1m distilled water, added to the process as a reducing agent, and the mix was incubated at room temperature in a vibration incubator. Core-shell NPs that were created were then cleaned by distilled water. Fe@Au were analysed by AFM (JPK Instruments, AG-NanoWizard® II, Germany), SEM (FE-SEM, Hitachi S-4160, Japan), TEM (Carl Zeiss AG - Zeiss EM900, Germany) and UV-Visible spectrophotometer (UV-Visible, Aquarius 7000, Italia).

3. Results and Discussion

The spectrum of absorption of colloid Fe@Au NP's was measured and compared with the pure Fe NP's and Au NP's colloid. Figure 3 shows the absorption spectra of (a)Fe NP's colloidal in distilled deionized water, (b) Au NP's colloidal in distilled deionized water, (c) Fe@Au NPs synthesized with HAuCl4.3H2 Volume 100 mg and different temperature and (d) Fe@Au NPs synthesized with HAuCl4.3H2O Volume 200 mg and different temperature. The Fe NP's colloid display an absorption band with a peak in 226 nm, while the Au NP's colloid displays an absorption band with a peak at 552 nm. Fe@Au NP's colloid with HAuCl4.3H2O Volume 100 mg shows two absorption bands with a peak in 283 nm and 560nm at 25°C, which it is the curve of the smallest particle size, the other curves showed a redshift toward a longer wavelength by increasing particle size [23]. Fe@Au NP's colloid with HAuCl4.3H2O Volume 200 mg shows two absorption band with a maximum at 283 nm and 552nm at 38°C which is the curve of the smallest particle size. The absorption band notices for gold colloid at 552 nm in the UV-visible range is because of surface plasmon resonance (SPR) which is a property of Au NPs [24, 25].

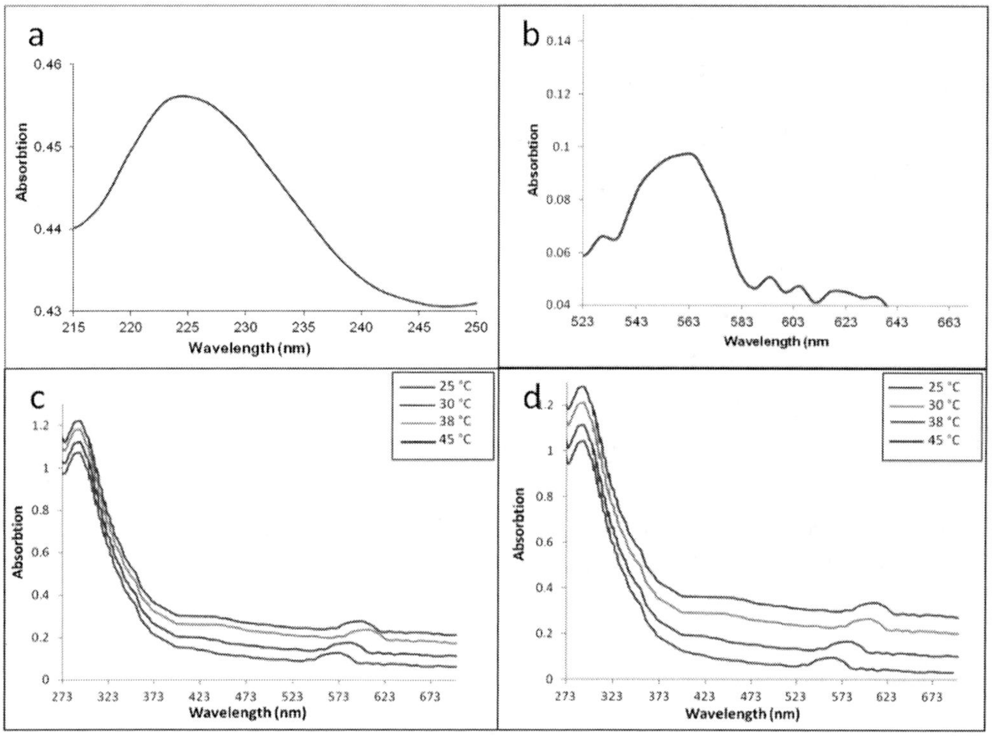

Figure 3. Absorption spectra of (a) Fe NP's colloidal in distilled deionized water, (b) Au NP's colloidal in distilled deionized water, (c) Fe@Au NP's synthesized with HAuCl4.3H2O Volume 100 mg and different temperatures and (d) Fe@Au NP's synthesized with HAuCl4.3H2O Volume 200 mg and different temperatures.

Figure 4 from the atomic force microscope (AFM) shows that the obtained nanoparticles have a spherical shape with suitable dispersity without remarkable aggregation. According to AFM measurements, Figure 4.a shows Fe NPs being analysed with the average diameter 32.47 nm with a uniform distribution of nanoparticles. Figure 4.b shows how Fe@Au NP's grain analysis was done with the average diameter 63.65 nm, the peak of size distribution narrows by intensity which means that most particles are of the same size.

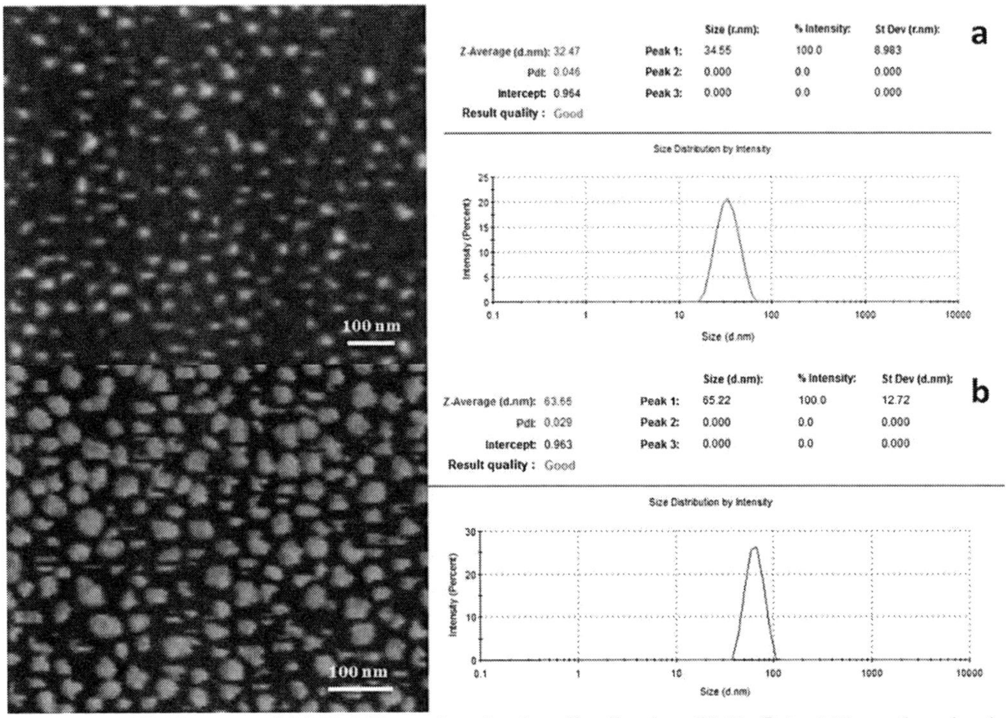

Figure 4. AFM images of (a) Fe NPs and grain size distribution (b) Fe@Au NPs and grain size distribution.

Figure 5 reveals the images of scanning electron microscopy displaying an excellently uniform distribution of spherical NPs in the synthesised nanoparticles [26]. Figure 5-a shows that the magnetite Fe NPs possess a nearly uniform size of around 32.47 nm and a smooth spherical form, and Figure 5-b reveals that Fe@Au NPs possess a nearly uniform size of around 63.65nm as well as a spherical rough form looking like Raffaello candy.

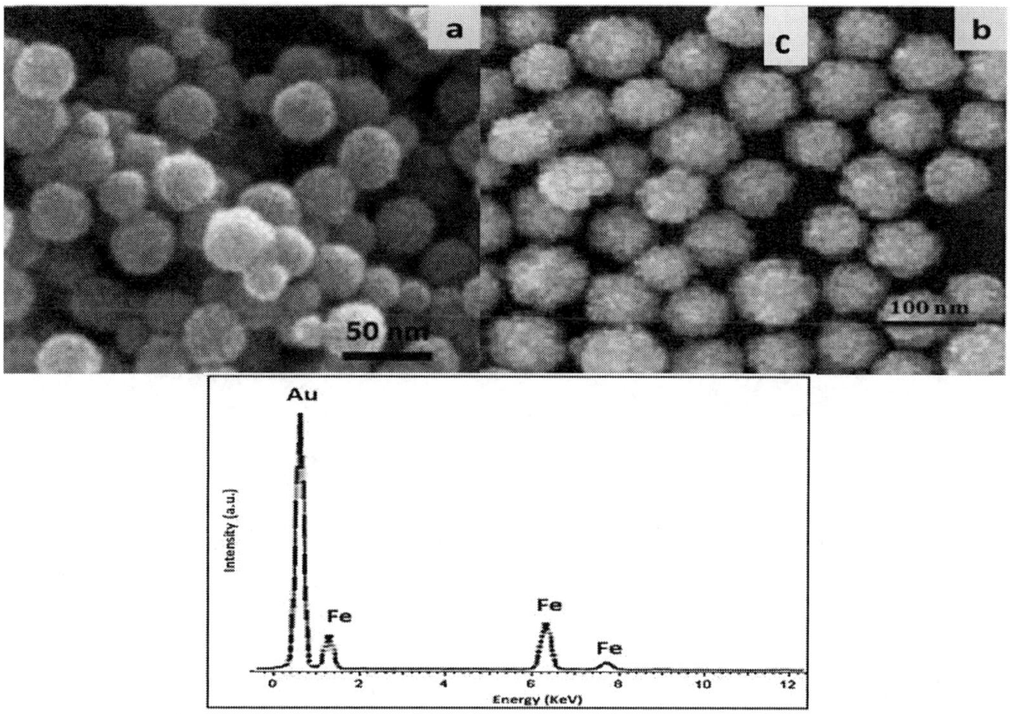

Figure 5. SEM image of (a) Fe NPs, (b) Fe@Au NPs and (c) EDAX of Fe@Au NPs.

The morphology and construction of the synthesised products were examined by transmission electron microscopy (TEM). Figure 6 illustrates the TEM micrographs of the nanoparticles obtained when synthesised at various temperatures and HAuCl4.3H2O Volume 100 mg. Figure 6-a shows that the magnetite Fe NPs possess a nearly uniform size of around 60nm as the core of the nanosystem. Fe@Au NPs possess a nearly uniform size of around 75nm, and spherical form Au shell is thin compared to the core. Figure 6-b demonstrates that the Fe NP core is small compared with the thick shell, while Fe@Au NP's size is about 88nm. Figure 6-c shows that the Fe NP core is very small compared with the thick Au NP shell, and the Fe@Au NP size is about 102 nm. Figure 6-d shows that Fe@Au NP possesses a nearly uniform size of around 99 nm, and spherical form Au NP's shell is thin compared with the core.

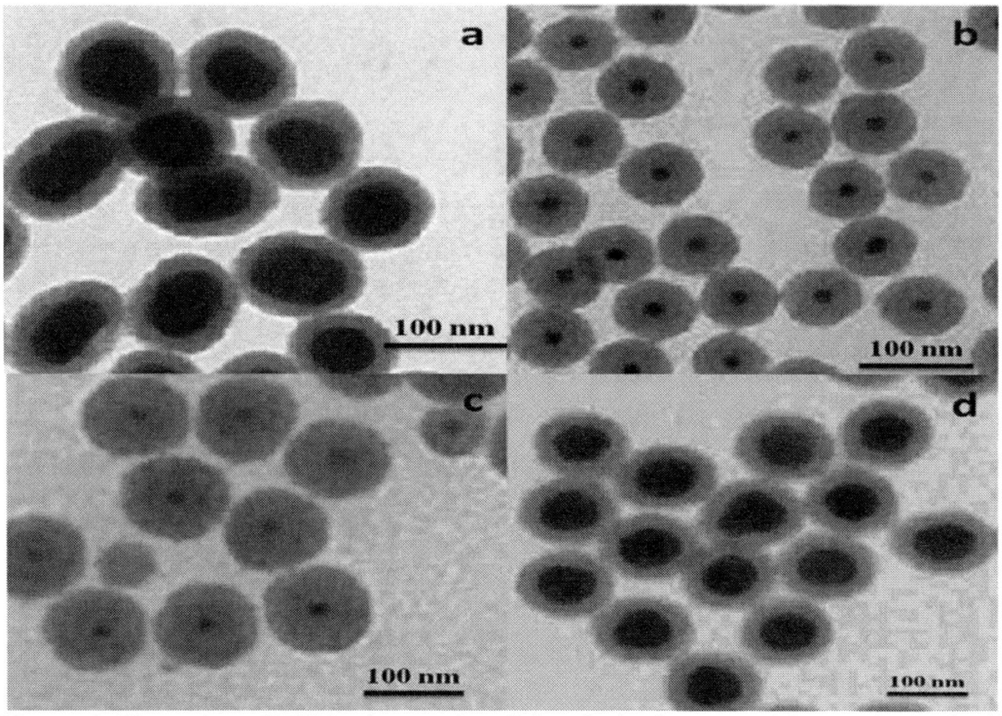

Figure 6. Illustration of TEM micrographs of the obtained nanoparticles synthesised with different temperatures and HAuCl4.3H2O Volume 100 mg (a) 25°C, (b)30°C, (c) 38°C, (d) 45°C. All scale bars are 100 nm.

Figure 7 illustrates the TEM micrographs of the obtained nanoparticles synthesised with various temperatures and HAuCl4.3H2O Volume 200 mg. Figure 7-a reveals that the Fe NP core is very small compared with the thick Au NP shell, and Fe@Au NP's size is about 120nm. Figure 7-b shows that the Fe NP core size is 20nm, and Fe@Au NP has a size of about 90nm. Figure 7-c shows that the magnetite Fe NP possesses an almost uniform size of around 32.47 nm, and a spherical form for the core of the nanosystem Fe@Au NPs possesses an almost uniform size of around 63.65 nm, as well as a spherical form as a core-shell, which is in perfect harmony with the result of (SEM). This is the smallest size obtained and is required for drug delivery applications. The Fe NP's core was much darker than the Au-coated nanoparticles [27]. Figure 7-d shows the Fe@Au NP's size to be about 83nm.

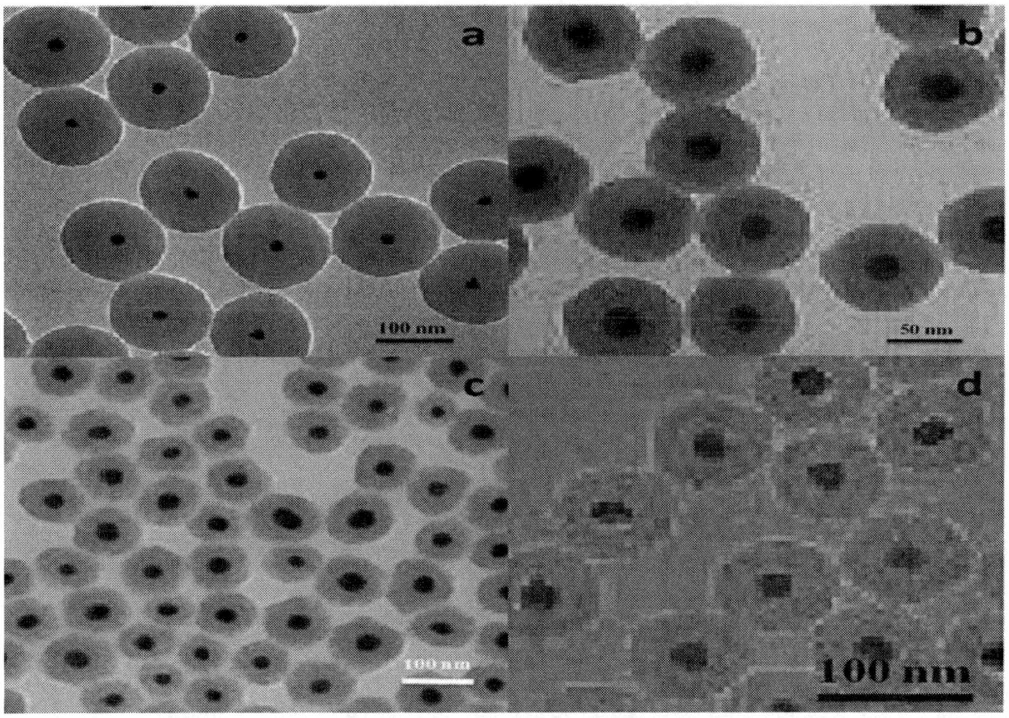

Figure 7. Illustration of the TEM micrographs of the obtained nanoparticles synthesised with different temperatures and HAuCl4.3H2O Volume 200 mg (a) 25°C, (b)30°C, (c) 38°C, (d) 45°C. All scale bars are 100 nm except (b scale bars 50nm).

Table 1 shows that, in the synthesising processes of a nanosystem, we find the best result of Fe@Au core-shell preparation in case (7) where HAuCl4.3H2O Volume 200mg at temperature 38 °C.

Table 1. Synthesis Processes of Fe@Au nanosystem

rank	$HAuCl_4.3H_2O$ Volume (mg)	Fe@Au solution mixing temperature (°C)	Fe@Au solution stirring (rpm)	Fe@Au Nanoparticle size (nm)
1	100	25	100	75
2	100	30	150	88
3	100	38	200	102
4	100	45	250	99
5	200	25	100	120
6	200	30	150	90
7	200	38	200	63.65
8	200	45	250	83

Conclusion

According to the results, the nanosystem of a Fe@Au NP possesses a uniform size of around 63.65nm and spherical form as core-shell at an optimum temperature of 38°C with a mass of 200mg. The smallest size obtained is beneficial, given its use in drug delivery applications. The TEM picture of the Fe@Au

NP confirmed the configuration of core-shell NPs, wherein the Fe core was much darker than the Au-coated nanoparticles.

Acknowledgments
The authors gratefully acknowledge the financial and technical support provided by the Applied Science Department, Laser Branch, University of Technology.

References

[1] Matijevic E 1994 *Langmuir Journal* **10(1)** 8–16.
[2] Jehan A Afraa A and Adawiya J 2018 *Global Pharma Technology Journal* **10(03)** 348-355.
[3] Maha A Adawiya J and Sharafaldin A 2020 *J Inorg Organomet Polym.*
[4] Al-Awady M Balakit A Al-Musawi S Alsultani M Ahmed Kamil and Alabbasi M 2019 *Nano Biomedicine and Engineering* **11(2)** 157-169.
[5] Al-Musawi S Hadi A Hadi S and Hindi N 2019 *J Global Pharma Tech.* **11(7)** 628-637.
[6] Al-Musawi S Kadhim M Hindi N 2018 *J Pharm Sci & Res.* **10(4)** 749-754.
[7] Chen M Yamamuro S Farrell D and Majetich 2003 *Applied Physics Journal* **93** 7551-7553.
[8] Eric D Michelle C Yue Z and Thelma S 2011 *Chemical Communications Journal*, 7.
[9] Ren J Shen S Pang Z Lu X Deng C and Jiang X 2011 *Chemical Communications Journal*, **47(42)** 11692.
[10] Kayal S and Ramanujan R 2010 *Nanoscience and Nanotechnology Journal* **10(9)** 5527–5539.
[11] Yi-Ran C Chao H Ying-Lin Z Yue L Xiao-Ming G and Xin-Xiang Z 2011 *Talanta Journal* **85** 1246–1252.
[12] Rawal R Chawla S and Pundir C 2012 *Biosens Bioelectron Journal* **31** 144–150.
[13] Baer R Neuhauser D and Weiss S 2004 *Nano Letters Journal* **4(1)** 85–88.
[14] Taton T 2000 Science Journal, 289(5485), 1757–1760.
[15] Dhelal A Adawiya J and Abdulalah T 2016 *International Journal of Humanities, Arts, Medicine and Sciences* **9** 101-108.
[16] Wang L et al. 2005 *Physical Chemistry Journal* **109** 21593–601.
[17] Wang L Luo J Maye M Fan Q Rendeng Q Engelhard M Wang C Lin Y and Zhong C 2005 *Materials Chemistry Journal* **15** 1821–32.
[18] Zhang J Post M Veres T Jakubek Z Guan J Wang D and Simard B 2006 *Physical Chemistry Journal* **110(14)** 7122–7128.
[19] Adawiya J Mohammad J and Mohammad S 2018 *American Institute of Physics Journal.*
[20] Hu A Yao Z and Yu X 2009 *Applied Polymer Science Journal* **113** 2202-2208.
[21] Boutonnet M Lögdberg S and Svensson E 2008 *Current Opinion in Colloid & Interface Science Journal* **13** 270–286.
[22] Zhang J et al. 2006 *Physical Chemistry Journal* **110(14)** 7122–7128.
[23] Bhatia P Verma S and Sinha M 2019 *Physics Letters A. Journal* **383** 2542–2550.
[24] Goon I Lai L Lim M Munroe P Gooding J and Amal R 2009 *Chemistry of Materials Journal* **21(4)** 673–681.
[25] Jos´e R Yolanda P and Manuel A 2013 *the Korean Physical Society Journal* **10** 1376-1381.
[26] Karamipour S Sadjadi M and Farhadyar N 2015 *Spectrochimica Acta Part A: Molecular and Biomolecular Spectroscopy Journal* **148** 146–155.
[27] Hossein S Elham H Meisam S and Kevin K 2012 *Nanobiotechnology Journal* 10.

Effects of New Modification and Refinement Techniques of Mechanical Properties and Corrosion Behaviours of Al-Si Alloys

Hussein. A. Hussein[1*], Mohammed. j. kahdim[1], Alaa. A. Atiyah[2]
[1]Department of Production Engineering and Metallurgy / University of Technology, Baghdad – Iraq.
[2]Department of Materials Engineering / University of Technology, Baghdad – Iraq.
[*]Corresponding author: 70082@uotechnology.edu.iq

Abstract. In this paper, the effects of new techniques of phosphor-doped elements on the microstructure and mechanical properties of Al-%Si alloy to produce modifier compounds were studied. The modifiers were produced by developing a new technique and additions field (nano- doped metals) under as-cast conditions. The results show that the addition of $CuPO_4$, $Ni_3(PO_4)_2$ and $Ca_3(PO_4)_2$ to the Al-Si alloys can change (refine) the primary Si phases and are less angular. In addition, the eutectic Si form has changed from a long needle-like to a fibrous shape. The modified Al-Si alloys have microstructures in the inter-dendrite region consisting of a uniformly distributed α-Al phase, eutectic Al-Si, and fine primary Si particles. The microstructure showed that the primary silicon fraction decreased 60% with increased phosphate elements content. The finest microstructure could be observed with a 0.03% addition of doped metals. Good enhancements in ultimate tensile strength and elongation with clear increases in the hardness were observed, reaching 55%, 9% and 30%, respectively. Moreover, additions give the possibility of changing corrosion behaviours from pitting corrosion to micro and general corrosion. Furthermore, the optimal holding and melting temperature is 780°C, with a holding time of 15 minutes.

Keywords: Aluminium Silicon, Corrosion Resistance, Wear Resistance, Fracture Behaviour.

1- Introduction

The Al-Si eutectic and primary Si alteration in hypoeutectic alloys play a significant role in developing mechanical properties and corrosion behaviours [1][2]. Grain size and morphology are the key microstructural factors that affect the processing and properties of alloys. A fine grain size leads, in general, to a decreased tendency to hot tearing [3], a more dispersed and refined porosity distribution [4] and improved directional feeding characteristics [5], although it can also reduce fluidity and cause defects in hot spots [5].

The improvement in the mechanical properties is believed to be due to the change in the size and morphology of eutectic silicon. It is well established that nucleation and growth during solidification play key roles in modification [6]. However, the detailed nucleation mechanism is still not well understood. Nucleation is difficult to study because of the complex role of the inherent impurities present in Al alloys. The novel entrained droplet technique first used by **Zarif et al.** [7] found that, unlike P, Sr refused to promote nucleation. However, the eutectic nucleation temperature is lowered by the contribution of Sr, which may be due to the formation of the intermetallic $\mathbf{Al_2Si_2Sr}$ layer, which may absorb or adversely affect the effective AlP nucleation sites. **Min Zuo et al.** [8] showed that the new Al-6Y-2P master alloy with cubic YP particles can also successfully develop complex modification properties for an Al-18Si alloy. By adding the intermediate alloy Al-6Y-2P, excellent dynamic changes of the Al-18Si alloy can be achieved. Under design conditions (780°C melting point, 1.5% additional melting point and 30–60 minutes holding time), the tensile strength and Brinell hardness of Al-18Si alloy can be significantly improved. It is proposed that the silicon atoms in the melt can improve the structural evolution of YP to AlP by releasing Y, which will lay the foundation

for the excellent performance of a t composite modification. **Hongseok Choi et al.** [9] conducted experiments to study how nano-particles affect Si particles during solidification in hyper-eutectic Al–Si–Cu alloys. For instance, 0.5% c-Al_2O_3 nanoparticles are added and further dispersed using an ultrasonic cavitation technique. Microstructural analysis with optical and scanning electron microscopy revealed significant refinement of the central and eutectic Si particles as well. The primary Si particles were simplified from star shapes to polygonal or blocky shapes, and their edges and corners were much smoother. The study reveals that c-nanoparticles simultaneously refine and alter Si particles as well as decrease porosity in Al–20Si–4.5Cu. **Jiehua Li et al.** [10] studied the role of traces of phosphorus on nucleation in high purity Al-Si alloys. They proposed that only 0.25–2 ppm phosphorus is sufficient to form AlP. **Brian Cantor et al.** [11] reported the poisoning effect of sodium on AlP, concluding that the formation of Na_3P might reduce the amount of AlP, thereby increasing the eutectic silicon nucleation and growth temperature, i.e. higher undercooling (ΔT). **S. Liang et al.** [12] confirmed this finding, noting that the ppm addition of sodium increased the eutectic droplet nucleation undercooling (ΔT). The AlP is believed to be the potent nucleation site for eutectic silicon[11]. A recent publication [14] proposed that the addition of strontium caused the formation of intermetallic **Al_2Si_2Sr**. These intermetallic compounds may consume the AlP phase, thereby reducing the number of nucleated eutectic grains [13].
From this moment, grain refinement research focused on determining the key issues of core particles and the role of alloy composition, and many attempts have been made to integrate these two factors into the grain refinement model. Research in the age of twin planes usually emphasises one effect, and another effect sometimes leads to the proposition of various theories, but it is now obvious that, if these two factors (and the effect of processing conditions) are taken into account at the same time, a comprehensive grain refining model can be developed to correctly describe the grain-formation process.
This paper initially introduces recent theoretical developments in grain refinement that incorporate both nucleant particles and solute, after which the details of the role of solute and of inoculants are discussed. The main objective of this paper is to investigate in detail the effect of different levels of (Nano-doped metals) on primary Si and eutectic Si nucleation, as well as the mechanical properties and corrosion behaviour in the Al-Si alloy. Moreover, the phase's reconstruction and phase structure are studied to illustrate the mechanism of intermetallic segments at some stage in the solidification process of the Al-Si alloy.

2. Experimental Method

2.1 Preparation of Nano-dopant elements

One means of synthesising nanoparticles is by use of solid liquid chemical reactions in several steps. Step one: add metals powder (μm-in-size) and mix with 10ml of phosphoric-acid (H_3PO_4) for 72h at room temperature. Table 1 shows the chemical formula of the phosphoric acid and elemental powder chemical reactions. The reaction will continue until stopping. Step two is another reaction followed by 130°C for 12h. Step three is calcination at 720°C for 2h. Step four is grinding and evaluation by X-ray diffraction (phases and grain size).

Table 1 Solid-liquid chemical reaction of elements that used in alloying experiments.

Material	Formulas
Cu	$2\ Cu + 2\ H_3PO_4 = 2\ CuPO_4 + 3\ H_2$
Ni	$3Ni + 2H_3PO_4 = Ni_3(PO_4)_2 + 3H_2$
CaO	$3\ CaO + 2\ H_3PO_4 = Ca_3(PO_4)_2 + 3\ H_2O$

2.2 Master alloy preparation

The pelleted form of a master alloy consisting of 0.3g, and grain size is 300μm pure aluminium powder mixed with Nano-Metal-Phosphate and compaction powders below 150 bars to improve oxidation resistance and flotation. Table 2 defines the elasticity of particles that have compressed together (nanoparticles and aluminium powder), and the pressure action leads to the contact points being fully bonded. Compact powder results in pellets which make handling easy.

Table 2 The composition of pellets.

Material	Wt.% added	Wt.% added	Al-Pure	Compact pressure
$CuPO_4$	*0.01 g*	*0.03 g*	*0.3 g*	*150 bar*
$Ni_3(PO_4)_2$	*0.01 g*	*0.03 g*	*0.3 g*	*150 bar*
$Ca_3(PO_4)_2$	*0.01 g*	*0.03 g*	*0.3 g*	*150 bar*

2.3 Casting techniques

The Al-13Si alloy was developed by liquid metallurgy. The chemical composition of the aluminium-silicon alloy used in this analysis is shown in Table 3. In the electric furnace (Carbolite, England), the alloy Al-13Si is heated to 780°C. Then, using an inert gas (Argon), degassing follows mechanical agitation with 500 rpm, as shown in Figure 1(a,b). At this point, the dopant-elements that were preheated to 350°C to eliminate any moisture content are added to the molten alloy. Particle-sized nano-metals (20 nm) were added (0.01and 0.03 %) to the molten alloy. The molten agitation was continued for about ten minutes to promote uniform dispersion of the wetting nuclant and uniform dispersion of the reinforcement. Then, the molten alloy was poured into the 1.5 cm x 15 cm preheated steel mould.

Table 3 Al-Si alloy chemical composition (wt. %) used in this study.

Bal.	Zn	Ni	Mg	Mn	Cu	Fe	Si
Al	*0.004*	*0.008*	*0.02*	*0.01*	*0.01*	*0.14*	*12.86*

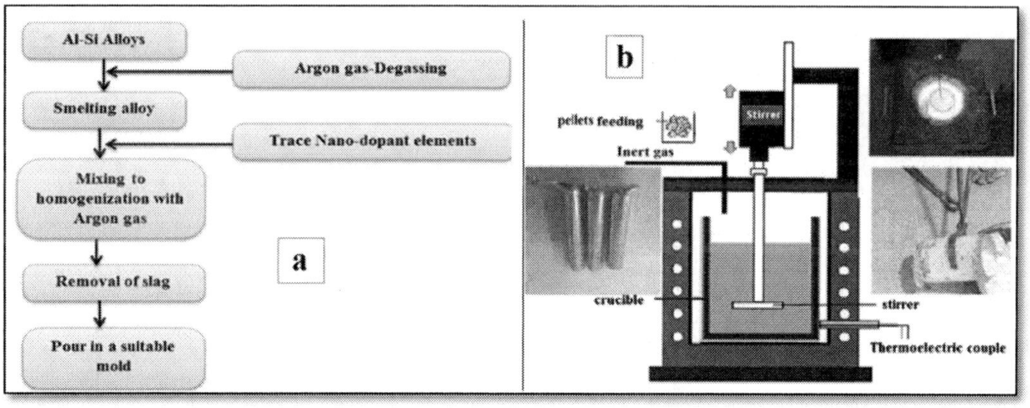

Figure 1. Process of stir casting

2.4. Microstructural Analysis

Al-13Si alloy SEM photos were prepared by grinding via 220, 320, 600, 800, 1,000 and 1,200 grit sheets, respectively. The specimens were then polished with diamond paste (0.1μm), and the killer solution washed and etched. Keller 's reagent is a mixture of 5ml NHO_3, 3ml HCl, and 2ml HF, and 190ml H_2O, the solution used for microstructure detection

2.5. Hardness Test

Vickers hardness tests were performed on the Al-13Si alloy (type ZWICK Z323 – Germany). For (10 seconds), the load was (0.5Kg). For each sample, three readings were used to obtain the average diameter value:

$$\mathbf{HV} = \mathbf{1.8544} * \frac{\mathbf{P}}{(\mathbf{D}_{\mathbf{AV}})^2} \text{------------} \mathbf{(1)}$$

where
HV: Vickers hardness number, **P:** applied load (N), $\mathbf{D_{AV}}$: diagonal length (mm)

2.6. Analysis of X-Ray Diffractions
The phases present in the alloy samples of modified Al-13Si were examined using X-ray diffraction via a Shimadzu X-ray diffractometer (XRD-6000) available at the Nano Technology Center University of Technology. X-ray tube: Cu, NF (1.54060 nm), radius of scanning: 185 mm, X-rays leakage: less than 2.5μSv / h at maximum output.

2.7 Corrosion Tests
The Al-Si alloy electrochemical analysis was concluded with cyclic polarisation. A computer-controlled device (PCI4/750, GAMRY, Inc., Warminster, PA) disbursed all tests in 3.5 % NaCl sol. In 25C. Ag / AgCl as the pole of reference and the auxiliary pole and working pole of noble metal (Pt) was used as the specimen keeper. The specimen was exposed to NaCl solution of 3.5 %. The area of exposure was 1 mm^2. Before each test, the samples were given a metallographic polish. Specimens were dipped into the solution until a steady open circuit (OCP) potential was obtained. The test specimens had an exposed area of about 10 mm^2. All data were neutralised by area of the surface and calculated using the Echem analyser to identify the test results evenly [15].

2.8. Wear rate measurement
Dry slip wear tests with ASTM F732-82 were conducted at room temperature using a laboratory pin on the disc wear test [11]. Nails, modified in Al-Si alloy o 10 mm and 20 mm in length, were polished with different sheets of grain carbide (220, 320, 500, 800, and 1,000) and cleaned with acetone and then dried. The wear device consists of a motor with a constant speed of rotation (510 rpm). The disk is made from tool steel with a 65 HRC hardness. Relative mass changes were tested by weighing the samples before and after wear using a sensitive-scale classification digital scale (DENVER Tool, Japan) (Max-210gm) of 0.0001 mg. The loads applied were 500 kN. After each test, the disk was cleaned.
The following equation was to calculate wear rate:

$$\textit{Wear rate} = \frac{\Delta W}{S} \text{----------} 2$$

$$S = V * t \text{------------------------} 3$$

$$V = \frac{\pi DN}{1000*60} \text{-------------------} 4$$

where
D: diameter (m) =0.14, **W**: The different in weight of sample before and after the test (gm), **S**: sliding distance (m), **t**: time (Sec), **N**: r.p.m=510, **V**: velocity (m/sec)

2.9. *Tensile tests*
Tensile tests on Al-Si alloy modifier specimens were performed with the dimensions shown in Fig. 2 (ASTM E8M) using Instron type 1195 machine made in England.

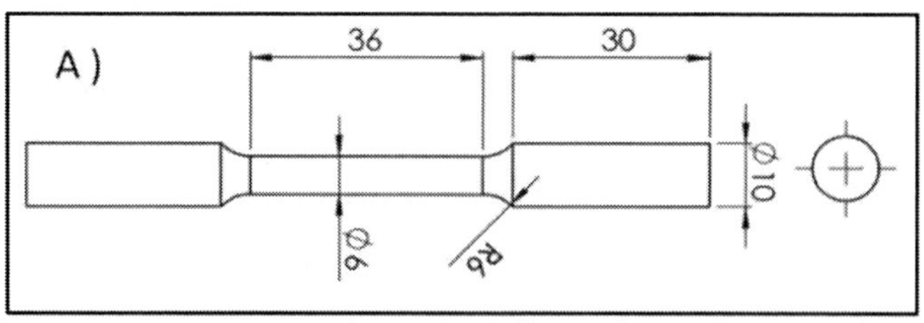

Figure 2. Geometry and Standard dimensions of test specimen

3. Results and Discussion

3.1 Microstructure of Experimental Materials

Fig. 3 illustrates unmodified Al-13Si alloy microstructures, initially composed by using J-Picture software of large grains of primary Si and platelet eutectic Silicon with a grain size equal to 37.32μm.

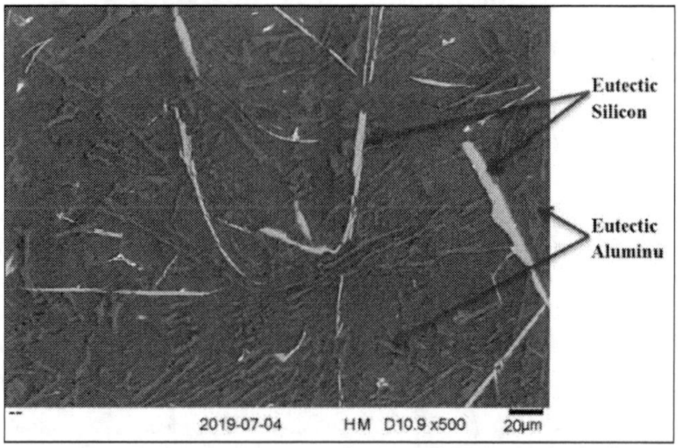

Figure 3. Microstructures of unmodified Al-13Si alloy.

Fig 4(a) shows the morphology of Al-13Si alloys with 0.01% $Ca_3(PO_4)_2$ which consists of coarse-primary-Si-particles. Fig 4 (b) shows the morphology of Al-13Si alloys with 0.01 % $Ca_3(PO_4)_2$ which consists of coarse-primary Si particles of **1.71**μm in size with a fine eutectic Si distribution. When the Al-13 % Si base alloy contains 0.03 % $Ca_3(PO_4)_2$, the primary Si particles of **1.54**μm in size are scattered in an eutectic configuration of fibres leading to substantial refinement.

The Al_2Ca phase and primary Si formed in intimate association with each other as well as in association with the oxide inclusions. Previous experimental evidence suggests that Si particles may nucleate and grow on oxide biofilms during solidification of hypereutectic Al-Si alloys [14]. The primary Si nucleation on oxide bifilms may be attributed to the oxide film, which provides a powerful substratum for primary Si heterogeneous nucleation. Potential alignment and related misfits of the measured lattice between Si and AlP, Al_2Ca and the appropriate oxides. The misfit between Si and AlP is very close to zero, resulting in an incredibly strong potential for Si nucleation by AlP. The misfits between Si and the two popular Al_2O_3 types are too large for Si to serve as heterogeneous nucleants for the oxides. Al_2O_3 nanoparticles can be used in hypereutectic Al-Si alloy melts for simultaneous refining and alteration of Si[15].

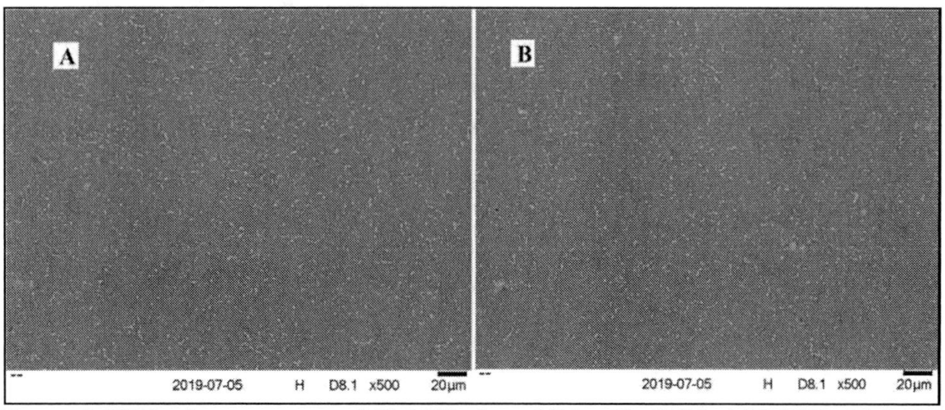

Figure 4. SEM microstructures of Al-13Si after additive $Ca_3(PO_4)_2$ (a) 0.01%, (b) 0.03%

IOP Conf. Series: Materials Science and Engineering **987** (2020) 012014 doi:10.1088/1757-899X/987/1/012014

Fig. 5 (a,b) shows the microstructure of Al-13%Si alloy with different content: 0.01% $NiPO_4$ master alloy addition; primary Si size has decreased while the most of the eutectic Si became **3.28** µm. In increasing the amount of $NiPO_4$ to 0.03%, the primary Si size decreases to 2.67 µm. **Fig. 4 (a,b)** shows the distribution and size of primary Si; meanwhile, the eutectic Si is modified to short rod or dot shapes. However, both of the primary and eutectic Si become fine during accretion of the amount of $NiPO_4$ master alloy.
The hypereutectic alloys with primary silicon crystals are modified by AlPO, and the silicon crystals will otherwise be detrimental to the mechanical properties. The modification of the primary silicon crystals with AlPO reduces the size. The intermetallic phases are well distributed, found in the vicinity of the aluminium dendrites and in the aluminium-silicon eutectic [8]. The size of the primary silicon crystals was decreased by ~73 % in the Al-13Si.
According to reaction equation, the AlNi intermetallic phase has a fibrous morphology, and the eutectic region consisted of Al_3Ni fibres and α-aluminium around the fibres per SEM micrographs of eutectic Si structures. These intermetallic compounds retard the grain growth, leading to grain refinement, and resulting in further strengthening effects[16].

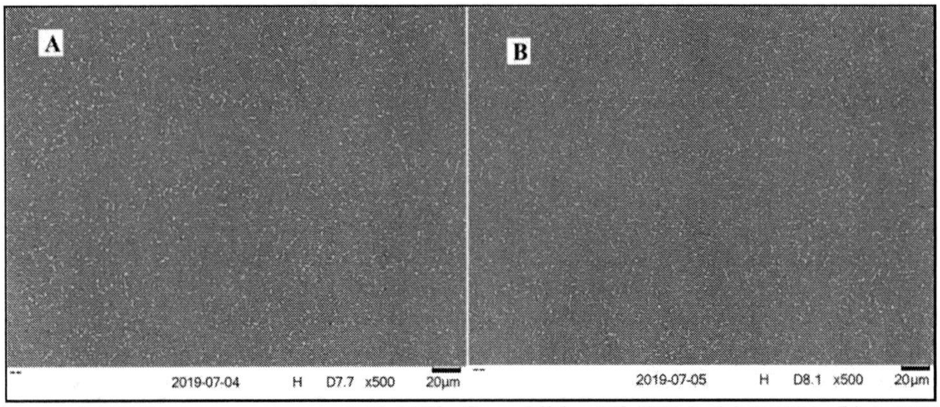

Figure 5 SEM microstructures of Al-13Si after additive $Ni_3(PO_4)_2$ (a) 0.01%, (b) 0.03%.

Fig. 6 shows a microstructural examination, revealing the polygonal primary silicon particles, with an increasing addition of (0.03%) CuP distributed uniformly in the matrix. This could be due to the fact of reactions with the liquid Al forms AlP, Al_2O_3 and Cu_3Al particles present in molten alloy, which may act as heterogeneous nucleating sites for the modification of primary silicon particles and refinement of eutectic silicon during solidification of hypereutectic alloys [17]. Rich phases (intermetallics) Al_3Cu found partially dissolved in the dendrite matrix. After the main Al-Si eutectic reaction, intermetallics will be solidified in one massive and one fine eutectic form. The primary Si particle size increased with the addition of CuP (0.01 and 0.03%), ranging from 4.52–3.74µm, respectively, which is about a 60% reduction in primary silicon size.

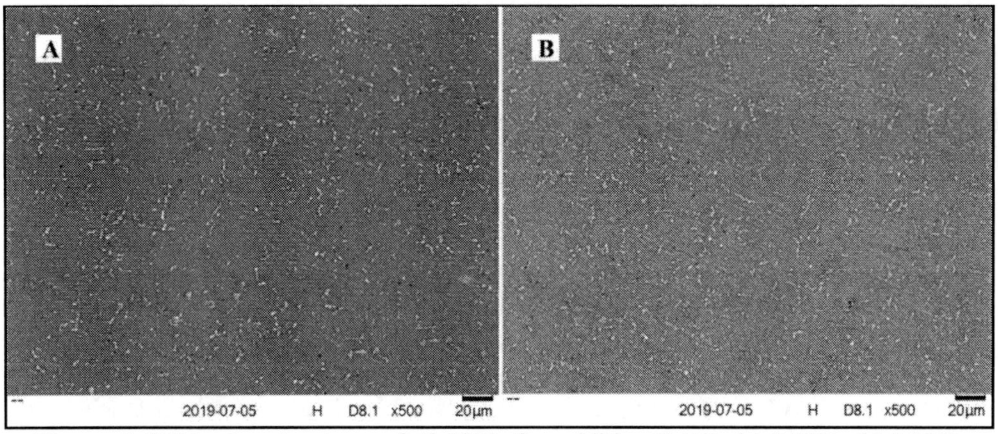

Figure 6. SEM microstructures of Al-13Si after additive $CuPO_4$ (a) 0.01%, (b) 0.03%.

3.2 Mechanical Properties

The results concerning mechanical characteristics show improved or comparable properties with the content of experimental material Nano-doped metals forms (Fig. 7). Vickers hardness is Hv=89.36 for the first sample without an additive. Vickers hardness are Hv=118.15 and 148.78, respectively in specimens with a content of 0.01–0.03% of Nano-copper-metals; moreover, Vickers hardness is Hv=102.24 and 120.71 respectively for specimens containing 0.01–0.03% of nano-nickel metals. The Vickers durability is Hv=95.64 and 114.71 for specimens with a content of 0.01–0.03% of nano-calcium metals. The difference in hardness is small, because the nano- doped metals are substituted, which means that the Al-matrix (fineness of each material) is very similar.

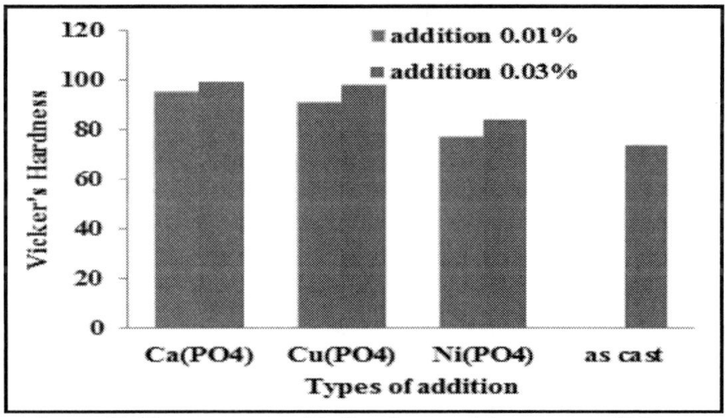

Figure 7. Show comparable hardness with content of Nano-doped metals types.

3.3 X-ray diffraction and crystallographic texture

The XRD pattern of Al-Si samples processed by a (**$Ca_3(PO_4)_2$**) master alloy is illustrated in Fig. 4.55. In order to investigate the extant form of Ca and its influence on the microstructure of aluminium alloy, Al-13Si samples are processed by $Ca_3(PO_4)_2$ and its phase composition analysis is shown in Fig 8 (a). It can be found that this alloy is composed of four phases, including α-Al, AlP, and Al_2O_3 primary Si. Al_2O_3 was observed through reflections on 2θ=12.19°, 21.17°, 38.54° and 47.86° and AlP was observed through reflections on 2θ=28.64° and 56.31° and Al_2Ca 19.15° and 44.95°. X-ray diffraction analysis shows the presence of $Ca_3(PO_4)_2$ in the synthesised sample as evidenced from reflections of low intensity [18]. The master alloy is composed of three elements: Ca, P and O. In this system, the atomic ratio of Al and P is about 1:1, indicating that all of the Ca atoms would react with Al to form Al_2Ca the atomic ratio of Al and Ca is about 2:1without P. In comparison, from the phase

two alloys, it can be deduced that, with the addition of $Ca_3(PO_4)_2$, the Al-Si alloy appears [19].
The XRD diffraction pattern of the Al-Si alloy processed by **$NiPO_4$** master alloy is illustrated in Fig. 4.59, showing the results of synthesis reactions using **$NiPO_4$** master alloy. In order to investigate the extant form of Ni and AlPO and its influence on the microstructure of aluminium alloy, Al-13Si samples processed by **$NiPO_4$** and the phase analysis are shown in Fig 8(b). This alloy is composed of four phases: α-Al, AlPO, primary Si and AlNi. AlNi was observed through reflections on 2θ=18.74°, 47.57°, 78.67° and 88.42° and AlP was observed through reflections on 2θ=47.54° and 69.24. X-ray diffraction analysis shows the presence of **$NiPO_4$** in the synthesised sample as evidenced from reflections of low intensity [20]. The master alloy is composed of three elements: Ni, P and O. The XRD pattern of Al-Si, indicating effective **$NiPO_4$** is much more reactive and is thus a very suitable source for AlPO. Fig 8(b) shows that the formation of AlPO is induced and appears as a single-phase of high crystallinity according to the strong signals in the XRD pattern reveal the presence of trigonal (rhombohedral axes) AlPO and cubic AlNi. The phase fractions are analysed using calibration standards card.
The XRD diffraction pattern of the Al-Si alloy is processed by a **$CuPO_4$** master alloy, as illustrated in Fig. 8(c). The microstructure was verified by means of X-ray diffraction (XRD). Apart from the obvious presence of aluminium, the diffraction pattern showed the presence of the Al_2Cu phase, as expected [21]. Subsequently, this method was chosen to characterise the microstructural evolution of the samples, since it allows the phase formation and dissolution to be correlated directly to the applied casting process. The introduced modification of the phase also permits the 2=20.70° to 78.31° region to be explored in the XRD pattern. It should be noted that the AlP and Al_2O_3 precipitates can be detected by the XRD approach, even if the small volume-fraction and small size of these precipitates are embedded in the Al matrix [22].
The XRD patterns from the extracted particles, as shown in 4.60, reveal the presence of cubic AlP, orthorhombic Al_2O_3 and cubic Al_2Cu in the sample. The phase fractions have been analysed using a calibration standards card.
Even after significant refinement and modification of the α-Al grain structure using intensive melt shearing, Al_2Cu was still the dominant phase. The XRD analysis further confirms that formation of AlP, Al_2O_3 and Al_2Cu, promoted by the addition of **$CuPO_4$** grain refiner. Overall, the present results confirm that the production of intermetallic phase is primarily governed by the addition of a **$CuPO_4$** grain refiner in Al-Si melts [23].

.

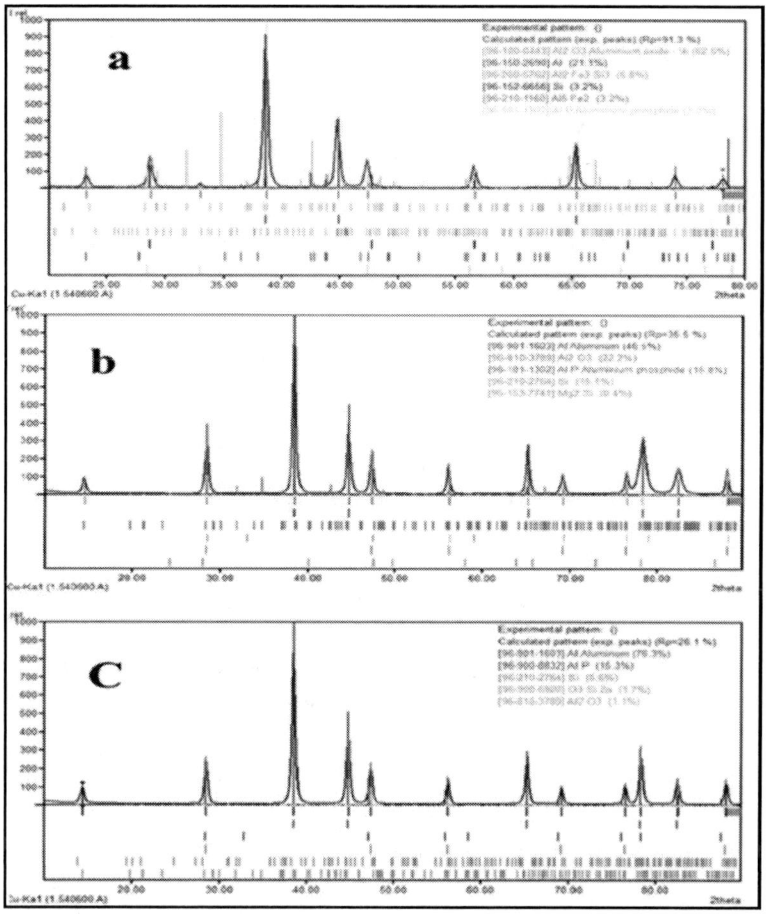

Figure 8. XRD pattern of Al-13Si Samples processed by: (a) additive of ($Ca_3(PO_4)_2$ (b) additive of $Ni_3(PO_4)_2$ (c) additive of $CuPO_4$.

3.4. Wear behaviour

The relationships between addition nano-metal-phosphate and wear rate of experiments alloys are presented in Fig. 9. It is clear the wear rate decreases with an increase in percentage of nano-doped metal particles. Furthermore, the relationship is linear. Hence, a decrease in grain size and good distribution of primary Si, eutectic and intermetallic phases signifies a wear rate decrease regardless of the reinforcement nano-Al_2O_3 size and percentage. Also, the wear and tear rate of processed composites decreases with increasing modifier elements (nano-metal-phosphate). The improved wear resistance to a casting procedure is referred to the distribution of particles within the base Al and refinement of grain.

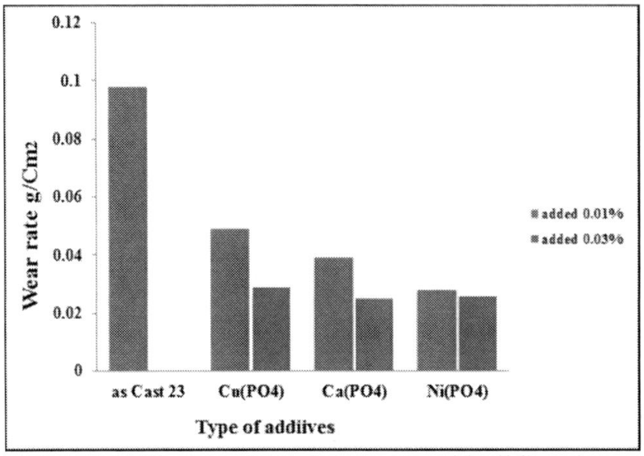

Figure 9. The variation of wear rate with sliding time with different nano-metal phosphates.

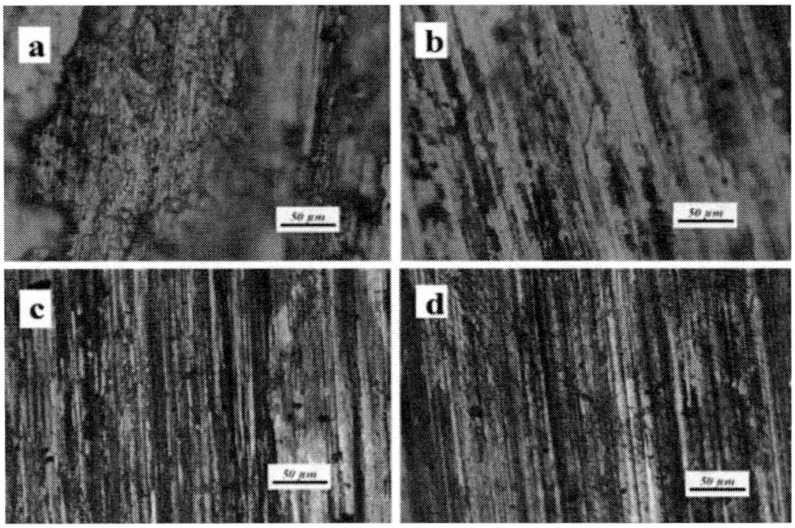

Figure 10. Optical micrographs of worn track: (a) as cast, (b) additive of $(Ca_3(PO_4)_2$ (c) additive of $CuPO_4$ (d) additive of $Ni_3(PO_4)_2$.

Fig. 10 shows an optical micrograph of worn surfaces of stir-processed composites. The micrograph reveals that the worn track breadth decreases in modified alloys compared to unmodified alloys. It is proof that the extent of wear and tear in a composite is below that of an unmodified alloy. The depth and breadth of the grooves generally imply an association in foster quantity of fabric from the specimen surface [26]. It is expected that uniform distribution of particles and modification of Si grains can cause a high wear resistance. Furthermore, the yield strength of the Al-13Si alloy is increased, which results in the strength of wear surface and the starting abrasive wear. Moreover, since Si is a hard material, it may plough out from the Al_2O_3 as the counter is slid across the face of the disc. In addition, the hard Si improves the resistance of a composite against plastic deformation and material flow by increasing the load-bearing capacity. The probability of adhesive wear is less in such Al_2O_3, and the dominant mechanism is abrasion and de-lamination.

3.5 Corrosion Behaviour

The cyclic polarisation curve was determined and shown in Fig.11 to evaluate the electrochemical stability of prepared samples in 3.5% NaCl; corrosion variations were evident in the alloys tested. Cyclic polarisation experiments have been carried out in order to assess the polarisation domains for the analysed samples. All three possible domains – catholic domain, passive domain and transpassive domain – were shown by cyclic polarisation curves. The large passive dominance and low-density values show higher resistance to corrosion in the analysed sample in the transpassive domain [24].

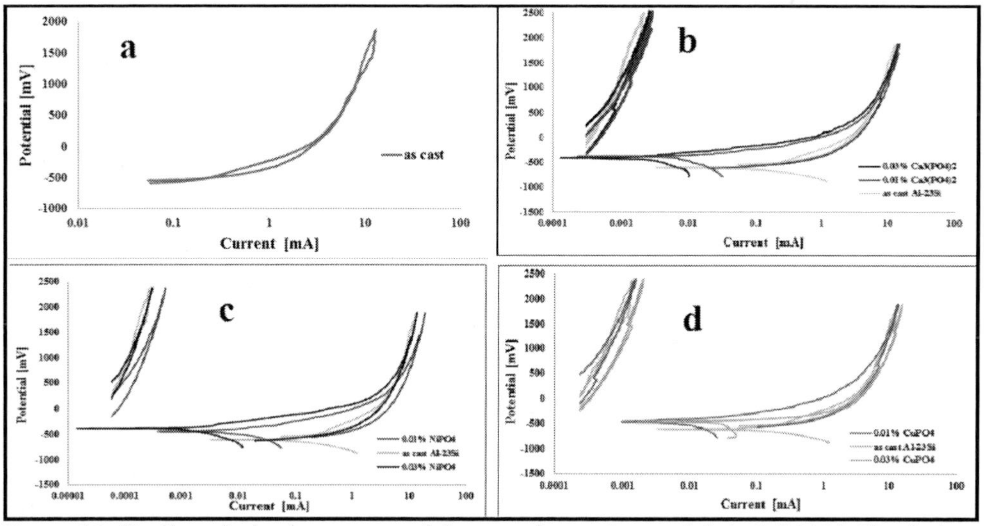

Figure 11. Cyclic polarisation curves of the experimental materials in 3.5 % NaCl solution. (a) as cast, (b) additive of $(Ca_3(PO_4)_2$ (c) additive of $Ni_3(PO_4)_2$ (d) additive of $CuPO_4$.

Indeed, despite the use of the Tafel extrapolation method for the determination of corrosion rates for Al-based alloys, the Tafel tables and their respective calculated uniform and pitting corrosion rates for metal covered with a semi-conductive passive film raised strong conviction [25]. Active dissolution near E_{corr} occurred in the polarisation curve of pure Al, followed by a major improvement of the current, with the potential added due to dilution and weakening of the passive layer, due to the attacks by Cl-anions shown in Fig. 12(a). The cyclic curves of polarisation also demonstrate the action of the board. This is obvious for the divisions of E_{corr} Al with the inclusion of phosphate elements, which displayed a certain curvature over the entire range of applications.

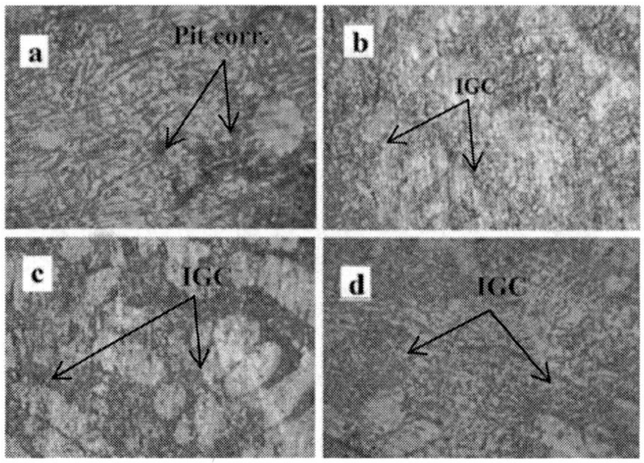

Figure 12. Microstructure characteristics of the tested alloys after corrosion test. (a) as cast; (b) additive of $(Ca_3(PO_4)_2$ (c) additive of $CuPO_4$ (d) additive of $Ni_3(PO_4)_2$.

As a result of microgalvanic corrosion processes, the effect of $(Ca_3(PO_4)_2$ on the micro pitting of alloys starts on the interface between the α-Al matrix. General irregular corrosion affected specimens, with grey corrosion products covering their surface. The anodic current densities were almost unchanged, and the passive scale decreased by $Ca_3(PO_4)_2$ to 0.03%. With both changes, the curves displayed the same noble change in corrosion ability, the strong passive range and the same anodic current intensity. Al-13Si alloys revealed reverse and close hysteresis, indicating that these alloys are sensitive to micro-galvanic and micro-pitting characteristics of the tested Al-13Si alloys, per Fig. 12(b).

Fig.12 (c) shows the microstructure of Al-13Si alloys processed by $CuPO_4$ after a polarisation test in 3.5% NaCl. The Al-13Si alloys possess a typical structure of dendrite. The surface corrosion of all specimens indicates that Cl^- ions severely attacked the surfaces of the alloys during polarisation. The major type of corrosion is mico-galvanic and localised corrosion. While nano-galvanic corrosion occurred at the junctions of dendrites, corrosion appeared at both the junctions of dendrites and the interfaces of Al_2Cu and matrix. Furthermore, considering the initiation and growth of pitting corrosion [26], the corrosion type at a lower potential is general corrosion, while localised and pitting corrosion dominate at higher potentials, especially in a transpassive region.

The resistance of IGC is referred to the refinement of and enhanced creation of the precipitates in the matrix by 0.03 %$NiPO_4$, where increased precipitation in the grain boundaries is consumed their solute material, thereby reducing the compositional disparity between the matrix and the grain boundary area, by a modifier and discrete platelets in the grain borders [27]. Fig. 12(d) shows that the pitting susceptibility in $NiPO_4$ can be correlated with the modification of AlPO and AlNi-phase precipitates in grain bodies as well as in grain-boundaries, resulting in a finer and regular distribution of noble cathodes on the surface, as compared to the prevalence of grain boundary precipitation obtained in the 0.03% $NiPO_4$ sample variants displaying IGC susceptibility.

3.6 Fracture Analysis

A cross-sectional view of the fracture surfaces shown in Fig. 13(a) may further explain the failure mechanism. As a cast material, the crack extends along the crossroads between the centre and the limit of the softer areas, which was consistent along the surfaces of the fracture. With the de-cohesion of the Si particles, the voids coalesced and spread until a complete separation occurred and the sample failed under tensile loading. With a 0.01 percent of $Ca(PO_4)$, Fig. 13(b) shows the results of eutectic Si being changed, acicular with only tiny regions of a modified fibrous Si current, increasing the concentration of $Ca(PO_4)$ to 0.03%. In addition, the Si morphology transition revealed that intermetallic particles were found with the introduction of 0.03% of $Ca(PO_4)$. These particles have a distinct polyhedral shape which makes them easy to separate from the eutectic Si, which tend to be located mainly at the interface between dendrites and eutectics. Ca-rich particles are distributed relatively uniformly in the sample [28], and the amounts of UTS, YS, % El including $Ca(PO_4)$ Al-13Si are 170.18–181.56MPa, 67.29–69.51MPa, respectively, with a related rise in ductility of 8.259–69.51%.

This appears to be a reasonable hypothesis since the nucleation level of Al-Si alloys represents extremely active nuclei, such as AlP and Al_2O_3. On the other side, $CaPO_4$ additions improved the coarse acicular morphology of eutectic Si, resulting in an increase in the eutectic-nucleation rate, resulting in a fibrous-eutectic Si[29]. The mechanism of this improvement can then be clarified because of the Si phase's increase in growth rate. As the P concentration and, therefore, the amount of AlP particles in the melt is small, most of the Al_2Si_2Ca that nucleated early during solidification may have infected much of the AlP particles, contributing to a higher-growth rate of the Al-Si eutectic. Such modification strengthens the mechanical properties, improves the overall tensile strength (UTS) and decreases ductility. The reinforcement process occurs from Al_2O_3 acting as a pinning of dislocation motion, being the amount of influence of multiple reinforcing systems as well as the composition of the alloy. The motion of dislocation happens by contact bond breakage and bond reformation, such that the tolerance for deformation can be strengthened by reducing the probability of movement [30].

Fig. 13(c) displays the impact of $Ni(PO_4)$ additions on the tensile properties of Al-13Si alloys (UTS, YS, and El). The amounts of UTS, YS, % El including $Ni(PO_4)$ Al-13Si are 192.11–214.66 and **75.2–78.42** MPa, respectively, with a related rise in ductility of **8.067–8.337**%. Eutectic and primary silicon particles are modified and refined with addition of $Ni(PO_4)$ (0.01–0.03%) and are distributed uniformly in the matrix. This could be due to the fact that Ni(PO_4) reacts with the liquid Al forms AlPO and Al_3Ni particles present, which can serve as heterogeneous nucleating sites during the solidification of Al-Si alloys, thus changing primary-silicon particles [10]. The Al_3Ni phase is formed in the immediate vicinity of eutectic-Si crystals. Associated temperatures of nucleation and growth of the eutectic and the solidus temperature of the alloy are shifted to lower values [28], displaying fine-sized dimples, instead of the evenly spaced dimples found. The cleavage seen on the surface of the fracture is typical of a fractural fracture. Fig. 13(c) indicates that the introduction of 0.03% Ni(PO_4) contributes to large dimples instead of randomly spaced dimples.

Fig. 13(d) indicates the yield-strength (YS), ultimate tensile-strength (UTS), strain (percentage El) and hardness (HV) of the aluminium alloys as a result of the Cu(PO_4) element. The strength resulting from cast condition in different alloys has varying values, including Ni(PO_4) Al-13Si as 195.31– 202.31 and **70.8–71.91**MPa, respectively, with a related rise in ductility of **7.837–9.837**%. In general, the tensile properties of aluminium alloys are heavily dependent on the DAS structure, which is a function of Cu(PO_4) content [31]. However, polygonal primary silicon particles are treated with increasing addition of Cu(PO_4) (0.01-0.03%) and get distributed uniformly in the matrix. This could be due to the fact that Cu(PO_4) reacts with the liquid Al forms AlP and Al_2O_3 particles present, which may act as heterogeneous nucleating sites for the modification of primary silicon particles during the solidification of hypereutectic alloys [10]. An increase in both yield and ultimate strength is observed in all samples. The aluminium-silicon alloys are heat-treatable in order to enhance the strength by the precipitation of the Al_2O_3 and Al_2Cu phases. At the peak-aging condition, the size of the Al_2Cu phase is nano-scale and aligned with <001> cube axis in the matrix. A detailed study of the effects of microstructure and crystal structure on the hardening precipitates [32] reveals that the size and distribution of the Al_2Cu phases precipitates do not show a significant difference in the alloys; this suggests that the yield strength is less dependent on the Al_2O_3 composition[33]. Arrangement reveals a major change in the 0.03% Cu(PO_4)-modified Al-13Si fracture descriptions where a regular dimple arrangement has generally substituted the fine particles. The fundamental cracks of collapsed eutectic Si particles have been sheared plastically, creating a ductile frame [34].

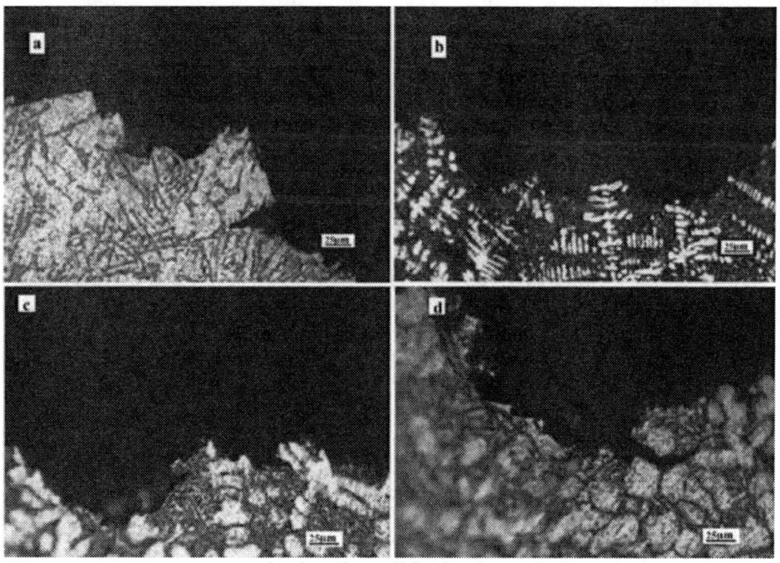

Figure 13. Fracture analysis of modified-Al-13Si alloys (a) as cast; (b) additive of ($Ca_3(PO_4)_2$ (c) additive of $CuPO_4$ (d) additive of $Ni_3(PO_4)_2$.

4. Conclusions

Based on the above, it can be concluded that, firstly, the addition of 0.01% nano-metal-phosphate had a slight change in the eutectic Si, but primary Si was moderately refined. Adding 0.03% of nano-metal-phosphate to Al-13Si alloy was necessary to reduce primary Si from an average particle size of 16μm to 2μm. Secondly, the addition of various quantities of nano-metal-phosphate in the Al-Si alloy has led to a significant variance in the wear behaviour of the measured Al-13Si. In addition, modification with additional nano-doped metal was observed, resulting in a fine fibrous structure of eutectic silicon, mainly in the addition of $CaPO_4$ and $NiPO_4$. After that, results included increased hardness and wear values of modified samples processed by nano-metal-phosphate. In addition, the XRD pattern revealed predominant aluminium, silicon, slight AlP and Al_2O_3 peaks. Finally, results of the cyclic polarisation test showed a susceptibility to micro-pitting corrosion was a dominant characteristic of the Al-Si alloys after being processed by 0.01% of nano-metal-phosphate, while micro-intergranular corrosion in the Al-Si alloys appeared after being processed by 0.03% of nano-metal-phosphate.

References:

[1] R. Haghayeghi; 2008 "Grain Refinement and Nucleation Processes in Aluminium Alloys through Liquid Shearing," Brunel University.

[2] H. Kim, J. Hong, S. Kim, Y. Park, H. Park, W. Park, J. Kim and K. Buem; 2017 "Effect of silicon on microstructure and mechanical properties of Cu-Fe alloys," J. Alloys Compd., vol. 707, no. 6, pp. 184–188.

[3] Y. Zuo, Z. Fan, Q. Zhu, L. Lei, and J. Cui; 2013 "Modification of a hypereutectic aluminium silicon alloy under the influence of intensive melt shearing," Mater. Sci. Forum, vol. 765, pp. 140–144.

[4] D. Li, C. Cui, X. Wang, Q. Wang, C. Chen, and S. Liu; 2016 "Microstructure evolution and enhanced mechanical properties of eutectic Al-Si die cast alloy by combined alloying Mg and La," Mater. Des., vol. 90, pp. 820–828.

[5] M. Easton, M. Qian, A. Prasad, and D. John; 2016 "Recent advances in grain refinement of light metals and alloys," Current Opinion in Solid State and Materials Science, vol. 20, no. 1, pp. 13–24.

[6] A. Mazahery and M. Shabani; 2014 "Modification mechanism and microstructural characteristics of eutectic si in casting Al-Si Alloys: A review on experimental and numerical studies," Jom, vol. 66, no. 5, pp. 726–738.

[7] S. Gebremeskel, S.Bhatnagar, N. Li, J. Jianping, M. Ludwig, T. Matsubara, Y. Hofer and F. Arnberg; 2016 "Effect of Modifier and Grain Refiner on Cast Al-7Si Aluminum Alloy : a Review," IOSR J. Mech. Civ. Eng., vol. 5, no. 1, pp. 1–13.

[8] M. Zuo, D. Zhao, Z. Wang, and H. Geng; 2015 "Complex modification of hypereutectic Al-Si alloy by a new Al-Y-P master alloy," Met. Mater. Int., vol. 21, no. 4, pp. 646–651.

[9] H. Choi and X. Li; 2012 "Refinement of primary Si and modification of eutectic Si for enhanced ductility of hypereutectic Al-20Si-4.5Cu alloy with addition of Al_2O_3 nanoparticles," J. Mater. Sci., vol. 47, no. 7, pp. 3096–3102.

[10] J. Li, F. Hage, X. Liu, Q. Ramasse, and P. Schumacher; 2016 "Revealing heterogeneous nucleation of primary Si and eutectic Si by AlP in hypereutectic Al-Si alloys," Sci. Rep., vol. 6, no. April, pp. 1–8.

[11] C. Ho and B. Cantor; 2004 "Modification of hypoeutectic Al-Si alloys," J. Mater. Sci. Vol., vol. 30, pp. pages1912–1920.

[12] S. Liang and R. Schmid; 2014 "Phosphorus in Al-Si cast alloys: Thermodynamic prediction of the AlP and eutectic (Si) solidification sequence validated by microstructure and nucleation undercooling data," Acta Mater., vol. 72, pp. 41–56.

[13] A. Haselhuhn, P. Sanders, and J. Pearce; 2017 "Hypoeutectic aluminum-silicon alloy development for gmaw-based 3-D printing using wedge castings," Int. J. Met., vol. 11, no. 4, pp. 843–856.

[14] W. Kyffin, W. Rainforth, and H. Jones; 2001 "Effect of phosphorus additions on the spacing between primary silicon particles in a Bridgman solidified hypereutectic Al-Si alloy," J. Mater. Sci., vol. 36, no. 11, pp. 2667–2672.

[15] P. Yan, W. Mao, J. Fan, and B. Wang; 2019 "Simultaneous refinement of primary Si and modification of eutectic Si in A390 alloy assisting by Sr-modifier and Serpentine Pouring Channel Process," Materials (Basel)., vol. 12, no. 19, pp. 1–13.

[16] A. Moffat; 2007 "Micromechanistic analysis of fatigue in aluminium silicon casting alloys, Phd Thesis," p. p274.

[17] L. Tian, Y. Guo, J. Li, F. Xia, M. Liang, and Y. Bai; 2018 "Effects of solidification cooling rate on the microstructure and mechanical properties of a cast Al-Si-Cu-Mg-Ni piston alloy," Materials (Basel)., vol. 11, no. 7, pp. 3–11.

[18] M. Wang; 2014 "Control of Morphology Al-Fe-Si Phase in Al-Si-Cu Hypoeutectic Alloy," p. 165.

[19] W. Ebhota and T. Jen; 2017 "Effects of Modification Techniques on Mechanical Properties of Al-Si Cast Alloys," Alum. Alloy. - Recent Trends Process. Charact. Mech. Behav. Appl.

[20] H. Bartsch, J. Mánuel, and R. Grieseler; 2017 "Influence of Nanoscaled Surface Modification on the Reaction of Al/Ni Multilayers," Technologies, vol. 5, no. 4, p. 79.

[21] H. Medrano, C. Garay, C. Gómez, J. Aguilar, M. Maldonado, and R. Martínez; 2016 "Evolution of microstructure in Al-Si-Cu system modified with a transition element addition and its effect on hardness," Mater. Res., vol. 19, pp. 59–66.

[22] A. Barros, A. Magno, I.Souza, F.Mota, C. Moreira, A. Silva, M. Rocha and O. Lima; 2015 "Measurements of microhardness during transient horizontal directional solidification of Al-Rich Al-Cu alloys: Effect of thermal parameters, primary dendrite arm spacing and Al2Cu intermetallic phase," Met. Mater. Int., vol. 21, no. 3, pp. 429-439.

[23] L. Wang, W. Lu, Q. Hu, M. Xia, Y. Wang, and J. guo Li; 2017 "Interfacial tuning for the nucleation of liquid AlCu alloy," Acta Mater., vol. 139, pp. 75–85.

[24] M. Reboul and B. Baroux; 2011 "Metallurgical aspects of corrosion resistance of aluminium alloys," Mater. Corros., vol. 62, no. 3, pp. 215–233.

[25] R. Melchers; 2018 "A Review of Trends for Corrosion Loss and Pit Depth in Longer-Term Exposures," Corros. Mater. Degrad., vol. 1, no. 1, pp. 42–58.

[26] M. Khushaim; 2015 "Investigation of the Precipitation Behavior in Aluminum Based Alloys," Materials Science and Engineering, King Abdullah University of Science and Technology , November.

[27] M. Kaiser, S. Sabbir, M. Rahman, M. Kabir, And M. Al Nur; 2018 "Effect of Fe , Ni and Cr on The Corrosion Behaviour of Hypereutectic Al-Si Automotive Alloy in 3 . 5 % Nacl Solution at Different Temperature and Velocity," Vol. M, No. December, Pp. 11–17.

[28] T. Ludwig, P. Schaffer, and L. Arnberg; 2013 "Influence of some trace elements on solidification path and microstructure of Al-Si foundry alloys," Metall. Mater. Trans. A Phys. Metall. Mater. Sci., vol. 44, no. 8, pp. 3783–3796.

[29] K. Al-Helal, Y. Wang, I. Stone, and Z. Fan; 2013 "Effect of Ca level on the formation of silicon phases during solidification of hypereutectic Al-Si alloys," Mater. Sci. Forum, vol. 765, pp. 117–122.

[30] D. Bacon, Y. Osetsky, and D. Rodney; 2009"Chapter 88 Dislocation-Obstacle Interactions at the Atomic Level," Dislocations in Solids, vol. 15, no. 09, pp. 1–90.

[31] M. Djurdjevič and M. Grzinčič; 2012 "The Effect of Major Alloying Elements on the Size of Secondary Dendrite Arm Spacing in the As-Cast Al-Si-Cu Alloys," Arch. Foundry Eng., vol. 12, no. 1, pp. 19–24.

[32] I. Sadeghi, M. Wells, and S. Esmaeili; 2018 "Effect of particle shape and size distribution on the dissolution behavior of Al2Cu particles during homogenization in aluminum casting alloy Al-Si-Cu-Mg," J. Mater. Process. Technol., vol. 251, no. 11, pp. 232–240.

[33] Y. Yang, R. Jiang, X. Li, Y. Zhang and R. Li; 2019 "Effect of Nanoparticle Content on the Microstructural and Mechanical Properties of Forged and Heat-Treated TiC/2219 Nanocomposites," J. Chem. Inf. Model., vol. 53, no. 9, pp. 1689–1699.

[34] G. Wignall, R. Rothon, G. Longman, and G. Woodward; 1977 "The structure of amorphous aluminium phosphate by radial distribution functions derived from X-ray diffraction," J. Mater. Sci., vol. 12, no. 5, pp. 1039–1049.

IOP Conf. Series: Materials Science and Engineering **987** (2020) 012015 doi:10.1088/1757-899X/987/1/012015

The Effect of the Nanoparticle SiO_2 on the Mechanical and Physical Properties of Rubber Composites

Ahmed Abdulameer Subeh[1], Mohammed H. Al-maamori[2] and Ahmed Fadhil Hamzah[3]

[1]Karbala Technical Institute, Al-Furat Al-Awsat Technical University, Karbala, Iraq.
[2,3]College of Materials Engineering, University of Babylon, Babylon, Iraq.

[1]ahmedeng6217@yahoo.com, [1]ahmedeng@ikr.atu.edu.iq,
[1]ahmed.abdulameersubeh@uobabylon.edu.iq,
[2]mat.mohammed.hamzah@uobabylon.edu.iq,
[3]mat.ahmed.fadhil@uobabylon.edu.iq

ABSTRACT. This paper investigates the effect of the nanoparticle SiO_2 on the mechanical and physical properties of a rubber blend of natural rubber (NR) and styrene-butadiene rubber (SBR). The proportions of the nanoparticle that are added are 0, 0.2, 0.4, 0.8, 1, 1.5, 2, 2.5 and 3 pphr to a fixed percentage of rubber blend (30% NR/70% SBR), which has been prepared with vulcanisation factors, accelerators and activators using a two-roll mill laboratory, moulds and a thermal piston; the pressure, temperature and time difference test conditions according to standards. The results of the laborator tests show an increase in the mechanical properties (i.e. tensile strength, modulus of elasticity, resilience and tear resistance) with increases in the percentage load (pphr) until 0.8 pphr, which is when there is a slight decrease; the properties of hardness, fatigue and compression set also increase. While the results for the physical properties show an increase in viscosity, the greater torque, scorch and cure times decrease with increases in the proportion of SiO_2 in the rubber recipe.

1. Introduction

Non-reinforced rubber composites have only a few use cases. The addition of ceramic materials such as silica is needed to enhance the rubber's mechanical and physical properties and to minimise the cost of production. Accordingly, studying the mechanical and physical properties of different engineering materials is important because of their engineering applications. The increased use of rubber in engineering has been the result of its unique properties, which include high extensibility, high strength, high energy absorption, high fatigue resistance, strong eco-resistance and high resilience [1]. Rubber by itself has poor properties and no advantage over other materials; in order to improve its performance, it needs to be vulcanised. Vulcanisation, named after Vulcan the Roman God of Fire, describes the process of transforming mechanically brittle raw rubber materials into durable goods. One of the most effective ways of producing high-quality rubber is to strengthen the rigid entities that enhance the mechanical and other properties of the rubber depending on the nature of the reinforcement, whether it is black carbon, silica, oxygen or clay. Integrating fillers into elastomer materials typically leads to a substantial improvement in their mechanical properties, resulting in increased abrasion and tear resistance. The interactions between the rubber fillers are an important parameter affecting these properties [2]. The term 'rubber blend' refers to a combination of two or more different rubbers resulting in a material with a range of properties that do not originate from either one of the components [3]. The elastomer original is NR, which acts as an imperative position among elastomers. NR is extracted as latex from the Pará rubber tree [4]. NR has a moderate teragram (around -73 °C), which is accompanied by flexible behaviour, especially when it contains a high level of cis, because the presence of methyl groups on the same side of the rubber chain prevents NR from being properly arranged, causing spaces between the particles to form and dissipate the passing wave vibration energy. However, due to the regularity of the molecular structure, NR can easily crystallise at low ambient temperatures, as well as increased in hysteresis. NR products exhibit high tensile strength and high resistance to tearing, cutting and abrasion [5]. NR also has very strong tolerance for low temperatures, down to -57 °C, at which point it becomes substantially stiffer. NR has a low glass transition temperature (around -73 ° C), which allows it to exist at lower temperatures than most other rubbers [6]. In addition to being used in tires and tubes, NR is used in conveyor belts, engine supports, structural bearing pads, flags, rubber bands, vehicle helmets, heels and gaskets [7].

The synthetic rubber SBR, which is made by polymerising styrene and butadiene, is the most common substitute for natural rubber [8]. Production-based polymerisation reactions can be divided into two forms: hot polymerisation and cold polymerisation. In hot polymerisation, the rubber comprises several branches in the chain and has greater dimensional stability and stronger extrusion compared to the rubber resulting from cold polymerisation. What is generated is linear chains with strong vulcanisation properties and with a low styrene content of up to 20–30% polymer [4]. Particulate fillers can make the strength of the rubber over ten times that of amorphous rubber. To trigger this effect, the fillers must have a highly specific surface area, i.e. the particles must be small. Small particles have a wide surface area to interact with the compound's rubber and similar particle-to-particle spacing. The two types of fillers that are most important for rubber reinforcement are black carbon and silica [9]. Nano-silica consists of spherical particles with a diameter of less than 100 nm. The hydrophilic character of silica also influences the cure properties of rubber compounds; vulcanised rubber properties and even compatible with non-polar rubber such as NR [10]. Silica retards vulcanisation as it reacts to the complex the zinc-accelerator - the sulphur [11]. The most widely used silane coupling agent is 3-(Trimethoxysilyl)propyl methacrylate (TMSPM). A silane coupling agent is a bi-functional molecule: one end has the ability to react chemically with silica, forming an intermediate silica–silane coupling agent, and the other end with rubber. The intermediate silica–silane coupling agent participates in the vulcanisation process, which improves the degree of compatibility between silica and non-polar rubber, thus increasing the reinforcing efficiency of the silica [10]. Interactions between rubber and silica can either happen through physical adsorption or chemical bonding.

This paragraph provides an overview of scientific studies in this area. Sun *et al.* [12] studied the improved performance of nano-silica-filled SBR composites through the surface modification of nano-silica using silane coupling agents. Nano-silica dispersion in SBR and the bonding power of nano-silica with SBR have been significantly improved, and the physical and mechanical properties of vulcanised rubber have also been greatly improved. Sombutosompop *et al.* [13] investigated the effect of untreated silica, precipitated silica (PSi) and fly ash silica (FASi) as fillers on the properties of NR and SBR. It was found that the cure properties at NR loading 30–75 pphr increased with PSi and decreased with FASi at the same rate. All properties of filled SBR were similar to filled NR except for tear strength and tensile, which makes NR the superior rubber. Salimi et al. [14] studied the effect of silica enhancements on NR and on NR/SBR compounds. To chemically bind the silica to the rubber, the silica surfaces were treated with a silane-coupling compound (bis(triethoxysilylpropyl)tetrasulfide). The tests of the NR/SBR compound with an optimal amount of silane demonstrate that the cure time decreases as the NR content increases. (The NR/SBR's tensile properties combine with rising SBR material). Mayyadaha [15] studied two types of rubbers (SBR and NR) combined with two types of binding fillers (Al_2O_3 and SiO_2). Her findings revealed that, rather than alumina fillers, silica fillers improved their properties. She also found that, with the introduction of all forms of reinforcing fillers and the improvement of their loading speed, their mechanical properties and even durability improved. SBR coated with 25 pphr of silica has the highest hardness qualities and lowest abrasion wear ratings. This research studies the impact of variable amounts of silica on different properties of rubber blends.

2. Materials and Method

2.1 Used Materials

The materials used in this research are natural rubber (SMR20) and synthetic rubber (SBR 1500), and the materials used in preparing the rubber recipe are CBS, zinc oxide (activated), stearic acid (activated), anox (HB), antiozone (WAX A 111), aromatic oil, carbon black (GPF 660), sulphur and IPPD (imported by the Diwaniyah tire company) and nanoparticle material (SiO_2).

2.2 Method of Preparation

The formulation of the rubber composites are given in Table 1.

Table 1. The composition of the rubber composites.

Item no.	Recipe ingredients	Loading level (pphr)								
1	SMR20					30				
2	SBR 1500					70				
3	Activator (zinc oxide)					4				
4	Sulphur					2				
5	CBS					0.5				
6	Anox (HB)					1				
7	IPPD					1.75				
8	Antiozone (WAX A 111)					2				
9	Activator (stearic acid)					2				
10	Aromatic oil					8.5				
11	Carbon black (GPF 660)					51				
12	SiO_2	0	0.2	0.4	0.8	1	1.5	2	2.5	3

Mixing was done in a two-roll mill (Figure 1); kneading and mixing operations were conducted on this machine as per ASTM D 15. The ingredients were added as follows, as per ASTM D3187-89:

1. Pass SMR20 and SBR 1500 between the two rolls several times with a reduction of 0.5–1 mm between the two rolls.
2. Add stearic acid and zinc oxide and then blend for material homogenisation.
3. Add half of the carbon black with operating oil evenly across the mill at a uniform rate.
4. When the carbon black has been completely incorporated, open the rolls to 1.65 mm.
5. Add the remaining carbon black and oil evenly across the rolls at a uniform rate.
6. Add reinforcement materials (nano-silica) while continuously mixing to start homogenising; the mixing process should continue until excellent homogenisation is achieved.
7. Add sulphur and mix to achieve the homogenisation of the materials under room temperature and then add the accelerators and sulfur.
8. Set the rolls to 0.8 mm and then combine the materials for homogenisation.
9. Open the mill to provide a minimum batch thickness of 6 mm and continue the mixing process until premature homogenisation is achieved.

After preparing the rubber composite recipes, test models were prepared to provide mouldings and state of test, such as temperature, pressure and time. Following that, according to the ASTM, which is a method of thermal piston vulcanisation, the temperature and time of the vulcanisation process variable according to examination.

Figure 1. The mixing machine.

3. Characterisation and Testing

3.1 Mechanical Properties

A tensile test was carried out on the samples prepared from the vulcanised slice (i.e. by cutting three dumbbell specimen). The cutting was done by hand press and by a Monsanto Tensometer T10 in conjunction with the ASTM d412-88, using special tensile test sample holding jaws that are mobile at a speed of 500 mm/min. A tearing examination was conducted in compliance with the standard ASTM D624-54 by using a system tensile (Tansometer10) and speed (500 mm/min) after model testing (dumbbell) by cutting machine models and by cutting machine (Nicking cutter) form (Wallace) incision work in the narrow area of the unit. A Tansometer10 was used to measure the results in N/mm. A hardness test was carried out using a durometer Hardness shore Aaccording to the standard ASTM-D2240. Standardised hardness-measuring equipment was applied directly to the surface of the specimens using a sharp needle to test the intensity. Six different positions were averaged. Resilience was tested using a Wallace Dunlop Tripsometer according to the specifications in ASTM D1054, and the data was linked to the digital computer information, Wrpra-wallace Rebound Resilience Data File, which records the percentage of resilience. The compression set test (ASTM D395) was conducted on ordinary 29 ± 0.5 mm diameter and 12.5 ± 0.5 mm thick cylindrical test specimens. Using a compression set tool, the samples were compressed to 40 percentage of their thickness for 72 hours and then left for 30 minutes until the final sample thickness was taken. Fatigue tests were carried out for the prepared samples in compliance with ASTM D430 using a flexibility fatigue instrument (Wallace-Dematta Flexing Machine), which included a monitor for the number of cycles at which a crack occurs.

3.2 Physical Properties Test

The viscosity and curing characteristics are conducted in compliance with ASTM D2084-89 using an oscillating disk rheometer (ODR) for the prepared samples (i.e. cut from the rubber composite recipes), in which the upper and lower jaw, restraint clutches for six minutes below 20 Pa and 185 °C. As shown by the following results (max. torque), min. torque , scorch time , cure time (min) and viscosity), the results are shown in a graph of torque vs. vulcanised time. The rehometer was connected to a machine with a precision control system.

4. Results and Discussions

4.1 Mechanical Properties

The load is increase from zero and even the breakdown happens rapidly (500 mm/min), to establish the properties of the material under the influence of axial tension load in one direction of the sample (longitudinal axis).Several models were tested, and an average wass taken to achieve the optimal outcome for get rid of human errors and to reduce the rate of error from the samples' heterogeneity. Figure 2 shows an increase in tensile strength with an increase in the percentage of silica in the rubber recipe; however, when the ratio of silica reaches a certain percentage, there is a noticable decrease in tensile strength due to the aggregation of the granules as a result of pressure and compression. The particles filled the spaces between the rubber chains and hardened the structure, while a large amount will result in an agglomeration in the compound material, which leads to defects in the formation of cracks that lead to weakened tension (10 pphr). This result is consistent with other research [16,17,18]. Figure 3 shows that, when a small amount of nano-silica is added to the rubber recipe, the elongation property of the rubber decreases, because the rubber extends when the fine particles fill the spaces, which reduces the elongation property by limiting the movement of the chains. On the other hand, when the quantity of nano-silica is increased, the elongation property also increases due to the fact that the rubber recipe does not absorb a high amount of filling; therefore, the accumulation of nanoparticles between the chains occurs and leads to the formation of cracks, thus reducing tensile strength and increasing the elongation property. This result is consistent with other research [17]. Figure 4 shows an inverse relationship between the modulus of elasticity and elongation: an increase in the modulus of elasticity was observed with only a small amount of modified nanoparticles. The distribution between the rubber chains facilitates the precision of the particles, and thus the elastic modulus increases. However, an increase in the amount of nanoparticles leads to a decrease in the modulus of elasticity; this is due to the heterogeneous distribution, which leads to the clumping of molecules between the chains. The entanglement then becomes weak, and thus cracks are generated, which reduces the modulus of elasticity. This result is consistent with other research [16,17,19].

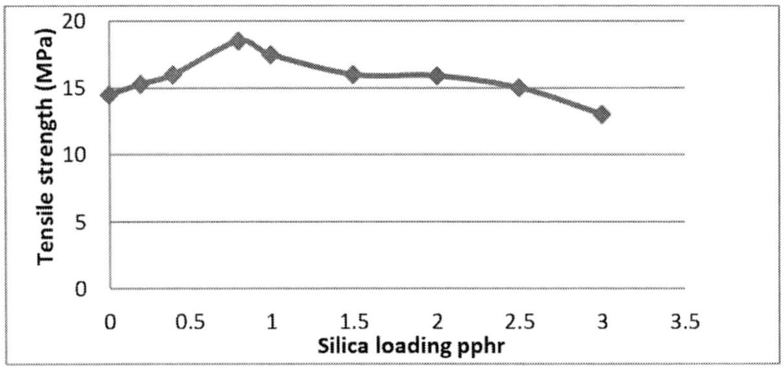

Figure 2. Effect of nano-silica (pphr) on the tensile strength of the NR/SBR composite.

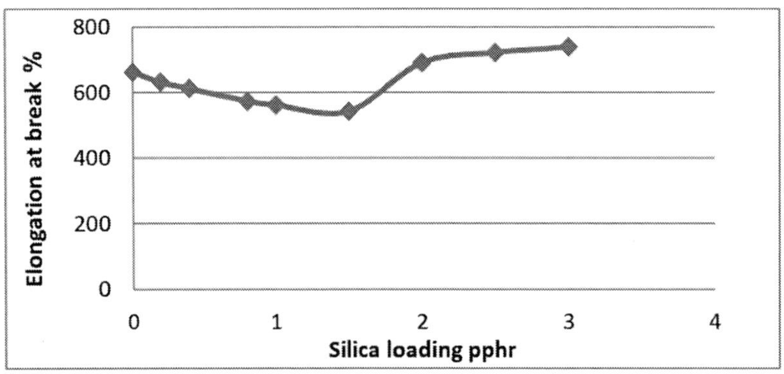

Figure 3. Effect of nano-silica (pphr) on the elongation of the NR/SBR composite.

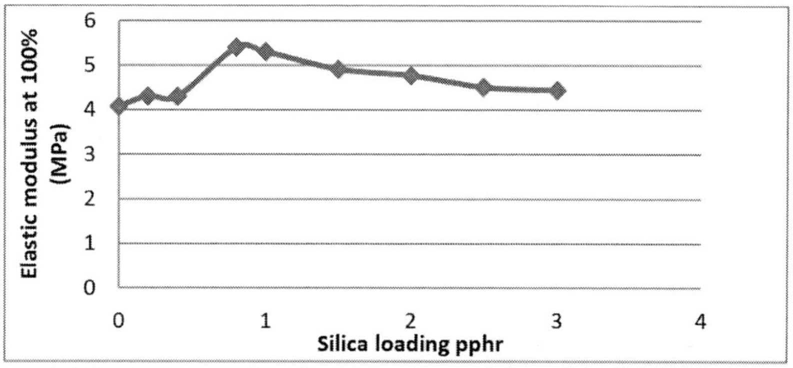

Figure 4. Effect of nano-silica (pphr) on the modulus of the NR/SBR composite.

Figure 5 shows the effect of additional materials on tear resistance, which is measured in N/mm. tear indicates the materials' resistance to the growth of any things when under stress. This feature is related to the tension feature. Accordingly, resistance increases when a small amount of silica is added. The maximum value of resistance is 8.2 MPa at 0.8 pphr of silica, because the particles fill the spaces between the rubber chains and increase the mechanical bond between them (i.e. the same reason as mentioned earlier in the case of the tensile property). This leads to the best resistance to tearing. In the case of increasing the amount of modified nano-silica to 3 pphr, resistance decreases to 7 MPa; the resistance decreases due to the irregular distribution of nano-silica particles in the rubber and the assemblies between the chains, as mentioned in the case of the tensile property. This result is consistent with other research [20].

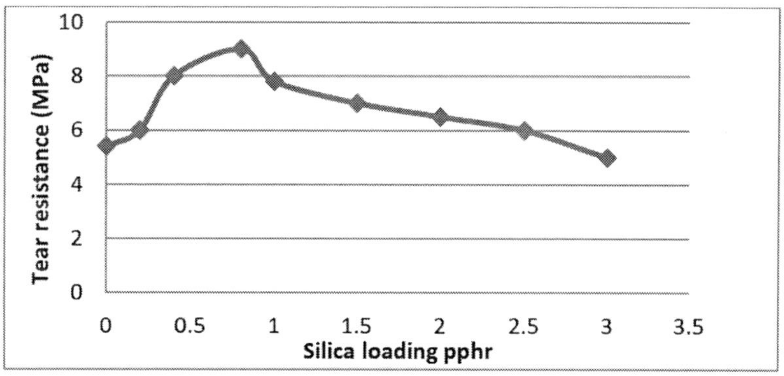

Figure 5. Effect of nano-silica (pphr) on the tear of the NR/SBR composite.

Test relies on testing the penetration of the rigid ball in a rubber sample, and releasing the force during load.The hardness was measured from the centre and sides for the samples, which were prepared with a cylindrical form, and took the range to get the maximum result and prevent errors. Figure 6 shows that, when adding nano-silica particles containing a particle size of 65–85 nm to the rubber in different proportions (i.e. 0, 0.2, 0.4, 0.8, 1, 1.5, 2, 2.5 and 3 pphr) in the rubber recipe, there is an increase in hardness, because silica particles have an active and irregular surface area which increases penetration between themselves and the rubber chains upon contact with the rubber, which increases the compound's outer body's resistance to penetration due to increased surface tension. The result is increased rigidity. This result is consistent with other research [19].

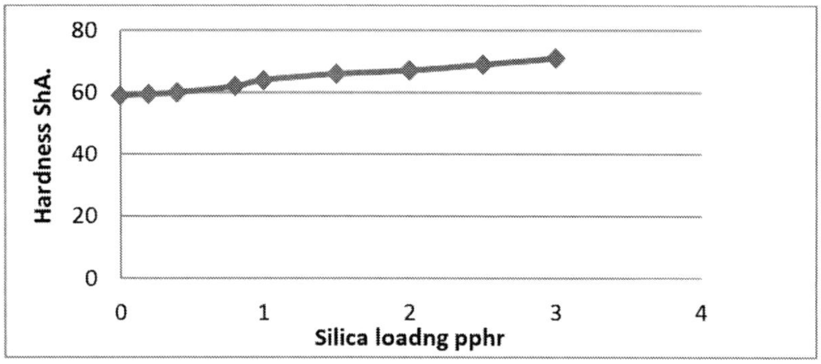

Figure 6. Effect of nano-silica (pphr) on the hardness of the NR/SBR composite.

Figure 7 shows an increase in the rebound; at 0.8 pphr, there is a jump in the values, because the nanoparticles act as fillers in the spaces and do not absorb energy. Then, there is a decline, even as the proportion of nanoparticles increases. At 3 pphr, the rebound elasticity increases due to particle aggregation and pressure in the compound rubber structure, which promotes the growth of cracks and reduces the hardness property. After that, agglomeration occurs, and a substance forms that absorbs and dissipates energy as heat, thereby reducing the strength of elasticity. Resilience is related to hardness and density.

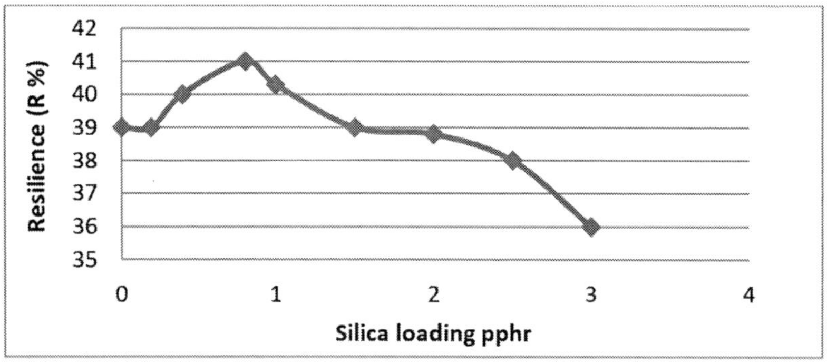

Figure 7. Effect of nano-silica (pphr) on the resilience of the NR/SBR composite.

Figure 8 shows that the compression set increases the ratio of silica; this is due to an increase in the surface area as a result of the interference between silica granules and rubber and an increase in hardness, i.e. increased crosslinking and reduced movement of chains, thus reducing Recoverability of the rubber compound. This result is consistent with other research [21].

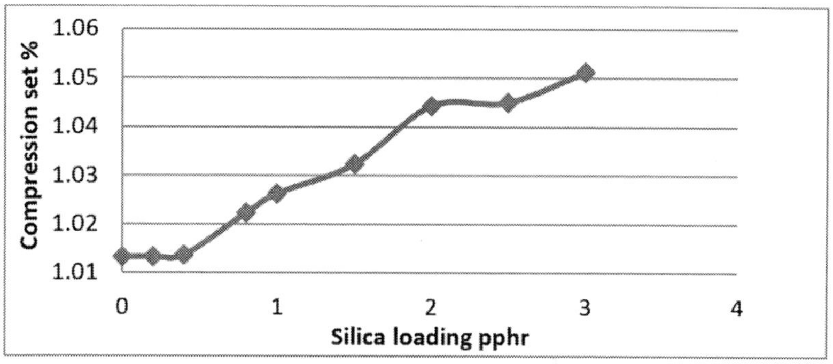

Figure 8. Effect of nano-silica (pphr) on the compression set of the NR/SBR composite.

Fatigue strength is represented by the materials that are resistant to the growth of cracks. Figure 9 shows that, in small quantities, silica provides resistance to crack growth. The higher the amount of nanoparticles, The greater the heterogeneous agglomeration and the distribution will make fatigue in the sample making the cracks invisible.

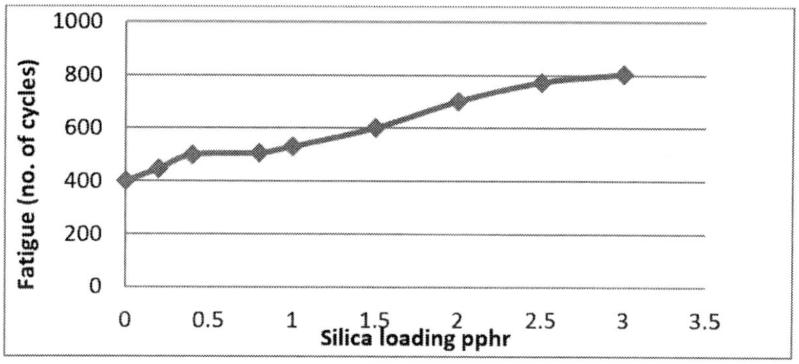

Figure 9. Effect of small quantities of nano-silica (pphr) on the fatigue of the NR/SBR composite.

4.2 Physical Properties

Addition nano-silica with limited percentage increases the force of the elastic mixture, indicating an increase in torque and viscosity. Figures 10 and 11 show that the presence of nano-particles means that the mobility of rubber chains is limited; therefore, these properties decrease. This result is consistent with other research [22].

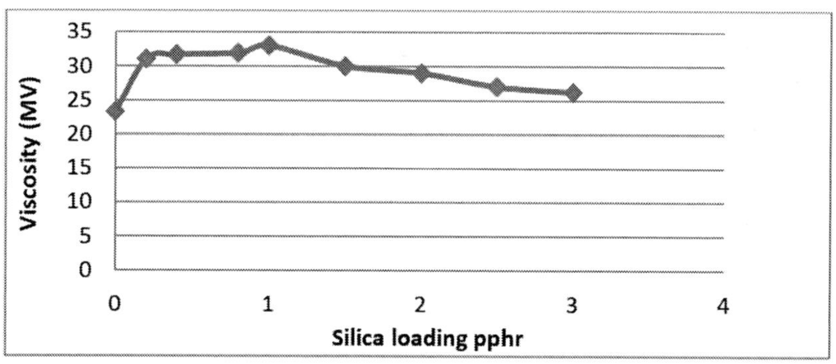

Figure 10. Effect of nano-silica (pphr) on the viscosity of the NR/SBR composite.

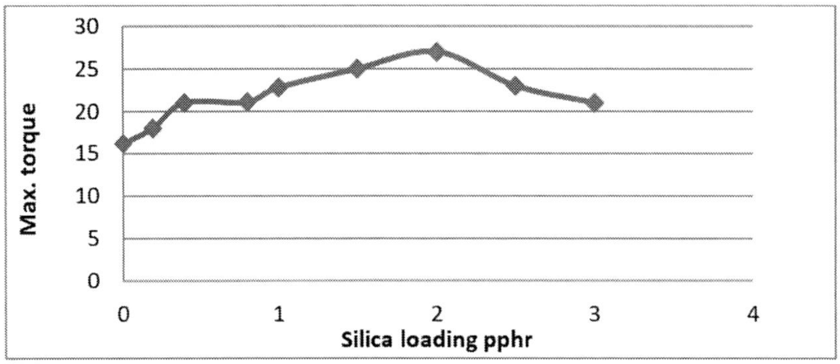

Figure 11. Effect of nano-silica (pphr) on the maximum torque of the NR/SBR composite.

Figures 12 and 13 show that, at low percentages, the modified nano-silica increases scorch and cure times. This is due to the fact that silica makes as a coat on the rubber molecules, which allows the reaction with sulphur partially. However, there is no subsequent effect on these times as the nano particles increase.

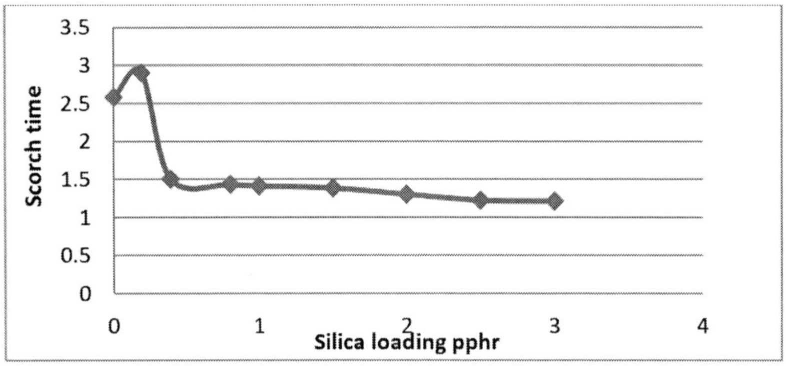

Figure 12. Effect of nano-silica (pphr) on the scorch time of the NR/SBR composite.

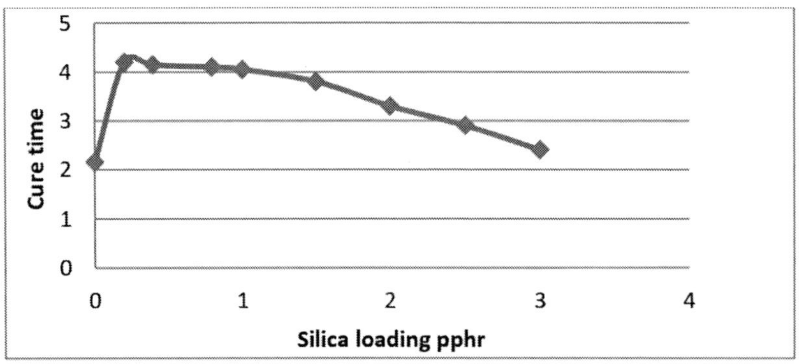

Figure 13. Effect of nano-silica (pphr) on the cure time of the NR/SBR composite.

IOP Conf. Series: Materials Science and Engineering **987** (2020) 012015 doi:10.1088/1757-899X/987/1/012015

5. CONCLUSIONS

These results show that increasing the silica content in a rubber blend of NR and SBR increases the rubber's tensile strength, modulus of elastcity, hardness, compression set and fatigue. Properties such as tear resistance and resilience decrease with increases in the silica. While the results for the physical properties of the rubber show an increase in viscosity, the greater torque, scorch time and cure time decrease when the proportion of SiO_2 in the rubber recipe is increased. The ratio of 0.8 pphr of nano silica achieved the best result in most of the tests. It is possible to take advantage of this new rubber compound in engineering applications that require the aforementioned advantages.

References

[1] J. G. Sommer 1996 *Rubber World*, December, p. **18**.

[2] N. Nakajima 1999 The Science and Practice of Rubber Mixin, Rapra Technology Ltd, pp. **12**.

[3] J. White, S. K. De and K. Naskar 2009 Rubber Technologist's Handbook Volume **2**, Smithers Rapra Technology, United Kingdom.

[4] http://www.energyjustice.net, Rubber Compound, Toyo Tire Talk.

[5] P. A. Ciullo 1996 Industrial Minerals and Their Uses, A Handbook & Formulary Copyright by Noyes Publications .

[6] M. A. B. Mat Junoh 2008 Studies on the Properties of Conductive Particulate Fillers Filled Rubber Composite, *M. Sc. Thesis*, University Sains Malaysia.

[7] M. J. Gregory 1979 Effect of Compounding on the Properties of Natural Rubber Engineering Vulcanizate, Paper Presented To Engineering Data Sheet (EDS) Issued By The Malaysian Rubber Research Association (ED53).

[8] Jennifer Renninger 2000 "Understanding Damping Techniques for Noise and Vibration Control", www.earsc.com.

[9] O. H. Campion, R. P., Ellul, M. D., Finney, R. H., Gent, A. N., Hamed, G. R., Hertz, Jr., D. L., James, F. O., Lake, G. J., Miller, T. S., Scott, K.W., Sheridan, P. M., Sommer, J. G., Stevenson, A., Sueyasu, T.,Thomas, A. G., Wang, C., Yeoh 2001 *Engineering with Rubber*.

[10] B. R. and W. Waddell 2005 in the science and technology of Rubber, in *Elsevier Academic Press*, 3rd ed., B. E. and F. R. E. J. E Mark, Ed. New York, pp. **421**–427.

[11] C. Hepburn 1984 "Plast," *Rubber Int*, vol. **9**, no. 11.

[12] Sun D., Li X., Zhang Y. and Li Y. 2011 Effect of Modified Nano-Silica on the Reinforcement of Styrene Butadiene Rubber Composites, Journal of Macromolecular Science, Part B: Physics Vol. **50**, Issue 9.

[13] Sombatsompop N., Thongsang S., Markpin T., Wimolmala E. 2004 Fly Ash Particles and Precipitated Silica as Fillers inRubbers. I. Untreated Fillers in Natural Rubber and Styrene–Butadiene Rubber Compounds, Wiley InterScience, DOI **10**.1002.

[14] Salimi D., Khorasani S. N., Abadchi M. R. and Veshare S. J. 2009 Optimization of Physico-Mechanical Properties of Silica-Filled NR/SBR Compounds, Advances in Polymer Technology, Wiley InterScience, Vol.**28**, No.4, pp 224–232 .

[15] Abed M. S. 2008 Improving the Properties of the Tire Tread by Adding SiO_2 and Al_2O_3 to SBR and NR Rubbers, *M.Sc. Thesis*, Materials Engineering Department, University of Technology , Baghdad.

[16] S. R. Thomas S. 2010 *Rubber nanocomposites: preparation, properties, and applications*. India: John Wiley & Sons (Asia) Pte Ltd.

[17] Fadil Abbas.H.Al-Husnawi 2014 A Study of the Effect of Zinc oxide on Physical Properties of NR/SBR Blends, Thesis M.Sc., College of Science, Physics Department, University of Kufa,.

[18] M. C. B. Jorgen S. B. 1999 Mechanical behavior of particle filled elastomers, *Rubber Chem. Technol*, vol. **72**, pp. 633–656.

[19] N. Parvin and M. Rahimian 2012 The Characteristics of Alumina Particle Reinforced Pure Al Matrix Composite, *ACTA Phys. Pol. A*, vol. **121**, no. 1, pp. 108–110,.

[20] B. Rodgers and W. Waddell 2005 in The Science and Technology of Rubber, *in The Science and Technology of Rubber*, B. E. and F. R. E. J. E. Mark, Ed. New York: Elsevier Academic Press, pp. 421– 427 .

[21] G. R. Schuur MVD 2005 Influence of chemical crosslinks on the elastic behaviour of segmented block copolymers, *Polymer (Guildf).*, vol. **46**, no. 6862, p. 8.

[22] T. Thomas, E. P. Ayswarya, and E. T. Thachil 2013 Nano Alumina as Reinforcement in Natural Rubber Composites, *Int. J. Innov. Res. Sci. Eng. Technol.*, vol. **21**, no. 6, pp. 2365–2370.

IOP Conf. Series: Materials Science and Engineering **987** (2020) 012016 doi:10.1088/1757-899X/987/1/012016

Optimising the chemical demulsification of water-in-crude oil emulsion using the Taguchi method

Mohammed Hussein[1,*] and Ibtehal Kareem[1]

[1]Department of Chemical Engineering, Engineering College, University of Baghdad, Baghdad, Iraq

[*]Corresponding author: mhjaaa95@gmail.com

Abstract. This study examines the effect of important factors such as the temperature, pH, freshwater, concentration, and type of demulsifier on chemical demulsification (CD) using the bottle test method. Three Iraqi heavy crude oils, each with different properties, were tested. Each of these samples contained different concentrations of salt solution, which need to be removed to improve the properties of the crude oil. Five different types of commercial chemical demulsifiers, namely oxyalkylated copolyether (D1), Polyethylene glycol octaadacyl ether (D2), polyether butoxy ethanol (D3), Ethylene-propylene copolymer (D4), and Cumene 1,2,4-trimethylbenzen (D5), were tested. The Taguchi method was used to design the experiments to obtain optimum separation. The data analysis was carried out using signal-to-noise (S/N) ratio and analysis of variance (ANOVA). The results of the best combination test show that the maximum removal efficiency of CD for the three samples are 94.2%, 95%, and 92%, respectively. In addition, there was a noticeable improvement in the original values of the API from 23.2 to 28.4, 14 to 20.9, and 4.6 to 24.1, respectively. The work plan and the results of this work could be used to improve crude oil processing in the Iraqi oil industry.

1. Introduction

Reservoir pressure drop is one of the most complicated problems in the oil industry, and it requires the use of artificial lift systems for drifting the crude oil from oil wells [1]. Artificial lift systems use electrical submersible pumping (ESP), steam injection, and chemical flooding [2,3]. The use of these techniques results in two main problems: i) increased water and salt content and ii) increased total cost of production and treatment of the oil barrel, with a cost of $10.57 per barrel in Iraq for 2019 [4][5][6]. Flow rates in the production process and continuous shacking form emulsion [7]. Hence, the presence of natural surface active agents (surfactants), such as asphaltenes, resins, organic acid, salts, and clay, stabilise the water droplets in a continuous oil phase, water-in-oil (W/O) emulsion or stabilise the oil droplets in a continuous water phase, oil-in-water (O/W) emulsion [8]. The emulsion is undesirable in the oil industry for the reasons of corrosion damage, increased viscosity, difficulties in oil transporting, decreased American Petroleum Institute (API), and catalyst poisoning [9,10,11]. Therefore, for the operational and economic goals of increasing pipeline usage and decreasing corrosion problems before the refining process, it is important to completely separate water from the oil [12,13]. The separation of water depends on destroying the interfacial film and draining the water or oil droplets; this process is called demulsification [14,15]. The demulsification process includes many types of methods, such as mechanical, electrical, and chemical [16,17]. The chemical demulsification (CD) method is a widespread method, which is indispensable even in combination with other methods, and relies on the use of chemical demulsifiers [18,19].

Chemical demulsifiers are amphiphilic surfactant polymers with interfacial activity more than the indigenous surfactants (natural surfactants) [20,21]. The amphiphilic characteristic enables them to adsorb at the oil/water or water/oil interface and destabilise emulsified droplets [22,23]. The stabilisation of natural surfactants depends on a reduction of interfacial tension (IFT) and the enhancement of emulsification and dispersion droplets [24]. The mechanism of demulsification depends on the substitution of natural surfactants at the interface and changes the gradients of IFT due to the Marangoni effect to promote the separation of oil and water [21,25,26].

There are many studies that have examined the effect of numerous factors on the demulsification process. Biniaz et al. [27] studied the effect of the temperature, pH, freshwater, and concentration of demulsifiers by using response surface methodology (RSM) that depended on central composite design (CCD) to obtain the optimum factor for different types of demulsifiers. The results show that the optimum value of a demulsifier depends on the type of demulsifier; however, in both types, efficiency declined after the optimum value. Freshwater had either a positive or negative effect depending on the solubility of each demulsifier. The demulsification process is proportional to increasing temperature and its effectiveness at the neutral range of pH. Roshan et al. [28] investigated the effect of temperature, demulsifier concentration, pH, and freshwater ratio on dehydration efficiency by using RSM. The results proved that high temperatures and freshwater contribute to increased demulsification efficiency, while the effects of pH and demulsifier centration are dependent on the type of demulsifier. Roostaie et al. [29] examined the effect of treating temperature, agent concentration and type, freshwater dose, and pH on separation efficiency by applying the central composite design (CCD), which is useful in RSM. The results indicate that separation efficiency is proportional to temperature, agent concentration, and freshwater dose, while the optimum value of pH is at a neutral state.

Based on the literature, it can be concluded that the temperature, pH, freshwater, concentration, and type of demulsifier have a critical effect on CD. However, unlike in previous studies, in this study, the Taguchi method was used to design the runs of experiments and to analyse the data. The Taguchi method is a powerful statistical tool for forming the best combination of the most influential factors with minimum time and trials then guessed the contribution for each factor to reach the optimum case [30]. Taguchi can be applied to signal-to-noise (S/N) ratio to predict the performance variability and to analysis of variance (ANOVA) to examine the contribution of each factor [31]. Five different types of demulsifiers were selected as a factor in the work plan to improve CD efficiency at the lowest possible cost and improve the characteristics of crude oil to increase the selling price. Five chemical demulsifiers, namely oxyalkylated copolyether (D1), Polyethylene glycol octaadacyl ether (D2), polyether butoxy ethanol (D3), Ethylene-propylene copolymer (D4), and Cumene 1,2,4-trimethylbenzen (D5), were chosen to be compared using the bottle test method.

2. Experimental methodology and data analysis

2.1. Experimental methodology

In this study, the plan of the experiments was designed based on the Taguchi method, which can be used to determine the optimal combination of factors to increase demulsification efficiency and to evaluate the effect of each individual factor. The steps of conducting the experimental evaluation were i) choosing the CD factors and their levels, ii) designing the experimental runs by using orthogonal array (OA), iii) testing the runs and recording responses, iv) analysing results by using S/N ratio to obtain the optimum levels for each factor and ANOVA to predict the contribution factors, and v) retesting the optimum case to confirm the results.

2.2. The control factors and factor levels

There are five controllable parameters affecting the efficiency of the CD process. The first parameter is the temperature of the crude oil, which must remain constant during the exam time, thus achieved by using a water bath. The water bath was used to regulate the temperature of the samples (model HH-S2/500W, operation range from room temperature to 99 ^oC). Second, the pH of the crude oil was controlled using HCl and NaOH depending on pH meter (Hanna: Model -HI98107, range 0 to 14 pH, 0 to 50 ^oC) [1–3]. Third, freshwater was used to decrease the concentration of salt content in the crude oil

[29]. Pipette HO-1630A (10ml) was used to add the exact amount of water. Fourth, the demulsifier concentration was controlled by a micropipette. The type of micropipette was Biozek mechanical pipette/A3024 (range 5–50 µl). Fifth, five types of commercial demulsifiers with unknown chemical composition were used to study their effects on the process of separating water from crude oil. The types of demulsifiers are referred to by their abbreviations (i.e. D1, D2, D3, D4, and D5). The control factors and their levels are showed in Table 1.

Table 1. Control factors and factor levels used in Taguchi experimental design.

Control Factors	Code	Levels				
		1	2	3	4	5
Temperature ^{0}C	A	30	40	50	60	70
pH value	B	3	5	7	9	11
Freshwater (%)	C	2	4	6	8	10
Concentration of demulsifier (ppm)	D	50	100	150	200	250
Type of demulsifier	E	D1	D2	D3	D4	D5

The temperature of the samples (i.e. 30, 40, 50, 60, and 70 ^{0}C) in the water bath were selected to include the widest range for testing and also to ensure that light hydrocarbons of crude oil would not be depleted during the treatment processes at high temperatures. Kokal [9] showed that the temperature of treatment must remain as low as possible, because high temperatures can cause i) increasing treatment cost, ii) a drop in American Petroleum Institute (API), or iii) scale deposition and corrosion of equipment and pipes. The range of pH included the transfer from acidity to the base case. The freshwater ratio was dependent on salt content. This ratio range was 2–10% in the crude oil sample [32]. Lastly, the concentration and type of demulsifiers were dependent on trial and error [19].

The standard of Taguchi design OA was L25 for Taguchi experimental plan. In this study, the treatment processes take place on crude oil like operational conditions in real plant, and the time spent in each stage was simulated by the bottle test method. The Taguchi experimental plan is shown in Table 2.

Table 2. Taguchi experimental plan.

Run	A	B	C	D	E
1	30	3	2	50	D1
2	40	5	4	100	D2
3	50	7	6	150	D3
4	60	9	8	200	D4
5	70	11	10	250	D5
6	30	5	6	200	D5
7	40	7	8	250	D1
8	50	9	10	50	D2
9	60	11	2	100	D3
10	70	3	4	150	D4

11	30	7	10	100	D4
12	40	9	2	150	D5
13	50	11	4	200	D1
14	60	3	6	250	D2
15	70	5	8	50	D3
16	30	9	4	250	D3
17	40	11	6	50	D4
18	50	3	8	100	D5
19	60	5	10	150	D1
20	70	7	2	200	D2
21	30	11	8	150	D2
22	40	3	10	200	D3
23	50	5	2	250	D4
24	60	7	4	50	D5
25	70	9	6	100	D1

2.3. Materials

Three Iraqi heavy crude oil samples with different water and salt contents and other physical properties were supplied from the Missan oil field. Table 3 shows the characteristics of these samples. The five types of commercial chemical demulsifiers (oxyalkylated copolyether (D1), Polyethylene glycol octaadacyl ether (D2), polyether butoxy ethanol (D3), Ethylene-propylene copolymer (D4), and Cumene 1,2,4-trimethylbenzen (D5)) that were used in this study were provided by the Missan oil company. Sodium hydroxide (NaOH) and hydrochloric acid (HCl) and the organic solvents such as ethanol, acetone, toluene, and xylene were provided by the Chem-Lab NV company.

Table 3. The characteristics of the Iraqi crude oil samples.

Property	Standard	Sample1	Sample2	Sample3
Specific gravity at 15 ºC	IP 235/69	0.9148	0.972	1.04
API gravity at 15 ºC	IP 235/69	23.2	14	4.6
Kinematic viscosity at 40 ºC (cSt)	ASTM D445	24.51	86.25	863.01
Asphaltenes (wt%)	IP 143	6.2	3.28	3.86
Sulphur content (wt%)	ASTM D4294	3.278	1.918	1.342
Salt content (mg/L)	IP 77	9425	87620	184800
BS&W (V%)	ASTM D4007	2	30	60
pH	PH meter	5.4	5.1	5.2

2.4. *Data analysis method*

Water separated during the settling time was recorded after 12 hours. The test tube was a centrifuge tube/T.CO-G. R ADARSH. The total amount of water in the sample of crude oil (W_T) was produced from i) the amount of water in the crude oil (water content (W_C)) and ii) fresh water (W_F) used to wash the crude oil and decrease the concentration of salt in the crude oil as much as possible. In this study, the efficiency of chemical demulsification (i.e. responses) is defined as the amount of water that is separated (W_S) compared to the total water (W_T) in the sample. The efficiency is calculated by using Equation 1.

$$\text{Efficiency\%} = \frac{W_S}{W_T} * 100 = \frac{W_S}{W_C + W_F} * 100 \quad (1)$$

The effect of studied factors and their levels on CD efficiency was predicted by using S/N ratio. The-larger-the-better of S/N ratio was applied to increase CD efficiency [31]. SNL is shown in Equation 2.

$$\text{SN}_\text{L} = -10 \log\left(\frac{1}{\text{n}} \sum_{i=1}^{n} \frac{1}{\text{Y}_i^2}\right) \quad (2)$$

Where n is the number of runs and Y_i is the response of each experiment at the ith run. ANOVA is a statistical tool used to find the significance and contribution of each factor on demulsification efficiency [30].

2.5. *Prepare the samples and bottle test method*

To save time and reduce effort, we examined every five runs at one time and each sample according to the specific experimental plan. For each run, the amount of crude oil tested was $100ml$. Hence, 25 runs had to facilitate $2.5l$ at least. The total amount of crude oil provided was $5l$. The total amount of sample was divided into five parts, each part with a volume of $1l$ to ensure that there was no shortage during the test period and that there was an additional amount of crude oil to examine the optimum case. As mention earlier, HCl and NaOH were used to regulate the pH value of the sample for each one of the five parts (i.e. 3, 5, 7, 9, and 11). According to the work plan, the temperature levels are 30, 40, 50, 60, and 70 °C; therefore, we were able to test all five samples simultaneously as follows: i) preparing the five centrifuge tubes so that they were clean and dry; ii) adding 100ml of sample crude oil for each tube with different pH values; iii) heating the samples for 5 minutes; iv) adding the specific demulsifier dose by using the micropipette; v) shaking the tube to reach homogeneity; vi) adding the specific freshwater by using the pipette; vii) shaking the tube to reach homogeneity again; viii) putting the samples in the water bath; ix) taking out the samples from the water bath after 12 hours and recording the amount of water separated; and xi) washing the tubes to use them again.

3. Result and discussion

3.1. *Effect of the five operating factors on CD efficiency*

Figure 1 shows the change in CD efficiency during the experimental runs at 12 hours. The change in CD efficiency is significant. The highest efficiencies were 87.5% (v/v) in Run 20 of sample 1, 90% (v/v) in Run 19 of sample 2, and 90% (v/v) in Run 11 of sample 3. The lowest efficiencies were 0% (v/v) in Runs 5, 9, 12, 13, 16, 17, and 21 of sample 1, 0% (v/v) in Runs 5, 6, 9, 12, 13, 16, 17, and 21 of sample 2, and 0% (v/v) in Runs 5, 9, 12, 13, 17, and 21 of sample 3.

The aim of this study is to evaluate the effect of each control factor and its contribution to the CD process. Hence, analysis of variance (ANOVA) was used to explain the effect of the operating parameters on CD. The contribution ratio means the amount that the factor influences the CD process within the specified level in Table 2. Figure 2 shows the contribution of the five control factors that were mentioned in Table 2. The first contribution values of temperature were 5% of sample 1, 19% of sample 2, and 14% of sample 3. The second contribution values of pH were 64.96% of sample 1, 36% of sample 2, and 57% of sample 3. The third contribution values of fresh water were 3.46% of sample 1, 5% of sample 2, and 16% of sample 3.

IOP Conf. Series: Materials Science and Engineering **987** (2020) 012016 doi:10.1088/1757-899X/987/1/012016

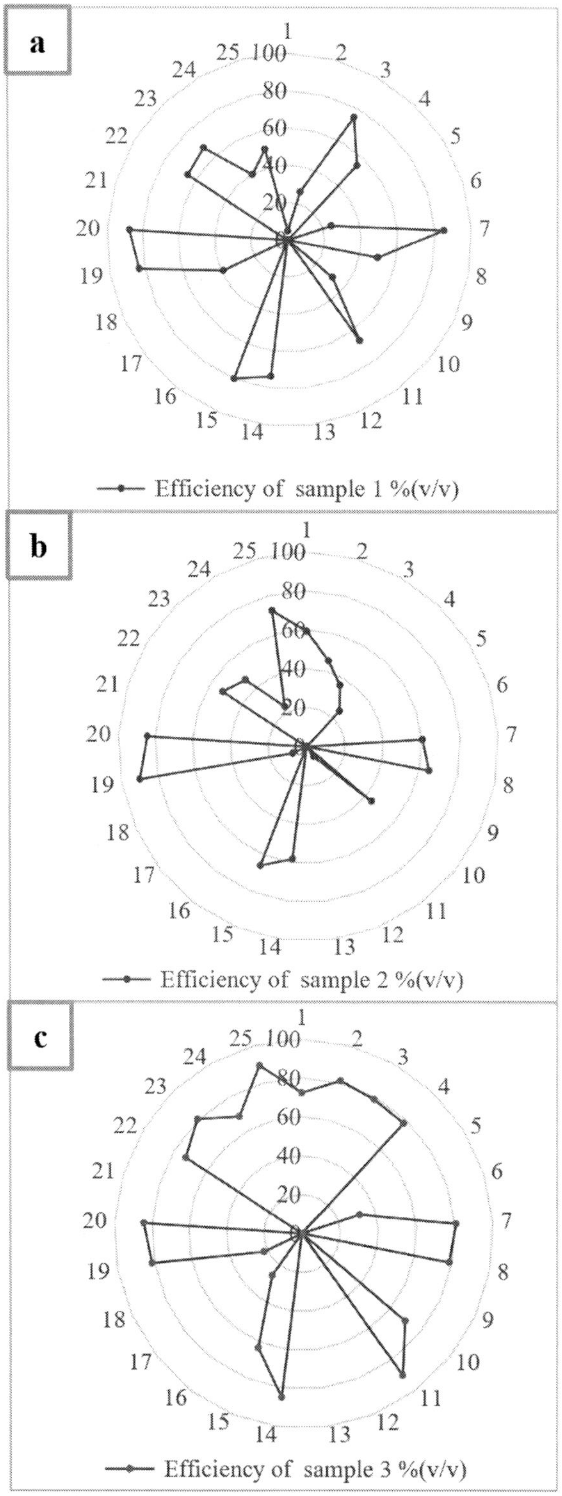

Figure 1. Efficiency of 25 experimental runs: a) sample 1, b) sample 2, and c) sample 3.

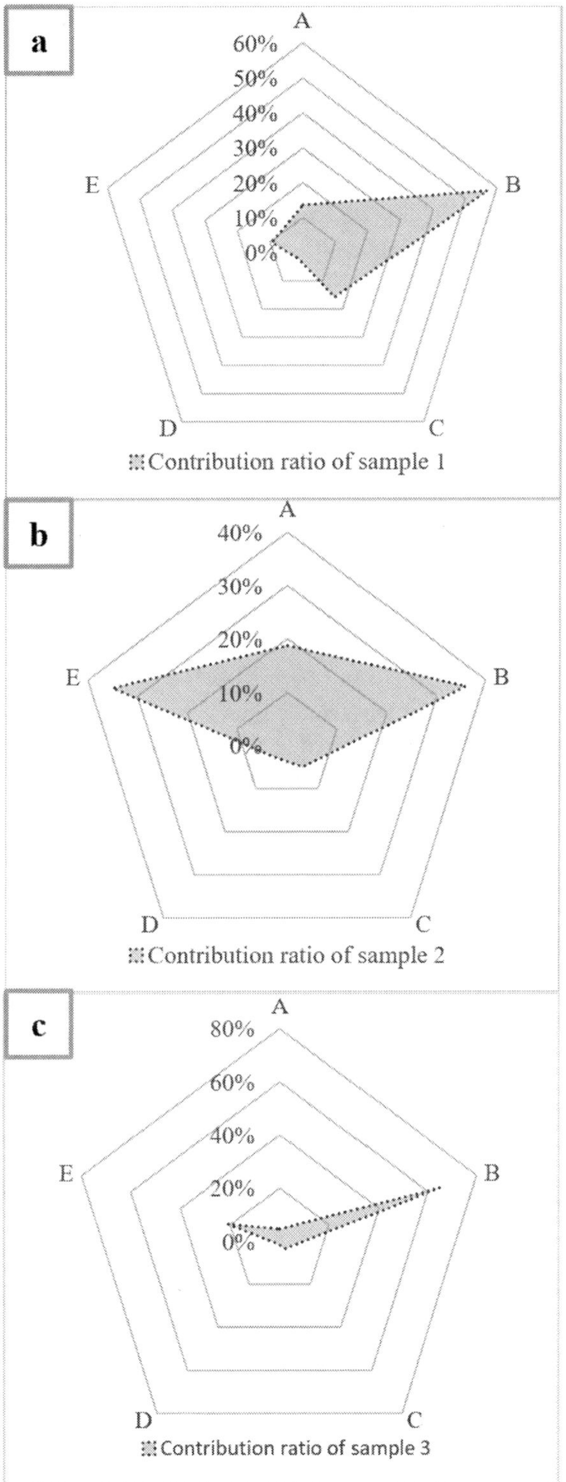

Figure 2. Contribution factors on the CD process: a) sample 1, b) sample 2, and c) sample 3.

The fourth contribution values of concentration of demulsifiers were 3.46% of sample 1, 5% of sample 2, and 16% of sample 3. The fifth contribution values of type of demulsifiers were 20.77% of sample 1, 35% of sample 2, and 9% of sample 3.

The results show that pH has the most significant influence on CD, because the stability strength of the emulsion is dependent on surfactant polarity. The weakest impact factor on CD is the concentration of the demulsifier. This leads to the conclusion that the performance of the demulsifier does not depend on the concentration in the levels chosen in Table 2. This opens the way for future research into the effect of demulsifiers in wider levels. The contribution of factors on samples 2 and 3 (i.e. type of demulsifier, temperature, and freshwater) is arranged in descending order. In contrast, the contribution of sample 1 is freshwater, temperature, and type of demulsifier. The difference between the contribution ratios is due to the difference in the sample content of water, asphaltenes, and sulphur, as mentioned in Table 2.

S/N ratio showed the optimum levels in the CD process and whether the factors were directed towards enhancing or discouraging the CD process. Figure 3 illustrates the S/N ratio analysis and shows the optimum case of the factors' level. First, the optimum case of temperature (A) is 60, 70, and 60 °C for the three samples, respectively. Second, pH (B) is 7 (i.e. neutral) for all samples. Third, freshwater (C) is 10% for all samples. Fourth, concentration of the demulsifier (D) is 200, 50, and 50 ppm, respectively. Fifth, type of demulsifier (E) is D2, D1, and D2, respectively.

Figure 3 also shows that the effect of temperature on CD is reflected in samples 1 and 3; however, in sample 2, the increasing temperature has a positive effect. Although the increasing temperature reduced the viscosity of interfacial film and increased the thermal energy of water droplets, the high temperature inverted the reaction that occurred at the oil–water interface from spontaneous to unspontaneous. This leads to conclusion that the optimum temperature depends on the type of demulsifier.

When pH is neutral, the ionisation of surfactants increases and leads to increased charge density so that the repulsion between molecules is enhanced to produce an increase in the CD process.

However, the efficiency of the CD process is directly proportional to increases in freshwater percent. The increase in CD efficiency is slow, as shown in Figure 3. As illustrated in Figure 2, the freshwater contribution to the CD process is inversely proportional to water content in the crude oil.

The optimum dose for the concentration of the demulsifier is proportional to the surfactants present in the crude oil samples (i.e. asphaltenes and sulphur), as shown in Table 1. Figure 3E shows that the concentration of the demulsifier must be accurate. Therefore, If the quantity is small, then separation efficiency is low; on the other hand, if it exceeds what is required, this results in so-called overdose.

As shown in Figure 3, the best type of demulsifier was dependent on the nature of the demulsifier and indigenous surfactants (i.e. anion, cation, and non-ionic). Consequently, the effectiveness of the demulsifier agent changed with variation in the type of crude oil, and this depends on several factors: i) the nature of the demulsifier and indigenous surfactants (i.e. anion, cation, and non-ionic), ii) solubility of demulsifier, and iii) the dose used. Due to the many parameters affecting the efficiency of the demulsifier, it is difficult to determine which of the previously mentioned factors is more effective, as mentioned in the introduction.

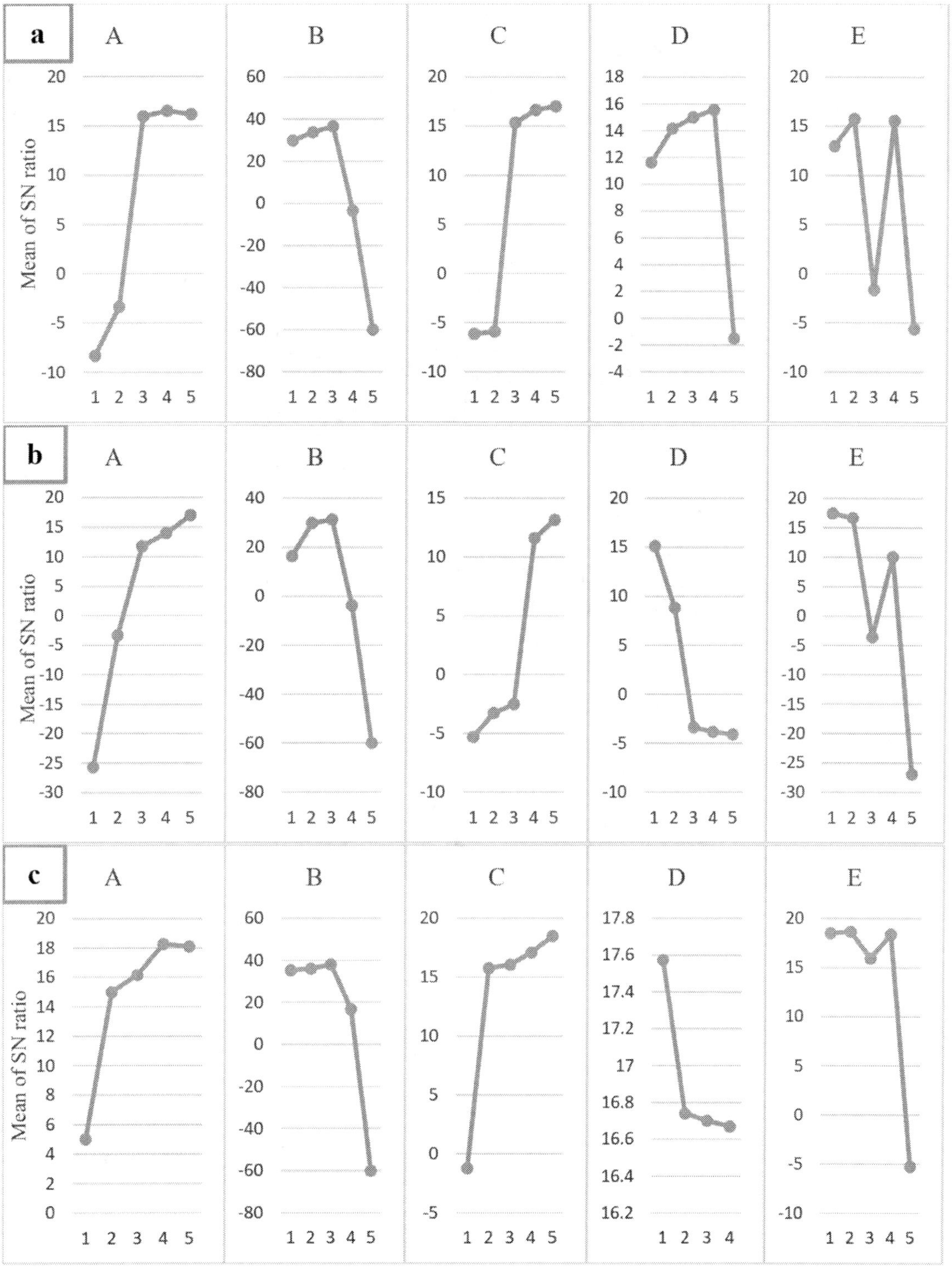

Figure 3. Mean of S/N ratio of the CD process: a) sample 1, b) sample 2, and c) sample 3.

3.2. Retesting the optimum levels

The performance of both the Taguchi and analysis methods were confirmed by testing the optimum levels obtained by using S/N ratio, as shown in Figure 3. Figure 4 illustrates the dehydration ratio and demulsification efficiency, which were calculated by using Equation 1. The results from this equation show that the highest CD efficiencies for the three samples were 94.2%, 95%, and 92%, respectively, as illustrated in the comparison in Figure 5.

It is concluded that the amount of contribution of the Taguchi method to reach the optimum situation with minimal effort and the fastest time and evidence of the method's success so the demulsification efficiency was increased.

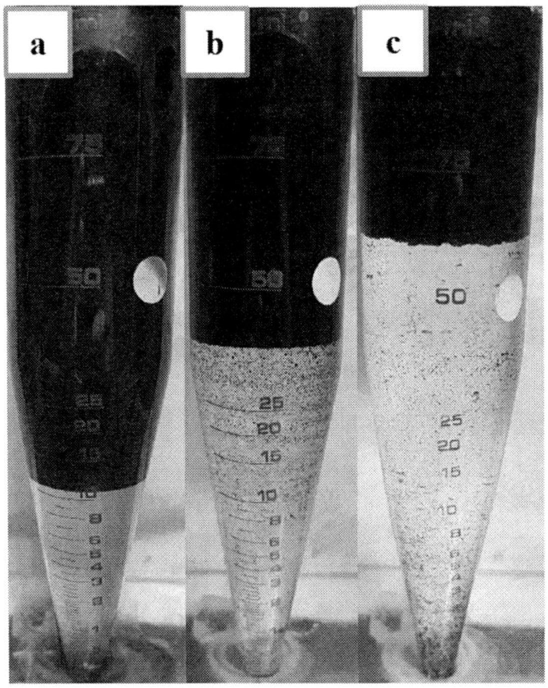

Figure 4. Optimum level results: a) sample 1, b) sample 2, and c) sample 3.

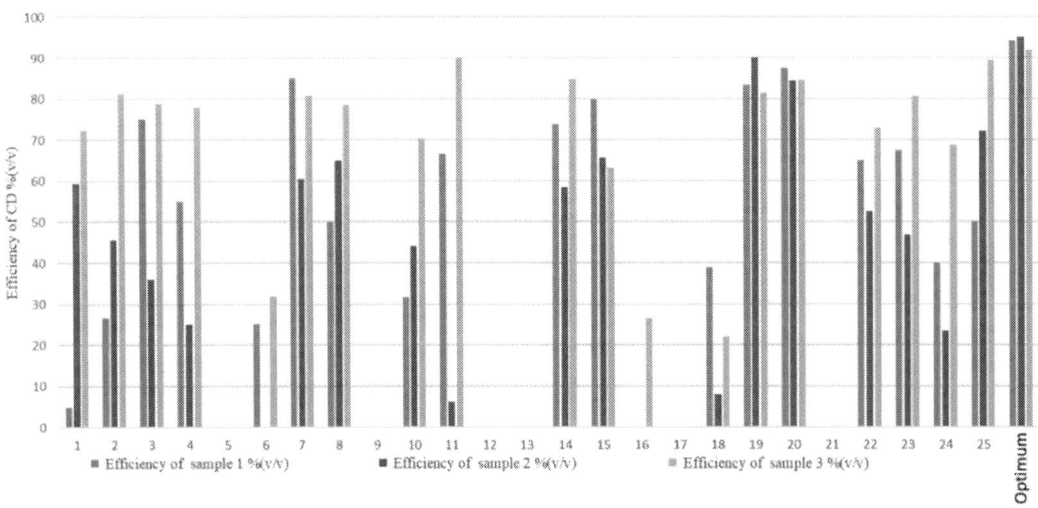

Figure 5. Comparison between efficiency of the three samples with the optimum test.

3.3. Improving the price of crude oil

As mentioned in Table 3, the API values for the three samples were 23.2, 14, and 4.6, respectively. After completing the optimum levels test, a sample was sucked from the three tubes using a pipette, and then the API value was measured depending on an IP 235/69 stander. The results show that the API values for the three samples were then 28.4, 20.9, and 24.1. Therefore, the essential parameters that decrease the value of the API are water and salt content, which lead to a drop crude oil price.

4. Conclusion

In the present paper, the results show that pH was the most influential factor on CD and that the weakest was the concentration of the demulsifier. The contribution ratio of temperature, freshwater, and type of demulsifier depended on the type of the crude oil sample. The plan of the experiments was designed based on the Taguchi method, which can be used to determine the optimal combination of factors to increase demulsification efficiency and to evaluate the effect of each individual factor. The results were analysed using S/N ratio to obtain the optimum levels for each factor and ANOVA to predict the contribution factors. The optimum levels for the three samples were respectively: i) temperature at 60, 70, and 60 °C; ii) pH at 7 for the three samples; iii) freshwater at 10% for the three samples; iv) demulsifier dose at 200, 50, and 50 ppm; and v) type of demulsifier of D2, D1, and D2. The results of the retest run at the optimum levels show improved CD efficiencies for the three samples with efficiencies of 94.2%, 95%, and 92%, respectively. In addition, there was a noticeable improvement in the original values of the API of the three Iraqi heavy crude oils tested in this study by reducing their water and salt content. The API increased from 23.2 to 28.4, 14 to 20.9, and 4.6 to 24.1, respectively.

5. References

[1] "Artificial Lift Systems and The 5 P's - Halliburton." [Online]. Available: https://halliburtonblog.com/artificial-lift-and-the-5-ps/. [Accessed: 21-Apr-2020].
[2] "Artificial lift - PetroWiki." [Online]. Available: https://petrowiki.org/Artificial_lift.
[3] J. G. Speight and J. G. Speight, "Upgrading During Recovery," *Heavy Oil Recover. Upgrad.*, pp. 231–257, Jan. 2019.
[4] "Cost of oil production in the world." [Online]. Available: https://kosatka.media/en/category/neft/analytics/sebestoimost-dobychi-nefti-v-mire. [Accessed: 08-Apr-2020].
[5] F. D. Martin and R. M. Colpitts, "Reservoir Engineering," *Stand. Handb. Pet. Nat. Gas Eng.*, pp. 1–362, Jan. 1996.
[6] S. A. M. Mohammed, "Microwave Assisted Demulsification of Iraqi Crude Oil Emulsions Using Tri-octyl Methyl Ammonium Chloride (TOMAC) Ionic Liquid," *Iraqi J. Chem. Pet. Eng.*, vol. 15, no. 3, pp. 27–35, 2014.
[7] H. Zaman *et al.*, "Effect of pH and salinity on stability and dynamic properties of magnetic composite amphiphilic demulsifier molecules at the oil-water interface," *J. Mol. Liq.*, vol. 290, p. 111186, Sep. 2019.
[8] M. Nikkhah, T. Tohidian, M. R. Rahimpour, and A. Jahanmiri, "Efficient demulsification of water-in-oil emulsion by a novel nano-titania modified chemical demulsifier," *Chem. Eng. Res. Des.*, vol. 94, pp. 164–172, Feb. 2015.
[9] S. L. Kokal, "Crude Oil Emulsions: A State-Of-The-Art Review," *SPE Prod. Facil.*, vol. 20, no. 01, pp. 5–13, 2005.
[10] D. Arla, A. Sinquin, T. Palermo, C. Hurtevent, A. Graciaa, and C. Dicharry, "Influence of pH and Water Content on the Type and Stability of Acidic Crude Oil Emulsions," *Energy & Fuels*, vol. 21, no. 3, pp. 1337–1342, May 2007.
[11] D. P. Ortiz, E. N. Baydak, and H. W. Yarranton, "Effect of surfactants on interfacial films and stability of water-in-oil emulsions stabilized by asphaltenes," *J. Colloid Interface Sci.*, vol. 351, no. 2, pp. 542–555, Nov. 2010.
[12] N. Hazrati, A. A. Miran Beigi, and M. Abdouss, "Demulsification of water in crude oil emulsion using long chain imidazolium ionic liquids and optimization of parameters," *Fuel*, vol. 229, pp.

126–134, 2018.
[13] L. Xia, S. Lu, and G. Cao, "Stability and demulsification of emulsions stabilized by asphaltenes or resins," *J. Colloid Interface Sci.*, vol. 271, no. 2, pp. 504–506, Mar. 2004.
[14] N. Ali *et al.*, "Interfacially active and magnetically responsive composite nanoparticles with raspberry like structure; synthesis and its applications for heavy crude oil/water separation," *Colloids Surfaces A Physicochem. Eng. Asp.*, vol. 472, pp. 38–49, May 2015.
[15] A. M. Al-Sabagh, N. G. Kandile, R. A. El-Ghazawy, and M. R. Noor El-Din, "Synthesis and evaluation of some new demulsifiers based on bisphenols for treating water-in-crude oil emulsions," *Egypt. J. Pet.*, vol. 20, no. 2, pp. 67–77, Jun. 2011.
[16] T. Sun *et al.*, "Influence of Demulsifiers of Different Structures on Interfacial Dilational Properties of an Oil–Water Interface Containing Surface-Active Fractions from Crude Oil," *J. Colloid Interface Sci.*, vol. 255, no. 2, pp. 241–247, Nov. 2002.
[17] V. K. Rajak, I. Singh, A. Kumar, and A. Mandal, "Optimization of separation of oil from oil-in-water emulsion by demulsification using different demulsifiers," *Pet. Sci. Technol.*, vol. 34, no. 11–12, pp. 1026–1032, Jun. 2016.
[18] A. M. Alsabagh, M. E. Hassan, S. E. M. Desouky, N. M. Nasser, E. A. Elsharaky, and M. M. Abdelhamid, "Demulsification of W/O emulsion at petroleum field and reservoir conditions using some demulsifiers based on polyethylene and propylene oxides," *Egypt. J. Pet.*, vol. 25, no. 4, pp. 585–595, Dec. 2016.
[19] J. Wu, Y. Xu, T. Dabros, and H. Hamza, "Effect of Demulsifier Properties on Destabilization of Water-in-Oil Emulsion," *Energy & Fuels*, vol. 17, no. 6, pp. 1554–1559, Nov. 2003.
[20] V. Kedar and S. S. Bhagwat, "Effect of polar head surfactants on the demulsification of crude oil," *Pet. Sci. Technol.*, vol. 36, no. 2, pp. 91–98, Jan. 2018.
[21] L. Zhang, H. Ying, S. Yan, N. Zhan, Y. Guo, and W. Fang, "Hyperbranched poly(amido amine) as an effective demulsifier for oil-in-water emulsions of microdroplets," *Fuel*, vol. 211, pp. 197–205, Jan. 2018.
[22] W. Kang *et al.*, "Demulsification performance, behavior and mechanism of different demulsifiers on the light crude oil emulsions," *Colloids Surfaces A Physicochem. Eng. Asp.*, vol. 545, pp. 197–204, May 2018.
[23] G. Kume, M. Gallotti, and G. Nunes, "Review on Anionic/Cationic Surfactant Mixtures," *J. Surfactants Deterg.*, vol. 11, no. 1, pp. 1–11, 2008.
[24] P.-C. Schorling, D. . Kessel, and I. Rahimian, "Influence of the crude oil resin/asphaltene ratio on the stability of oil/water emulsions," *Colloids Surfaces A Physicochem. Eng. Asp.*, vol. 152, no. 1–2, pp. 95–102, Jul. 1999.
[25] N. Ali *et al.*, "Novel Janus magnetic micro particle synthesis and its applications as a demulsifier for breaking heavy crude oil and water emulsion," *Fuel*, vol. 141, pp. 258–267, Feb. 2015.
[26] M. M. Abdulredha, S. A. Hussain, and L. C. Abdullah, "Optimization of the demulsification of water in oil emulsion via non-ionic surfactant by the response surface methods," *J. Pet. Sci. Eng.*, vol. 184, p. 106463, 2020.
[27] P. Biniaz, M. Farsi, and M. R. Rahimpour, "Demulsification of water in oil emulsion using ionic liquids: Statistical modeling and optimization," *Fuel*, vol. 184, pp. 325–333, Nov. 2016.
[28] N. Roshan, S. Ghader, and M. R. Rahimpour, "Application of the response surface methodology for modeling demulsification of crude oil emulsion using a demulsifier," *J. Dispers. Sci. Technol.*, vol. 39, no. 5, pp. 700–710, May 2018.
[29] T. Roostaie, M. Farsi, and M. R. Rahimpour, "Dehydration of crude oil by cellulose-based demulsifiers: statistical modeling, sensitivity analysis and optimization," *Pet. Sci. Technol.*, vol. 37, no. 20, pp. 2134–2141, Oct. 2019.
[30] A. Al-Jubainawi, Z. Ma, Y. Guo, and L. D. Nghiem, "Effect of regulating main governing factors on the selectivity membranes of electrodialysis used for LiCl liquid desiccant regeneration," *J. Build. Eng.*, vol. 28, p. 101022, Mar. 2020.
[31] B. D. Baloni, Y. Pathak, and S. A. Channiwala, "Centrifugal blower volute optimization based on Taguchi method," *Comput. Fluids*, vol. 112, pp. 72–78, May 2015.
[32] J.-P. W. (Editor), *Petroleum Refining, Volume 2, Separation Processes*. Editions Technip, 2000.

Theoretical Model to Investigate the Heat Transfer Mechanism through a Heat Pipe with Graphene Oxide/Distilled Water as Working Fluid

Jihad Kadhim AbdAli[1], **Adil Abbas Alwan**[1]

[1]Department of Mechanical Engineering, College of Engineering, University of Babylon, Babylon, Iraq
jihad@agre.uoqasim.edu.iq

Abstract. This study investigates heat transfer through a heat pipe by analysing the evaporation and condensation of working fluid (distilled water) in two parts of the heat pipe to find the phase change phenomenon. A computational methodology was developed to simulate the phase change processes and a thermal model was simulated to determine the spatial and temporal temperature profile of the heat pipe and phase change location. The velocity and location of the phase change in the evaporator and condenser sections of the pipe were calculated at each stage, as was the temperature across the domain. The thermal efficiency of the heat pipe was improved by adding graphene oxide nanoparticles at an increase of 15%. When nanoparticles were added to the distilled water, the time of the phase change phenomenon was reduced in the evaporator and condenser sections. The capillary property was enhanced when nanoparticles were deposited in the wall of the evaporator's wick structure. The temperature distribution of the phase change phenomenon was evaluated via the evaporation and condensation processes and the location of the interface line in the two main parts of the heat pipe.

Keywords: Heat transfer mechanism; evaporation; condensation processes; graphene oxide nanoparticles.

1. Introduction

A heat pipe is a passive cooling device; high amounts of heat can be transferred by evaporation and condensation across a range of temperatures in a short period. The pipe consists of a copper container with a capillary wick structure and three main components: evaporator, adiabatic and condenser. The wick structure works as a capillary property and generates the constraint of liquid to reduce velocity and pressure when it is poured into the evaporator. The consistency, structure and size of the wick materials also contribute to controlling the performance of the heat pipe. Different wicks are used depending on the application of the heat pipe, such as dissipation or heat transfer from one heat source to another.

The passive device (heat pipe) can be used to remove heat generated by friction and drag force from a layer of air on the frame of an aircraft or from heat generated by electronic devices. The heat pipe can also be used a heating cycle; for example, pre-heating the air in a steam boiler to improve the thermal efficiency of the boiler. The most recent use of heat pipes to transfer heat is to move geothermal heat from underground to the inside of greenhouses without using traditional means.

The shut-down systems in nuclear reactors are composed of neutron absorbers (rod assembly) where heat decay can be removed using passive safety or containment systems. A flat heat pipe that combines the functions of current control rod assembly and passive cooling systems could avoid future accidents. Solar energy technologies, such as panels and solar water heaters, could use heat pipes as cooling devices to prevent the freezing and backflow of working fluid during the night.

A numerical procedure was developed to create predictive behaviour to analyse the heat transfer mechanism coming into contact with a phase change problem in the evaporation and condensation processes. The plan was executable in a nonlinear limited component programme for mathematical assessment. The outcomes clarified reasonable information within the accessible expository arrangement. Kim et al. investigated the vaporisation and condensation processes when water was used as the working fluid in heat pipe components [1]. Zainal et al. studied the heat transfer mechanism on working fluid temperature and the effect of latent heat on vaporisation and condensation. They studied energy release and the proportion and time taken for a liquid to vaporise when it was put in the evaporator section of a hollow heat pipe. The experiment was conducted in observational conditions to determine and control the charge of the working fluid and the release of energy in a liquid-vapour change phenomenon [2]. Siriwan et al. created a numerical model of a helical oscillating heat pipe to predict the time required to get to steady-state temperature under transient conditions. The temperature of the pipe wall and the heat transfer rate were evaluated [3].

Computer codes were developed for the design and simulation of a high-temperature heat pipe. The codes applied empirical equations, such as modifying the finite element technique. Computational fluid dynamics were applied to study the heat pipe working in different steady and unsteady conditions. Panda et al. analysed the saturated temperature of liquid-vapour interface conditions using a commercial computational fluid dynamic code. A three-dimensional unsteady numerical model was proposed to study the vapour zone, the temperature of the wall, the pressure of vapour and vapour velocity in a screen-mesh wick in a sodium heat pipe [4]. Senthilkumar et al. analysed heat pipe operation characteristics by applying the Taguchi approach. The study included calculating flow properties for the working fluid when it started from laminar, and it predicted the active nodes in the domain [5].

Ding et al. investigated three types of heat sinks using a 180 W high-powered diode lamp using the finite element analysis process. Thermal performance at different working angles was investigated in the optimised heat sinks. The working angles of the lamps greatly impacted the heat sink's thermal performance. The angle of operation was recommended for a heat sink with the best thermal efficiency. The findings played a significant role in the design of high-power diode lamps [6]. Mall evaluated the maximum heat transfer rate using the C++ program for different operating temperatures of a particular heat pipe. The selected fluid was water. The dimension was constant and the temperature was varied; the heat pipe was tested in a horizontal position. The heat transferred through the pipe was approximately 15 times the heat transferred through pure conduction in the heat pipe. Pure conduction occurs when the working fluid absorbs the latent heat and is converted to vapour. This could be considered low heat transfer. In contrast, the overall heat transfer for a large amount of energy from the heat sink to other parts was a lot higher [7].

Hari et al. studied flow specifications in the wick of a flat heat pipe when the working fluid inside the wick structure was subjected to boiling and condensing. The two-phase mixture model was applied.

The laminar boundary-layer flow was steady in the wick structure and was examined in both vertical and horizontal orientations. The governing equations were the partial differential of the boundary layer simplified using transformation. The transformed equations were then solved numerically using an iterative shooting technique. The study confirmed that orientation plays a significant role in the flow and temperature fields [8]. Beautician et al. developed a three-dimensional model to simulate the heat transfer rate of a heat pipe in a transient state. They calculated the temperatures for a domain consisting of a wall, wick and vapour core using the finite element method; numeric equations were simulated. The model attempted to predict the domain's transient temperature, velocity and heat transfer rate profiles. The estimated values were then compared with the experimental results of the studies. The experiment showed that the time needed to achieve steady-state agreed with the model [9].
Rashidian developed a computer simulation using MATLAB in what could be considered a simple tool for modelling and designing heat pipe heat exchangers. The results of the programme were compared to a pilot plant and demonstrated agreement with the experimental data. The programme was capable of figuring out economic evaluations for designing a heat exchanger, such as cash flow, payback and fuel-saving [10]. Chen et al. determined the influence of wick structure parameters using the theory of capillary limitation for the porosity powder diameter and thickness of the wick structure. A miniature heat pipe with sintered dendritic copper powder wicks was fabricated and tested. The maximum heat transfer rate reached over 13 W with an outer diameter of 3 mm and an effective heat pipe length of 20 cm. A simple correlation equation was proposed for the prediction of the maximum heat transfer rate of the sintered miniature heat pipe with dendritic powder [11].
Suh and Par investigated the effects of interfacial shear stress on the thermal performance of a micro-heat pipe with an axial flat-grooved channel. Shear stress was caused by the difference in velocity of the liquid and vapour phases. The heat pipe was tested using ammonia as the working fluid. The results were obtained mainly at the 300 K operating temperature, i.e. liquid and vapour flow passed through trapezoidal grooves [12]. Torii analysed convective heat transfer, adding graphene oxide nanoparticles that were coursed through a low level of cylindrical heat under heat transition conditions. A multi-heat pipe cooling device was studied when filled in various proportions (40%, 60%, 80% and 100%) in steady heat transition conditions. The results found that heat transfer was improved when the nanoparticles were suspended, so the most extreme estimation of the Nusselt number was over twice that of the pure working fluid. The heat pipe incrementally expanded the grouping of graphene oxide nanoparticles in the base working fluid and the greatest heat transfer upgrade yielded 0.075% of the fixed weight [13]. The graphene oxide and water nanofluid demonstrated lower thermal opposition compared with pure water. The ideal thermal obstruction was acquired when filled at 100%, and the heat transfer coefficient of the heat pipe expanded with the fixation of graphene oxide nanoparticles.
Rui et al. theoretically investigated fully developed turbulent heat transfer characteristics of liquid sodium inside a high-temperature heat pipe. A model was built to predict the heat transfer coefficient and was dependent on universal velocity, turbulent eddy diffusivity, linear law and the inter-thermal boundary layer [14].
In this study, the phase change phenomenon was analysed by evaporating and condensing working fluid in two parts of a heat pipe. A computational methodology was developed to simulate the spatial and temporal temperature profile through the heat pipe. The velocity and location of the phase change in the evaporator and condenser sections were calculated. Graphene oxide nanoparticles were added to the base working fluid (distilled water) to improve the thermal efficiency of the heat pipe by reducing the

phase change time and location along the r-z axes. When nanoparticles were deposited in the wall of the evaporator, the capillary property of the wick was enhanced.

2. Mathematical Model

The assumptions of this case study are:

1. The temperature of the adjacent phase is equal to the prescribed constant temperature tube, which is normally the boiling temperature of the working fluid.
2. An energy balance must be satisfied at the interface.
3. The thermophysical properties of each phase are uniform and remain constant.
4. The distinction of density of liquid and vapour does not create a considerable nearby movement of fluid.
5. Heat transfer is controlled by an energy equation.
6. The material is homogenous and isotropic.

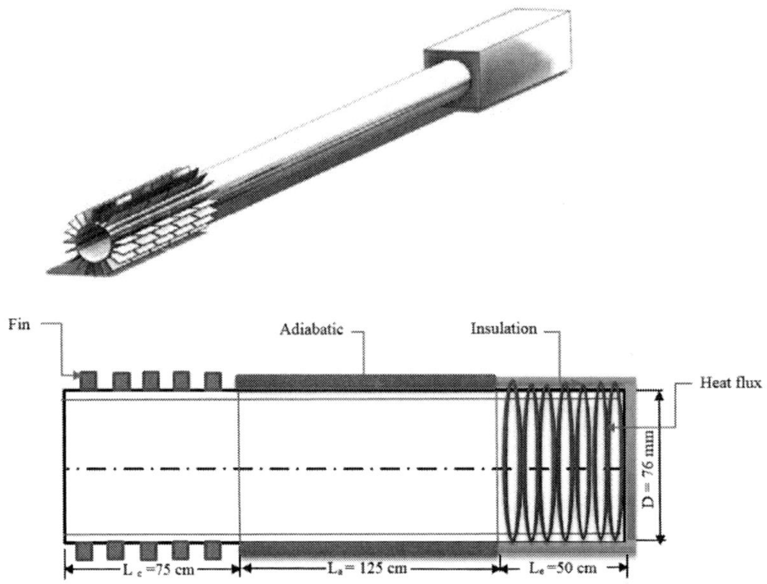

Figure 1 Schematic of the heat pipe

A cylindrical coordinates approach has been used to predict the temperature distribution through the heat pipe and explicate the phase change phenomenon of the working fluid. The working fluid changed from liquid to vapour due to the location of the boiling point in the evaporator and vice versa from a vapour to a liquid in the condenser. This is called a moving boundary problem because of the unexplored moving liquid–vapour interface. The moving boundary problem is nonlinear and rendered more intricate.

2.1 Evaporator section

$$\frac{1}{\alpha}\frac{\partial T}{\partial t}+\frac{v_l}{\alpha_l}\frac{\partial T}{\partial z}=\frac{\partial^2 T}{\partial r^2}+\frac{1}{r}\frac{\partial T}{\partial r}+\frac{1}{r^2}\frac{\partial^2 T}{\partial \theta^2}+\frac{\partial^2 T}{\partial z^2} \quad (1)$$

when $\alpha = \frac{k}{\rho c_p}$.

All of the focuses have domain-limit (a-b-c-d) dimensions, as shown in Figure 2. It has been assumed that a half-domain of the cylinder has been used to design the model because of its symmetrical shape. Numerical simulations have been performed using an explicit technique.

$$\frac{T^n_{i+1,j}-2T^n_{i,j}+T^n_{i-1,j}}{\Delta r^2}+\frac{1}{r_i}\frac{T^n_{i+1,j}-T^n_{i-1,j}}{2\Delta r}+\frac{T^n_{i,j+1}-2T^n_{i,j}+T^n_{i,j-1}}{\Delta z^2}=\frac{1}{\alpha}\frac{T^{n+1}_{i,j}-T^n_{i,j}}{\Delta t}+\frac{v_l}{\alpha_l}\frac{T^n_{i,j+1}-T^n_{i,j-1}}{2\Delta z} \tag{2}$$

The boundary conditions are at z=0 and r=R, $-k\frac{\partial T}{\partial r} = q^{=}$(heat flux) at z=50, T=$T_v$

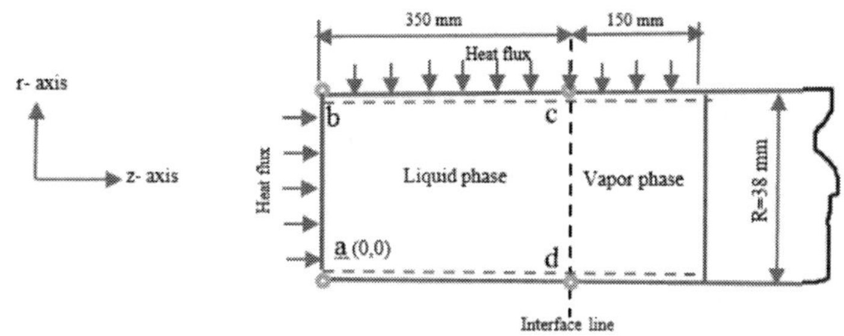

Figure 2 The two-dimensional computational mesh of the evaporator

2.1.1 Temperature distribution through the liquid phase

In the liquid phase, a two-dimensional transient energy equation with polar coordinates allows the temperature distribution in an interface region of limited space to be determined before a change of phase occurs on an interface line, as shown in Figure 2. The interface line is located 35 cm from the left side of the evaporator. The liquid temperature was increased to a saturated temperature (boiling point) at the interface line; this records the vaporisation process. At this time, the remaining fluid has stagnated and its velocity is equal to zero. The computer program, depending on the boundary conditions, has a greater capacity to measure temperature profiles through the r and z plane when the explicit scheme is applied.

1. Temperature at Point a

Point a represents the starting location of the liquid phase in the evaporator. At this point, r and z coordinates are 0, 0, where r=0 and z=0 after applying the energy balance to this boundary node, which is a corner node.

$$T^{n+1}_{i,j} = 4\lambda_1 T^n_{i+1,j} + (1-6\lambda_1)T^n_{i,j} + 2\lambda_1 T^n_{i,j+1} - 2\frac{\lambda_1 q^{=}\Delta r}{K} \tag{3}$$

where $\lambda = \frac{\alpha_l \Delta t}{\Delta r^2} \leq 0.5$ is the Fourier number and $\lambda_1 = \frac{\alpha_l \Delta t}{r_i \Delta r}$.

IOP Conf. Series: Materials Science and Engineering **987** (2020) 012017 doi:10.1088/1757-899X/987/1/012017

2. Temperature distribution through the r-axis

To find the temperature through the r-axis in the liquid phase, an energy balance on the boundary line was applied. For r>0 and z=0, the explicit scheme was constructed as:

$$T_{i,j}^{n+1} = \lambda\, T_{i+1,j}^{n} + \lambda\, T_{i-1,j}^{n} + (1 - 4\lambda_1)T_{i,j}^{n} + 2\lambda_1 T_{i,j+1}^{n} - \frac{2\lambda q^{=}\Delta r}{K} \tag{4}$$

3. Temperature distribution at corner point b

The energy balance was applied for this boundary node, and the temperature was estimated as:

$$T_{i,j}^{n+1} = 2\lambda\, T_{i,j+1}^{n} + (1 - 4\lambda\,)T_{i,j}^{n} + 2\lambda\, T_{i-1,j}^{n} - 4\frac{\lambda q^{=}\Delta r}{K} \tag{5}$$

4. Temperature distribution through line bc in the liquid phase

This is also a boundary line; therefore, the energy balance was needed. Temperature distribution was obtained when r=R and z>0, i.e. i=R to R and j=0 to 349.

$$T_{i,j}^{n+1} = 2\lambda\, T_{i-1,j}^{n} + (1 - 4\lambda\,)T_{i,j}^{n} + \lambda\, T_{i,j+1}^{n} + \lambda\, T_{i,j-1}^{n} - 2\frac{\lambda q^{=}\Delta r}{K} \tag{6}$$

5. Temperature distribution through the z-axis in the liquid phase

The temperature was calculated by applying the two dimensions of the energy equation when r=0 and z>0, i.e. i=1 to 1, j=2 to 349 and $T_{i+1,j} = T_{i-1,j}$.

$$T_{i,j}^{n+1} = 4\lambda\, T_{i+1,j}^{n} + (1 - 6\lambda\,)T_{i,j}^{n} + \lambda\, T_{i,j+1}^{n} + \lambda\, T_{i,j-1}^{n} \tag{7}$$

6. Temperature distribution through the r-z plane in the liquid phase

The temperature was calculated by applying the two dimensions of the energy equation where r>0, i=1 to L-1, z>0 and j=2 to 349.

$$T_{i,j}^{n+1} = (\lambda + \tfrac{\lambda_1}{2})T_{i+1,j}^{n} + (1 - 4\lambda\,)T_{i,j}^{n} + (\lambda - \tfrac{\lambda_1}{2})T_{i-1,j}^{n} + \lambda\, T_{i,j+1}^{n} + \lambda\, T_{i,j-1}^{n} \tag{8}$$

2.1.2 Temperature distribution at the interface line cd

To satisfy the fundamental relationship at an interface, the temperature of adjacent phases should be equal to the boiling temperature at the interface line. $T_v = T_l = T_{boiling}$ at $z = \delta_{(t)}$, where T_v is the vapour temperature, T_l is the liquid temperature and $T_{boiling}$ is the boiling temperature of the liquid.

The energy balance must be satisfied at the interface; therefore, the energy equation at the liquid–vapour interface should be expressed as:

$$k_L \frac{\partial T}{\partial z} - k_v \frac{\partial T}{\partial z} = \rho L_a \frac{\partial \delta(t)}{\partial t} \text{ at } z=\delta(t) \tag{9}$$

$$k_L \frac{\partial T}{\partial z} = \rho L_a \frac{\partial \delta(t)}{\partial t} \text{ at } z=\delta(t) \tag{10}$$

This is a moving boundary condition that supplies the energy equation to evaluate the temperature distribution through a phase change. Alwan has outlined the details of the velocity of the phase change [18].

$$\frac{d\delta(t)}{dt} = V_a = \frac{\text{heat flux}}{[\rho_l L_a + \rho_l c_l (T_V - T_L)]} \tag{11}$$

1. Phase change at point d

This point on the domain surface of the phase change is noted in Figure 2. The interface line between the liquid and vapour phase change is assumed to be propagated through the r-axis. The finite-difference form is written as follows:

$$k_L \frac{T_{i,j+1} - T_{i,j-1}}{2\Delta z} = \rho L_a V_a \tag{12}$$

$$T_{i,j-1} = T_{i,j+1} - \frac{\rho L_a V_a 2\Delta z}{k_L} \tag{13}$$

where r=0 and z=350. The heat equation is then used to analyse the temperature distribution at point d, which is the evaporating point.

$$T_{i,j}^{n+1} = 4\lambda\, T_{i+1,j}^{n} + (1 - 6\lambda_1) T_{i,j}^{n} + 2\lambda\, T_{i,j+1}^{n} - \frac{2\lambda}{K_L} \rho_L L_L V_a \Delta z \tag{14}$$

2. Phase change through the r-axis

For r>0 and z=350, the following equation was used to demonstrate the phase-change mechanism through the r-axis.

$$T_{i,j}^{n+1} = \left(\lambda + \frac{\lambda_1}{2}\right) T_{i+1,j}^{n} + (1 - 4\lambda_1) T_{i,j}^{n} + \left(\lambda - \frac{\lambda_1}{2}\right) T_{i-1,j}^{n} + 2\lambda T_{i,j+1}^{n} - \frac{2\lambda}{K_L} \rho_L L_L V_a \Delta z \tag{15}$$

3. Phase change at point c

This point is a boundary node; therefore, a heat balance must be performed when r=38 and z=350.

$$T_{i,j}^{n+1} = 2\lambda\, T_{i,j+1}^{n} + (1 - 4\lambda) T_{i,j}^{n} + 2\lambda\, T_{i-1,j}^{n} - 2\lambda \frac{q^{=}\Delta z}{K} - \lambda T_{i,j+1}^{n} - \frac{2\lambda}{K_L} \rho_L L_a V_a \Delta z \tag{16}$$

2.1.3 Temperature distribution through the vapour phase in the evaporator

1. Temperature distribution through the z-axis

At the z-axis where z>350 , r=0 and $c = \frac{\Delta t v_a}{\Delta z}$ is the courant number $0 \le c \le 1$.
the following energy equation should be used:

$$T_{i,j}^{n+1} = 4\lambda\, T_{i+1,j}^{n} + (1 - 6\lambda) T_{i,j}^{n} + \left(\lambda - \frac{c}{2}\right) T_{i,j+1}^{n} + \left(\lambda + \frac{c}{2}\right) T_{i,j-1}^{n} \tag{17}$$

2. Temperature distribution in the r-z plane in the vapour phase

$$T_{i,j}^{n+1} = \left(\lambda + \frac{\lambda_1}{2}\right) T_{i+1,j}^{n} + (1 - 4\lambda_3) T_{i,j}^{n} + \left(\lambda - \frac{\lambda_1}{2}\right) T_{i-1,j}^{n} + \left(\lambda - \frac{c}{2}\right) T_{i,j+1}^{n} + \left(\lambda + \frac{c}{2}\right) T_{i,j-1}^{n} \tag{18}$$

2.2 Adiabatic Section

The adiabatic section was designed using suitable dimensions, materials and insulation properties so that the heat pipe would work as a cooling device. The working fluid vapour was passed through this section with no loss of temperature. The vapour temperature of the working fluid was considered the initial value of the temperature in the condenser section because there was no heat transfer between the system and its surroundings in the adiabatic section when it was insulated by a layer of glass wool. The phase change phenomenon did not occur in this section [22].

2.3 Condenser section

The working fluid entered the condenser section as a vapour for 120 min. The vapour changed to liquid at the interface line and phase change occurred at 45 cm from the z-axis. Heat was rejected by the high-fins that were specifically designed for this section. The liquid phase continued to the last node of the condenser's dimensions. The energy equation was applied to predict the temperature distribution for the condenser section in two zones (vapour–liquid), as shown in Figure 3.

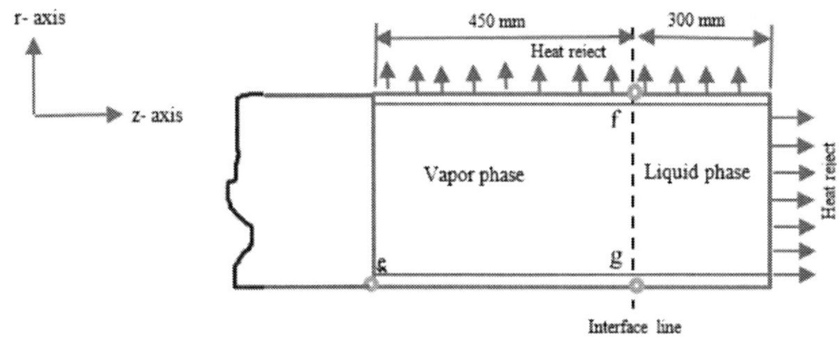

Figure 3 The two-dimensional computational mesh of the condenser

2.3.1 Temperature distribution through the vapour phase in the condenser

The domain area of the vapour phase and the interface line was evaluated at 45 cm from the left, as illustrated in Figure 3. The initial temperature in the condenser was the vapour temperature when it left the evaporator after two hours in the heat pipe.

1. Temperature distribution through the z-axis

Where r=0 and z>0, i.e. i=1 to 1 and j=0 to 449, then the energy equation was used. The equation was represented with an explicit technique when r=0 as :

$$T_{i,j}^{n+1} = 4\lambda\, T_{i+1,j}^{n} + (1 - 6\lambda)T_{i,j}^{n} + (\lambda - \tfrac{c}{2})T_{i,j+1}^{n} + (\lambda + \tfrac{c}{2})T_{i,j-1}^{n} \quad (19)$$

2. Temperature distribution through the r-z plane in the vapour phase

Temperatures of the vapour phase in the r-z plane were calculated at r>0 and z>0, i.e. i=2 to L-1 and j=2 to 449. The energy equation was used to estimate the temperatures through the r-z plane:

$$T_{i,j}^{n+1} = (\lambda + \tfrac{\lambda_1}{2})T_{i+1,j}^{n} + (\lambda - \tfrac{\lambda_1}{2})T_{i-1,j}^{n} + (1 - 4\lambda_3)T_{i,j}^{n} + (\lambda - \tfrac{c}{2})T_{i,j+1}^{n} + (\lambda + \tfrac{c}{2})T_{i,j-1}^{n} \quad (20)$$

2.3.2 Temperature distribution at the interface line fg

The interface energy at z=zδ(t) was represented in the equation 9 , for the condensation process it was expressed as:

$$-k_v\frac{\partial T}{\partial z} = \rho L_a \frac{\partial \delta(t)}{\partial t} \tag{21}$$

$$\frac{d\delta(t)}{dt} = V_* = \frac{\text{heat flux}}{[\rho_v L_a + \rho_v c_v (T_V - T_L)]} \tag{22}$$

1. Temperature distribution at point g

Where r=0 and z=450, the phase change from vapour to liquid began.

$$T_{i,j-1} = T_{i,j+1} + \frac{2}{K_v}\rho_v L_v V_* \Delta z \tag{23}$$

$$T_{i,j}^{n+1} = 4\lambda\, T_{i+1,j}^{n} + (1-6\lambda)T_{i,j}^{n} + 2\lambda\, T_{i,j+1}^{n} + \frac{2\lambda}{K_v}\rho_v L_v V_* \Delta z \tag{24}$$

2. Temperature distribution at line gf

Where r>0 and z=450, for sweep i=2 to L-1 and j=450 to 450.

$$T_{i,j}^{n+1} = (\lambda + \tfrac{\lambda_1}{2})T_{i+1,j}^{n} + (1-4\lambda)T_{i,j}^{n} + \left(\lambda - \tfrac{\lambda_1}{2}\right)T_{i-1,j}^{n} + 2\lambda\, T_{i,j+1}^{n} + \frac{2\lambda}{K_v}\rho_v L_a V_* \Delta z \tag{25}$$

2.3.3 Temperature distribution through the liquid phase in the condenser

It was assumed that the vapour would change to liquid after interface line gf. Then, the temperature distribution of the liquid domain in the condenser was simulated using the energy equation.

1. Temperature distribution through the z-axis in the liquid phase

When r =0, z>0, i.e. i=1 to 1 , j=451 to m-1and $c_1 = \frac{v_*\Delta t}{\Delta z}$,the energy equation was presented explicitly as:

$$T_{i,j}^{n+1} = 4\lambda\, T_{i+1,j}^{n} + (1-6\lambda)T_{i,j}^{n} + \left(\lambda - \tfrac{c_1}{2}\right)T_{i,j+1}^{n} + \left(\lambda + \tfrac{c_1}{2}\right)T_{i,j-1}^{n} \tag{26}$$

2. Temperature distribution through the r-z plane in the liquid phase

When r>0 and z>0, i.e. i=2 to L-1 and j=451 to m-1, the energy equation was formed explicitly as:

$$T_{i,j}^{n+1} = (\lambda + \tfrac{\lambda_1}{2})T_{i+1,j}^{n} + (1-4\lambda)T_{i,j}^{n} + \left(\lambda - \tfrac{\lambda_1}{2}\right)T_{i-1,j}^{n} + \left(\lambda - \tfrac{c_1}{2}\right)T_{i,j+1}^{n} + \left(\lambda + \tfrac{c_1}{2}\right)T_{i,j-1}^{n} \tag{27}$$

3. Results and Discussion

Figure 4 shows the temperature distribution through the liquid phase in the evaporator with respect to the r-z axis. The temperature ranged from 26 ˚C to 58 ˚C during the 60 min when the heat pipe was in operation. Phase change did not occur in the evaporator at this time because of the heat flux, and the latent heat supplied was not enough to cause the liquid to change to vapour. Figure 5 shows that the temperature of the phase change was considered at 96–96.5 ˚C. The phase change position was displayed in the cylindrical enclosure (evaporator) at 180 min when z=35 cm. The latent heat of the distilled water was estimated at 337 KJ/kg in this model, which calculated that the phase change position shifted.The interface line of the phase change varied according to the latent heat supply and the heat current; the speed was equal to 0.0003431 m/s in the step position. The isothermal contour map in Figure 6 identifies the temperature range of the vapour phase in the r-z plane at 180 min. The working fluid acquired more latent heat and energy due to the phase change from liquid to vapour when the time reached two hours. The maximum vapour temperature was 108 °C in the domain at the interface line of the evaporator.

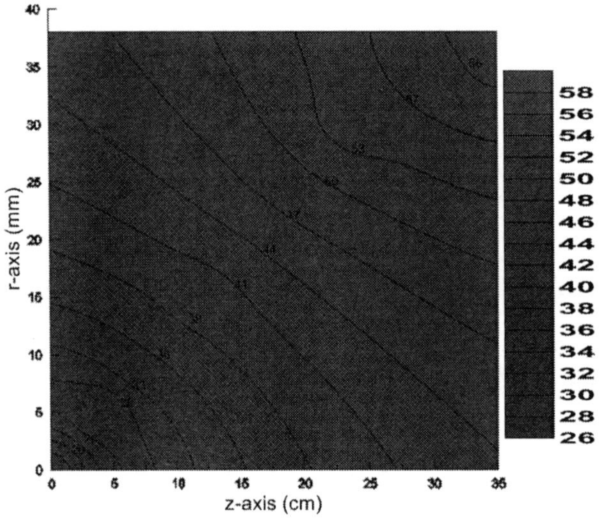

Figure 4 Temperature distribution through the liquid phase at 60 min in the evaporator

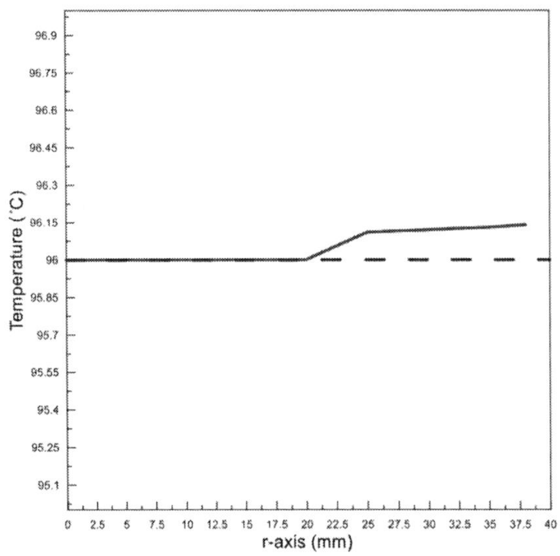

Figure 5 Temperature distribution of the phase change at z=35 cm and 180 min in the evaporator

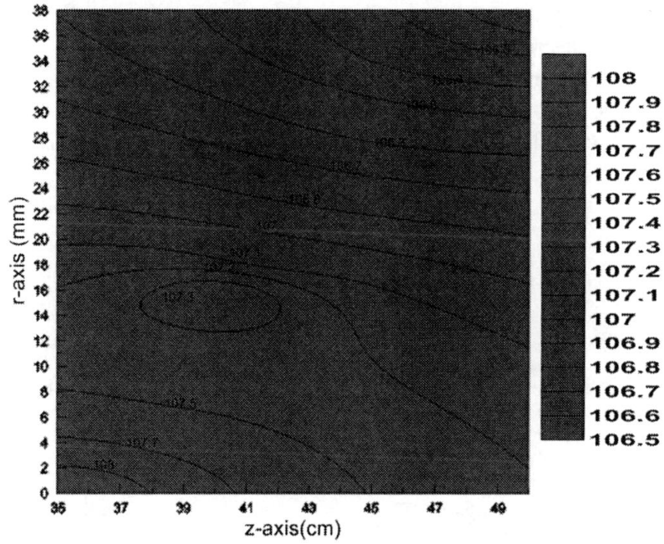

Figure 6 The isothermal contour of the temperature distribution through the r-z plane in the vapour phase at 180 min in the evaporator

Figure 7 indicates the temporal temperature distribution in the area of the cylindrical enclosure in different regions during evaporation. When the temperature increased after three hours, the liquid–vapour interface line shifted to another position, especially in the upper portion. The shift happened because the heat flux was near the upper layer. When the working fluid reached the evaporation temperature of 62 °C at 120 min, the interface line of the liquid–vapour phase was calculated at 35 cm. When the temperature increased up to evaporation point, the working fluid was not immediately vaporising, but it was needed for long enough to convert all of the liquid to vapour by increasing the temperature of the working liquid molecules and latent heat. After the working fluid evaporated, the latent heat increased and the structure of atoms was less stable in lattice positions.

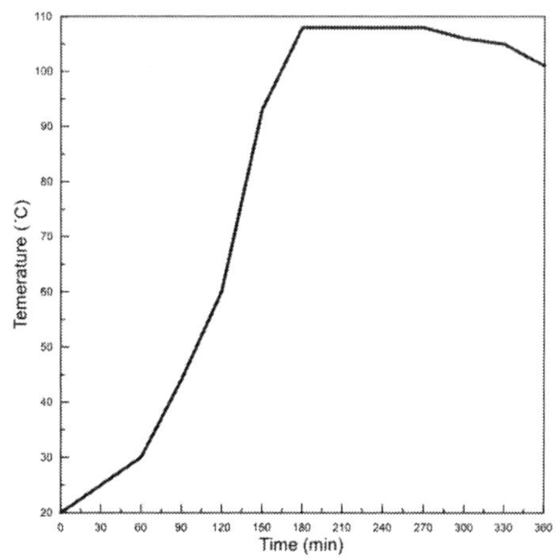

Figure 7 Temporal temperature distribution through the evaporator at r=19 mm and z=50 cm

Figure 8 displays the temperature distribution at the location r=0 at different evaporation times. The temperature response at a positioning layer less than 35 cm showed that the working fluid remained liquid after 60 min, which implies that low-temperature values and latent heat were not enough to cause the phase change phenomenon. This means that the phase change did not occur until 120 min, as shown by the temperature response function at 60 and 120 min. This figure demonstrates that the phase change position rises when the working fluid heated layer moves forward very slowly. The maximum temperature was 108 °C after three hours and reached a steady-state of device operation.

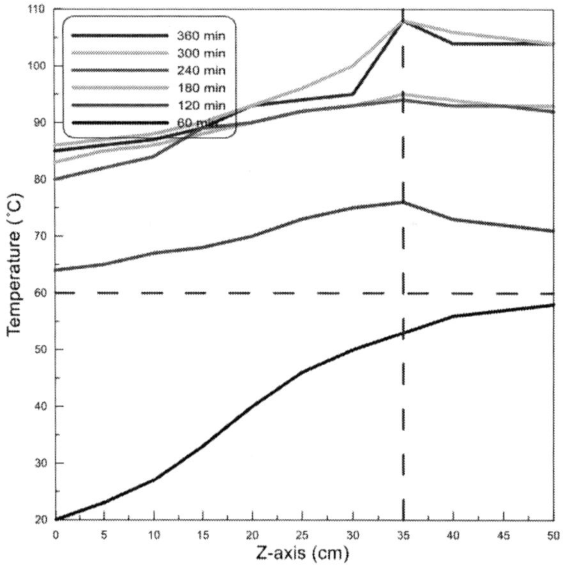

Figure 8 Temperature distribution at the z-axis at r=0 in the evaporator at different periods

Figure 9 explains the isothermal temperature distribution of the phase change phenomenon at 180 min into the evaporation process. It depicts the liquid phase in the first section of the heat pipe and before the interface line at 35 cm. When the working fluid reached an evaporation temperature of 60 °C under a vacuumed pressure of 15 KPa inside the enclosure, the interface line of the phase change was located at 35 cm. During this investigation, the estimation of temperature at an interface location was around 94 °C. The working fluid at this point was needed to acquire latent heat to change to vapour, as referenced by Muhieddine et al. [20].

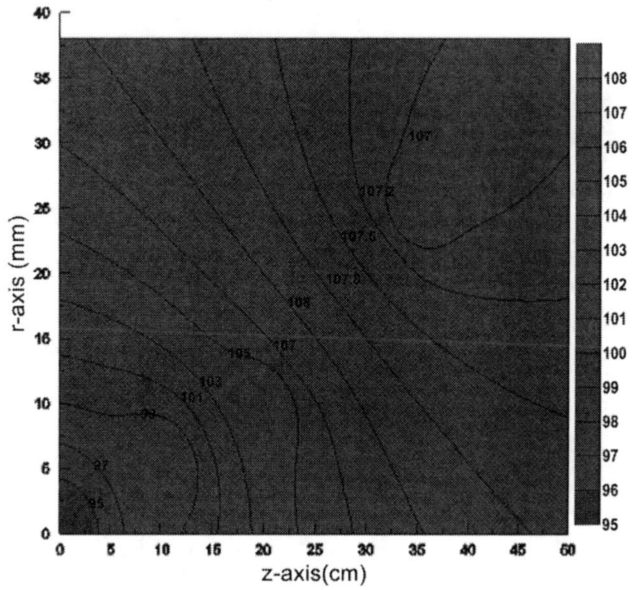

Figure 9 Isothermal contour map of the temperature distribution of evaporation at 180 min

The temperature distribution in the condenser was different from the evaporator. The condensation process started 120 min after vapour entered the condenser at 68 °C. The vapour was lost as latent heat and converted to liquid, and the phase change occurred at 45 cm. Figure 10 shows the isothermal contour map of temperature distribution through the r-z plane in the vapour phase at 120 min in the condenser. Figure 11 indicates the phase change temperature at 180 min when the interface line was at 45 cm.

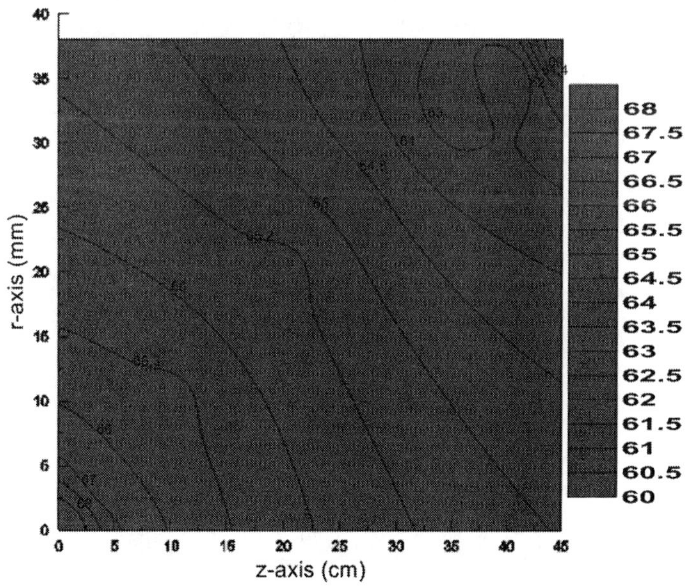

Figure 10 The isothermal contour of temperature distribution through the r-z plane in the vapour phase after 120 min in the condenser

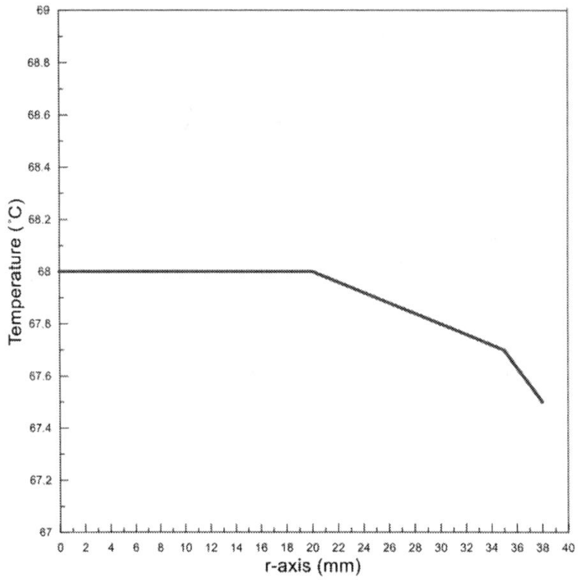

Figure 11 Temperature distribution of the phase change at z=45 cm after 120 min in the condenser

Figure 12 represents the isothermal contour of temperature distribution through the r-z plane for the liquid phase at 120 min in the condenser when the temperature of the liquid decreased to 49.5 °C. During this investigation, the phase change temperature was around 94 °C at 180 min. The condensation and conversion of vapour to liquid did not happen quickly. When it reached the phase change temperature, it continued to expend energy and needed more time for heat transition. The temperature decreased; therefore, the design of the adiabatic and condenser sections of the heat pipe with suitable dimensions and materials to work as a cooling device was effective.

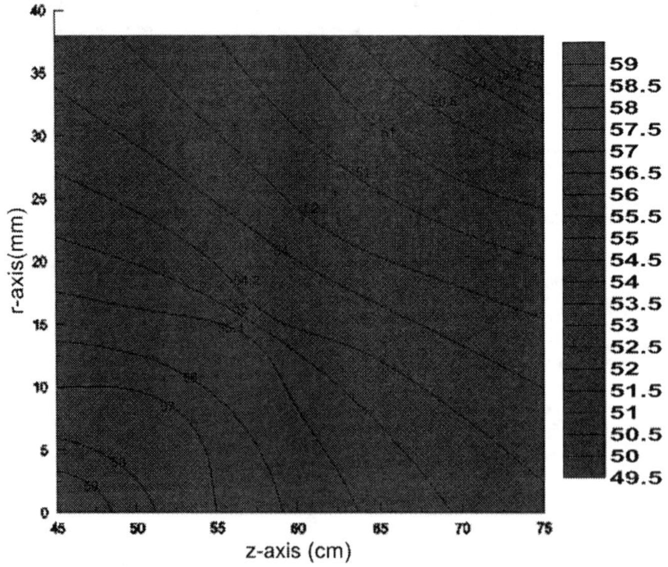

Figure 12 The isothermal contour of temperature distribution through the r-z plane for the liquid phase at 120 min in the condenser

Figure 13 explains the temperature distribution through the z-axis at r=0 of the condenser at different periods. The condenser was operational for over two hours, and the vapour of the working fluid changed to liquid when the interface line for the z-axis was at 45 cm. The temperature values decreased from 68–54 °C over two hours. When the temperature increased in the evaporator, the dissipation of heat increased and operated the cooling process. Figure 14 shows the temporal temperature values through the centre of the condenser; the final nodes at r=0 and z=75 cm were satisfied throughout the working time.

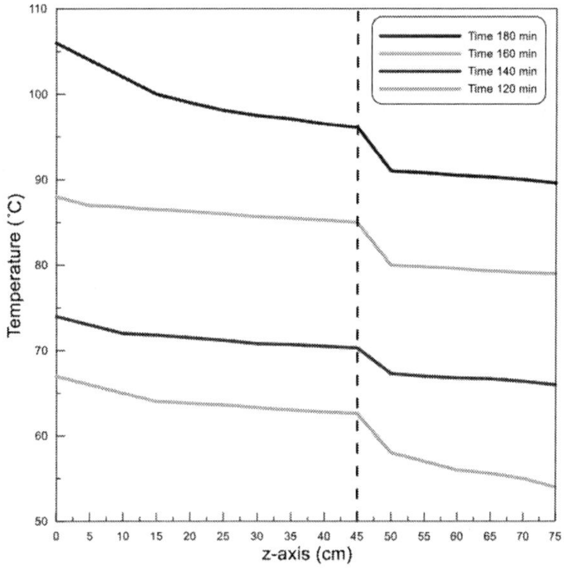

Figure 13 Temperature distribution through the z-axis at r=0 of the condenser section at different periods

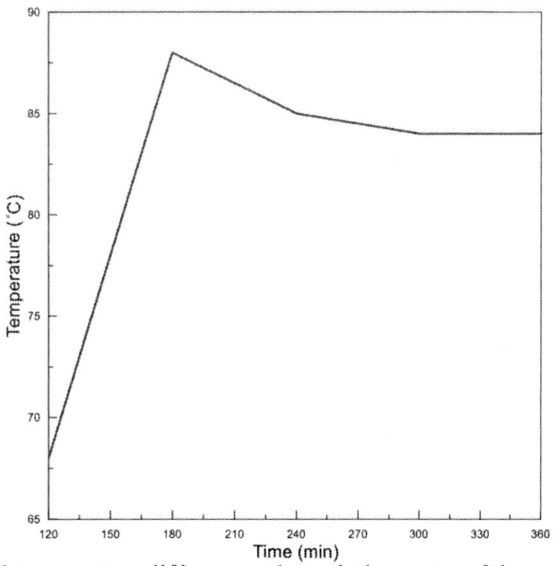

Figure 14 Temporal temperature differences through the centre of the condenser at z=75 cm

IOP Conf. Series: Materials Science and Engineering **987** (2020) 012017 doi:10.1088/1757-899X/987/1/012017

When graphene oxide nanoparticles were added to the base working fluid (distilled water), the thermal performance of the heat pipe improved and the phase change occurred in 60 min at 30 cm on the z-axis of the evaporator. Figure 15 shows that the evaporation temperature was less than 60 °C when graphene nanofluid was used. The weight fraction percentages of graphene nanoparticles were added to the distilled water in a range from 0.01–0.075 wt.%. Phase changes time occurrence and location of its enhanced caused by transfer high amount of heat at least period of operating time. The nanoparticles deposited on the wick near the evaporator wall improved its capillary property and reduced the pressure of the liquid entering the evaporator. The effect of nanoparticles in this analysis is in agreement with results from a study by Zhao [19].

The efficiency of the heat pipe was enhanced by increasing the absorption heat from the heat source caused by the addition of graphene oxide nanoparticles to the base working fluid. This enhanced the dissipation of heat in the condensation process when compared to the work by Alijani [21]. The heat pipe activity increased with the heat load as thermal resistance is lower at higher heat loads. The vapour movement from the evaporator to the condenser was primarily due to density. Figure 16 explains the effect of adding graphene oxide nanoparticles to the base working fluid on the condensation process. This is where heat was transferred at a high temperature from the condenser wall and the drop in temperature from the evaporator to the condenser enabled the phase change to occur in less time at 40 cm on the z-axis.

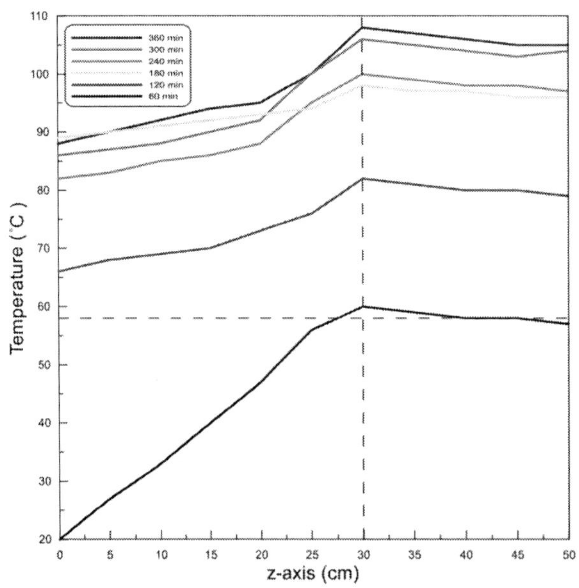

Figure 15 Temperature distribution of the nanofluid during evaporation at different time intervals at r=0 mm

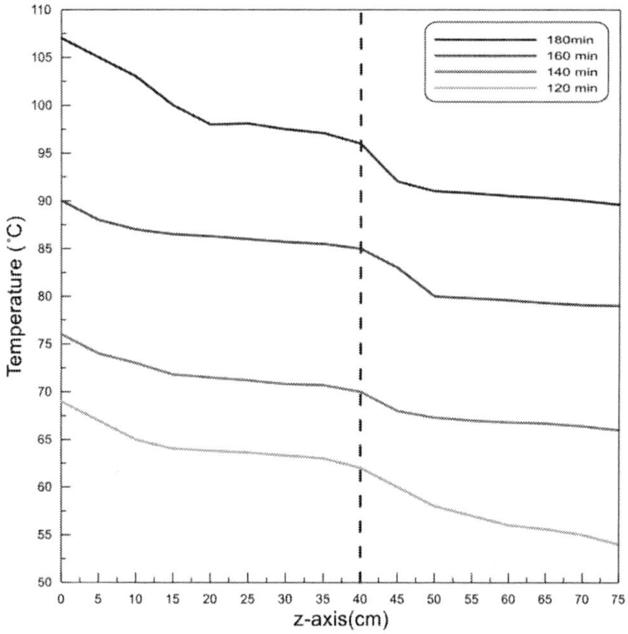

Figure 16 Temperature distribution of the nanofluid during condensation at different time intervals

4. Conclusions

A mathematical model predicted the heat transfer mechanism inside a flat heat pipe by analysing the spatial and temporal temperature during the evaporation and condensation processes. The prediction was dependent on the design of the three main parts of the heat pipe and the position and amount of heat flux in the evaporator section. This model was simulated numerically using an explicit technique, which provided more accurate results and damped oscillation errors than other long-term calculation methods. The new assumption that the boundary layer moved was used to forecast distance in the cluster domain with time to learn the phenomenon of the phase change in the evaporator and condenser.

The interface line of the liquid–vapour phases jumped into an adjacent position by acquiring latent heat in the evaporator and releasing energy in the condenser when vapour converted to liquid. The interface line moved with time in the z-direction. It was proven that the phase change occurred at 35 cm from the left of the evaporator, where heat flux was fixed. The maximum temperature in the evaporator was 381 K and 370 K in the condenser after 180 min from the start of operation. After 180 min, steady-state was reached, the thermodynamic cycle started and the evaporator temperature decreased.

In the condenser, temperature distribution was evaluated through the z-axis and phase change location at 45 cm. This was achieved when the heat pipe had been operational for two hours. Graphene nanoparticles affected the time the phase change occurred as it happened in less than two hours. The phase change location moved to 30 cm from the z-axis of the evaporator and 40 cm from the condenser axis. Heat pipe thermal efficiency was enhanced by 15% because of the addition of the graphene oxide nanoparticles.

Nomenclature

cp_L Specific heat capacity of liquid at constant pressure J/kg.k
cp_v Specific heat capacity of vapour at constant pressure J/kg.k
k_L Thermal conductivity of liquid phase W/m.k
k_v Thermal conductivity of vapour phase W/m.k
k_{nf} Thermal conductivity of nanofluid W/m.k
k_{bf} Thermal conductivity of base fluid W/m.k
Va,V_* Velocities of interface m/s
T Temperature k
t Time s
ρ Density kg/m^3
α_L Thermal diffusivity of liquid phase m^2/s
α_v Thermal diffusivity of vapour phase m^2/s
ρ_L Density of liquid phase kg/m^3
ρ_v Density of vapour phase kg/m^3
$\delta(t)$ Location of liquid-vapour interface m
Δr Increment distance through r-axis mm
Δz Increment distance through z-axis cm
$\lambda_{1,2,3,4}$ Convergent

References

[1] Kim Y, Hossain A and Nakamura Y 2013 *International Journal of Heat and Mass Transfer.* **63** 101–122

[2] Zainal N, Sulaiman S, Shamsudin M and Waini I 2017 *MATEC Web of Conferences.* **97** 01077

[3] Siriwan N, Chompookham T, Ding Y and Rittidech S 2017 *Songklanakarin J.Sci. Technol.* **39(6)** 765–772

[4] Panda K, Basak A and Dulera I 2015 *Thorium Energy Conference.* **15** 12–15

[5] Senthilkumar R, Vaidyanathan S and Sivaraman B 2010 *International Journal of Engineering Science and Technology.* **2(4)** 564–569

[6] Ding X, Li Y, Li Z, Tang Y, Yu B and Yuan D 2014 *Applied Mechanics and Materials.* **602–605** 2713–2716

[7] Mall P 2015 *International Journal of Research.* **2(08)** 708–720

[8] Hari R, Jolly T and Muraleedharan C 2015 *Journal of Computational Multiphase Flows.* **7(4)** 259–270

[9] Boothaisong S, Rittidech S, Chompookham T, Thongmoon M, Ding Y and Li Y 2015 *Advances in Mechanical Engineering.* **7(2)** 1–11

[10] Rashidian B 2015 *Proceedings of the 2nd WSEAS International Conference on Engineering Mechanics Structures and Engineering Geology. Process Engineer.* 114–119

[11] Chen Y, Wu S and Chu C 2001 *Heat and mass transfer.* **37** 611–616

[12] Suh J and Par Y 2003 *Tamkang Journal of Science and Engineering.* **6(4)** 201–206

[13] Torii S 2018 *IOP Conf. Series: Materials Science and Engineering.* **376** 012016

[14] Rui Z, Zhenying W, Zhenhong W, Tingting R and Jiangwu S 2020 *Frontiers in Energy Research.* **8(10)** 1–8

[15] Karlekar B and Desmond R 1982 Engineering Heat Transfer, Book, West-Publishing Company, St.Paul, Minn

[16] Adams J and Rogers D 1973 *Tokyo* Computer-Aided Heat Transfer Analysis, Book, McGraw-Hill, Kogakasha Ltd

[17] Anderson D, Tannehill J and Pletcher R 1984 Computational Fluid Mechanics and Heat Transfer, Book, Second edition McGraw-Hill

[18] Alwan A 2002 Modeling of Thermal History of Solid Materials Processed by Laser PhD Thesis Department of Mechanical Engineering University of Technology

[19] Zhao S, Xu G, Wang N and Zhang X 2018 *Nanomaterials.* **8(2)** 6–16

[20] Muhieddine M and Canot EMarch R 2018 *International Journal on Finite Volumes.* **6(1)** 1–19

[21] Alijani H, Cetin B, Akkus Y and Dursunkaya Z 2017 ICCHM^{2}T2017May 28–June 1.**160221** 1–9

[22] AbdAli J, Alwan A and Hameed R 2019 *Test Engineering and Management.* **83(3)** 13182–13198

IOP Conf. Series: Materials Science and Engineering **987** (2020) 012018 doi:10.1088/1757-899X/987/1/012018

Effect of Alloying Elements on the Erosion–Corrosion Behaviour of Cu-Based Alloys

Noor Sabah Aslan[1], Abdalwahid K. Rajih[1], Ali Hubi Haleem[1]

[1]University of Babylon, nooraslan1985@gmail.com, alialihhhobi@yahoo.com

Abstract. This study used a powder metallurgy (P/M) to prepare a sample of a Cu / 9.5 % Al alloy and add alloying elements (9.5 % Al five % Ni 0.9 % Mn 4.5 % Fe 0.2 % Ge) to it. The Cu / five % Ni alloy has been used for comparison with other research alloys. The powders were mixed for six hours and pressed under 750 MPa. The samples were sintered at 550 °C for two hours, and then the temperature was raised to 950 °C for a period of 180 min with a heating rate of ten °C / min under an inert gas environment with a continuous stream of argon. It was then left to cool inside the furnace. After this, several heat treatments were used. Some tests were carried out, such the scanning electron microscopy (SEM), energy dispersion spectrometer (EDS), X-ray diffraction (XRD), Vickers micro-hardness and the erosion–corrosion test. The alloy Cu / five % Ni showed the least hardness. The addition of Ni, Mn and Fe to the Cu / 9.5 % Al alloy increased the hardness, and these additions caused a relative increase in the resistance to erosion–corrosion. However, it also caused a large increase in the resistance to erosion–corrosion when adding Ge to the alloy Cu / 9.5 % Al / five % Ni 0.9 % Mn 4.5 % Fe 0.2 % Ge when immersed in seawater. Additionally, it demonstrates a high hardness. These improvements were attributed to an enhancement of the properties of alloy-based Cu – 9.5 % Al, which is used in the marine industry.

1. Introduction

Copper, as well as alloys, is a metal that is recognisable by the public. It is easily recognised by its colour, and it has been used since the Bronze Age for tools, arms, jewels and housewares at the reference point of metallurgy. These were commonly used for their thermal properties. Nowadays, it is still the main metal for electrical connections, as well as for wires in devices, appliances and microelectronics. Regarding the mechanical properties, copper is also an fcc-type metal [1]. Aluminium bronzes (Cu / Al) are best known for their high strength and excellent resistance to corrosion. Their resistance to stress-corrosion fatigue exceeds that of austenitic stainless steels. They can be easily welded and can be machined or ground, although good lubrication and cooling are necessary for a fine surface finishing. Common uses involve marine hardware, shafts, pumps and valve components for seawater handling and sour mine water, as it has no oxidising acids or fluids from the manufacturing process. The strong wear resistance of the alloys makes it the perfect choice for heavy-duty sleeve bearings and machine tools [2]. Its properties are strongly dependent on the microstructure, as an understanding of the phase transformations that occur is important to ensure optimum properties. The binary diagram is complicated, but the most important reaction for commercially applicable binary alloys is the eutectoid at 565 °C. Alloys that contain Fe and Ni are also of great commercial significance. Each of these elements have an important effect on the process relationships and affect the alloy properties [3]. Alloys containing less than eight % aluminium are single-phase α alloys. Solidification starts when α dendrites form. The range of freezing is short, but there is no evident segregation, and the alloy solidifies that as a single phase. As β solidifies, it creates an alloy of ten % Al. Therefore, when the temperature falls by approximately 930 ° C, the β, as shown in Figure 1, starts to deposit. This is shown by the binary phase diagram of Cu-Al [2]. When the temperature falls, the amount increases. A Widmanstätten structure is created, which is directly comparable to that found in brass of 60 / 40. The remaining β starts to decompose at 565 °C into a lamellar eutectoid, $\alpha + \gamma_2$. The reaction is regulated by diffusion, and only stops when the alloy is kept at temperatures far below that of the eutectoid for a significant amount of time, as may occur during slow cooling. To allow that reaction to occur, normal casting rates are too rapid. Therefore, β is retained. The retained β transforms the martensite into a hard, acicular form by a method that does not require diffusion. As it is a phase of non-equilibrium, the martensitic changes are not included in the equilibrium diagram. However, it does have important implications for the mechanical properties of the diagram and the alloy. With respect to the mechanical properties and corrosion resistance, the eutectoid α+ γ2 structure has some unacceptable characteristics. Therefore, it is often avoided as it is not practicable for most industrial applications [4].

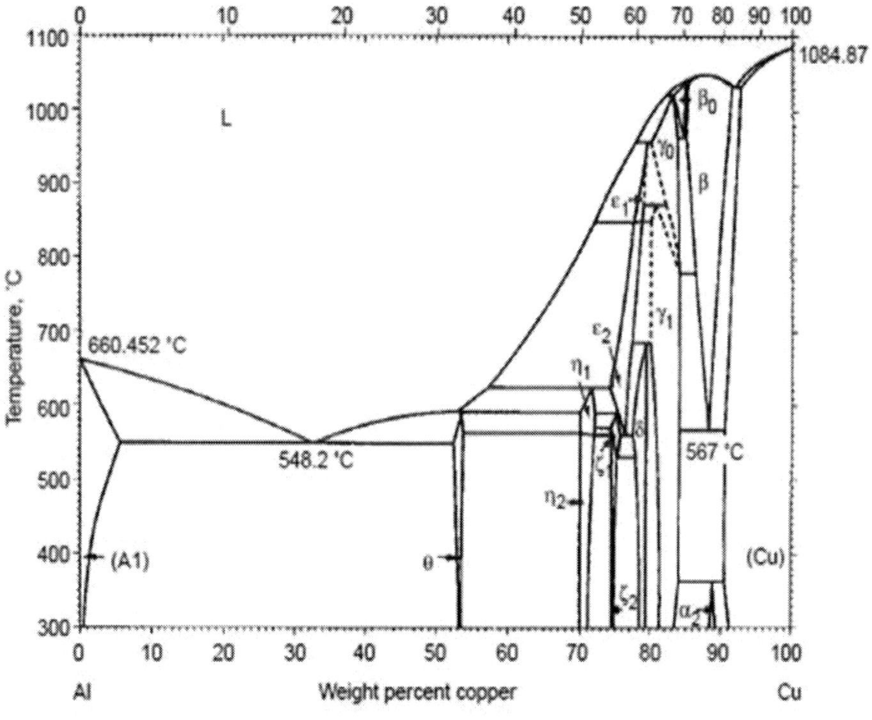

Figure 1. The copper-aluminium (Cu-Al) binary phase diagram [2].

In 2005, Al-Haidary J. T. [5] studied the erosion–corrosion behaviours of α-brass with some additives (0.2 % Al, 0.2 % Te and two % Ge) while subjected to a jet of tap water + 3.5 % NaCl. The results suggested the roles of these elements in improving the erosion–corrosion resistance. In 2009, Faisel et al. [6] prepared alloys (Cu – 9.5 % Al) that were used in marine propellers using the powder metallurgy technique. They added alloying elements to the base alloy (five % Ni / one % Mn / four % Fe / three % Si) and carried out several tests, such as hardness and corrosion / erosion tests. The results showed that the alloying elements increase the corrosion / erosion when added to Cu 9.5 % Al. The improvement ratio was 488.9 % when the heat treatment was done, and the hardness value was 185 HV. In 2014, W.A. Badawya et al. [7] studied adding Al, Zn and Ni to copper as a single alloying element in binary alloys, such as bronzes and brasses. This was to research the effect of mixing two alloying elements with Cu, as well as to investigate the electro-chemical behaviour of these alloys in solutions consisting of chloride. Cu- / 10.0 Al / 10.0 Ni, Cu / 10.0 Al / 10.0 Zn and Cu / 10.0 Ni / 10.0 Zn were used, and these alloys' surfaces were examined using the EDAX and SEM analyses. The surface analysis showed the participation of the various alloying elements in the passive film according to the alloy constituents. It also showed that the chloride ions were deeply permeated in the alloy. In 2019, Vu Anh Tuan et al. [8] studied the effect of Ni on the mechanical and microstructure of the Cu alloy. This study shows the results of nickel components on the mechanical and microstructure properties of the Al / Cu alloys. With the heat treatment process, it was heated to 850 °C, quenched in water, and then tempered at 350 °C for two hours. As the sample had added Ni, it had phases of nickel with aluminium and iron. These phases made the mechanical properties of the sample better than the sample without nickel. In the samples that had nickel added, the hardness was higher than those without Ni, as it was 10.92 HRB. The results of abrasion showed that the Ni alloyed samples presented better than those without Ni. This study aims to produce an alloy of Cu / 9.5 Al using powder technology and other alloys by adding alloying elements to study the impact of seawater for use in marine industries.

2. Experimental Programme

2.1. *Preparation of the Samples' Stages*

The samples were prepared using the powders shown in Table 1, where the particle size of the powders, the purity and source are shown as follows:

Table 1. Powders used in the samples.

Powder	Average particle size (μm)	Purity %	Source
Copper	5.03	99.97	Sky, Sipring Nanomaterials, Inc / USA
Aluminium	20.07	99.913	Sky, Sipring Nanomaterials, Inc / USA
Nickel	4.736	99.87	Sky, Sipring Nanomaterials, Inc / USA
Iron	17.74	99.926	Sky, Sipring Nanomaterials, Inc / USA
Manganese	29.49	99.882	Sky, Sipring Nanomaterials, Inc / USA
Germanium	3.782	99.938	BDH Chemicals Ltd / England

2.1.1. Preparation of the Blended Mixtures

Using a small amount of acetone (ten ml per 30g of the mixture) to avoid agglomeration of the mineral powders and oxidation problems, the elementary powders were blended using the electric roll mixer. The maximum rotating speed was 90 rpm, while the total mixing time was approximately six hours. The mixtures included Cu + five % Ni and Cu + 9.5 % Al, and Ni, Mn, Fe and Ge were added to the Cu + 9.5 % Al mixture with different weight percentages (0.9 %, 4.5 % and 0.2 %) to explain the influence of alloying elements on the hardness and erosion–corrosion resistance of the prepared alloys. The prepared alloys that were used in this study are shown in Table 2.

IOP Conf. Series: Materials Science and Engineering **987** (2020) 012018 doi:10.1088/1757-899X/987/1/012018

Table 2. Mixtures and their compositions, as prepared by this study.

Alloy No.	Composition (weight parent of powders)
A	Cu + 5% Ni
B	Cu + 9.5% Al
C	Cu + 9.5 % Al + 5 % Ni + 0.9 % Mn + 4.5 % Fe
D	Cu + 9.5 % Al + 5 % Ni + 0.9 % Mn + 4.5 % Fe + 0.2 % Ge

2.1.2. Compaction of Powders

The hydraulic press for uniaxial pressing was used to compact the green samples. The cold uniaxial pressing cylindrical die was used to prepare samples with dimensions of 20 mm diameter and eight mm height for use in the erosion–corrosion test. The steel die interior walls were graphite lubricated. Exactly 750 MPa compression pressure was used to determine the optimum compression pressure to give a lower green porosity and higher green density.

2.1.3. Sintering

After the compaction procedure, the green compacts of all samples were placed in a furnace (type VBF-1200X-H8-USA) under an inert gas environment with a continuous stream of argon. The pressure of the gas inside the furnace was controlled by a graduated gage from one bar to 25 bar. This was done to isolate the samples from oxygen to prevent oxidation at elevated temperatures during the sintering process. This used the thermal cycle shown in Figure 2 and included the following steps:

1- From room temperature, the heating green compacts up to 550 °C with a heating rate of ten °C / min.

2-The duration is 120 min.

3- The temperature of the furnace is increased to 950 °C with the same heating rate.

4- The duration is 180 min.

5- The samples are left to slow cool in the furnace.

The sintering of all samples was successful with low porosity. No distortions occurred in the samples' shapes, and no fusion in any elements occurred. This is due to the high purity of the powders that were used, as well as the accurate thermal cycle [9].

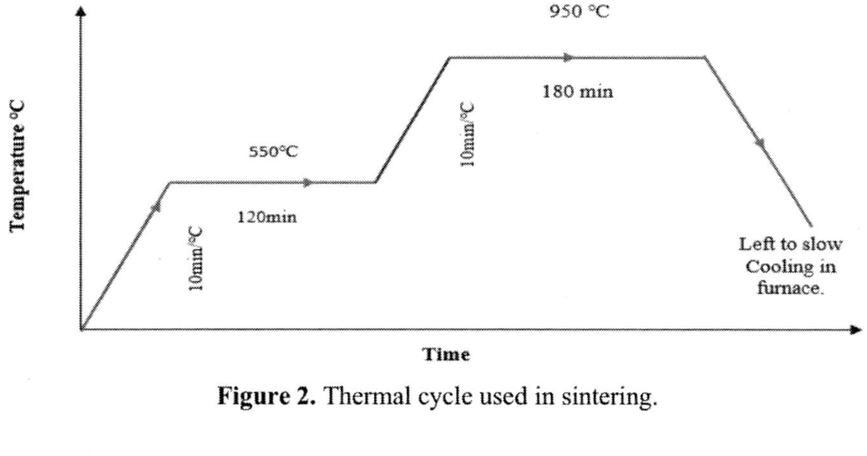

Figure 2. Thermal cycle used in sintering.

2.1.4 Heat Treatment After Sintering

Several heat treatments have been used in this study, including solution heat treatment (betatising) and aging heat treatment (post-quench). The details of this treatment are provided in the following paragraphs:

2.1.4.1. Solution Heat Treatment (Betatising)

After sintering, the samples are solution treated by heating the samples up to 850 °C for 60 min. Next, the samples (B, C, D) are rapidly quenched into ice water at three °C–six °C to avoid decomposition from the β phase into the equilibrium phases, including $\alpha + \gamma_2$, which is the eutectoid reaction at 565 °C for aluminium bronze in the alloy phase diagram. The solution heat treatment thermal cycle is shown in Figure 3.

2.1.4.2. Aging Heat Treatment

Aging heat treatment was done after the quenching heat treatment to age several intermetallic compounds, such as γ_2, in the microstructure to modify their properties. The aging heat treatment temperature was 200 °C for 30 hrs for the prepared samples (B, C, D), followed by cooling in iced water so as to not begin the decomposition phone during slow cooling [9]. Increasing the ageing time leads to an increase in the γ_2 precipitates size and a loss of coherency [10]. The thermal cycle is shown in Figure 4.

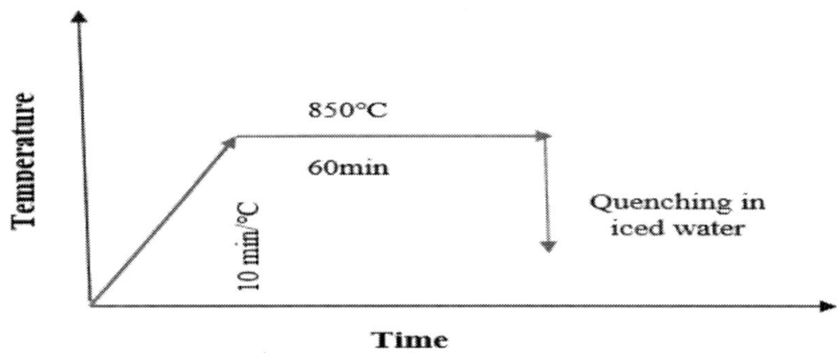

Figure 3. Solution heat treatment (betatising) thermal cycle.

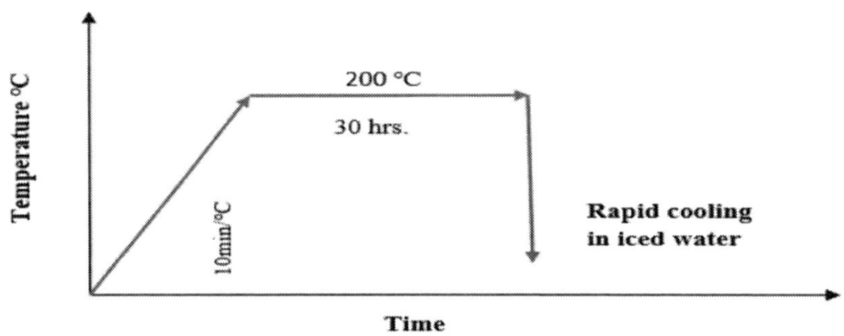

Figure 4. Thermal cycle of aging heat treatment.

2.2. Samples Tests

2.2.1. Chemical Compositions

The chemical composition examination is shown in Table 3 below. It analyses the chemical compositions and concentrations of impurities for all samples.

Table 3. Chemical compositions for all samples with impurities in the concentration.

Alloy Code	Cu %	Al %	Ni %	Fe %	Mn %	Ge %	Mg %	Mo %	Ti %	W %	Co %	Si %
A	Bal.	0.0755	4.554	0.0537	0.0434	---	0.034	0.025	0.0130	0.015	0.0313	0.016
B	Bal.	9.1308	0.0888	0.0609	0.0406	---	0.035	0.05	0.0111	---	0.0318	0.015
C	Bal.	9.1356	5.762	4.367	1.311	---	0.037	0.045	0.0155	0.015	0.0295	---
D	Bal.	9.2546	6.009	4.864	1.999	0.179	0.035	0.0621	0.0110	---	0.0176	0.013

2.2.2. Scanning Electron Microscopy SEM and EDS analyses

The SEM observation was used to detect the microstructure of the etched samples (sample A after sintering and samples B, C and D after the aging process). The images were captured with magnifications 200X, 208X, 180X and 20 KV. Wet grinding was done on the samples that were prepared using 320, 400, 600, 800, 1000, 1200, 1500, 2000, 2500, 3000 and 5000 grit silicon carbide papers with a grinding wheel machine. They were then polished using a diamond with one to three μm as the particle size and polishing paper. After that, they were cleaned with water and dried using hot air with an etching solution of 2.5 g $FeCl_3$ $.6H_2O$ + 48 ml CH_3OH + ten ml HCl ferric chloride acid / methanol / hydrochloric acid [11] for approximately ten sec etching for the samples. Then, the samples were ready for the scanning electron microscope observation. The SEM examination was done in the University of Technology Production Engineering and Metallurgy.

2.2.3. X-Ray Diffraction Analysis (XRD)

The phase's analysis of the prepared samples was based on the X-ray diffraction technique and microstructures examination using an optical microscope with different magnifications. The test was done after the aging treatment was carried out on all samples (A, B–H) to see the effect of heat treatment. The X-ray diffraction tests were carried out at the Faculty of Materials Engineering and in the General Company for Engineering Inspection and Qualification in Baghdad in Iraq. The Mini flex2 device was held at angles of (2θ) = ten–90 deg. with a step (0.02 deg.) This gave more precise details. In addition, the test was done for all samples after the erosion–corrosion test. The X-ray diffraction tests were carried out at the Iran / Mashhad / Tavous Research institute POW3040/60 with XPertSoftware type Philips / Holland.

IOP Conf. Series: Materials Science and Engineering **987** (2020) 012018 doi:10.1088/1757-899X/987/1/012018

2.2.4. Micro-Hardness Test

The Vickers micro-hardness test was done for all samples with a load of 500 g and holding time of 20 sec [9], along with X200 as the magnification for the light optical microscope that was attached using the Vickers instrument. Four readings were taken for each sample because there could have been pores that were located on the sample's surface that would affect the hardness. The measurements were done automatically and taken directly from the digital screen of the device.

2.2.5. Erosion–Corrosion Test

Erosion is known as a mechanical wear process, i.e. the loss of materials from solid surfaces by the impacting or impingement of liquid, solids or gases. The erosion–corrosion testing device was manufactured and designed during this work according to (G 73) ASTM [12], as shown in Figure 5 .The erosion–corrosion device consists of the chamber and comprises a cylindrical plastic tank. The dimensions of the tank are 50 cm in height and 30 cm in diameter. The various media is directly supplied at high pressures, and the vertical position of the sample's face through the jet nozzle (d = 0.75 mm) uses a pump to suck the different media from the plastic tank chamber with $^3/_4$horse power. From Teflon, the single-phase electric motor (chemical pump) was made, so the distance between the sample and nozzle was ten mm. The chemical pump is resistant to the chemical solution and slurry. The valves and pipe joints that are connected to the chamber with the chemical pump are made from PVC plastic.

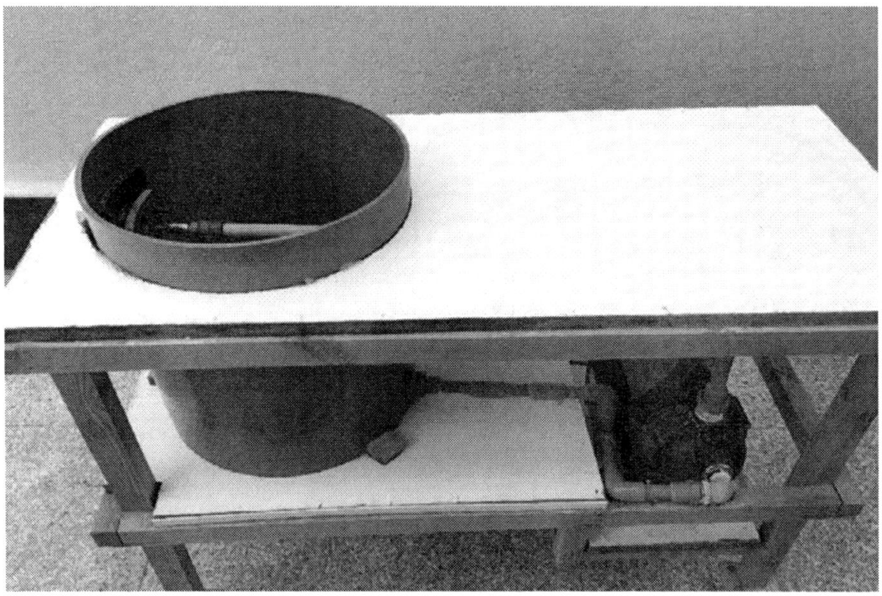

Figure 5. The erosion–corrosion device.

3. Results and Discussion

An SEM observation was done for alloy A after sintering at 550 °C for 120 minutes and 950 °C for 180 minutes. It was also done for alloys B, C and D after the aging heat treatment, where the samples were heated up to 200 °C for approximately 30 hours, and then rapidly quenched in iced water. According to Figures 6, 7, 8 and 9, sample A contains the α phase, while in samples B, C and D, the phases are identified as the α phase and contain the intermetallic compound γ_2. This conclusion was also corroborated by an

XRD analysis. The elemental compositions of these samples were obtained using SEM–EDS. According to Figure 4, the elemental compositions in sample A are Cu and Ni. Figures 5, 6 and 7 show the alloy composition in three samples that are made of the bronze alloy. The main alloying elements of all samples are Cu and Al. Nickel is detected as the main concentration in samples C and D. Manganese is considered to be an alloying addition in samples C and D, while iron is present in the elemental compositions for samples C and D as an alloying addition. However, C and O are considered impurities in the alloy composition. On the other hand, low amounts of Ge in sample D are absent in the elemental compositions of this sample, which may be due to the limitations of EDS as an analytical tool. However, a clear improvement in the resistance to erosion–corrosion for this sample is evidence of germanium.

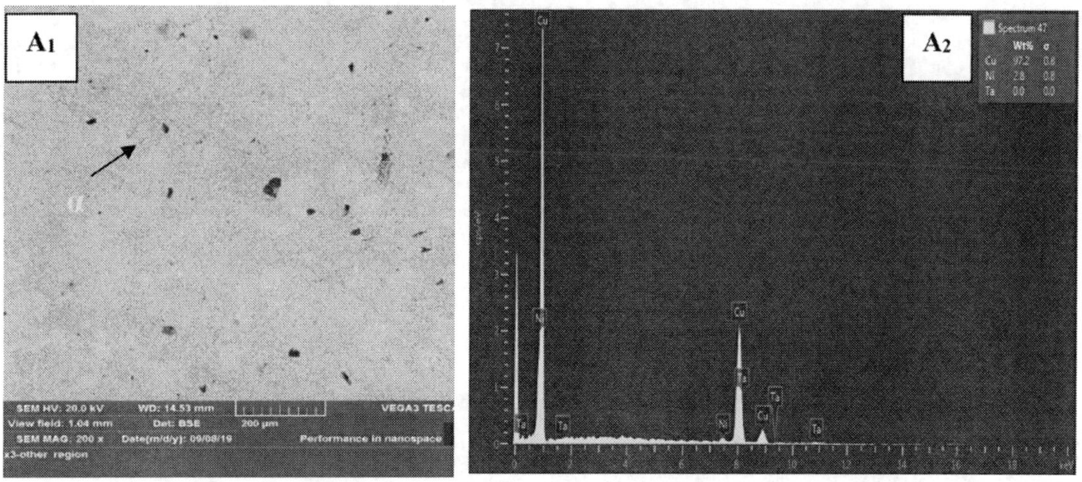

Figure 6. A_1. SEM image, A_2. EDS analysis for sample A after the sintering process.

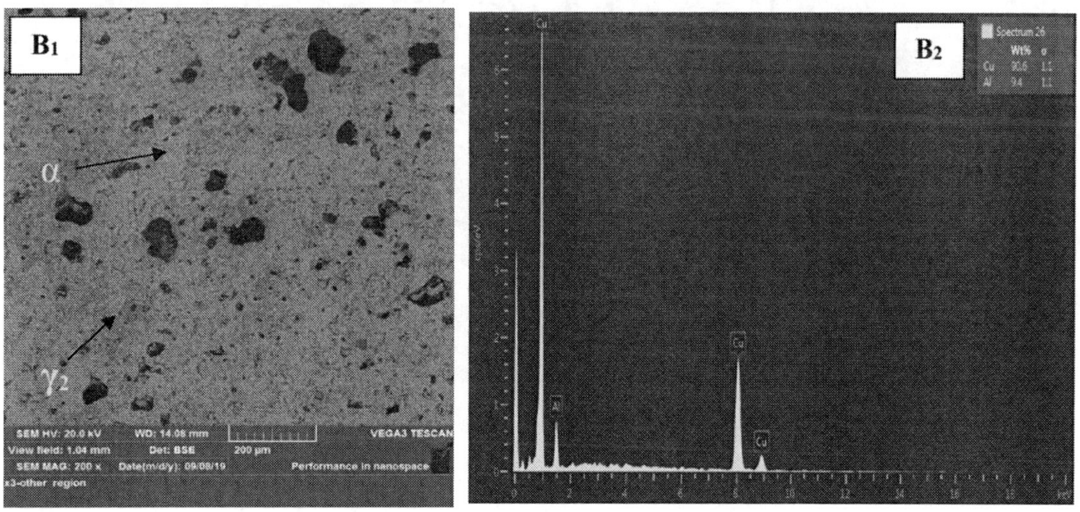

Figure 7. B_1. SEM image, B_2. EDS analysis for sample B after the ageing process.

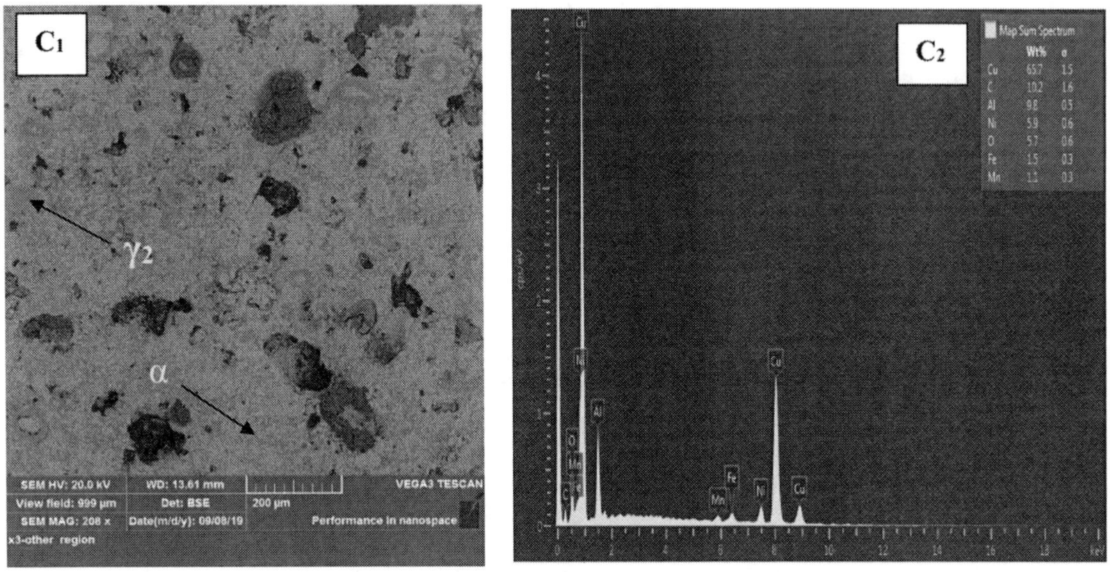

Figure 8. C_1. SEM image, C_2 . EDS analysis for sample C after the ageing process.

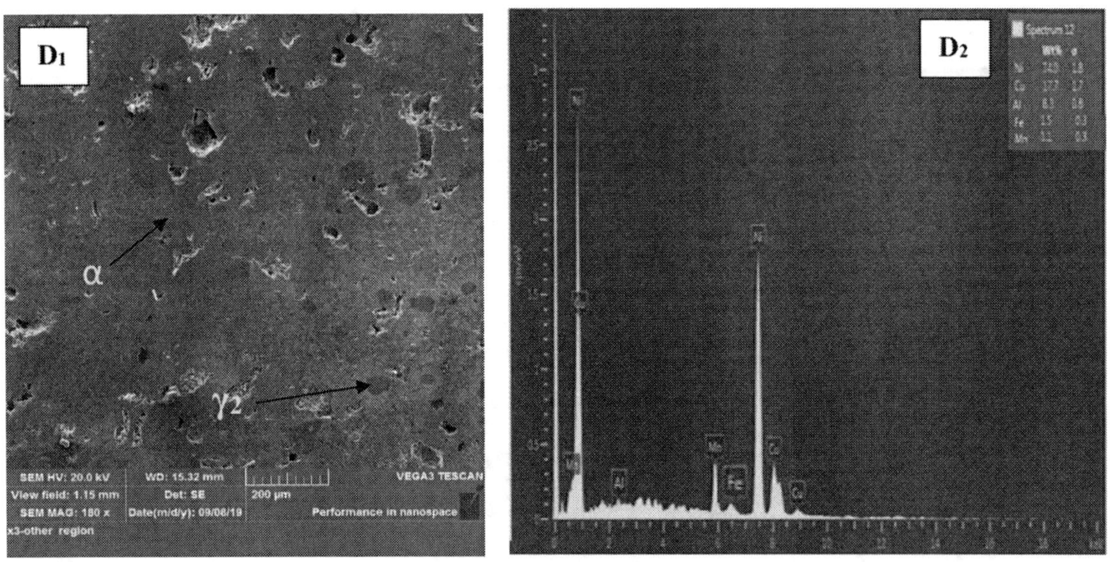

Figure 9. D_1. SEM image, D_2 . EDS analysis for sample D after the ageing process.

The XRD pattern of alloy A, as shown in Figure 10, shows that the existing phases were α-Cu, Ni and $Cu_{3.8}Ni$ after the sintering process at 550 °C for 120 minutes and 950 °C for 180 minutes. Figures 11 to 13 show the XRD pattern for alloys B, C and D after the aging heat treatment in which the samples were heated up to 200 °C for approximately 30 hours, and then rapidly quenched in iced water. It was found that the aging heat treatment led to precipitation in the intermetallic compound γ_2 $AlCu_3$ and α phase. Regarding the other phases that can be formed between copper, aluminium, nickel, iron and germanium, they can have a significant influence on the alloys' properties. However, they did not appear in the X-ray diffraction pattern, as which suggests that they may be outside the limits of the XRD detection.

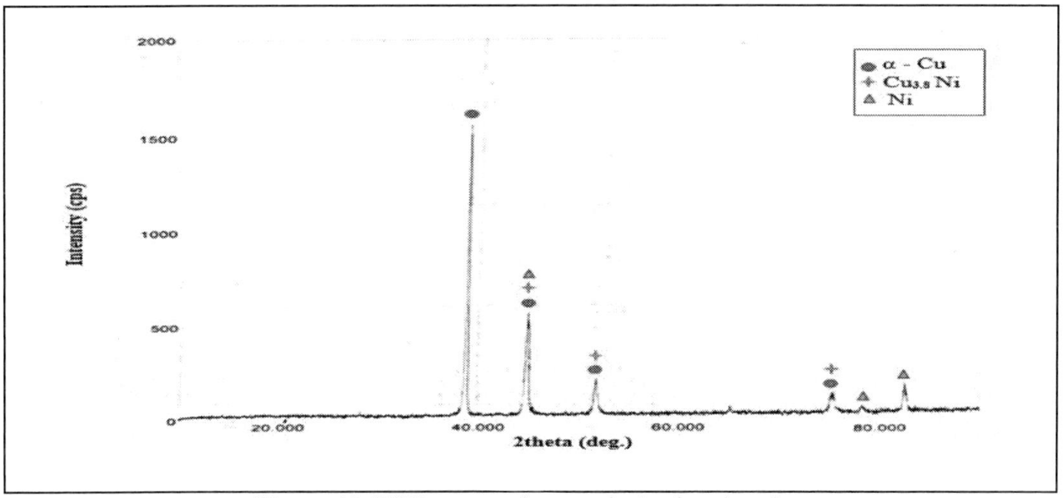

Figure 10. XRD pattern for sample A after sintering.

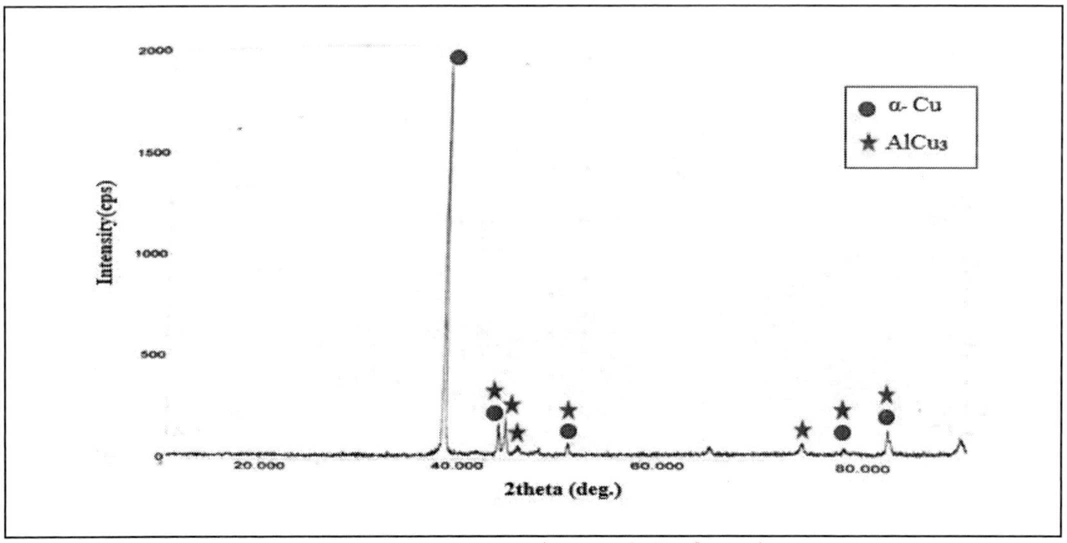

Figure 11. XRD pattern for sample B after aging.

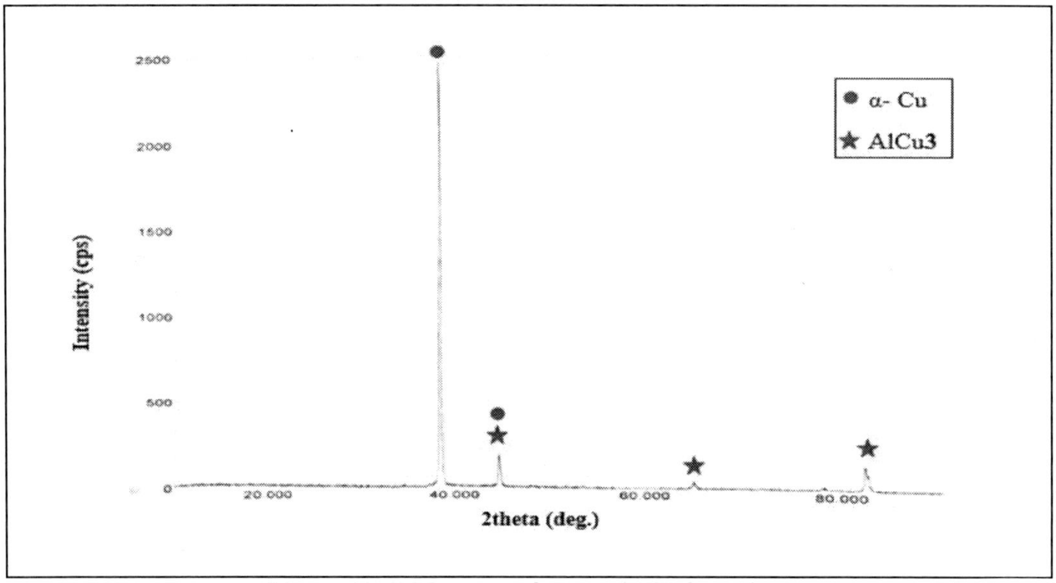

Figure 12. XRD pattern for sample C after aging.

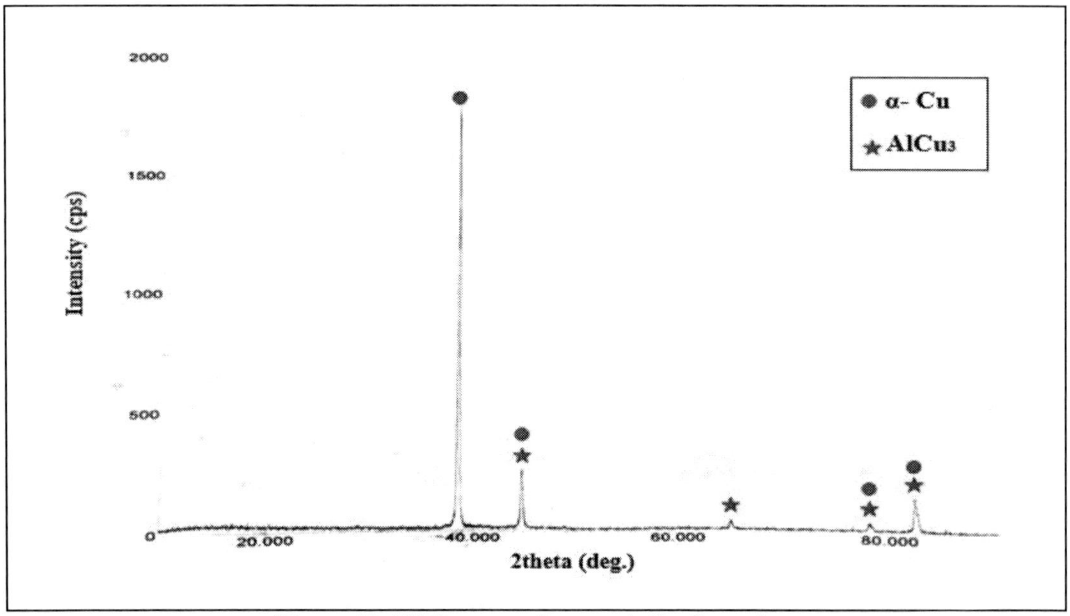

Figure 13. XRD pattern for sample D after aging.

Table 4 provides the hardness values of each individual sample. Sample A has the lowest hardness value compared to B, C and D. This is expected because the relationship between copper and nickel is complete solubility, and at any percentage, they completely dissolve each other. As is known at five % nickel, the tensile strength is less likely to be present for this alloy, which increases after this value. Therefore, we see that this alloy has lower hardness [15]. Adding the alloying elements Ni, Mn and Fe will increase the strength because heat treatment increases hardness. The addition of Ge has the same effect as the previous results, as the addition of germanium will be deposited freely, thereby helping to refine the grains. Therefore, it will increase the hardness compared to alloys A, B and C. At the same time, it will improve the plasticity of the alloy [16].

Table 4. Vickers micro-hardness for the samples.

Sample	Vickers Micro-hardness g /μm2	Improvement %
A	65	....
B	139	113.85
C	189	190.77
D	200	207.69

The effect of the chemical composition and the duration of exposure to the sea water jet on the weight loss of the samples A–D are shown in Figure 14. Figure 15 shows the erosion–corrosion rate test in solution 3.5 % NaCl of the samples A–D for ten hours. For samples A, B and C, we note in Table 5 that the weight loss of sample B (0.088943 g.cm^{-2}) is the lowest compared with A (0.063663 g .cm^{-2}) and C (0.080945 g .cm^{-2}) in the steady state. This was expected because sample B contains 9.5 % Al; therefore, there is a rich oxide layer of Al_2O_3. The rate of weight loss for samples A and B changed with the duration of exposure to the seawater jet, where the incubation period was not observed. When the test began, the surface of the samples hardness by drops of water. Following that, spalling occurs, which causes an acceleration in the weight loss. This reached a great value in the first quarter hour of the test. It was followed by a deceleration period, which represents a decrease in the rate of weight loss because the sample surface is rough in this period with grooves. These are then filled with salty water, and the jet of salty water will contact the water rather than the alloy. This was followed by a steady state, which has a stable weight loss. On the other hand, sample C was characterised by the appearance of the incubator period and a lower rate of weight loss. It reached the steady-state period faster than samples A and B. This sample showed a good resistance to erosion–corrosion, as there was no weight loss during the first half of its exposure to seawater due to the combined effect of iron with manganese. It also showed a clear increase in erosion–corrosion resistance, and this result is in agreement with Callcut [15]. Alloy D showed excellent resistance to both the erosion–corrosion and the weight loss compared with the others. The longest incubation period resulted in the lowest rate of weight loss and the least time to achieve the steady-state period compared to the other samples. This is due to the presence of a germanium element that can improve mechanical properties, give high ductility to the oxide layer [16] and improve corrosion resistance, which is in good agreement with Rosalbino [17].

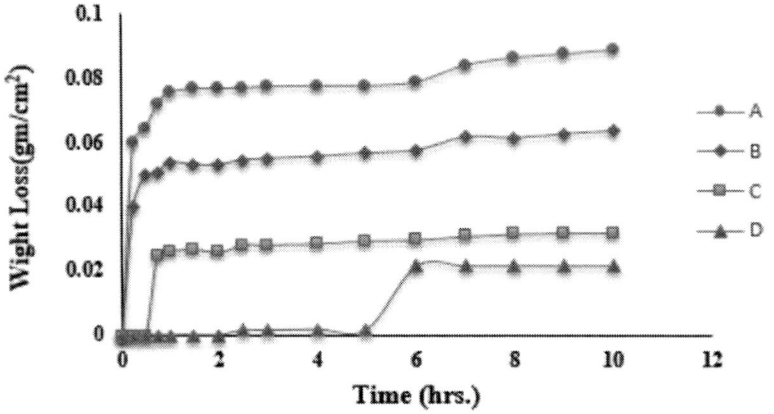

Figure 14. The effect of chemical composition and the exposure period of the seawater jet on the weight loss of samples A–D for ten hours.

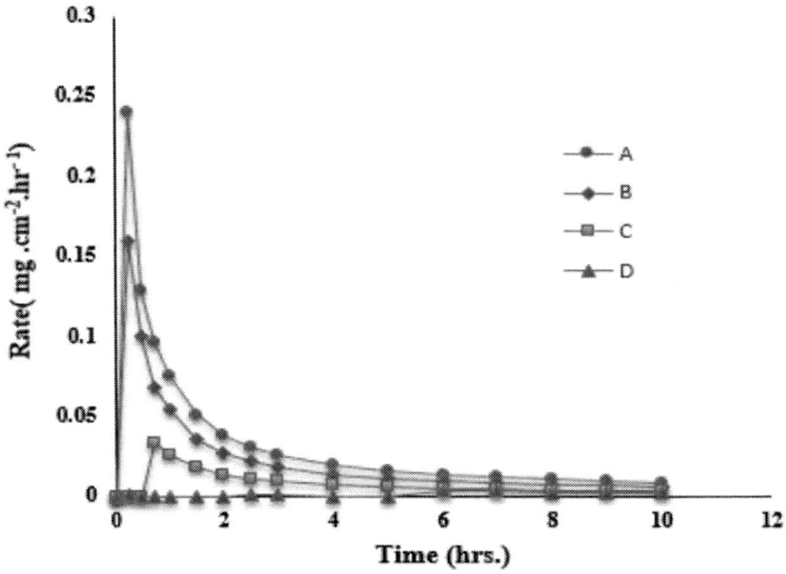

Figure 15. The erosion–corrosion rate test in the solution 3.5 % NaCl of the samples A–D for 10 hours.

Table 5. Weight loss for all samples and their improvement.

Sample	Weight loss g .cm^{-2}	Improvement %
A	0.089	-----
B	0.064	39.71
C	0.032	177.97
D	0.022	308.84

4. Conclusion

The XRD analysis and SEM observation showed that the formed phase in the sintered sample (Cu-five % Ni) was (α), whereas in the aged samples of Cu- 9.5 % Al with the alloying addition of elements Ni, Mn, Fe and Ge, the formed phases were $\alpha + \gamma_2$. The alloy D showed an increase in hardness by 207 % as compared to the alloy A which showed the lowest hardness. Alloy D showed a decrease in weight loss by 308.84 % compared to alloy A, which showed the highest weight loss after being subjected to an erosion–corrosion test in laboratory-prepared seawater for ten hours. The incubation period (latency) for the sample D was long, whereas for the rest of the research alloys, it was either absent or short. The addition of germanium to alloy D represented a significant shift in the erosion–corrosion resistance compared to the other alloys.

References

[1] J.-M. Welter.2006 *Copper Proceedings of the International Conference Copper '06*, France.
[2] J. R. Davis, Davis 2001 *Copper and Copper Alloys* ASM International Handbook Committee.
[3] Dimitri. K. 1991 *Graphite Manufacturing process* substances and technologies, USA.
[4] M.E.Wilms,V.J.Gadgil,J.M.Krougman, F.P.Ijsseling 1994 *The effect of σ-phase precipitation at 800°C on the corrosion resistance in sea-water of a high alloyed duplex stainless steel* Corrosion science , Vol **36** pp. 871-875, 877-881.
[5] 5-Al-Haidary J. T. 2005 *Erosion-Corrosion Resistance Improvement of α-Brass Dept. of Materials and Metallurgical Engineering* Al-Balqa' Applied University, Al-Salt, Proceedings of The First Jordanian International Conference of Materials Science and Engineering.
[6] Faisel A.M.Al-Bassam, Abd Al-Wahed K.R. Al-Bakry and Saif M.H.Al-Sharify 2009 *Effect of Alloying Elements and Heat Treatment on Behavior of the Alloy (Cu,Al) Used in Marine Industries* The Iraqi Journal For Mechanical And Material Engineering, Vol.**10**, No.1.
[7] W.A. Badawya, , M.M. El-Rabieib, H. Nadyb 2014 *Corrigendum to Synergistic effects of alloying elements in Cu-ternaryalloys in chloride solutions* Electrochimica Acta, Vol **120**, pp. 39-45 .
[8] Minh Quang Chau 2019 *Effect of different quenching meadia on mechanical properties of AISI 1018 low carbon stee*l Journal of Mechanical Engineering Research and Developments pp 1024-1752.
[9] Reed-Hill , R. 1964 *Physical metallurgy principles* ,4th edition, Library of Congress, USA.

[10] Elef, T. 2004 *Aging Characteristics of Cu Based Shape Memory Alloys* PhD Thesis, metallurgical and materials engineering, The Middle East Technical University, Ankara, Turkey,

[11] Safaa N., Hamzah, E. , Bakhsheshi-Rad, H. and Abubakar, T. 2014 *Microstructure and corrosion behaviour of Cu- Al-Ni shape memory alloys with Ag nanoparticles* Materials and Corrosion vol. **66**, No. 6.

[12] A. Mohammed Ahmed 2007 *Evaluation of Erosion and Erosion-Corrosion in 3.5% NaCl for Steel Pipe* MSc. Thesis, Dept. of Production Engineering and Metallurgy, University of Technology,

[13] *Copper-Nickel alloys :props,processing application*,German,copper Institute(DKI).

[14] Ali H. Haleem and Hussein A. Hashim 2018 *Investigate the influence of Ge addition on the corrosion resistance and microstructure of Cu-Al-Ni shape memory prepared via powder metallurgy* ,Journal of Engineering and Applied Science , pp 10882-10887 .

[15] Callcut,V.A 1989 *Aluminium bronzes for industrial use* Metals and materials vol.**5** No. 3, pp 128-132.

[16] Kwai S. Chan and David L. Davidson 2003 *Improving the fracture toughness of constituent phases and Nb-based in-situ composites by a computational alloy design approach* Metallurgical and Materials Transactions . pp 1833–1849.

[17] F. Rosalbino, R. Carlini , R. Parodi,, G. Zanicchi 2012 *Effect of silicon and germanium alloying additions on the passivation characteristics of Fe3Al intermetallic in sulphuric acid solution* Electrochimica Acta pp.62 305– 312 .

IOP Conf. Series: Materials Science and Engineering **987** (2020) 012019 doi:10.1088/1757-899X/987/1/012019

Pomegranate Peel Plant Extract as Potential Corrosion Inhibitor for Mild Carbon Steel in a 1 M HCl Solution

Asia Mishaal Salim[1a], Nawal Mohammed Dawood [2,b], Rasha Ghazi [2]
[1] AL-FURAT AL-AWSAT TECHNICAL UNIVERSITY, Diwaniyah Technical Institute
[2] University of Babylon, Babylon, Iraq
[a]dw.asia@aut.edu.iq; [b]mat.newal.mohammed@uobabylon.edu.iq

Abstract Herein, *Pomegranate Peel* plant extract was assessed as a low-cost, green and efficient corrosion inhibitor for mild carbon steel in 1 M hydrochloric acid solution using weight loss, potentiodynamic polarisation, an atomic absorption Spectroscopy test, surface roughness test and surface studies by optical microscopy (LOM). Increasing *Pomegranate Peel* extract concentration increases the charge transfer resistance as a result of the adsorption of *Pomegranate Peel* extract on the mild carbon steel surface. The polarisation results showed the mixed-type inhibition behaviour of the studied extract with an inhibition efficiency (%IE) of 90.193% achieved with the addition of 4% vol. of *Pomegranate Peel* plant extract. The adsorption isotherm model of *Pomegranate Peel* plant extract on a mild carbon steel surface followed the Langmuir adsorption isotherm in a HCl solution. The inhibition mechanism of *Pomegranate Peel* plant extract was explained according to the increase in the IE% with concentration and the activation parameters which suggested significant chemisorption of the *Pomegranate Peel* plant extract on the mild carbon steel surface. LOM studies confirmed the formation of a protective layer of *Pomegranate Peel* plant extract on a mild carbon steel surface. Results obtained by different techniques showed good agreement that confirm the potential use of Pomegranate Peel plant extract as a corrosion inhibitor for mild carbon steel in an acidic media.

Keywords: *Pomegranate Peel extract · Corrosion inhibition · mild-carbon steel · HCl · LOM*

1.0. Introduction

Corrosion is still a significant problem for many industrial systems that use water or acids in their applications, such as cooling systems, condensers, heat exchangers and engines, all of which need water more than any other purpose. Moreover, acids play a big role in pickling processes or acid-cleaning processes in oil refining or manufacturing industries [1]. The maintenance of these systems is very expensive unless the water or acid is treated to protect the systems from erosion, sediment or biological growth. One of the most important and widely used techniques for reducing or preventing metal corrosion is the use of inhibitors, which are chemical compounds that, when added in small amounts to a medium (usually a liquid phase) in which the metal corrodes, impede or eliminate this corrosion. Organic inhibitors are a good method widely used to protect iron and steel from corrosion in acidic media. At present, modern organic inhibitors that have been experimentally discovered have shown remarkable superiority in neutral solutions or salt solutions [2].

Organic compounds, especially adsorbent ones, reduce the rate of corrosion, either by interfering with the corrosion reaction or by forming an adsorbent uniform layer that impedes the effective sites on the metal surface and acts as a protective layer [3]. The toxic nature and high costs of many of the existing organic inhibitors have prompted researchers to develop corrosion inhibitors that are more acceptable and less expensive; natural products can be considered a good source for this purpose. Natural products of plant origin contain many organic compounds such as pigments, tannins, alkaloids and, finally, organic and amino acids. Most of these substances possess a suitable inhibitory action [4].

Hydrochloric acid (HCl) solutions are widely used for acid cleaning, industrial acid pickling, oil well acidizing and acid descaling [5–8], but a hydrochloric acid medium has aggressive effect on substances. Plant extracts are considered cheap, safe, on-hand and renewable sources of inhibitors. Such plant extracts include different organic compounds containing heteroatoms, and some [9–12] have been used to act as metal dissolution inhibitors in various corrosive mediums. "So, we need the using of substances that have inhibiting effect to protect the metals which used in different fields of applications. The formation of thin layer resulting from interaction of organic compounds on the metal (mixed inhibitors) while in case of inorganic inhibitors may block anodic or/and cathodic sites [13–15]. Due to the currently imposed environmental requirements for eco-friendly corrosion inhibitors, there is a growing interest in the use of natural products such as leaves or seed extracts. The terms *eco-friendly corrosion inhibitor* or *green inhibitor* refer to the substances that are biocompatible such as plant extracts since they are of biological origin. Thus, such natural products are being studied by several researchers for their corrosion-inhibition potential as they are more environmentally friendly, showing good inhibition efficiency with the low risk of environmental pollution [16]. The use of extracts of plant as inhibitors in acidic solutions, as Nigella sativa L.[17], Apium graveolens L. [18], Glycyrrhiza glabra [19], pomelo peel extract [20], Dryopteris cochleata leaf extracts [21], Myrmecodia pendans extract [22], Morinda citrifolia [23], Thymus vulgaris plant extract [24], Mentha spicata L. extract [25], phytoconstituent of Ervatamia coronaria [26], myrrh extract [27], Cucurbita maxima [28], Adhatoda vasica, Eclipta alba and Centella asiatica [29] have been investigated. Pomegranate peel have shown several *in vitro* pharmacological effects such as antibacterial, antifungal and anti-inflammatory properties. It is also known for its ethno-medicinal properties, mainly used for boils, rheumatic pain and digestive disturbances. The current research is an attempt to study the inhibitory performance of a pomegranate peel extract as an inhibitor of carbon steel corrosion in 1M HCl. To evaluate the performance of this inhibitor, simple immersion tests were performed; moreover, a corrosion-erosion test and an electrical conductivity test of solutions were performed, and, finally, a microstructure test.

1.1. Corrosion Rate

Corrosion rate may be determined by mil per year (mpy), with a value expressed in (mpy), either from weight loss or from corrosion current density according to Equations 1 and 2. The weight loss can be converted into a corrosion rate (C.R.) **[30].**

C.R=(K*W) / (A*t*ρ)..........................1

Where :

K: constant indicate the corrosion rate units=$3.45*10^6$
t : time of exposure in hours
A: area in cm^2
W: mass loss in grams
ρ: density in g/ cm^3

The corrosion rate can also be determined via an electrochemical method by (corrosion current density) and translated into other expressions by Equation 2:

$$\textbf{\textit{Corrosion rate}} = \frac{0.13\ \text{icorr(Ew)}}{A\rho} \quad ....2$$

Where:

0.13 = metric and time conversion factor
Icorr = corrosion current (μA)
E.W = equivalent weight (g/eq.)
ρ = density of alloy (g/cm^3)

A = Sample Area (cm^2)
mpy = Corrosion rate (mils per year).

1.2. Inhibition Efficiency

Inhibition efficiency was calculated for different inhibitors at different concentrations using the following formula [31]:

$$IE = (C.R^0 - C.R) / (C.R^0) * 100\% \quad ....3$$

Where : IE is the inhibitor efficiency
$C.R^0$ is corrosion rate without inhibitor.
C.R is corrosion rate with an inhibitor.

1.3. Polarisation Resistance

The polarisation resistance (Rp) was calculated using Stern-Geary equation [32, 33]:
$Rp = (ßa * ßc) / [2.303 (ßa+ßc)\ i_{corr}]$ 4

Where ßa, ßc are the anodic and cathodic Tafel slopes respectively, and i_{corr} is the corrosion current density. Current density i_{corr} was determined by the extrapolation of the Tafel lines to E= E_{corr} or according to the Stern-Geay equation, resulting in, where Rp was the polarisation resistance, as defined as the tangent of a polarisation curve at E_{corr} [4]. Corrosion-current values may be obtained from polarisation measurement, and the first step is to convert the measured or estimated current value to current density. This is accomplished by dividing the total current by the exposed area of the anodic specimen used in this equation. This calculation may be expressed as follows [33]:

$$I_{corr} = i_{corr} / A$$

Where I_{corr}: the corrosion current density (μA/ cm^2)
i_{corr} : corrosion current μA
A: exposed specimen area, cm^2

1.4. Adsorption Isotherms

Adsorption isotherms are usually used to describe the adsorption process. The establishment of adsorption isotherms that describe the adsorption of a corrosion inhibitor can provide important clues as to the nature of the metal-inhibitor interaction. In order to obtain the adsorption isotherm, the degree of surface coverage (θ) for various concentration of the inhibitor has been calculated according to the following equation [35]:

$$\theta = [1 - i_{inh} / i_{un\ inh}] \quad ....4$$

The Langmuir isotherm was carried out for its fit to the experimental data. The Langmuir adsorption isotherm is given by the following equation [36]:

$$C_{inh} / \theta = (1/ K_{ads}) + C_{inh} \quad ....5$$

Where

K_{ads}: the equilibrium constant of the adsorption-desorption process.

θ: the degree of surface coverage

IOP Conf. Series: Materials Science and Engineering **987** (2020) 012019 doi:10.1088/1757-899X/987/1/012019

C_{inh} : concentration of inhibitors in the bulk solution

The Langmuir isotherm is based on the presumption that every site of metal surface carries one adsorbed species H_20. Therefore, one adsorbed H_20 molecule is replaced by one molecule of the inhibitor adsorbed on the metal surface. The apparent free energy of adsorption (ΔG°_{ads}) can be calculated from the following equation [35]:

$$\Delta G^{\circ}_{ads} = - 2.303 \, RT \log 55.5 \, K_{ads} \;\ldots\ldots\ldots\; 6$$

Where

$$K_{ads} = \theta / c\,(1- \theta) \;\ldots\ldots\ldots\; 7$$

2.0. Experimental Work

2.1 Pomegranate Peel Plant Extracts

A pomegranate peel used as a green corrosion inhibitor. The pomegranate peel was collected, and then it was washed with distilled water, dried and ground into a powder. The process of extraction was achieved by two types of solvents in order to show the solvent effect on inhibition process: distilled water and diethyl ether.

An amount of 10g was dissolved in 200ml of solvent and put in the stirrer for four hours, left to soak for 24 hours and then filtered with a piece of gauze to separate the solids. After filtration, the liquid was further filtered by filter papers two times and then dried from both the distilled water and the diethyl ether extract to achieve a black powder and brown paste respectively several times to achieve a sufficient quantity. When needed as an inhibitor, the amount was dissolved from this extract to get a sample with a 10% concentration.

2.2. Preparation of Mild Steel Samples

The approximate chemical composition of the tested mild steel is explained in Table 1.

Table 1. The composition of mild steel sample (wt.%)

C	Si	Mn	P	S	Cr	Al	V	Cu	Mo	Ni	Fe
0.056	0.034	0.303	0.0008	0.007	0.018	0.049	0.0005	0.016	0.002	0.010	Bal

The mild steel samples were cut by a CNC machine into 1.5*1-cm pieces, and made into working electrodes for electrochemical measurements, as well as 2*2*0.4-cm pieces for weight loss studies, with a 2-mm bore to employ a cooling fluid during the cutting procedure to minimise mechanical damage. Prior to all corrosion experiments, the surface of mild steel was abraded with emery paper (80 to 2,000). Thereafter, the samples were polished with a cloth diamond paste with a size of 1 μm. Afterward, these samples were rinsed and degreased with ultrapure water and ethanol, and then dried and preserved in cans with a silica gel in order to keep them dry until use for corrosion investigation.

2.3. Microscopy Test

Mild steel samples were examined before exposure to HCl by using a light microscopy so as to explore the surface topography of the sample.

2.4. Inhibitor Evaluation
2.4.1. FTIR Spectra
Fourier transform infrared (FT-IR) spectra were recorded to detect the active groups. A 0.002 gm of pomegranate peel plant powder was mixed with 0.02 gm KBr; the mixture was triturated well, and then put in a spatial mould with a strong press to obtain a pellet to put in the device to achieve the FT-IR spectra.

2.5. Corrosion Tests
2.5.1. The Immersion Test
The weight loss examination by immersion test determined the rate of corrosion in addition to the percent of inhibition efficiency. This measurement will furnish a direct response to how corrosion environments affect the samples. The immersion test for polished mild steel engaged samples in HCl with and without different volume percentages of pomegranate peel plant extract (1, 2, 3 and 4% vol.) at laboratory temperature. The samples were submerged in 70 ml of 1MHCl in the presence of inhibitors for a duration of 20 h. Previous soak in HCl weights were achieved for the polishing, eviscerating as well as drying of the samples. Later, a settled interim of an exploratory examination of every test was brought out, washed completely in ultrapure water and ethanol and desiccated in warm air, whereupon the weight reduction was computed.

2.5.2. Tafel Extrapolation Measurements
A Tafel plot was extrapolated to determine corrosion current density and corrosion rate. A three-electrode cell was employed in electrochemical measurements, with mild steel as a working electrode, a calomel electrode as a reference electrode and a platinum electrode as an auxiliary. The exposed part to HCl is 1.33 cm^2 and the working electrode was immersed in 1M HCl to 70 ml for 30 minutes to open circuit potential. Before electrochemical tests, substantial variations of rest potential did not confirm the attainment of steady state. Then, the device was set to a scan rate of 0.1 mV/s. from an O.C.P over and below the potential with ±200 mV. All experiments were completed in laboratory temperature, fresh prepared solution and under-stirred conditions before tests.

2.6. The Absorption Spectroscopy
Atomic absorption spectroscopy (AAS) was used to determine the concentration of Fe^{+2} when the mild steel had been immersed for 3 hrs in HCl only and with pomegranate peel plant extract at laboratory temperature. The absorbance was concluded to the standard elaboration, and set up an adjustment curve from got esteems. Then, the researchers determined Fe^{+2} from the calibration curve for HCl with and without pomegranate peel plant extraction.

2.7. pH Measurement
The pH of HCl has been measured wit and without pomegranate peel plant extract before the immersion test at laboratory temperature, and after 3 h of immersion in order to identify its effect on inhibition.

2.8. Surface Roughness Test
A surface roughness test was used to measure the surface roughness of mild steel before and after the immersion test. The samples tested with and without pomegranate peel plant extract run over the surface of mild steel to register roughness.

3.0. Results and Discussion
3.1. The Microstructure Examination for Mild Steel Surface
A deliberate choice of a mild steel sample to experience microstructure examination. This progression is trailed by segmenting, pounding, cleaning and carving in Nital solution to uncover a precise microstructure and substrate. The specification survey of the mild steel is

finished through the light microscope. The microstructure of mild steel, shown in Fig. 1, consists of ferrite in addition to pearlite.

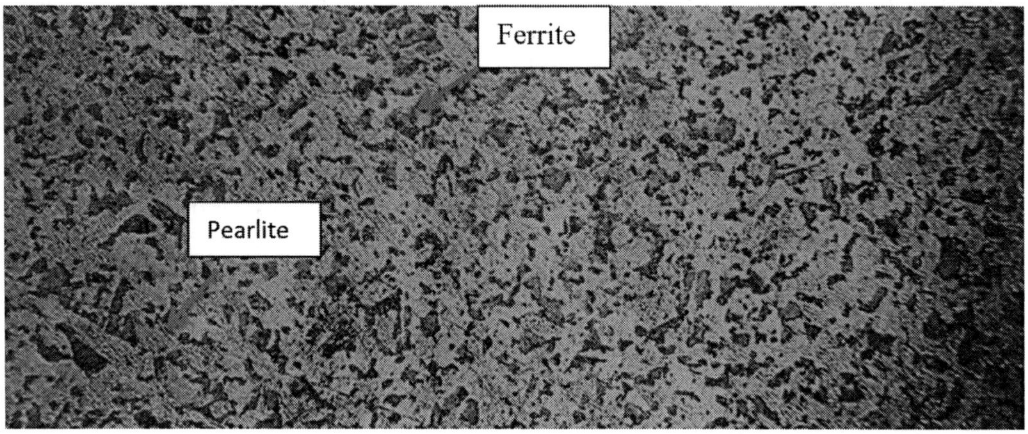

Fig. 1: The microstructure of mild carbon steel (100X)

3.2. Fourier Transform Infrared Spectroscopy Measurement (FTIR)

FTIR Spectra were utilised to assess the defensive film created on the mild steel surface to determine the sort of bonding for organic inhibitors adsorbed on the surface. FTIR to organic inhibitor, as illustrated in Figure 2, showed the results of this test for pomegranate peel powder, and Table 2 shows the effective groups and the corresponding wave numbers in the pomegranate peel plant powder. According to the table, the powder contains many active groups, which are mostly aldehyde compounds, ketones, amines, amides, alcohols, or aromatic or phenolic compounds; all of these compounds have inhibitory properties. Also, the presence of double and triple bonds and aromatic rings in this powder will enhance its inhibitory action, and this is consistent with [37].

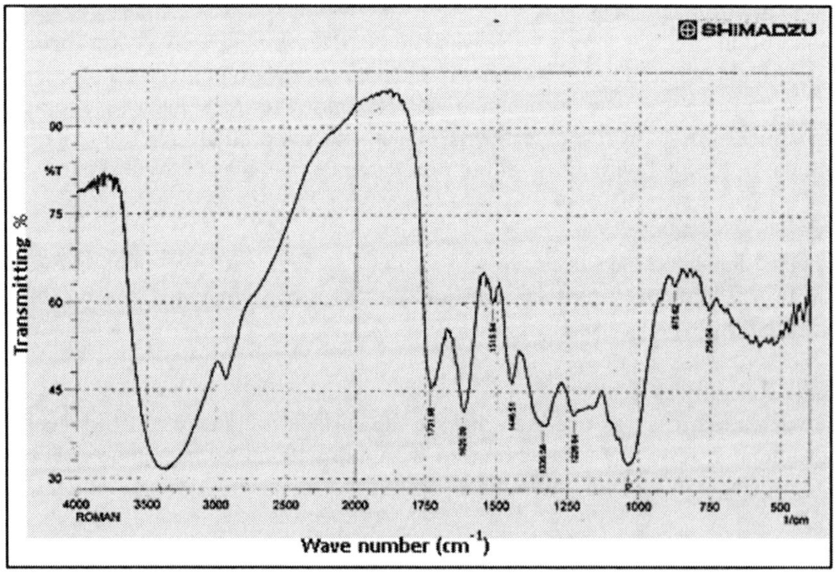

Fig.2: Fourier Transform Infrared (FT-IR) spectroscopy of pomegranate peel plant powder

Table 2: FTIR detection of functional groups of pomegranate peel plant

	Functional Groups	Wave Numbers (cm^{-1})
1	C-H aromatic	756.04
2	C-C	875.62
3	C-H bending	1035.7
4	C-N (amide) 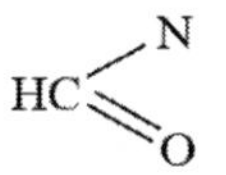	1226.64
5	C-H stretching	1336.58
6	CH_2	1446.51
7	CH_3	1515.94
8	C=C	1620.09
9	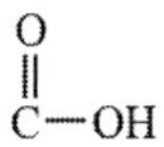	1731.96
10	$\equiv CH_2$	3365.55
11	C≡C-O	3392.55

3.3. Corrosion Test

3.3.1. Immersion Test

Figure 3 shows the effect of the period of exposure to a solution of 1M HCl on weight loss in the presence and absence of the inhibitor, wherein the mild carbon steel submerged in the acid solution free of the inhibitor showed a continuous and significant loss in weight with an increase in the period of exposure to the acid solution. The reason is due to the presence of hydrogen ions that work on discharging electrons and speeding up the reaction [38]. As for the loss of weight of the samples submerged in the acidic solution that contains the inhibitor and its different concentrations, it decreased significantly compared to the absence of the inhibitor. This completely reduces the corrosion rate [39].

According to Figure 3, the mild steel sample is subjected to a higher weight loss in 1M HCl without organic inhibitors to 0.37 mg/ cm2 , while, in presence of 1, 2, 3 and 4% vol. organic inhibitors, the lower loss is 0.101 mg/cm2 for organic inhibitor in a concentration 4% vol.

The weight loss, corrosion rate and efficiency of pomegranate peel plant extract have been calculated from the immersion test samples in the absence and presence of 1, 2, 3 and 4% vol. in 1M HCl, as given in Table 3. The corrosion rate in 1M HCl is 186.78 mpy without the addition of an inhibitor. The rate of corrosion is significantly diminished respectively by the addition of an inhibitor to 1MHCl, which exhibited lower corrosion rates of 64.85, 57.54, 46.43 and 31.41 mpy according to organic inhibitors in 1, 2, 3, and 4% vol. concentrations. The corrosion rate without and with different vol.% of organic inhibitor is illustrated in Figure 4.

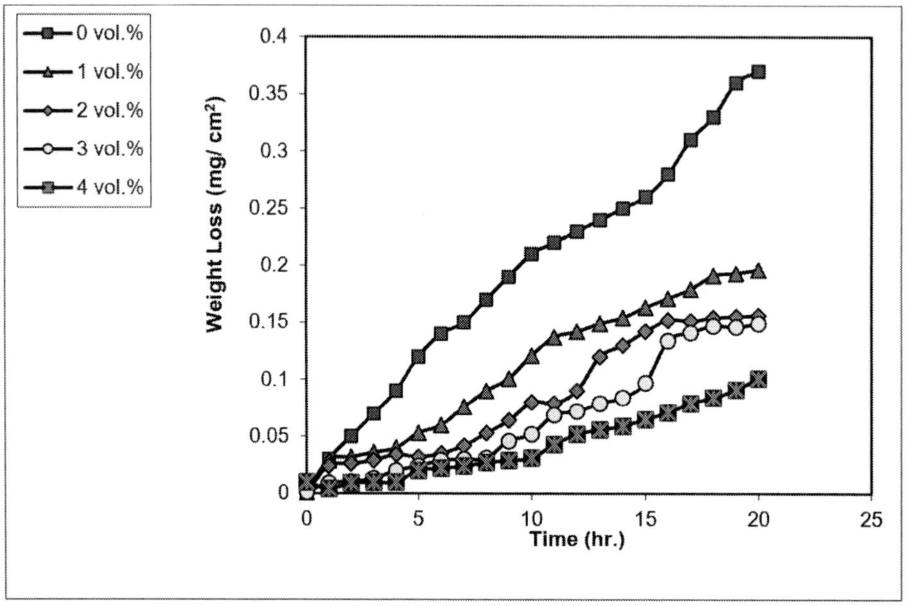

Fig. 3. Weight loss change over time of mild carbon steel samples in a 1M HCL solution containing different concentrations of inhibitor: 0, 1, 2, 3 and 4% vol.

Table 3. Corrosion Parameters Obtained from Immersion Test for Mild Steel

Concentration vol.%	W (mg/ Δcm^2)	CR(mpy)	η(%)
0	0.37	186.78	……
1	0.196	64.85	65.28
2	0.156	57.54	69.19
3	0.149	46.43	75.14
4	0.101	31.41	83.18

The addition of an organic inhibitor to 1 M HCl plays a favourable role in reducing the rate of corrosion through forming a hydrophobic film barrier on the surface, which could prevent the dissolution of the mild steel grains in a HCl solution [40]. A higher efficiency of an organic inhibitor is 83.18% at a concentration of 4% vol., as illustrated in Figure 5, which reveals the efficiency with various concentrations of organic inhibitors in 1 M HCl for the mild steel. The efficiency of an organic inhibitor is a significant evaluation because of the existence of (-C=N-) and other active groups in the molecules [41]. Active groups are adsorbed on the surface, spontaneously consisting of a monolayer. The planer in organic inhibitor includes electronegative atoms as N as well as O as active centres; moreover, having solitary electron pairs plus conjugated two-fold bonds in the aromatic rings hinders corrosion via adsorption [42].

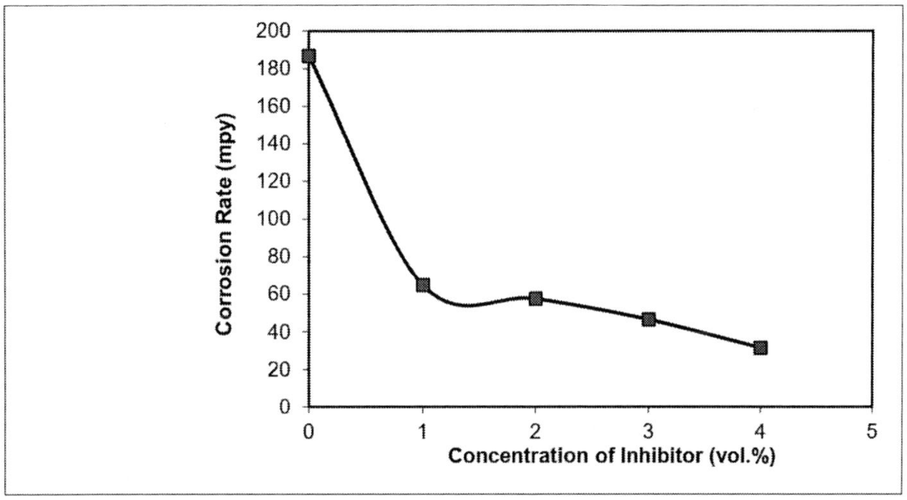

Fig.4. Corrosion rate (mpy) changes without and with 1, 2, 3 and 4% vol. inhibitor concentration immersed in a 1 M HCl solution at 30 °C.

3.3.2. Adsorption Isotherm

The surface coverage values (***θ***) according to Langmuir model for different inhibitor concentrations have been calculated by immersion test measurements. Plots of inhibitor concentrations (C) versus (C/ ***θ***) are shown in Figure 6. The adsorption for standard free energy of (ΔG°_{ads}) values for the investigated inhibitor is reported in Table 4. Outcomes reveal a spontaneous estimation of the free energy of adsorption in addition to an extreme interface among inhibitor molecules and surface of mild steel. In the organic inhibitor (ΔG°_{ads}) smaller than (-20 KJ/ mol), this construes a physical adsorption happening by electrostatic cooperation. The surface comprises a settled number of adsorption destinations, and every site holds a unique adsorbed. A large adsorption equilibrium is constantly proportional with the interface to the inhibitor with the surface of mild steel. The (K ads) equilibrium constant signifies a bigger adsorption as an organic inhibitor [43]

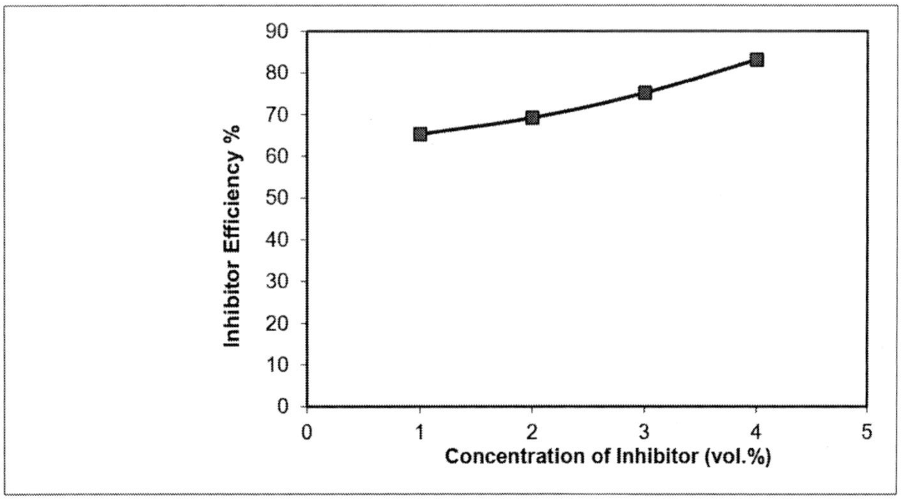

Fig. 5. Inhibition efficiency changes without and with 1, 2, 3 and 4% vol. inhibitor concentration immersed in 1 M HCl solution at 30 °C.

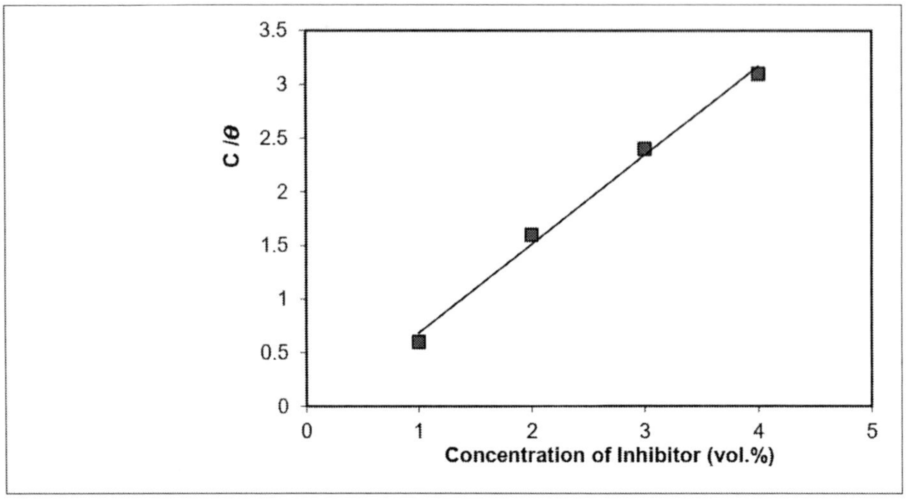

Fig. 6. Langmuir adsorption isotherm of organic inhibitor

Table 4: Standard free energy

Concentration of inhibiter (vol.%)	***θ***	**C / *θ***	**K_{ads} (mg/ l)$^{-1}$**	**ΔG^{o}_{ads} (KJ/mol)**
1	1.66	0.5	0.089	-4.856
2	1.26	1.6		
3	1.25	2,4		
4	1.29	3.1		

3.3.3. Tafel Extrapolation Curves

Polarisation offers important data regarding the dynamic of reactions. The limitations, for instance, the rate of corrosion, anodic in addition to the cathodic Tafel slopes, corrosion current density, potential and polarisation resistance are mentioned in Tables 5 and 6. The results show that, in incremental concentrations of an organic inhibitor, the corrosion current density declines as shown in Figure 7. Also, corrosion potential (E_{corr}) values do not change with incremental organic inhibitor concentrations. The change in the corrosion potential is slight, given that the addition of organic inhibitors can be interpreted as an inhibition of both the anodic and cathodic processes [44]. Organic inhibitors are mixed-type inhibitors, which inhibit the corrosion process by geometric obstructive of both cathodic and anodic sites [45]. The increase in polarisation resistance (R_P) magnitude is because of the creation of a protecting layer on the mild steel / HCl interface. The impediment influence emerges from the interface of π –electrons of the inhibitor rings as well as the existence of C-N (amide) and **C=C** forming bonds per mild steel [46]. The efficiency attained in both immersion tests, besides Tafel investigations, shows analogous performance through slight variance in the efficiencies of organic inhibitor.

IOP Conf. Series: Materials Science and Engineering **987** (2020) 012019 doi:10.1088/1757-899X/987/1/012019

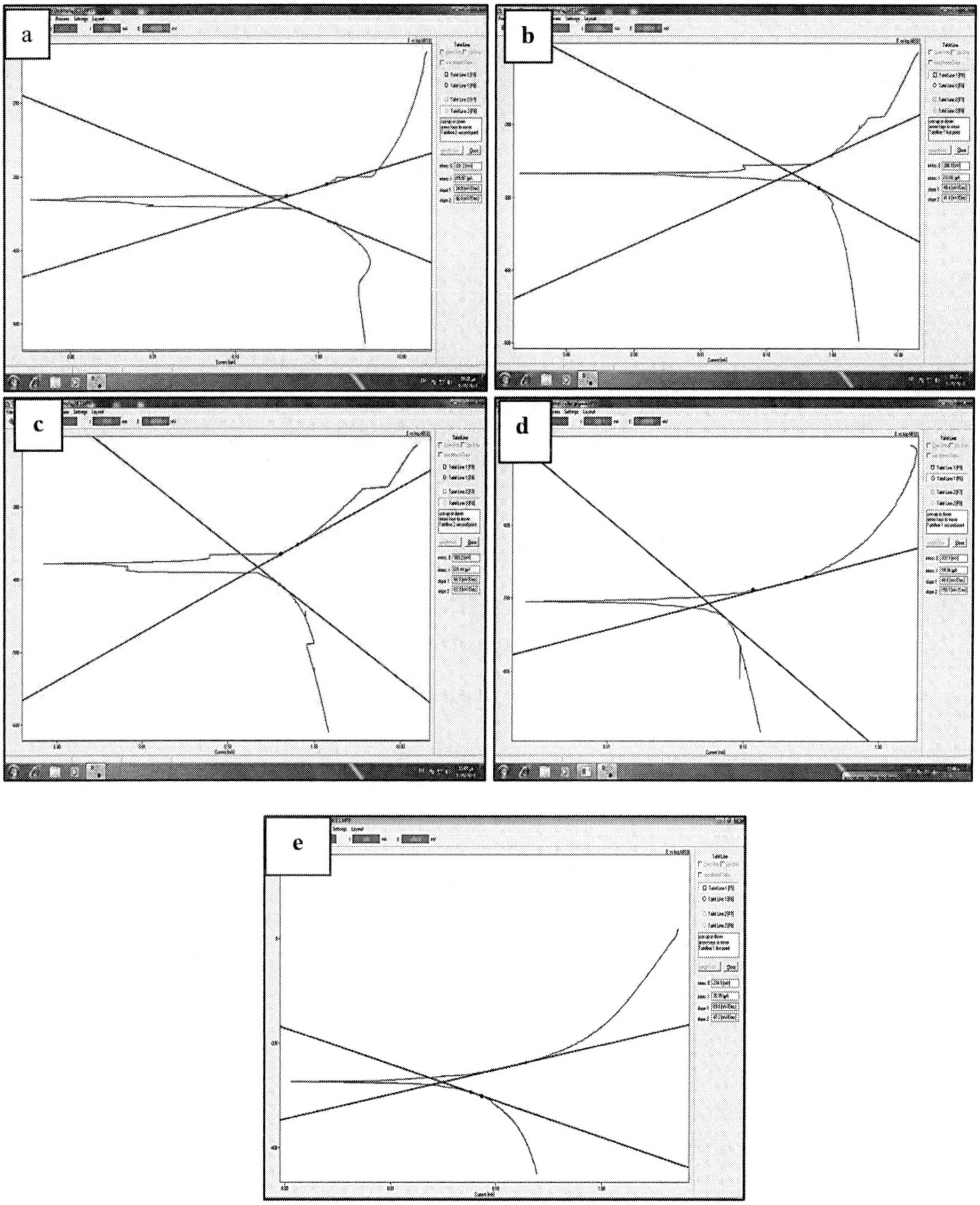

Fig. 7. Polarisation curves for mild carbon steel in 1M HCl without and with (a) 0% vol., (b) 1% vol., (c) 2% vol., (d) 3% vol. and (e) 4% vol. organic inhibitor.

IOP Conf. Series: Materials Science and Engineering **987** (2020) 012019 doi:10.1088/1757-899X/987/1/012019

Table 5. The Polarisation Constraints in 1 M HCl for mild steel

Concentration of inhibiter (vol.%)	i_c (μA)	Ic (μA/ cm^2)	CR(mpy)	η(%)
0	315.58	238.173	111.425	…….
1	233.86	176.498	82.571	25.985
2	221.44	167.124	78.186	29.830
3	54.94	41.464	19.398	82.813
4	30.95	23.358	10.927	90.193

Table 6. The Polarisation Constraints in 1 M HCl for the mild steel

Concentration of inhibiter (vol.%)	E (mV)	Rp (Ohm.cm^2)	ßa(Mv/Dec)	ßc(Mv/Dec)
0	-331.2	0.08	34.0	- 46.0
1	- 266.9	0.83	41.4	- 48.4
2	-383.2	0.61	66.5	- 93.3
3	- 707.1	0.42	49.4	-158.1
4	-274.9	0.29	47.2	-69.6

3.4. Atomic Absorption Spectroscopy (AAS)

The analysis of AAS results obtained shown in Table 7 could explain the dissolution concentration of Fe^{+2} in 1 M HCl only, without and with organic inhibitor for three hours.

Table 7. Atomic Absorption Spectroscopy Results

Concentration of inhibiter (vol.%)	(Fe^{+2}) concentration dissolved in 1 M HCl (mg/ l)
0	79.6
1	56.7
2	44.2
3	34.9
4	29.9

IOP Conf. Series: Materials Science and Engineering **987** (2020) 012019 doi:10.1088/1757-899X/987/1/012019

Results obtained indicate that organic inhibitors reduced in Fe^{+2} amount compared to the HCl without inhibition in 1 M HCl only Fe^{+2} dissolved is 79.6 mg/ l. This value is reduced to 29.9 mg/ l for an organic inhibitor with a concentration of 5% vol. This could impute adsorption of organic inhibitor on the surface, producing an obstruction that separates the surface from the corrosion [47]. The result is consistent with the observed trend from weight loss and Tafel polarisation measurements.

3.5. The surface roughness test

From the results of surface roughness test, it appears that the surface roughness of the pure mild steel sample is 0.147 μm, but, when the sample is immersed in 1M HCl for just 3 hr, it is 0.510 μm. When adding an organic inhibitor to HCl , the surface roughness is 0.094 μm in a 5% vol. concentration. There is a convergence of value in surface roughness between the sample immersed with organic inhibitors. The same goes for the pure sample. This significant improvement in the surface roughness for the sample with an organic inhibitor reveals that organic molecules cover the protective surface, isolating the mild steel from HCl and saving it from corrosion [48] .

3.6. Measurement of PH

The pH of HCl has been measured with and without organic inhibitors before the immersion test, as well as after 3 h of immersion. The pH of 1 M HCl is 0.38 before immersion, while, after 3 hr, the pH is 0.47. The pH values with organic inhibitor before the immersion test are the same (0.35), unchanged after 3 hrs from the immersion test. The lack of sensitivity to pH refers only to corrosion control, and the organic inhibitor remains active in HCl [49].

3.7. Microscopic Examination Test

The optical microscopy was used to scan the surface with and without organic inhibitors in 1M HCl as shown in Figure 8. Fig. 8a shows the pure mild steel surface after being immersed without an organic inhibitor for 20 hrs, which is highly corroded with cracks and pits due to rapid corrosion attack. Thus, the mild steel surface is exceedingly susceptible to damage from HCl attack in the absence of an organic inhibitor. However, in the presence of an inhibitor, a relatively smoother and less-corroded morphology of the mild steel surface can be observed as shown in Figure 8b, which is indicative of the construction of a passive layer on the surface of the mild steel [50].

4.0 Conclusions

Pomegranate peel plant extract, as an organic inhibitor, exhibits excellent inhibition performance when exposed to a series of techniques. In an immersion test, the inhibiting efficiency increases to (83.18%) with organic inhibitor at a concentration of 4% vol. Whereas, from the Tafel polarisation outcomes, the least corrosion current density is 23.358 μA/ cm^2 with an organic inhibitor at a concentration of 4% vol. The adsorption according to Langmuir isotherm is a preferable deposition of an organic inhibitor, including physical adsorption. From the results of atomic absorption spectroscopy, the organic inhibitor produces a barrier which isolates from the HCl medium, and less Fe^{+2} is dissolved in the organic inhibitor: 29.9 mg/ l. Results of surface roughness test reveal that the roughness is reduced to 0.049 μm with an organic inhibitor at a concentration of 4% vol.

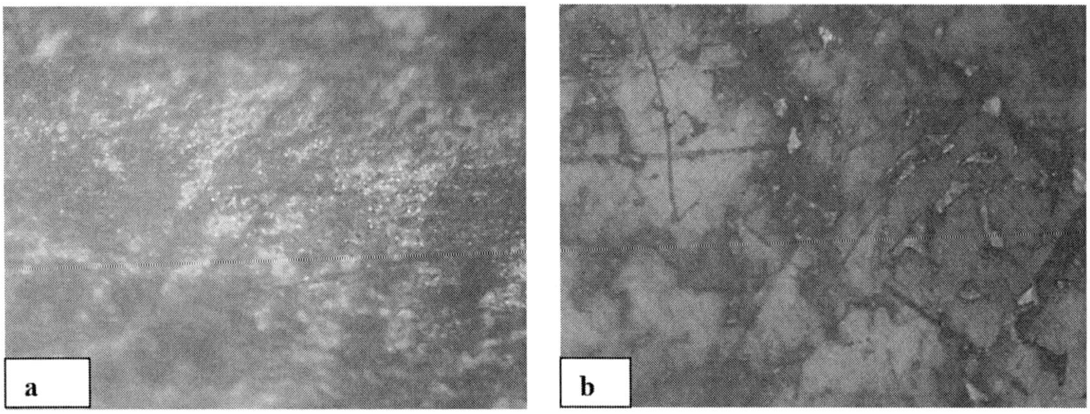

Fig. 8. Microscopy of submerged mild carbon steel samples in: (a) uninhibited 1 M HCl solution, and (b) inhibited 1 M HCl solution: 5% v. organic inhibitor.

References

[1] C.B. Pradeep Kumar, K.N. Mohana, 2014,Phytochemical screening and corrosion inhibitive behavior of Pterolobium hexapetalum and Celosia argentea plant extracts on mild steel in industrial water medium, Egyptian Journal of Petroleum 23, 201–211.

[2] Evrim Baran , Ahmet Cakir, Birgu Yazici , 2019, Inhibitory effect of Gentiana olivieri extracts on the corrosion of mild steel in 0.5 M HCl: Electrochemical and phytochemical evaluation, Arabian Journal of Chemistry 12, 4303–4319.

[3] Kamalon Rajitha, Kikkeri Narasimha Shetty Mohana, Saurav Ramesh Nayak, Mahesh Bhaskar Hegde, Ambale Murthy Madhusudhana, 2020,An efficient and eco-friendly anti-corrosive system based on Beeswax-Graphene oxide nanocomposites on mild steel in saline medium, Surfaces and Interfaces 18 100393.

[4] R. Saedah. Al-Mhyawi, 2014, Inhibition of mild steel corrosion using Juniperus plants as green inhibitior, African Journal of Pure and Applied Chemistry, Vol. 8(1), pp. 9-22, January, Academic Journals DOI: 10.5897/AJPAC2013.0497 ,ISSN 1996 - 0840 © .

[5] Li LF, P. Caenen, JP Celis , 2008,Effect of hydrochloric acid on pickling of hot-rolled 304 stainless steel in iron chloride-based electrolytes. Corros Sci, 50:804–810.

[6] MA Quraishi, R. Sardar , 2002, Aromatic triazoles as corrosion inhibitors for low carbon steel in acidic environments. Corrosion, 58:748–755.

[7] M Karakus, M Sahin, S Bilgic , 2005,An investigation on the inhibitioneffects of some new dithiophosphonic acid monoesthers on the corrosion of the steel in 1 M HCl medium. Mater Chem Phys, 92:565–571.

[8] EM Sherif, SM Park, 2006, Inhibition of copper corrosion in acidic pickling solutions by N-phenyl-1,4-phenylenediamine. Electrochim Acta, 51:4665–4673.

[9] F Zucchi, IH Omar, 1985, Plant extracts as corrosion inhibitors of mild steel in HCl solutions. Surf Technol 24(4):391–399.

[10] G. Gunasekaran, LR. Chauhan, 2004, Eco friendly inhibitor for corrosion inhibition of mild steel in phosphoric acid medium. Electrochim Acta,49:4387–4395.

[11] Li Y, P. Zhao, Q. Liaqng, B. Hou , 2005, Berberine as a natural source inhibitor for mild steel in 1 M H2SO4. Appl Surf Sci , 252:1245–1253.

[12] AY El-Etre, M Abdallah, ZE El-Tantawy, 2005, Corrosion inhibition of some metals using lawsonia extract. Corros Sci47:385–395.

[13] AY El-Etre, 2006, Extract as inhibitor for acid corrosion of SX 316 steel. Appl Surf Sci 252:8521–8525.

[14] KO Orubite, NC Oforka, 2004, Inhibition of the corrosion of mild steel in hydrochloric acid solutions by the extracts of leaves of Nypafruticans Wurmb. Mater Lett 58:1768–1772 .

[15] GO Avwiri, FO Igho, 2001 Inhibitive action of Vernonia amygdalina on the corrosion of aluminum alloys in acidic media. Mater Lett 57:3705–3711.

[16] A Boujakhrout, I Hamdani, O. Krim, A Bouyanzer, RV Santana, A Zarrouk, B Hammouti, H Oudda , 2005, Kimbiolongo extract as corrosion inhibitor for mild steel in 1.0 M HCl. Der Pharmacia Lettre, 8(2):180–187.

[17] AH Al-Moubaraki, AA Al-Howiti, Al-Dailami, MM Al-Ghamdi EA, 2017, Role of aqueous extract of celery (*Apium graveolens* L.) seeds against the corrosion of aluminum/sodium hydroxide systems. J Environ Chem Eng 5(5):4194–4205.

[18] V Koundal, R Haldhar, A Saxena, D Prasad ,2017,AIP natural nonpoisonous green inhibitor of *Glycyrrhiza glabra* for mild steel in 3.5% NaCl. In: Conference proceedings, 1860, 020063.

[19] Z Sun, A Singh, Xu X, Liu W, Lin Y ,2017, Inhibition effect of pomelo peel extract for N80 steel in 3.5% NaCl saturated with CO2 solution. Res Chem Intermed 43:1–18.

[20] RS Nathiya, Raj V ,2017, Evaluation of *Dryopteris cochleata* leaf extracts as green inhibitor for corrosion of aluminum in 1 M H2SO4. Egypt , J Pet 26(2):313–323.

[21] A Pradityana, W. Sulistijono, , B. Luwar, M. Mursid, 2017, Effect of temperature on the application of Myrmecodia pendans extract for environmentally friendly corrosion inhibitor. In: AIP CONFERENCE proceedings, 1840, 030001.

[22] R Kusumastuti, RI Pramana, JW Soedarsono, 2017,The use of *Morinda citrifolia* as a green corrosion inhibitor for low carbon steel in 3.5% NaCl solution. In: AIP Conference proceedings 1823, 020012.

[23] A. Ehsani, MG. Mahjani, M Hosseini, R Moshrefi, H Mohammad Shiri, ,2017,Evaluation of Thymus vulgaris plant extract as an eco-friendly corrosion inhibitor for stainless steel 304 in acidic solution by means of electrochemical impedance spectroscopy, electrochemical noise analysis and density functional theory. J Colloid Interface Sci 490:444–451 .

[24] M Shahidi, G Golestani, MR Gholamhosseinzadeh, 2017, Mentha spicata L. extract as a green corrosion inhibitor for carbon steel in HCl solution. Phys Chem Res 5(2):293–307.

[25] MG Sethuraman, V Aishwarya, C. Kamal, T. J. I.Edison,2017, Studies on Ervatinine—the anticorrosive phytoconstituent of *Ervatamia coronaria*. Arab J Chem, 10:S522–S530. .

[26] HS Gadow, MM Motawea, HM Elabbasy ,2017, Investigation of myrrh extract as a new corrosion inhibitor for α-brass in 3.5% NaCl solution polluted by 16 ppm sulfide. RSC Adv, 7(47):29883–29898 .

[27] K. Anbarasi, VG. Vasudha, 2017, Influence of ecofriendly plant material (Cucurbita maxima) on mild steel corrosion. Anti-Corros Methods Mater, 64(5):492–498 .

[28] M. Shyamala, A. Arulananthan ,2017,A comparative study on the inhibitory action of some green inhibitors on the corrosion of mild steel in hydrochloric acid medium. Malays , J Anal Sci , 21(2):346–355 .

[29] R. Venegas, F Figueredo, G Carvallo, A Molinari, R. Vera, 2016,Evaluation of *Eulychnia acida* Phil. (Cactaceae) extracts as corrosion inhibitors for carbon steel in acidic media. Int J Electrochem Sci 11:3651–3663 .

[30] Standard , ASTM, G1-90 (Re approved 1999),Standard Practice for Preparing, Cleaning and Evaluating Corrosion Test Specimens, Philadelphia, PA, ASTM.

[31] A. Schweitzer Philip, 2010, A Fundamental of corrosion: mechanism, causes, and preventative methods, Handbook, CRC Press.

[32] R. Roberge, Pierre, 2000, Hand book of corrosion engineering, Handbook, McGraw- Hill.

[33] Standard ASTM, G102-89,1994,Standard practice for calculation of corrosion rates and related information from electrochemical measurement, Note section 6.

[34] M. Aliofkhazraei, 2014, Developments in corrosion protection handbook, Publisher: In Tech.

[35] R. Laamari, J. Benzakour, F. Berrekhis, A. Abouelfida, A. Derja, and D. Villemin,2011, Corrosion inhibition of carbon steel in hydrochloric acid 0.5 M by hexa methylene diamine tetramethyl- phosphonic acid, Arabian journal of chemistry 4, No. 3, pp. 271-277 .

[36] A. S. Fouda, , A. M. EI- desoky and H. M Hassan., 2013, Quinazoline derivatives as green corrosion inhibitors for carbon steel in hydrochloric acid solutions, Int. J. Electrochem. Sci 8:pp.5866- 5885 .

[37] M.B. Harb, S. Abubshait, N. Etteyeb, M. Kamoun, 2020, Olive leaf extract as a green corrosion inhibitor of reinforced concrete contaminated with seawater, Arabian Journal of Chemistry 13, 4846–4856 .

[38] A. D. Aprael S. Yaro, A. Anees B. Khadom, R. K. Wael , 2013, Apricot juice as green corrosion inhibitor of mild steel in phosphoric acid, Alexandria Engineering Journal 52, 129–135 .

[39] A. Lebe Nnanna, I. O. Owate, E. E. Oguzie, 2014,Inhibition of Mild Steel Corrosion in HCl Solution by Pentaclethra macrophylla Bentham Extract, International Journal of Materials Engineering, 4(5): 171-179 DOI: 10.5923/j.ijme.20140405.02 .

[40] E. Aveen, , B. Rammath, , C. Elanchezhian, , M. Naziruden, 2016, Influence of organic corrosion inhibitors on pickling corrosion behavior of sinter forged C45 steel and 2% Cu alloyed C45 steel, Journal of alloys and compounds, vol. 695, pp. 3299-3309.

[41] H. Jafari, , I. Danaee, H. Eskandari, , M. Rashv, and Avei, 2014,Combined Computational and Experimental Study on the Adsorption and Inhibition Effects of N_2O_4 Schiff Base on the Corrosion of API 5L Grade B Steel in 1 mol/ HCl, Journal Material Science Technology, vol. 30, No.3, pp.239-252 .

[42] S. Benabid, , T.Douadi, , S.Issaadi, , C.Penveme, , and S. Chafaa, 2016,Electrochemical and DFT studies of a new synthesized Schiff base as corrosion inhibitor in 1 M HCl , Journal of the International Measurement Confedration, vol. 99, pp.55-63 . .

[43] G. Khan, , W Basirun, S.Kazi, , P.Ahmed, , L. Magaji, , S.Ahmed, G.Khan, and M.Rehman, 2017, Electrochemical investigation on the corrosion inhibition of mild steel by quinazoline Schiff base compounds in hydrochloric acid solution, Journal of colloid and interface science, vol. 502, pp.134-145.

[44] V.S. Sastri, , 2012, Green corrosion inhibition: theory and practice, John Wiley& sons, New York .

[45] E.Ituen, , O.Akaranta, , A.James, , and S. Sun, , 2016,Green and sustainable local biomaterials for oil field chemicals Griffonia simplicifolia extract as steel corrosion inhibitor in hydraulic acid, Journal of Sustainable Materials and Tehnologies, pp.1-15 .

[46] D. Daoud , T. Douudi, , S.Issaadi, , and S.Chafaa, 2014,Adsorption and corrosion inhibition of new synthesized thiophene Schiff base on mild steel X52 in HCl and H_2SO_4 solution , corrosion science, vol.79, pp.50-58 .

[47] S Umoren, U. Eduok, , M.Solomon, and A. Udoh , 2016, Corrosion inhibition by leaves and stem extracts of sida acuta for mild steel in 1M H_2SO_4 solutions investigated by chemical and spectroscopic techniques ,Arabian Journal of Chemistry, King Saud University, vol. 9, pp.S209-S224 .

[48] C.Verma, M.Quraishi, , K.Kluza, , M.Janusik, , L.Olasunkanmi, E. Ebenso, 2017,Corrosion inhibition of mild steel in 1 M HCl by D-glucose derivatives of dihydropyrido [2,3-d:6,5-d] dipyrimidine-2,4,6,8(1H,3H,5H,7H)- tetraone, , www.nature.com/scientificreports .

[49] L.Garverick,1994, Corrosion in the petrochemical industry, , online,http://www.asminternational.org .

[50] V.Kumar, ,and B.Venkatraman,2016, Influence of polythiophene –Anthraqinone Sulphonic Acid (PTh-aqsa) Nanopowder on corrosion inhibition of mild steel in HCl solution, pp 1-6 .

IOP Conf. Series: Materials Science and Engineering **987** (2020) 012020 doi:10.1088/1757-899X/987/1/012020

Influence of Gas Flow Rate on Plasma Parameters Produced by a Plasma Jet and its Spectroscopic Diagnosis Using the OES Technique

Kadhim A Aadim[1], Sabah N. Mazhir[2 *], Nisreen Kh. Abdalameer[2], Alyaa H. Ali[2]

[1]Dept. of Physics, College of Science, University of Baghdad
[2]Dept. of Physics, College of Science for Women, University of Baghdad
*Corresponding author: sabahnm_phys@csw.uobaghdad.edu.iq

ABSTRACT: In this study, the emission spectra of plasma generated from the argon gas in a plasma jet system were measured under normal atmospheric pressure, at constant voltage and for different flow rates from 1–5L/min. The plasma parameters were calculated based on electron density, frequency of plasma, the temperature of an electron, Debye length and the number of particles in the Debye sphere. We employed optical emission spectrometer (OES) technology, which captured the spectrum resulting from the plasma at various flow rates of argon gas. While the flow rate of argon gas to the plasma generated from the discharge current (D.C.) increased, the ranges of the temperature and density of the electron (T_e) were 0.075–0.1eV and 6.15–9.75x10^{17}cm^{-3}, respectively. In contrast, a rise in the intensity of spectral lines was observed.

Keywords: argon gas; Debye length, OES, plasma jet, plasma parameters

1. Introduction:

Plasma is usually a quasi-neutral gas of charged and neutral particles that exhibits collective behaviour, which incorporates charged particles (electrons, ions, and molecules). The word 'ionised' refers to the life of a free-electron but does not refer to an atom or molecule. Plasma has free charging particles with both the negative and positive charges storing roughly at the macroscopic stage. This occurs when the elements for a particular state of matter are heated to a temperature that is higher than thermal energy and above binding energies [1]. A diagnostic of the plasma aims to get information about plasma parameters using various experimental methods. It is necessary to know the characteristics of plasma to understand the impacts of the various physical proceedings that take place and to determine plasma properties from these proceedings. Such diagnostics include electrical samples and optical spectroscopy of emissions [2].

Typically, only probes are regularly used to measure the density and temperature of ion and electrons. Nevertheless, in magnetised or deposited plasmas, analysing the characteristics of the probe is challenging and may lead to incorrect results [3].

Spectroscopic methods, for instance, dispersion of laser, emission, absorption or spectrometry of fluorescence are specific methods of sampling different pieces of plasma in the absence of disrupting its status and structure [4]. Optical emission spectroscopy (OES) is one of these techniques [5]; it depends on the calculation of the optical radiation released from the plasma because it represents plasma properties in the chemical, molecular and ionic radiator's immediate environment [6]. The electrons or ions interact with the other particles of the plasma that give radiation. Electron interactions tend to take control of the procedures and the excitation of collision and ionisation because of their high velocity. Three electron transitions can take place during these interactions: boundary transitions, significant transitions and free transitions. As a consequence of these transformations, emitted light forms a band, line or progressive spectra. When the electrons pass with various levels of energy into the atomic system, the ions and atoms for active gas and trace defects emit radiation. Contrary to continuum radiation in the free-electron, for example, bremsstrahlung, the emission is in the form of short spectral lines [7].

Under atmospheric pressure discharge conditions, non-equilibrium plasmas can be produced, most of which can be handled significantly. These discharge plasmas can be generated at the liquid state, where the numbers of applied fields increase, as in the medical, food, agricultural and textile industries [8–11].

McWhirter [12] suggested four plasma models based on the mechanism of electron interaction. These models are local-thermal-equilibrium (LTE), stable-state corona, time-dependent corona and collision radiative. For low-density plasmas, spontaneous decay balances the collisional excitation and radiative recombination balances collisional ionisation. The population densities of the new rates are calculated for such low-density plasmas from a balance between radiative decay to ground state and collisional excitations [13].

The temperature of the plasma is a significant thermodynamic property because of its ability to identify and forecast specific characteristics of plasma; for example, populations with particle velocity distribution and relative energy levels. The two-line hydrogen ratio approach is utilised in this laboratory experiment. The ratio is a common way of measuring the temperature of plasma that can be determined by the intensity ratio for a pair of atom spectral lines or ions of a similar level of ionisation [14].

For LTE, the electron temperature (T)eV is determined from the equation below [15]:

$$T_e = \frac{-(E1-E2)}{KLn\left(\frac{\lambda 2\ I2 g1\ A1}{\lambda 1\ I1\ g2\ A2}\right)} \ldots\ldots\ldots\ldots\ldots\ldots\ldots\ldots (1)$$

where g represents the statistical weight, λ represents the wavelength, E represents the excited state energy in eV, I represents the intensity, A represents the transition probability and k represents the Boltzmann constant.

There are quite a few reliable methods for the determination of electron density, which include the spectroscopy of plasma, interferometry of the laser, Thomson scattering and microwave. A well-known technique for determining electron density uses a linear Stark spectral line extension, which is used in this study. Broadening the line in a plasma jet is fundamentally brought about by the effect of Stark and Doppler's width; the latter only depends on the temperature and atomic mass of the emitting species. This form of expansion is ignored in this test as the width of Doppler's usually ranges between 0.04–0.07 nm [16, 17]. Broadening the pressure can be achieved by the Stark effect as it embraces the interactions for both neighbouring particles and radiators. These interactions are the product of ion collisions between fewer electrons. The primary responsibility of the Stark effect represents the enlarging of the hydrogen line that is utilised in this study [18].

The Saha–Boltzmann formula uses spectral lines for the element itself and sequential stages of ionisation and is introduced as [19, 20]:

$$n_e = \frac{I_1}{I_2^*} 6.04\times 10^{21} (T)^{3/2} e^{\frac{E1-E2-X_Z}{KT}} cm^{-3} \ldots\ldots\ldots (2)$$

where

$$I_2^* = \frac{I_2\,\lambda_2}{g_2\,A_2} \ldots\ldots\ldots\ldots\ldots (3)$$

X_z is the organism's ionising potential at ionising level Z in eV, I_2 represents the intensity of line for a transfer from top-level -2 to level -1, λ is the congruous transition wavelength from level -2 to level -1, g_2 is the statistical move weight from level -2, A_2 is the transition eventuality from level -2 to level -1 and the subscript z denotes the species ionisation stage for the referred species [21].

The frequency of plasma is determined from the following equation [22]:

$$f_p = 8.98\sqrt{n_e}\,(Hz) \ldots\ldots\ldots (4)$$

This frequency, which only depends on density, is one of the basic parameters of plasma. Plasma frequency is usually very high due to the small mass of electrons [22].

A charged particle reaction that decreases the impact of electric on local fields is named Debye shielding (λ_D), where the shielding accords the plasma its property of quasi-neutrality. The distance (λ_D) can be named as the length of Debye, which was previously described as [23]:

$$\lambda_D = \sqrt{\frac{\varepsilon_\circ K_B T_e}{n_e e^2}} = 743 \times \sqrt{\frac{T_e}{n_e}} \ldots (5)$$

Debye length should be minimal when compared to the system dimension. This first condition for plasma existence is described as follows [24]:

$\lambda_D <<< L$

L is the system dimension (cm) and $N_{D.}$ represents the particles' number density on the surface of Debye based on the density and temperature of the electron. This second condition for plasma existence $N_{D.} >>> 1$ is as follows [25]:

$$N_D = \frac{4\pi}{3} n_e \lambda_D^3 \ldots\ldots\ldots (6)$$

The specific number of particles in a Debye sphere is given by plasma parameter [26]:

$$\Lambda = 4\pi n \lambda_D^3 = \frac{1.38 \times 10^6 T_e^{\frac{3}{2}}}{n^{\frac{1}{2}}} \ldots\ldots (7)$$

where Λ is the plasma parameter?

In this paper, Ar plasma is described in terms of argon fraction and operating parameters utilising OES to investigate the production of active species. The electron temperature, which also influences the plasma reactivity through inelastic collisions producing the active species, is evaluated from ArI spectral line intensities using the Boltzmann plot method. The key aims of this paper are to underpin the plasma reactivity species and to achieve a better understanding of the mechanisms regulating the development of these species.

2. Experimental section

2.1 Plasma jet system

An atmospheric plasma jet was used to produce cold plasma with various flow rates of argon gas at a constant voltage. The plasma system consists of the following parts:

- Argon gas: to generate cold plasma.
- Gas flow meter: to monitor the gas intake that connects to the hollow metal tube with a calibrator of 1–5 L/min.

- Hollow metal tube: stainless steel tube that is 3cm long with an internal diameter of 2 mm.
- Power supply (D.C.): high voltage between 10 and 20 kV and a cut frequency of 25 kHz.

2.2 Experiment setup

Figure 1 illustrates the work steps for this research.

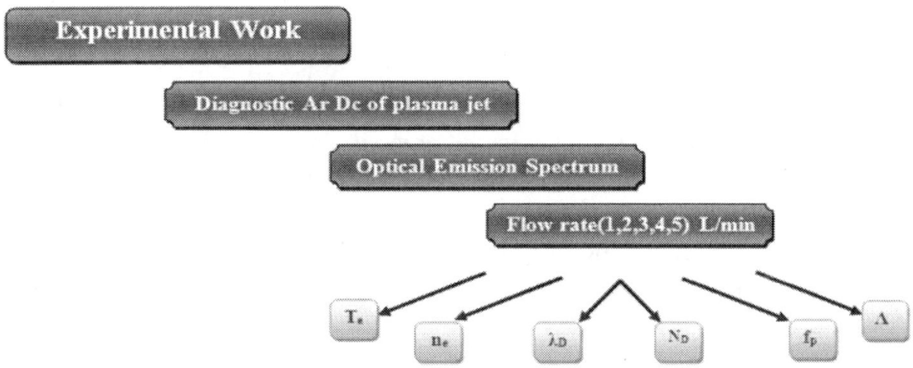

Figure 1 Diagram of the experiential framework of the plasma jet

The 3 cm-long hollow metal tube is made of stainless steel with an internal diameter of 2 mm. This metal tube is connected to the anode electrode of the high-voltage power supply. It has a flat end measuring 1 × 1 cm, and the spectrum is recorded for different flow rates of argon gas (1–5 litres per minute). Figure 2 shows a schematic diagram of the plasma jet configuration.

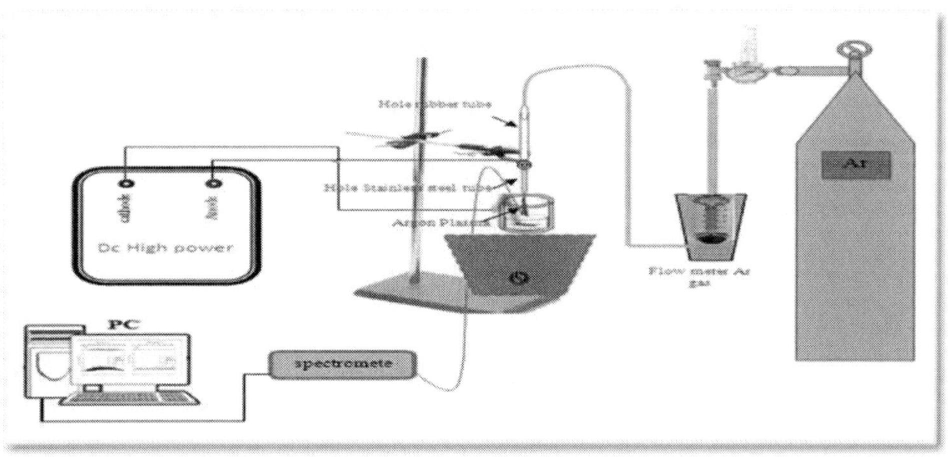

Figure 2 Device design for the plasma jet and OES technique

The spectrometer has a high resolution and responds with 3,648 pixels to a wavelength between 200–900 nm. The spectrum of Ar plasma was obtained within a wavelength of 690–780 nm for each flow rate of gas (1–5 L/min).

Eventually, the data from the National Institute of Technology and Standards (NIST) database was analysed and measured [27]. The parameters and characteristics of the plasma were determined afterwards.

3. Results and discussion

3.1 Evaluation of plasma parameters

3.1.1 Optical emission spectroscopy

Plasma emissions with different flow rates produced by atmospheric pressure in the plasma jet system were recorded. The resulting distribution of the spectrum was then plotted as amplitude against wavelength. The spectra of the optical emission of argon gas to generate cold plasma were recorded using the OES technique. A spectrum consists of several characteristic spectral lines of a particular atom and ions of Ar. Figure 3 shows the emission spectra of argon gas to generate cold plasma in a fundamental wavelength with a spectral range of 690–780 nm and flow rates of 1–5 L/min, in comparison with sharp standard lines for ArI and ArII [28]. The optical emission spectra of Ar were measured using the OES technique for the cold plasma.
Figure 3 demonstrates that the peak flow rate of argon has an essential and robust effect on emission line intensities, where the intensities of the spectral lines increase with the increasing gas flow rate as more molecules are passing through the tube. Most of the gas molecules passing through the plasma needle ionise the gas passing through the state of the plasma, which leads to an increasing number of excited atoms and, hence, the peaks of spectral line intensities.

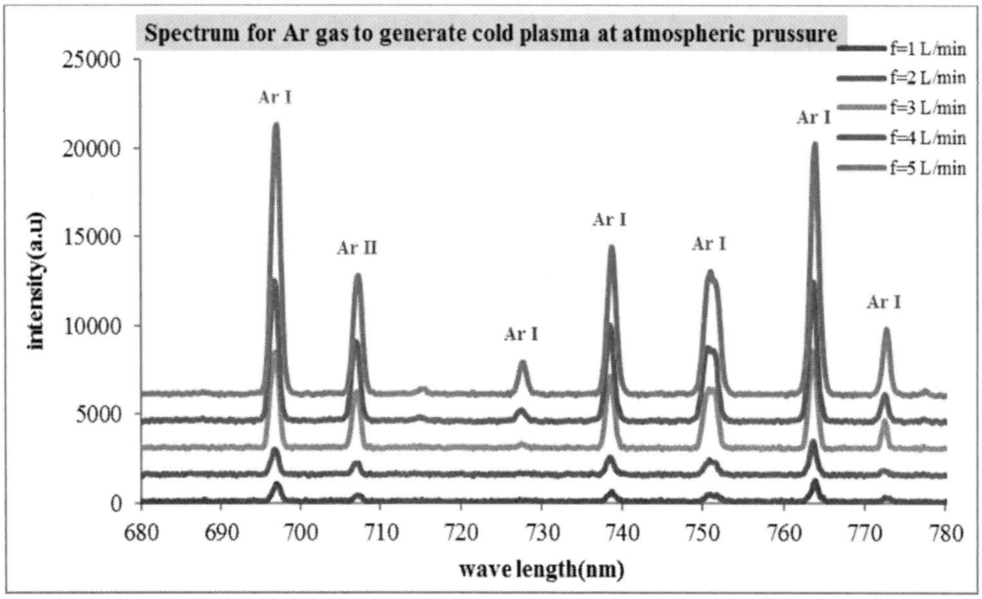

Figure 3 Emission spectra Ar gas to generate cold plasma at atmospheric pressure with different gas flow rates

For temperature measurements, the lines were taken at 696.73, 738.31 and 774.50 nm. Figure 4 shows the Boltzmann plot where the electron temperature is equal to the reverse of the slope of the fitting line. R^2 is a statistical coefficient that indicates the interest of the linear fit and takes a value between 0 and 1, with the best having an R^2 value closer to 1. Figure 5 demonstrates the Gaussian curve fitting of the ArI line and the variation of the FWHM broadening with different flow rates.

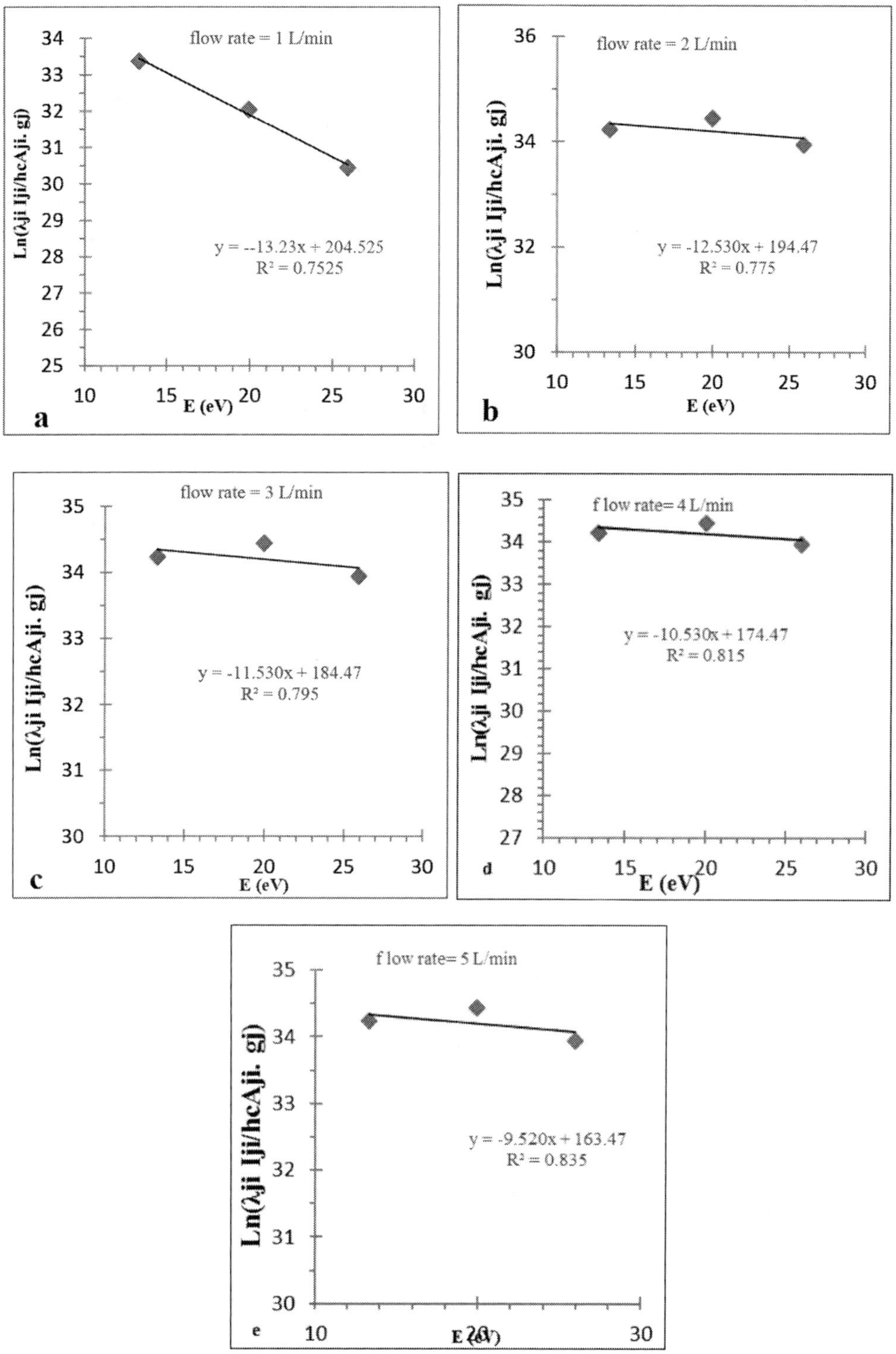

Figure 4 Boltzmann plot for argon gas with different flow rates: (a) 1 L/min, (b) 2 L/min, (c) 3 L/min, (d) 4 L/min, (e) 5 L/min

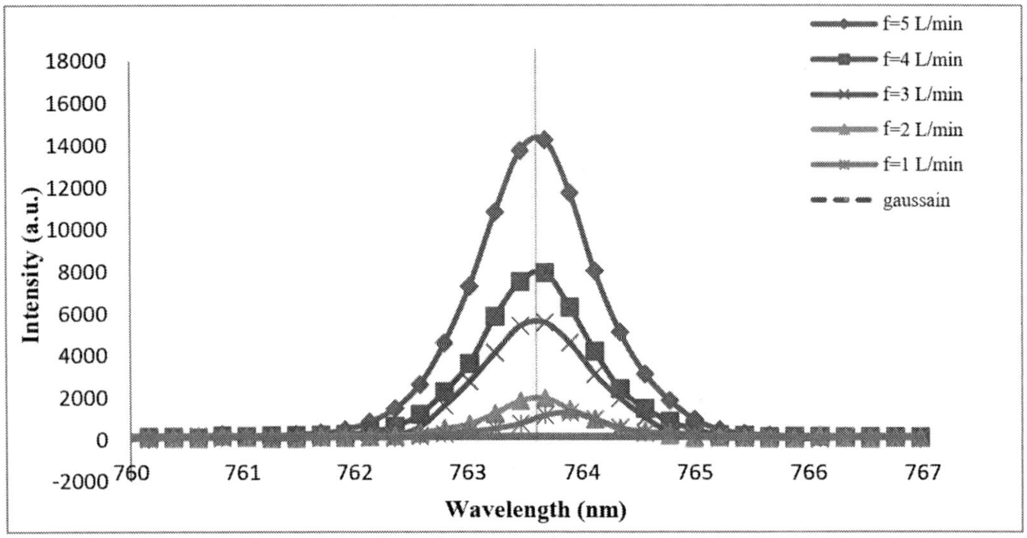

Figure 5 Stark broadening and the Gaussian curve fitting to argon discharge from 760 to 767 nm

3.1.2 Determination of temperature and density of the electron

The temperature of the electron (T_e) was measured from the line strength ratio using Eq (1); spectra from the OES were considered and the corresponding values of different parameters were taken from the NIST database [27]. The densities of the electrons were determined using Saha–Boltzmann in Eq (2) and the measured electron temperature.

The line graph in Figure 6 illustrates the influence of the change of gas flow rate on the temperature (T_e) and the density (n_e) of the electron. It is clear from this line graph that T_e and n_e increase gradually with an increased flow rate in the cold plasma system [28]. It is also shown that electron temperature and density at 1 L/min flow rate are 0.075 eV and $6.15x10^{17}$ cm^{-3}, whereas at 5 L/min the values are 0.10 eV and$9.75x10^{17}cm^{-3}$, respectively.

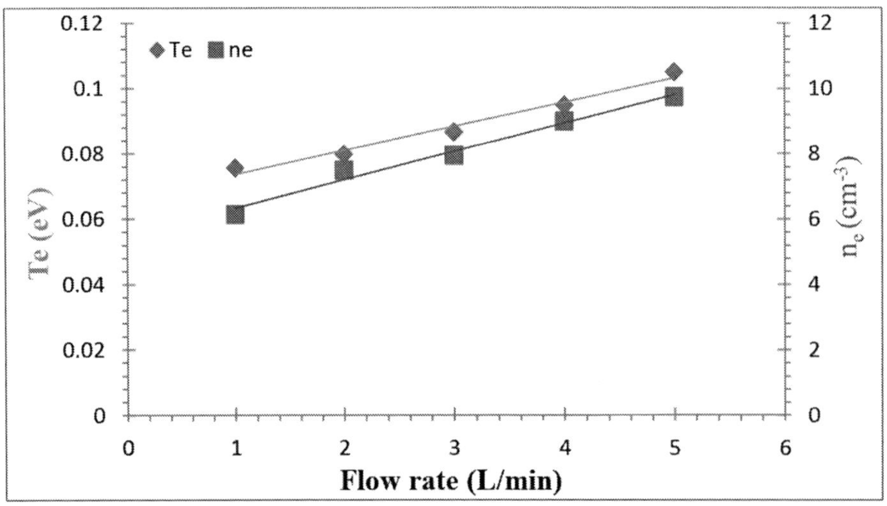

Figure 6 Change in electron temperature and density as a function of the gas flow rate

IOP Conf. Series: Materials Science and Engineering **987** (2020) 012020 doi:10.1088/1757-899X/987/1/012020

Figure 7 a, b and c show that the Debye length (λ_D) and plasma parameters (Λ) decrease. In contrast, the plasma frequency (f_p), and Debye number (N_d), increases with an increase in Ar gas flow rate because it is proportional with n_e.

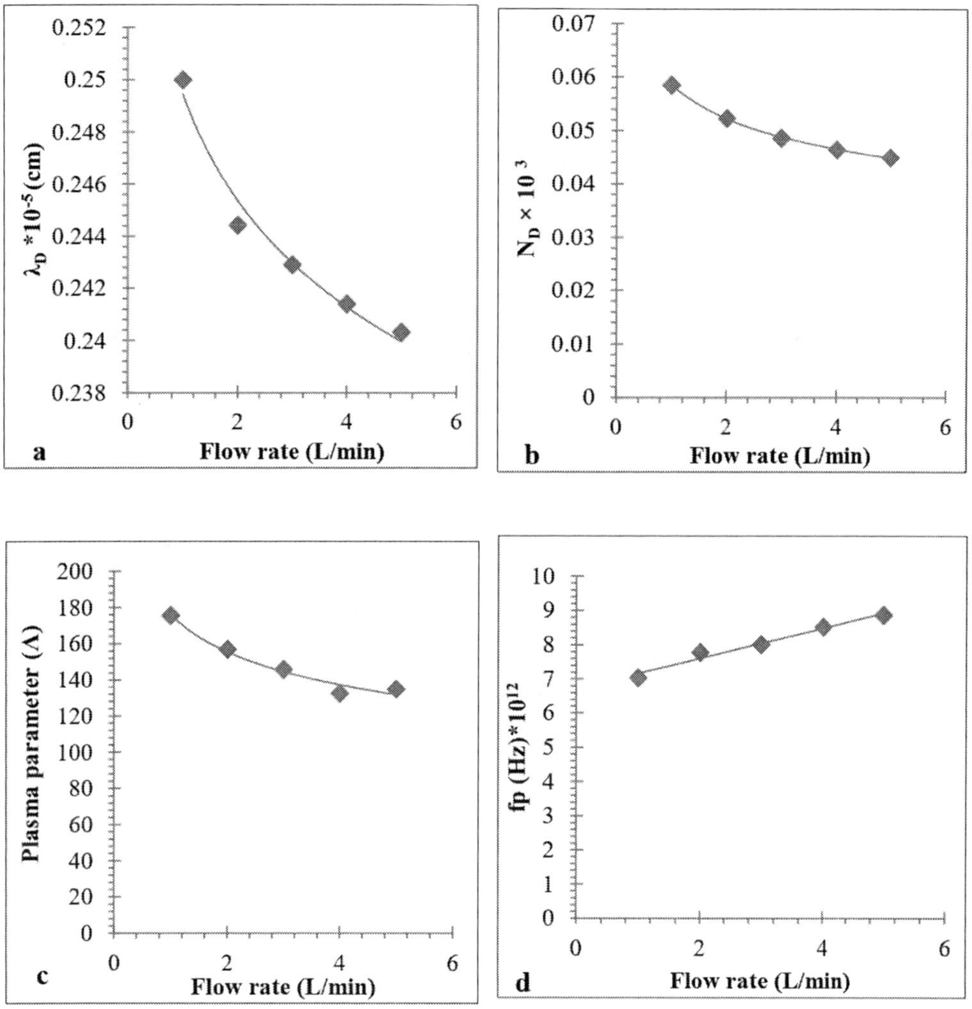

Figure 7 Determination of (a) Debye length(λ_D), (b) Debye number (N_d), (c) plasma parameters (Λ), (d) plasma frequency (f_p) as a function of different gas flow rates

Table 1 shows the T_e, n_e, λ_D, f_p, N_D and Λ for Ar gas at different flow rates. Criteria plasma were achieved through the results of the plasma parameters (λ_D, Λ), which showed a decrease with increasing flow rate because it is proportional to n_e. In contrast, f_p and N_D had an opposite (decreasing) trend. This agrees with the results reported by a previous study [29].

Table 1 Plasma parameters for Ar gas with different flow rates (L/min)

Flow rate (L/min)	FWHM	T_e (eV)	$n_e \times 10^{17}$ (cm^{-3})	$f_p \times 10^{12}$ (Hz)	$\lambda_D \times 10^{-5}$ (cm)	$N_d \times 10^3$	Λ
1	0.820	0.07558579	6.15	7.042289826	0.259426442	4.5E+01	175.4798
2	1.000	0.07980846	7.5	7.776908126	0.244418178	4.6E+01	156.8736
3	1.060	0.086730269	7.95	8.006816971	0.242890729	4.9E+01	145.7999
4	1.200	0.09491268	9	8.519176016	0.241393489	5.2E+01	132.5036
5	1.300	0.105042017	9.75	8.867039529	0.24031041	5.8E+01	134.8676

4. Conclusion

It has been concluded that the intensity of the emission spectrum lines of cold plasma generated by a plasma jet highly depends on environmental conditions. The investigation showed that the increase in the flow of argon gas leads to an increase in the intensity of emission, implying an increase in the number of gas molecules. This indicates that the energy supplied to the particles by the electrolyte field is sufficient to cause secondary ionisation of the molecules, thus, ionising most of the gas molecules passing through the plasma tube. Therefore, the passing gas turns into a plasma state, so the intensity appears at the lowest peak at a flow rate of 1 L/min, whereas it seems that its highest peak is at a flow rate of 5 L/min. We conclude from this that an increase in the gas flow rate contributes significantly to the effects on the plasma parameters of the density and temperature of an electron, Debye length, the frequency of plasma and the number of particles on the surface of Debye, which increases with an increased gas flow rate.

Reference:

[1] Najarian M L, Chinni R C 2013 Temperature and Electron Density Determination on Laser-Induced Breakdown Spectroscopy (LIBS) Plasmas, *J. Chem. Educ.* 90(2)244–247.https://doi.org/10.1021/ed3003385.

[2] Laux C O, Spence T G, Kruger C H and Zare R N 2003 Optical diagnostics of atmospheric pressure plasmas, *Plasma source sci. Technol.*12,25-138.

[3] Bech A, Hemmers D, Schweer H B and Uhlenbusch J 2000 Comparative measurement of electron density and temperature profiles in low-temperature ECR discharges by a lithium atom beam and Thomson scattering, *J. Phys. Appl. Phys*, 33,360–366.

[4] Hoppstock K, Harroson W W 1995 Spatial distribution of atoms in a dc glow discharge, *Anal. Chem.* 67:3167–3171.

[5] Meulenbroek R FG, Steenbakkers M F M, Qing Z, Van deSanden M C M and Schram D C 1994 four ways to determine the electron density in low-temperature plasmas, *Phys. Rev. E*, 49(3) 2272–2275.

[6] Behringer K and Fantz Y, Spectroscopic diagnostics of glow discharge plasmas with non-Maxwellian electron energy distributions 1994 *J. Phys. Appl. Phys.* 27,2128–2135.

[7] Hutchinson I H 2002 Principles of Plasma Diagnostics. Cambridge Univ. Press, U.K. 1-434

[8] Toyokuni S, Ikehara Y and Kikkawa F 2018 *Plasma medical science*. Cambridge, MA: Academic Press.

[9] Metelmann H R, Von Woedtke T and Weltmann K D 2018 Comprehensive clinical plasma medicine. New York, NY: Springer.

[10] Yang Y, Cho Y I and Fridman A 2017 Plasma discharge in liquid: water treatment and applications, *Boca Raton*, FL: CRC Press.

[11] Akatsuka H 2019 Advances in Physics: X, 4:1. DOI: 10.1080/23746149.2019.1592707

[12] McWhirter R W P, Plasma Diagnostics Techniques, NewYork: Academic. (1965).

[13] Edg A and Murat C 2019 Investigation of the plasma parameters of a laboratory argon plasma source using a collisional radiative model with the comparison of experimental and simulated spectra, *Spectrochimica Acta*, Part B.

[14] Yuan H, A. Gojani A, Gornushkin I and Wang X 2018 Investigation of laser-induced plasma at varying pressure and laser focusing. *Spectrochimica Acta*, Part B Atomic Spectroscopy. 150,33-37.

[15] Cremers A D and Radziemski L J 2013 Handbook of Laser-Induced Breakdown Spectroscopy, 2nd Edi. John Wiley & Sons, Ltd USA.

[16] Harilal S S, Bindhu C V, Issac R C, Nampoori V P N and Vallabhan C P G 1997 *J. Appl. Phys.* 82(5) 2140-2146.

[17] Khalaf M K, Mahdi M S, Taha S K and Bououdina M 2018 Influence of R.F. sputtering power on surface properties and biocompatibility of 316L stainless steel alloy by deposition of TiO_2 thin films, *Materials Research Express*. 6(3)035401. https://doi.org/10.1088/2053-1591/aaf2e9.

[18] Mazhir S N, Taha S K, Harb N H and Khalaf M K 2020 A Study of Plasma parameters in gold sputtering System by Means of Optical Emission Spectroscopy, IOP Conf. Series: *Materials Science and Engineering*. 871, 012081.

[19] Naeema N, Kudher A, Mohammed G H 2020 Study of the Spectroscopic Performance of Laser Produced CdTe, and CdTe:Ag Plasma, *IOP Conf. Ser.: Mater. Sci. Eng.* 757, 012025.

[20] Pae J Y, Medwal R, Vimal Vas J, MathamM V and Rawat R S 2019 Remote plasma-assisted low temperature large-area graphene synthesis, *J. Vac. Sci. Technol.* B 37, 041201. https://doi.org/10.1116/1.5093241

[21] Chen M C and Chen E C 1983 Introduction to Plasma Physics and Controlled Fusion, 1. Los Angeles.

[22] AlAamer A S and ElSherbini A M 2012 *World Journal of Nano Science and Engineering*. 2 (4)206-212.

[23] Smirnov B M 2001 *John Wiley & Sons*, Inc, 16(1) (2001).

[24] Chandra S 2010 Textbook of plasma physics, 1ed. India.

[25] Connell D O, Cox L J, Hyland W B, McMahon S J, Rreuter S, Graham W G and Gans T, J. C 2011 *Applied Physics letters*. 98, 043701. doi:10.10631/1.3521502.

[26] National Institute of Standards and Technology (NIST) Atomic spectra database, 5(2017).

[27] Sansonetti J E and Martin W C 2005 Spectroscopic data handbook of basic atomic, 34(4).

[28] Mazhir S N, Abdullah N A, Rauuf A F, Ali A H and Al-Ahmed H 2018 Effects of Gas Flow on Spectral Properties of Plasma Jet Induced by Microwave. *Baghdad Science Journal* Vol.15(1)81-86. http://dx.doi.org/10.21123/bsj.2018.15.1.0081.

[29] Muna E A and Kadhim A A 2019 Spectroscopic studying of plasma parameters for SnO_2 doped ZnO prepared by pulse Nd:YAG laser deposition, *Iraqi Journal of Physics* 17(42)125-135. Doi:10.20723/ijp.17.42.125-135.

Synthesis and Hemozoin Inhibitor of Side-Chain Modified Copper-Chloroquine Derivatives

*R A K Al-refaia[1] and A A Alkarimi[2]

[1] Lecturer in Biochemistry, Department of Chemistry, College of Science, University of Babylon, Iraq

[2]Assistant Professor in Analytical Chemistry, Department of Chemistry, College of Science, University of Babylon, Iraq

*E-mail: sci.rana.abdul@uobabylon.edu.iq

Abstract:

This study uses copper (I) as a transition metal to improve the activity of 4-aminoquinoline as an antimalarial agent. This chloroquine derivative was synthesised and tested for in vitro antimalarial activity using a simple colourimetric method compared to the conventional purification method to measure hemozoin formation. This compound has been characterised by the combination of NMR and IR spectroscopic methods. Copper-chloroquine (Cu-CQp) might strongly exhibit antimalarial activity after showing significant inhibition of hemozoin formation compared to commercial chloroquine (CQ). This is possibly due to its lipophilicity, which enhances cell permeation. The highest activity was shown by the Cu-CQp complex in comparison to that of commercial CQ. Cu-CQp complex and CQ were used in a range of concentrations from 10–50 μM.

Keywords: chloroquine, hemozoin, malaria, trans-metals

1. Introduction:

Malaria is a common problem around the world, particularly in developing countries. The parasite causes haemoglobin degradation in the host erythrocytes so that it can consume the final product, which includes amino acids to build its protein during proliferation, which is necessary for the parasite to survive [1, 2]. The parasite can detoxify free heme in different ways, such as degradation with reduced glutathione and neutralisation with protein-rich histidine [3–6]. It has been shown that hemozoin is identical to b-hematin (BH) both chemically and structurally [7]. Recently, many reports show that the best target for antimalarial selection is destroying or reducing BH formation.

Currently, 4-aminoquinoline derivatives in drug form, for example, chloroquine (CQ), have been successfully used for malarial treatment, but the most dangerous parasite, called *Plasmodium falciparum (P. falciparum),* has improved its resistance against these types of drugs [8]. Therefore, new antimalarial drugs are needed to control malaria [9], and the most encouraging method includes the use of transition metals [10]. Research has been conducted to develop the effectiveness of known drugs like chloroquine, one of those involved in the modification of drug structures with transition metals [11]. However, many studies have demonstrated that numerous transition metals can inhibit the resistance of malaria against chloroquine-sensitive and chloroquine-resistant strains of *P. falciparum*. It is well known that several metal complexations of chloroquine and its derivatives have remarkable properties as a drug compared to chloroquine (CQ), which has been used recently for chemotherapy treatment. A reliable improvement in strengthened metal-chloroquine conjugation suggests that metals conjugated with chloroquine are an essential source for generating a new area of antimalarial development structurally and/or chemically

[11, 12]. Recently, a good example of metal-chloroquine conjugation as an antimalarial agent is gold chloroquine (Au-CQ), which is more active than CQ against chloroquine-resistant strains of *P. berghei* and *P. falciparum* in vivo and in vitro [23]. It was also published that the main reason for the increase in antiplasmodial activity in vitro against two different *P. falciparum* strains (D10 and W2) in comparison to the activity of the CQ alone was the coordination of gold (I) and thiosemicarbazone [13]. To remove or reduce the CQ-resistance of *P. falciparum* parasites, copper (Cu) was used to modify the short side chine of CQp (chloroquine-propionic acid), and the role of Cu to develop the activity of CQp was investigated against the activity of malaria [14]. As heme polymerisation is an important step in the formation of BH, a quantity of heme is needed to study the mechanism action of BH formation. Several papers have described how BH formation can be measured in vitro and how it can be characterised using IR or X-ray diffraction [15]. Another method is to use toxic material, such as pyridine combined with laborious purification steps [16]. In this study, simple colourimetric and conventional purification methods have been used to identify whether or not CQ-metal has any speciality in the field of malaria treatment.

2. Material and methods:

2.1 Reagents

All chemicals were purchased from Sigma-Aldrich UK, except for CQp, which was synthesised and fully characterised in the labs at Hull University as part of a PhD project.

2.1.1 Chemical synthesis of copper-chloroquine complex (Cu-CQp)

An ethanolic solution of anhydrous copper (I) chloride (0.19 g, 1.99 mmol) was added to an ethanolic solution of the compound 3-((7-chloroquinolin-4-yl) amino) propionic acid (CQp) (0.5 g, 1.99 mmol). The mixture was refluxed for 2 h until the solution became brown. Then, the precipitate was filtered and dried under vacuum to form Cu-CQp compound as a dark brown solid (0.39 g, 43%).
IR: ionised carboxylic COO^- cm^{-1}, NH stretch at 3,340 cm^{-1} and C=O stretch at 1,730 cm^{-1}. ^{1}H NMR (400 MHz, DMSO-d6): δH/ppm 3.45 (t, 2H, CH2), 4.3 (t, 2H, CH2), 6.25 (d, J=6.12 Hz, 1H, ClQ-C3-H), 7.60 (dd, , J=7.2 Hz, 1H, ClQ-C6-H), 7.55 (d, J=7.8 Hz, 1H, ClQ-C5-H), 7.90 (d, J=6.8 Hz, 1H, ClQ-C8-H), 8.12 (d, J=7.2 Hz, 1H, ClQ-C2-H) and 8.15 (s, 1H, NH).

2.1.2 Beta-hematin preparation

A stock solution of heme was freshly prepared by dissolving hemin chloride in 0.2 M NaOH; to remove the remaining hemin crystal a stock solution centrifuged for 15 min at 7,000 g. The concentration was evaluated from the absorbance at 385 nm using ε_{mM} = 58,400 in 0.1 M NaOH [20]. All absorption spectra were recorded on a double-beam spectrophotometer. The formation of BH was performed following the published method with slight modifications [17, 18]. Briefly, heme (10 ml) was heated in an acetate buffer (0.56 M, pH 5.5) in the absence of CQ or CQp at 70 °C or at 75 °C with a range of concentrations of CQp and Cu-CQp (10–50 μM). At the identified times, 2 ml of heme solution was withdrawn and BH formation was calculated using two methods:

1- Simple colourimetric method

After heating the mixture of heme solution and heme control separately in an acetate buffer, the absorbance of each sample was investigated at 400 nm and 700 nm to confirm hemozoin formation. Consequently, heme fractions were calculated after they were altered to BH using Eq (1).

$$Fractions = \frac{(A\ at\ 400nm - A\ at\ 700nm)control - (A\ at\ 400nm - A\ at\ 700nm)\ sample}{(A\ at\ 400nm - A\ at\ 700nm)control} ----(1)$$

2- Conventional purification method

After 2 hrs of heating, the heme solution was centrifuged at 7,000 g for 15 min. A 2 ml mixture of 0.2 M, pH 9.2 sodium bicarbonate and 2.5% sodium dodecyl sulphate were used to suspend the pellets. The insoluble material was improved by centrifugation after 1 h shaking at room temperature. This step was repeated twice. The final product of BH was decrystalised after incubation in 2% sodium dodecyl sulphate and 0.1 M NaOH. To evaluate the crystalised heme (BH), heme concentration (C) was calculated from the absorbance of the solution at 400 nm with an extinction coefficient of 10^5 [19]. Fractions were calculated using Eq (2):

$$Fractions = C\ before\ coverting\ to\ BH - C\ after\ converting\ to\ BH - - - - - - - - (2)$$

3. Results and discussion

In this paper, a copper-chloroquine was obtained by conjugating CQp with an anhydrous copper (I) chloride to give a new compound of Cu-CQp, which is more active than commercial CQ and CQp alone in in vitro treatment as a BH inhibitor. The disappearance of characteristic ^{1}HNMR of a singlet proton at 11.66 ppm, corresponding to 1H and COOH of CQp and OH broadband of IR spectra for the same compound at 3,010 cm^{-1}, provided good evidence for the production of the Cu-CQp compound. Additionally, the functional group of CQp with copper (I) ion, which was determined using the IR spectra, contained an intense band of asymmetric and symmetric stretching vibrations of the ionised carboxylic COO group involved in the coordination of Cu^{+1} (the bands at 1,545.4 and 1,326.1 cm^{-1}). This is in agreement with other published values [20]. After synthesis and characterisation, Cu-CQp was tested as a BH inhibitor using two different methods as described in the experimental section. In vitro tests using colourimetric and traditional purification methods were used to confirm whether or not Cu-CQp can work as the same mechanism of CQ that Saroj et al. suggested [21]. Before testing CQp and metal-CQp against the growth of heme crystallisation compared with commercial chloroquine, the effect of temperature on the rate of BH formation was studied. In the colourimetric method, all samples were withdrawn after 30-second intervals, and their absorbance at 400 and 700 nm was recorded using the double beam spectrophotometer. The results display the best incubation time that indicates the beginning of BH formation in each sample, nucleation time and the growth rate of heme crystals. A noticeable difference was found between CQp and Cu-CQp activity as a BH inhibitor compared to commercial CQ, as shown in Figure 1.

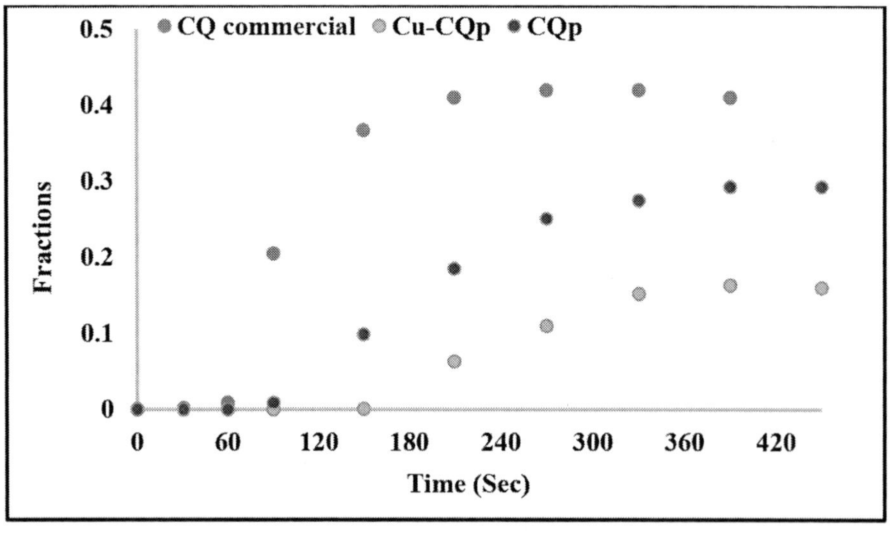

Figure 1: Effect of time on BH growth at 73 °C in the presence of 10 μM of CQ, CQp and Cu-CQp

As shown above, the quantity of heme-converted BH significantly changed when CQp was used as a heme crystallisation inhibitor compared to CQ, while the most significant inhibition was indicated after using Cu-CQp as an inhibitor agent of BH formation in comparison with both CQ and CQp. Moreover, there was a remarkable difference in the amount of BH crystallisation after treatment with Cu-CQp, and a noticeable change was observed after using commercial CQ as a BH inhibitor compared with untreated BH. This change increased when the concentration of each CQ-derivative increased from 10–50 μM. The reason might be associated with the ability of CQ-derivatives to conjugate with heme and block its active site to produce BH [22], as shown in Figure 2. The absorbance at 400 and 700 nm was recorded, and the fraction (f) of heme converted to BH was calculated. The results represent data from three independent experiments.

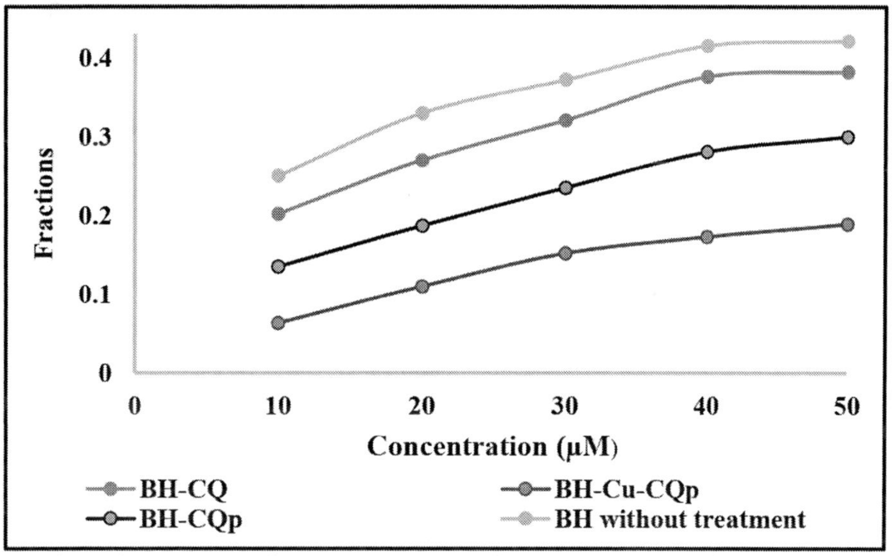

Figure 2: Effect of CQ, CQp and Cu-CQp concentrations on BH crystal growth at 75°C using the simple colourimetric method

Similarly, the results are a very good indicator of the role of Cu-CQp as an inhibitor of heme polymerisation growth, as shown in Figure 3. The fact that the free Cu of CQp obtained fairly low antimalarial activity compared to metal-CQp, suggested that the ability of the metal to bind CQp might play an important role in exhibiting antimalarial activity, rather than CQp or Cu alone [14].

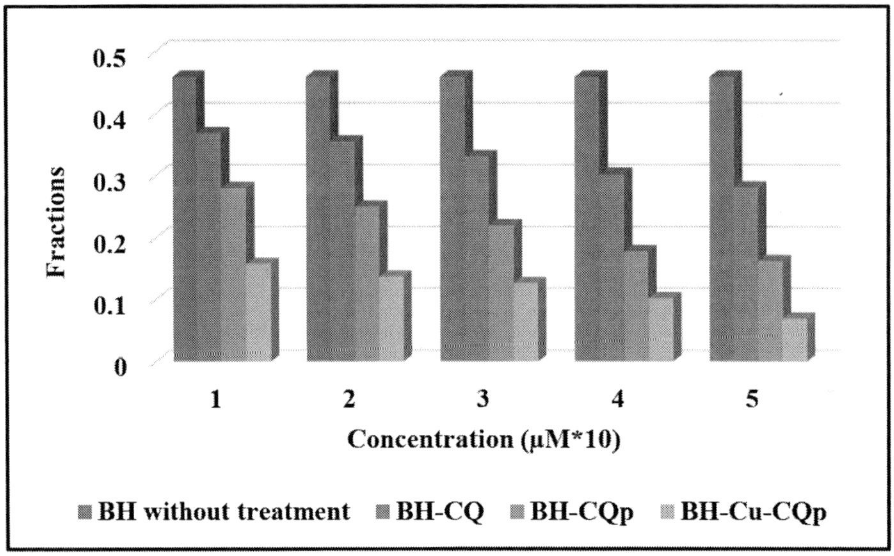

Figure 3: Effect of CQ, CQp and Cu-CQp concentrations on BH crystal growth at 75 °C using the conventional purification method

The growth rate of BH approximately matches in the two methods, thus, providing an acceptable relationship between the colourimetric and purification methods. Each experiment was performed in triplicate upon addition of CQ, CQp and Cu-CQp. As expected, when increasing the dosage of each compound, crystal growth began late, especially with Cu-CQp, thus, suggesting it is dosage-dependent.

4. Conclusion

In conclusion, the resistance of malaria against traditional antimalarial drugs has put pressure on public health systems to develop new therapeutics. The use of metal chloroquine as a new agent against BH formation may eliminate chloroquine resistance to traditional drugs. Additionally, using the colourimetric method provides useful information assimilated from BH crystal growth experiments and suggests that using an acetate buffer at high temperatures is convenient, safe and reproducible; it also outperforms other methods for monitoring BH formation.

5. References

1- Henk S, Piet A.K, Gerard J.S, and Petra F.M., 2006, *Malaria Journal*, **5(1):80.**
2- Eckman J.R., Modler S., Eaton J.W., Berger E., and Engel R.R., 1977, *J. Lab. Clin. Med.* **90**, 767–770.
3- Sullivan Jr. D.J., Gluzman I.Y., and Goldberg D.E., 1996, *Science*, **271**, 219–222.
4- Huy N.T., Serada S., Trang D.T., Takano R., Kondo Y., Kanaori K., Tajima K., Hara S., and Kamei K., 2003, *J. Biochem.*, **133**, 693–698.
5- Atamna H., AND Ginsburg H., 1995, *J. Biol. Chem.* **270**, 24876– 24883.
6- Huy N.T., Kamei K., Yamamoto T., Kondo Y., Kanaori K., Takano R., Tajima K., and Hara S., 2002, *J. Biol. Chem.* **277,** 4152–4158.
7- Francis S.E., Sullivan Jr D.J., and Goldberg D.E., 1997, *Annu. Rev. Microbiol.* **51**, 97–123
8- Kirandeep K., Meenakshi J., Ravi P. R., and Rahul J., 2010, *European Journal of Medicinal Chemistry*, **45**, 3245-3264
9- Marat K., Edwin G.T., and Mathew H.T, 2019, *Malaria Journal*, **18(1):93.**
10- Alim A., 2015, *Journal of Antimicrobial Chemistry,* **70(6)**.

11- Navarro M, Perez H, and Sanchez-Delgado RA. J. 1997, *Med. Chem.*, **40,** 1937–1937.
12- Adams, M., Li, Y., Khot, H., De Kock, C., Smith, P. J., Land, K., Chibale, K. and Smith, G. S., 2013, *Dalton Transactions*, **42,** 4677–4685.
13- Khanye SD, Smith GS, Lategan C, Smith PJ, Gut J, Rosenthal PJ, and Chibale K. J. 2010, *Inorg. Biochem.*, **104,** 1079–1083.
14- Timothy J. H., Prince N. A. A., Kimberly D. R., Natalie C. S., Randall D. M., Tarynn N. C. F., Amy N. C., Justin G. L., Stephen J. A., Shabana I. K., Tekwani B.L., and Faruk Khan M.O., 2014, *Bioorg Med Chem.*, **1; 22(13),** 3239–3244.
15- Egan T.J., Mavuso W.W., and Ncokazi K.K., 2001, *Biochemistry*, **40**, 204–213.
16- Chong C.R., and Sullivan Jr. D.J., 2003, *Biochem. Pharmacol.*, **66,** 2201–2212.
17- Taramelli D., Recalcati S., Basilico N., Olliaro P., and Cairo G., 2000, *Lab. Invest.*, **80,** 1781–1788.
18- Slater A.F., Swiggard W.J., Orton B.R., Flitter W.D., Goldberg D.E., Cerami A., and Henderson G.B., 1991, *Proc. Natl. Acad Sci. USA*, **88**, 325–329.
19- Sullivan Jr. D.J., Gluzman I.Y., Russell D.G., and Goldberg D.E., 1996, *Natl. Acad Sci. USA*, **93**, 11865–11870
20- Skorik N. A., Filippova M. M., Bukhol'tseva E. I., Mal'kov V. S., and Kurzina I. A., 2015, *Russian Journal of Inorganic Chemistry,* **60(6)**.
21- Verma S., Pandey S., Agarwal P., Verma P., Deshpande S., Saxena J.K., Srivastava K., Chauhan P. M. and Prabhakar Y. S., 20016, *RSC Advances*, **6,** 25584-25593.
22- Egan T. J., 2002, *Journal of Inorganic Biochemistry,* **91,** 19-26.

The effect of different at.% Ag elements on the wear rate of nitinol alloys fabricated using the casting method

Saja Mohammed Hussein[1], Khansaa D Salman[1] and Ahmed Abdulqader Hussein[1]

[1]Electromechanical Engineering Department, University of Technology, Baghdad, Iraq

50748@student.uotechnology.edu.iq, drkhansaa_ds@yahoo.com, ahmedabdulqaderhussein@gmail.com

Abstract. Shape memory alloys are considered smart alloys due to their properties, such as biocompatibility, superelasticity (SE) and shape memory effect (SME). These alloys are used in many applications (actuators, MEMS, biomedical and aerospace). In this work, the effect of adding the silver element to the NiTi alloy with different atomic percentages (zero, one, two and three at.% Ag) on mechanical properties, such as macrohardness and wear, was studied. The samples of this work were prepared by the casting method using a VAR furnace to obtain homogeneous shape memory alloys. The purity of the elements was Ni 99.2 %, Ti 99.7 % and Ag 99.99 %. In this work, many examinations were carried out to define the characteristics of the manufactured shape memory alloys, such as FESEM and OM, for microstructural analysis. Meanwhile, the mechanical properties were defined using macrohardness and wear tests. The results of the mechanical tests showed a slight impact in the percentage of the wear resistance and macrohardness at the point of increasing the silver element ratio to the binary alloy (NiTi). The photomicrographs of the OM and FESEM examinations showed that the Ag element is homogeneously distributed in the NiTi matrix. In addition, the FESEM examination showed the emergence of the martensite phase, austenite phase and some impurities. The results of the DSC test showed the degrees of the phase transformation for each alloy, and the SME recovery rate of 89.99% was achieved. Furthermore, the surface topography was analysed using an optical microscope for all the tested samples. The obtained results emphasised that the movement between the disc and the pins leads to a rise in the friction temperature and the formation of the layer of oxidation. These oxidation layers will be ploughed and uprooted to create many grooves at the surface of the samples. Hence, these grooves are similar for all samples and increased with increasing loads.

Keywords: Shape Memory Alloys, NiTiAg, Casting, VAR, Macrohardness, Wear.

1. Introduction

Smart alloys are considered a class of active materials (smart materials), which are used in applications requiring high precision. In particular, aerospace and medicine applications use the smart alloys widely, as these alloys possess the unique property of recovering their original shape after large deformation, which is known as the shape memory effect (SME) and pseudo elasticity (PE). Both of these properties constitute a phase transformation from the low temperature and high pressure (martensite phase) to the high temperature and low pressure (austenite phase) [1]. When a certain stress is applied to all metal materials, this stress is removed. Most of the metals return to their original shape, but when the applied stress exceeds the force of carrying the material, a plastic deformation occurs, which leads to a permanent deformation. Most metallic materials have a recovery rate that does not exceed 0.1 %, whereas shape memory alloys (SMAs) possess a seven % – eight % recovery rate [2].

The main problem is the manufacturing process of the SMAs as well as the type of alloy adopted, such as NiTi-based and the added element to this alloy of Cu, Ag, Pd or V. Each element has a certain effect on the mechanical and microstructure properties.

The main types of SMAs that have been used industrially are Ni-Ti-based, Fe-based and Cu-based. Among these three alloys, the NiTi-based alloys are the most widely used due to their excellent properties [3]. Nickel titanium binary alloys (nitinol) are probably the best-known materials in this class. However, these alloys are not always the most suitable for every purpose. One solution for improving this alloy is to add an element through the production process [4]. Each element has a specific impact on the properties of alloys; for example, adding Fe, V and Co causes the phase transformation temperatures to decrease, while elements, such as Hf, Zr, Au, Pd and Pt, have the opposite effect. Likewise, hysteresis decreases when adding the element Cu, whereas it increases when adding Nb and Fe [5]. Ag is one of the best electrically conductive materials that possesses high electrical conductivity and antibacterial effect biocompatibility and satisfactory stability. Therefore, it is predicted that the composite made with the NiTi and Ag component will have an excellent shape-memory effect, superelasticity, high electrical conductivity and good corrosion resistance [6–10].

The manufacturing process of SMAs is critical and has a significant impact on homogeneity, machinability, ductility, microstructure, biocompatibility and transformation temperatures. All of these properties are important in material applications. In general, there are two main methods of the manufacturing, as follows: powder metallurgy and casting methods, which are followed by several processes, such as cold and hot working, heat-treatments and surface treatments. Heat treatments have an important role in shaping the SMA [11]. The aim of this research is to manufacture an alloy (NiTiAg) using a VAR furnace with the addition of the silver element in atomic percentages (zero, one, two and three at.% Ag) to study its impact on macrohardness and wear resistance, as well as to examine the microstructure of the prepared alloy.

The rest of this paper consists of the following: Section 2 describes related works from previous studies; Section 3 focuses on experimental work, the manufacturing process, annealing, ingot cutting process and preparing samples for tests and examinations; Section 4 describes the results and discussion for all examinations and tests; and finally, Section 5 consists of the conclusions of this work.

2. Related Work

In this section, previous studies related to the NITI alloys-based casting techniques will be presented. Additionally, other relevant works concentrating on the impact of adding the silver element to the NITi alloys under different conditions and scenarios will be discussed.

The materials that were used for the casting method have important characteristics, such as high purity due to the presence of non-metallic elements. However, non-pure elements are greatly affected by the produced alloys. Non homogeneity leads to an irregular distribution and a change the properties of the products [11]. Vacuum induction melting (VIM) requires a graphite crucible. This method will contaminate the alloys with the carbon element. The carbon is considered a non-pure element; therefore, it will produce an alloy that does not have high purity [12]. On the other hand, the mechanism relies on the electron beam melting process, which produces high-purity alloys. The vacuum pressure rises during the melting process and leads to evaporation of some elements [13, 14]. Regarding the vacuum arc remelting (VAR), this method does not contain a graphite crucible and indicates a low percentage of the carbon element. Therefore, the alloy that is produced via this method is considered high purity, high quality and homogeneous [15]. Adding the silver element to alloys that are NiTi-based is presented in [16]. This method was prepared by using the electron beam melting (EBM) technique. The results of this study showed that it is difficult to add silver to the binary alloy due to the high vapor pressure of silver. Another addition of the silver element to the binary NiTi alloy using arc melting has been proposed in [17]. The purpose of manufacturing this alloy was to use this alloy in the medical and orthodontic fields. The results showed that the silver solubility and silver recovery rate are low due to silver element evaporation. However, the wear resistance improved with the increasing silver percentage, and the range of phase transformation was also achieved. Manufacturing a thin film NiTiAg-based on the sputtering technique is presented in [18,19]. In this method, the thin films were manufactured from NiTiAg using the sputtering technique to increase the silver content (the silver component that was added is almost pure). The results showed that the high percentage of silver does not affect the phase transformations of alloy (NiTi). An Arc melting technique was used to introduce pure silver deposits into the matrix TiNi, which is proposed in [6]. The results of this technique showed that the melting process is quite difficult due to the evaporation of the silver element, and the silver element is distributed randomly inside the matrix TiNi. These alloys showed a good corrosion resistance, higher tensile strength and greater elongation than binary alloys. Adding the silver element to the binary alloys NiTi using the VAR method is presented in [20]. The results of this approach showed that the prepared alloys are homogeneous and stable. However, the silver element was lost with repeated melting due to the high steam pressure.

3. Experimental Work

3.1. Materials and Methods

The ternary TiNiAg ingot with an actual composition of $Ni_{55}Ti_{45}Ag_0$, $Ni_{55}Ti_{44}Ag_1$, $Ni_{55}Ti_{43}Ag_2$ and $Ni_{55}Ti_{42}Ag_3$ was prepared from 99.2 % purity Ni, 99.7 % purity Ti and 99.9 % purity Ag through VAR at 1350 °C. There are two types of VAR technologies that are used to produce SMAs. The first method uses a non-consumable electrode, while the second method uses a consumable electrode. In this work, the VAR technique without a consumable electrode is preferred for many types of SMAs. Figure 1 illustrates the VAR furnace. At the beginning, the elements Ni, Ti and Ag were fixed on a copper mould and heated via radiation with the argon. After completing the melting process, the ingot was cooled with ice water. This process was repeated four times to obtain the best homogeneity. This process was followed by the annealing process at 660 °C for 24 hours. The process was followed by the cutting stage using a cutting machine to prepare the samples for the microstructural examination and mechanical tests.

Figure 1. Vacuum arc melting furnace.

3.2. Microstructural Examinations

The samples were examined by the field emission scanning electron microscopy (FESEM) before and after adding the silver element with different atomic proportions to the alloy (NiTi). Before starting the test, the samples were grinded using emery papers with particle sizes of 320, 500 and 1000 µm. After that, the samples were polished with a polishing cloth (chloride paste) and etched with etching solution

(95 mL H2O + 3.5 mL HNO3 + 1.5 mL HF) for ten minutes [17]. After completing the preparation of the samples, the samples were coated with palladium and gold for two minutes. Finally, the examination was tested via using FESEM (cam scan Mv 2300), which is shown in Figure 2.

Figure 2. FESEM examination device.

3.3. Phase Transformation

This first test was performed using the differential scanning calorimetry (DSC) device for the samples with different atomic percentages of the Ag element (zero, one, two and three a.t% Ag) to determine the degrees of phase transformation (As, Af, Ms and Mf). The second test was the SME. An examination was performed for the sample containing (zero, one, two and three a.t% of the silver element addition.

3.4. Mechanical Tests

3.4.1. Macrohardness Test

Vickers hardness apparatus was carried out for all the samples of SMAs before and after adding the Ag element, according to the ASTM E384 standards. Before the Vickers hardness testing, the samples (ten mm in diameter and seven mm in height) were prepared using the grinding process and emery papers (320, 500 and 1000 μm) and followed by the polishing process to obtain smooth surfaces. This test was performed via applying a load of three Kg for 15 sec. At least four readings were obtained, and then the average diameter of the indenter for all the samples was calculated. This test was performed using the Vickers macrohardness apparatus model (Universal Hardness Dia-Testor 722), as shown in Figure 3a. Figure 3b shows the macrohardness test samples.

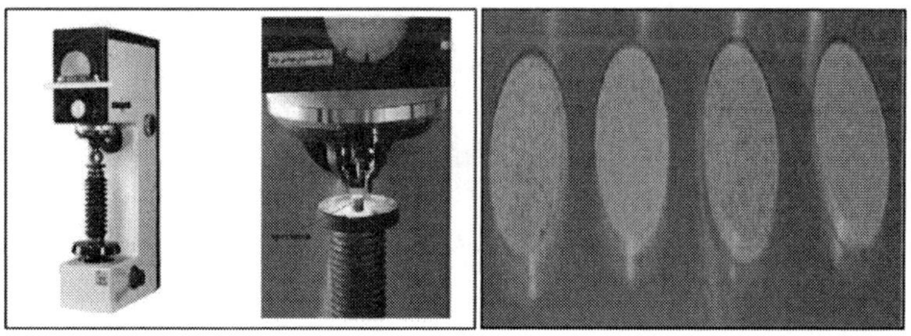

Figure 3. a: Hardness Dia-Testor, b: Hardness test samples.

3.4.2 Wear Test

In this work, the wear test was performed using the pin-on-disc technique for all SMAs before and after adding the Ag element, according to the ASTM-G99 standards. The samples of the wear test were two cm in height and one cm in diameter. Figure 4 shows the wear samples. Before the testing stage, the samples were prepared by grinding at 320, 500 and 1000 grit size and polishing using a suitable polishing cloth with chloride paste. After polishing, the samples were tested by the wear machine, changing the loads at five, ten, 15, 20 and 25 N at a constant time of 15 min and sliding speed of six cm / min. The wear rate was measured using the weighing method, and the weight loss for all shape memory samples was calculated using the sensitive electronic balance (the accuracy of the balance is approximately 0.01 mg). The wear rate of the proposed method was calculated using the following formula [21]:

$$wear\ rate = \frac{\Delta w}{S.D}\left(\frac{g}{cm}\right) \quad (1)$$

$$\Delta w = w_1 - w_2 \quad (2)$$

Where:
Δw : change in the weight of the sample (g).
w_1, w_2 : weight of sample before and after the wear test (g).

$$S.D = 2\pi . r . n . t \quad (3)$$

S.D: sliding distance (cm).
r: radius of the rotating disc (cm).
n: number of rotations per min (rpm).
t: sliding time (min).

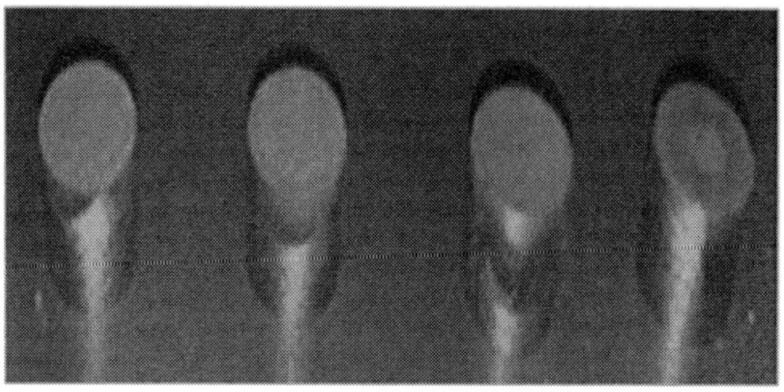

Figure 4. Wear test samples.

4. Results and Discussion

4.1. FESEM Results

Figure 5 shows the micrographs of the FESEM examination and the photomicrographs of samples before and after adding the silver element (zero, one, two and three a.t%) to the alloy NiTi. These photomicrographs show the appearance of the austenite phase (Ti_2Ni) and martensite phase (Ti002) with some impurities because of the processes (casting, cutting and polishing).

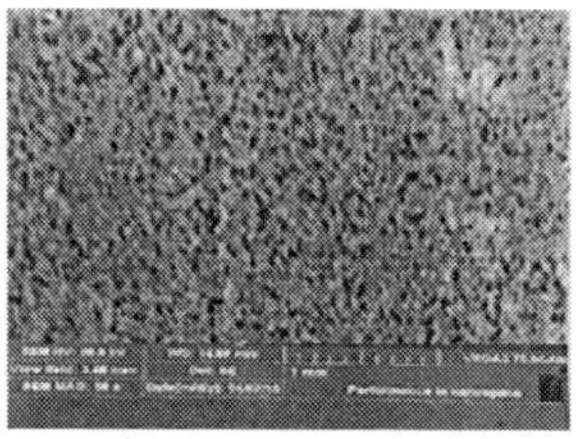

Zero at.% Ag

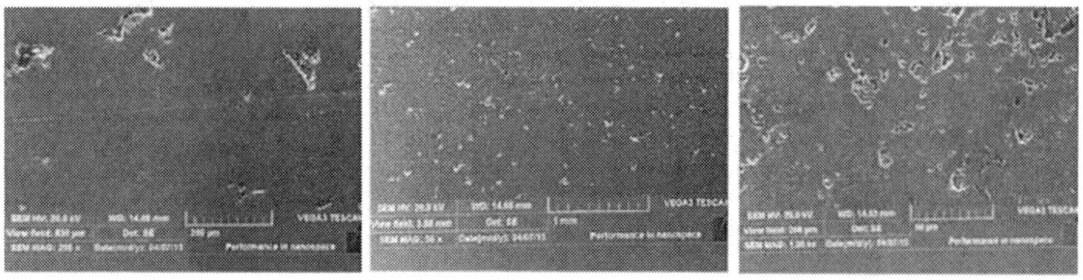

One at. % Ag Two at.% Ag Three at.% Ag

Figure 5. Photomicrographs of the samples for different at.% Ag.

IOP Conf. Series: Materials Science and Engineering **987** (2020) 012022 doi:10.1088/1757-899X/987/1/012022

4.2. OM Results

Figure 6 shows the results of SMAs before and after heat treatment by annealing at 660 *°C* for 24 hr at one at.% Ag, two at.% Ag and three at.% Ag. These results revealed that the Ag element is distributed homogenously in the NiTi-based alloy, and all the grains are equiaxed. This is attributed to a uniform mixing of the components of manufactured shape memory alloys with different atomic percentages of the Ag element. As a result of the melting process, which is followed by annealing, the grains are smaller in size (refined grain size) as the amount of the Ag element increases.

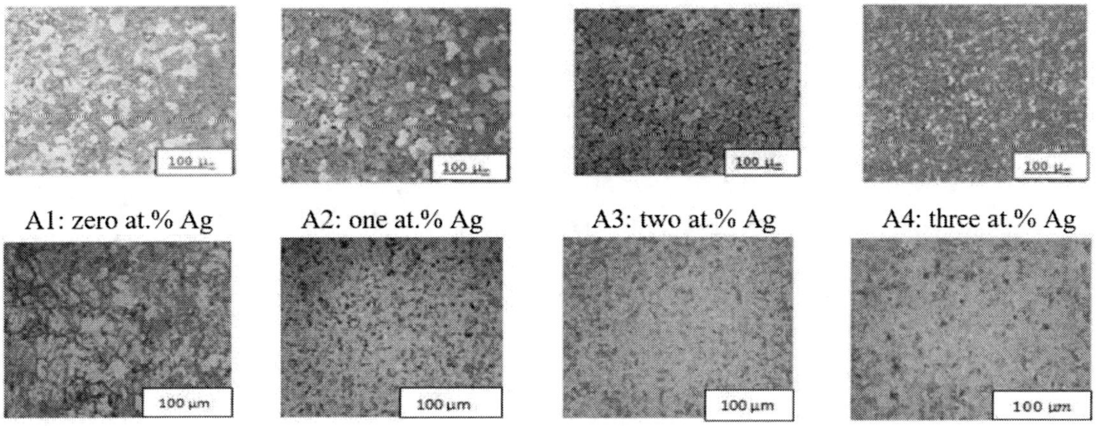

A1: zero at.% Ag   A2: one at.% Ag   A3: two at.% Ag   A4: three at.% Ag

B1: zero at.%Ag   B2: one at.% Ag   B3: two at.% Ag   B4: three at.% Ag

Figure 6. Photomicrographs of OM microstructures samples.
A: Before annealing heat treatment with at.% Ag; B: After annealing heat treatment with at.% Ag.

4.3. DSC Results

The results of the differential scanning calorimeter examination showed that the degrees of thermal transformation (A_s, A_f, M_s and M_f) explain the start and finish of both the austenite phase (B2) and the martensite phase (B19). The phase transformations were slightly affected by the increase in the silver element addition.

The results of the DSC examinations for (zero, one, two and three a.t%) is shown in Figure 7. The creation of two peaks during the cooling and heating process indicates the transition from the austenite phase to the martensite phase [22] with a shape memory effect of approximately 89.35 %–89.99 %.

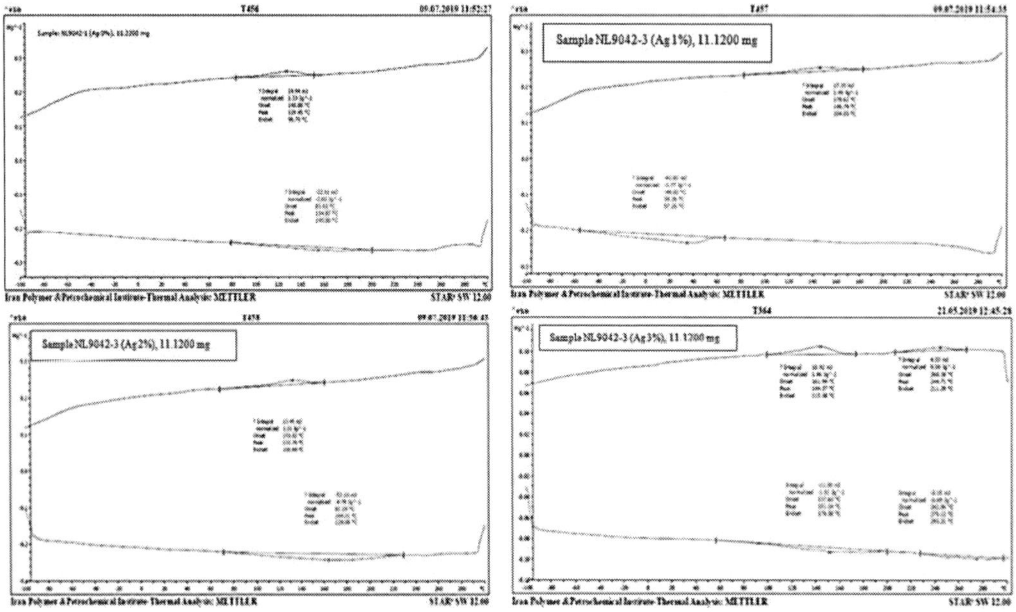

Figure 7. DSC test for all specimens.

4.4. Macrohardness Results

The macrohardness test was carried out for all samples before and after adding the Ag element at different atomic percentages. For each sample, at least four readings were taken, and then the average value was calculated. The obtained results of this test showed a slight increase in hardness with an increase of the atomic percentage of the Ag element. The slight increase in the macrohardness brought the grains close together and slightly refined the grains due to the heat treatment.
Figure 8 shows the effect of increasing the atomic percentage of the Ag element on Vickers macrohardness.

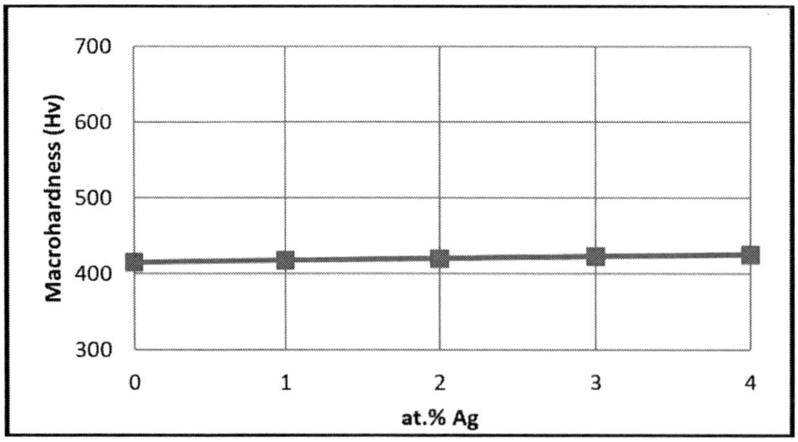

Figure 8. Macrohardness vs at.% Ag.

4.5. Wear Results

Figure 9 shows the wear behaviour of the samples before and after adding the Ag element when changing the applied load at a constant time of 20 minutes. The wear rate of the samples depends strongly on the conditions of the casting process. During the casting process, many imperfections were created, which affected the properties of the manufactured ingots. An important imperfection is voids. These voids will fill the additive element and improve the wear resistance. Increasing the applied load leads to an increase in the wear rates for all specimens due to raising the temperature between the samples and rotating disc and then forming oxidation layers between them. Moreover, increasing the applied load will increase the plastic deformation of the samples and decrease the wear rate. The difference between the melting temperatures of Ni, Ti and Ag will encourage the formation of thermal stresses, increase the dislocations density and create a loop of dislocation to obstacle the fragment of the worn surface and decrease the wear rate. This is in agreement with [23]. There is a slight impact of adding Ag on the wear behaviour of the NiTi-based SMA because the NiTi alloy is harder than the Ag element. The dry sliding wear behaviour depends on the macrohardness of the sample. The added element will obstacle or pin the movement of the dislocations and increase the wear resistance (decreasing wear rate).

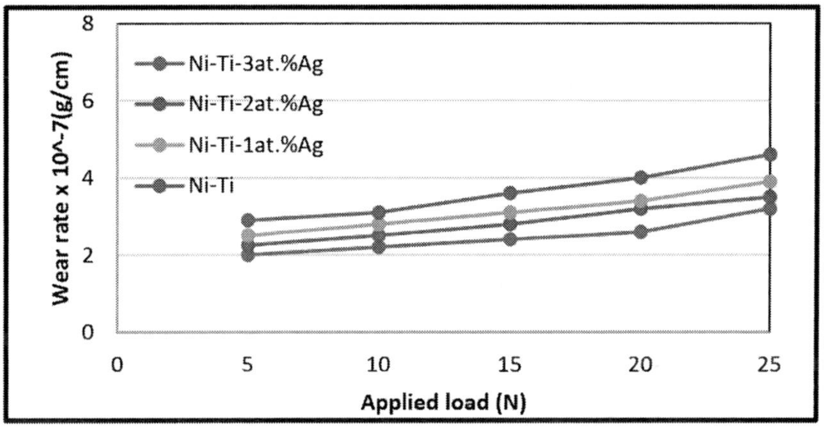

Figure 9. Wear rate vs at.% Ag.

Surface Topography

The surface topography for all the samples before and after adding the Ag element in different atomic percentages was carried out. Surface topography is defined using an optical microscope for the samples, which are tested using the wear technique. Surface topography is a result of the changes to the loads at a constant sliding time and sliding speed. The relative motion between the rubbing disc (made of hardened steel at 60 HRC) and the pins (samples at different atomic percentages of the Ag element) causes a rise in the friction temperature and creates oxidation layers between the disc and pins. Afterwards, these oxidation layers will be ploughed and uprooted to create many grooves on the surface of the samples. These grooves are similar for the samples of different atomic percentages of the Ag element. Hence, increasing the applied loads causes more fragments of sample surfaces and creates deeper worn grooves. Regarding the increasing applied loads, it ploughs an additional oxidation layer at the surfaces of the samples, as shown in Figure 10.

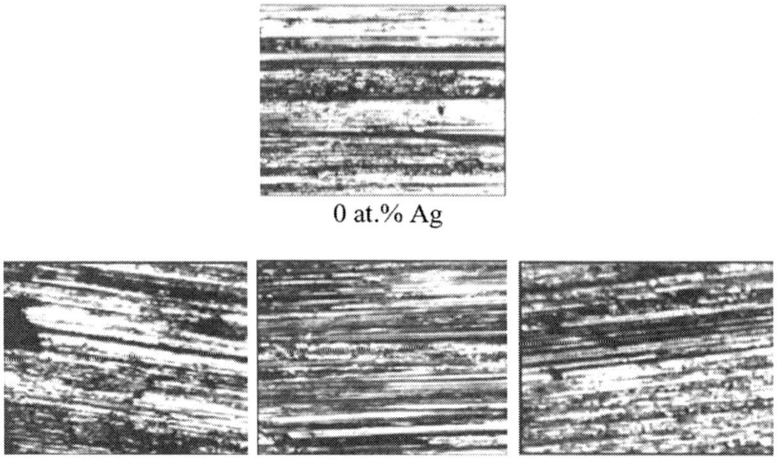

Figure 10. Worn surfaces for the samples at 15 N.

5. Conclusion

SMAs (NiTiAg) have been successfully fabricated by the casting process using a VAR furnace with different atomic percentages of the silver element as zero, one, two and three a.t%. This type of melting shows the production of the homogeneous alloy, and the silver element is uniformly distributed. The proposed method is evidenced by the examinations OM and FESEM. The results of FESEM showed the emergence of the martensite phase (Ti002) and austenite phase (Ti_2Ni), in addition to some impurities as a result of the manufacturing process and the subsequent processes. The results showed that the macrohardness value increased slightly when the percentage of the silver element in the base alloy (NiTi) increased. The wear results showed a slight increase in the wear resistance (decreasing wear rate) with increasing silver due to the high hardness of the NiTi alloy (high hardness of Ti). The maximum value of the shape memory effect of 89.99 % at three at.% Ag was achieved. Meanwhile, the starting and finishing phases of austenite and martensite were defined by the DSC test. The proposed method emphasised that the width of the hysteresis loop was slightly affected by the increasing addition of the silver element. Moreover, the surface topography was analysed using an optical microscope for all tested specimens using the wear technique. The obtained results showed that increasing the applied loads directly produces more fragments of the sample surfaces and creates deeper eroded grooves.

References

[1] Otsuka K, Kakeshita T. Science and technology of shape-memory alloys: new developments. mrs bulletin. 2002 Feb;27(2):91-100.

[2] Wayman CM. Deformation, mechanisms and other characteristics of shape memory alloys. In Shape Memory Effects in Alloys 1975 (pp. 1-27). Springer, Boston, MA.

[3] Petrini L, Migliavacca F. Biomedical applications of shape memory alloys. Journal of Metallurgy.Vol.2011;2011.Hindawi.

[4] Badakhshan Raz S, Sadrnezhaad SK. Effects of VIM frequency on chemical composition, homogeneity and microstructure of NiTi shape memory alloy. Materials Science and Technology. 2004 May 1;20(5):593-8.

[5] Ramaiah KV, Saikrishna CN, Bhaumik SK. Processing of Ni-Ti shape memory alloy wires. In International Conference of Smart Materials Structures and Systems 2005.

[6] Zheng YF, Zhang BB, Wang BL, Wang YB, Li L, Yang QB, Cui LS. Introduction of antibacterial

function into biomedical TiNi shape memory alloy by the addition of element Ag. Acta biomaterialia. 2011 Jun 1;7(6):2758-67.
[7] Zhao L, Chu PK, Zhang Y, Wu Z. Antibacterial coatings on titanium implants. Journal of Biomedical Materials Research Part B: Applied Biomaterials. 2009 Oct;91(1):470-80.
[8] Babu R, Zhang J, Beckman EJ, Virji M, Pasculle WA, Wells A. Antimicrobial activities of silver used as a polymerization catalyst for a wound-healing matrix. Biomaterials. 2006 Aug 1;27(24):4304-14.
[9] Bitzer M, Bram M, Buchkremer HP, Stöver D. Phase transformation behavior of hot isostatically pressed NiTi-X (X= Ag, Nb, W) alloys for functional engineering applications. Journal of materials engineering and performance. 2012 Dec 1;21(12):2535-45.
[10] Singhal R, Pivin JC, Chandra R, Avasthi DK. Ion irradiation studies of silver/amorphous carbon nanocomposite thin _lm. Surface and Coatings Technology. 2013 Aug 25;229:50-4.
[11] Xenos-Despina S, Gregory N. The unique properties, manufacturing processes and applications of near equatomic Ni-Ti alloys.2016.
[12] Kocich R, Szurman I, Kursa M. The methods of preparation of Ti{Ni{X alloys and their forming. Shape Memory Alloys-Processing, Characterization and Applications. 2013 Apr 3:28-35.
[13] Elahinia MH, Hashemi M, Tabesh M, Bhaduri SB. Manufacturing and processing of NiTi implants: A review. Progress in materials science. 2012 Jun 1;57(5):911-46.
[14] Otubo J, Rigo OD, Moura Neto CD, Kaufman MJ, Mei PR. Low carbon content NiTi shape memory alloy produced by electron beam melting. Materials Research. 2004 Jun;7(2):263-7.
[15] Olier P, Barcelo F, Bechade JL, Brachet JC, Lefevre E, Guenin G. Effects of impurities content (oxygen, carbon, nitrogen) on microstructure and phase transformation temperatures of near equiatomic TiNi shape memory alloys. Le Journal de Physique IV. 1997 Nov 1;7(C5):C5-143.
[16] Matsumoto H. Addition of an element to NiTi alloy by an electron-beam melting method. Journal of materials science letters. 1991 Jan 1;10(7):417-9.
[17] Oh KT, Joo UH, Park GH, Hwang CJ, Kim KN. Effect of silver addition on the properties of nickel–titanium alloys for dental application. Journal of Biomedical Materials Research Part B: Applied Biomaterials: An Official Journal of The Society for Biomaterials, The Japanese Society for Biomaterials, and The Australian Society for Biomaterials and the Korean Society for Biomaterials. 2006 Feb;76(2):306-14.
[18] Zamponi C, Wuttig M, Quandt E. Ni–Ti–Ag shape memory thin films. Scripta materialia. 2007 Jun 1;56(12):1075-7.
[19] Quandt E, Zamponi C. Superelastic NiTi thin films for medical applications. InAdvances in Science and Technology 2008 (Vol. 59, pp. 190-197). Trans Tech Publications Ltd.
[20] da Silva GH, Otubo J. Designing NiTiAg Shape Memory Alloys by Vacuum Arc Remelting: First Practical Insights on Melting and Casting. Shape Memory and Superelasticity. 2018 Dec 15;4(4):402-10.
[21] Lorenzo-Martin C, Ajayi OO, Singh D, Routbort JL. Evaluation of scuffing behaviour of single-crystal zirconia ceramic materials. Wear. 2007 Sep 10;263(7-12):872-7.
[22] Liu Y, Blanc M, Tan G, Kim JI, Miyazaki S. Effect of ageing on the transformation behavior of Ti{49.5 at.% Ni. Materials Science and Engineering: A. 2006 Nov 25;438:617-21.
[23] Neupane R. Indentation and Wear Behavior of Superelastic TiNi Shape Memory Alloy (Doctoral dissertation).2014.

IOP Conf. Series: Materials Science and Engineering **987** (2020) 012023 doi:10.1088/1757-899X/987/1/012023

The synthesis and study of the electrical properties of composites ($FeAl_2O_3$) with paraffin wax and the determination of the mechanism of electrical conductivity

Ahmed J H Almaliky[1], Hussein F Hussein[1] and Khalid I Ajeel[1]

[1]Department of Physics, College of Education for Pure Science, University of Basra, Iraq.

Corresponding author: ahmed.jasem@mail.ru

Abstract. Thick films of composites of iron–aluminium oxide ($FeAl_2O_3$) were prepared by mixing paraffin wax with a powder of aluminium oxide (Al_2O_3) and iron granular (Fe) (600 µm in size). The current–voltage (I-V) characteristic was measured at different temperatures (293–328 K). The electrical conductivity was also measured at those temperature levels. It was noted that an increase in temperature leads to an increase in electrical conductivity. The activation energies (Ea) were determined to be 0.357–0.125 ev; the electrical conductivity mechanism was identified as the Schottky and Poole–Frenkel effects.

1. Introduction

Composite materials have great potential for many technological and industrial applications due to their unique properties, such as low density, ability to form intricate shapes, versatile electrical applications and low manufacturing cost, as well as having a very high dielectric strength, good change storage capacity and dopant-dependent electrical and optical properties. The already wide range of composite applications can be extended by using different materials with inorganic fillers, because dispersed fillers may enhance different physical properties of the host material. The electrical applications of composite solids can be controllably modified depending on many factors, such as the type of the filler used, its concentration, the conductivity of the filler and the method in which it penetrates and interacts with the chains of the polymer [1,2]. To achieve both electrically insulating and thermally conducting polymer composites, alumina is used. Metallic fillers such as graphite and carbon black are used to enhance electrical and thermal properties [3,4]. A nanoparticle is the most basic component in the fabrication of a nanostructure. Metallic nanoparticles have different physical and chemical properties from bulk metals (lower melting points, higher specific surface areas, electronic and optical characteristics, mechanical strength and magnetisation properties that could be used in various industrial applications). Nonconductive fillers increase dielectric permittivity due to interfacial polarisation. The effectiveness of composites depends on their ability to disperse the fillers homogeneously throughout the material [5]. In recent years, studies on the optical and electrical characteristics of composite materials have been the subject of considerable research efforts due to their wide applications in optoelectronic devices. Thin films of the composite $FeAl_2O_3$ with the nano dimension have been studied intensively [6,7]. The physical properties and determination of carrier concentration as well as the mobility of Sb2(Te1-xSex)3 thin films have been investigated [8,9]. The dielectric properties of manganese phthalocyanine (MnPc) thin films have been studied [10]. The influence of sodium zirconate nanoparticles on the structural and

electrical characteristics of polyvinyl alcohol (PVA) nanocomposite films have also been studied [11]. Optical properties of pure and iron-doped cellulose acetate (CA) or PVA have been studied in the 200–800 um wavelength region [12]. The effect of cobalt(II) chloride on the electrical properties of poly(o-toluidine) have been investigated [13]. The synthesis, characteristics and optical properties of polyaniline (PANI) filled by graphene nano-films have also been investigated [14,15].
In the present study, we report on the synthesis and electrical characterisation of $FeAl_2O_3$/paraffin wax composite film of aluminium oxide (Al_2O_3 = 0.07 gm) and iron (Fe = 0.01 and 0.03 gm). The I-V characteristic was measured at different temperatures (283–328 K). Electrical conductivity was also measured at those temperature levels. The activation energies (Ea) and the voltage barrier (ϕ) were determined, and the electrical conductivity mechanism was also studied. At present, a significant scientific concern is the study of the electrical properties of blend and composites [16,17] and also the effect of the insulating matrix material on the electrical and magnetic properties of metal–insulators [18,19,20]. It is considered one of the most successful and sensitive methods in studying the composition of material and also to be important in studying the properties of layer films [21]. The process of determining conduction mechanisms that control the transmission of current through composite materials and physical composites depends on the measurements of the curves (current–voltage and conductivity–temperature) [22]. It may be that there is more than one conduction mechanism in the compound metal–insulator and in the thick film consisting of a specific physical composite, as it depends on the amount of thickness of the prepared film, the intensity of the projected electric field or the degree of deformation.

Table (1): The weight ratio of composites ($FeAl_2O_3$) and activation energies.

Aluminium oxide (AL_2O_3)	0.07 gm	0.07 gm
Iron (Fe)	0.01 gm	0.03 gm
Activation energy (Ea)	0.125 ev	0.357 ev

2. Experimental procedure

The films' composite $FeAl_2O_3$ was prepared by mixing paraffin wax with a powder of aluminium oxide (Al_2O_3) and iron (Fe). The iron element powder was sifted using a granular sieve with a size of 600 μm, and then 0.07 gm of aluminium oxide was mixed with 0.01 and 0.03 gm of a granular iron element with a size of 600 μm, as shown in Table 1. The mixing process was performed using an evaporating dish and a mixing spatula. After the process of mixing was complete, 0.1 gm of paraffin wax was added, which was melted at a temperature of 328 K, while continuing manual mixing as in the previous method. The aluminium slices (2.5*4 cm^2) were washed with acetone and distilled water and then dried in an oven with a temperature of 323 K for 15 minutes. The composition consisting of paraffin paste and aluminium oxide with iron was deposited on aluminium slices. The slices were left in laboratory conditions for 24 hours. The thickness of the samples was measured, and it was found to be 0.448 mm. The aluminium electrodes were deposited in the form of circles with a radius of 1 mm by using the evaporation apparatus under evacuated pressure (10–5 torr). When the electrode deposition process finished, the samples were placed in a special electrode measuring circuit, which consisted of a chamber containing electrodes made of copper. These electrodes were separated by the electric measuring circuit, which consisted of regulated voltages ranging from 1–30 V, where supplied by PHYWE 2592 power. The current was measured by an electrometer type measuring amplifier D53200 L.H.Co. The sample temperature was measured with a thermocouple placed near the sample.

IOP Conf. Series: Materials Science and Engineering **987** (2020) 012023 doi:10.1088/1757-899X/987/1/012023

3. Results and discussion

3.1. Measurement of electrical properties

The method of measuring electrical properties plays an important role in providing an accurate description of the change in current with voltages [22,23]. All data were taken after the current stabilised, in which case the current is independent of time. The current–voltage characteristics were measured at different temperatures (293–323 K) and at different concentrations (Al_2O_3 = 0.07 gm; Fe = 0.01 and 0.03), as shown in the Figures 1 and 2. In the two figures, it can be noted that an increase in temperature leads to an increase in current, where both curves display the same behaviour (the ohmic characteristic) at high temperatures, but Figure 1 differs from that characteristic at low temperatures (20–25 K) because of the difference in Fe metal concentration, which in turn leads to the difference in the mechanism of electrical conductivity. When recording values of current and voltage (corresponding to the current), it is followed by raising the temperature of the sample to a temperature higher than its predecessor, which is left for a period of half an hour to reach a degree of thermal stability, and then the process of re-recording the feature (current–voltage) begins; these measurements were repeated until the highest permissible temperature range was reached. Electrical conductivity was measured as a function of temperature of compound samples ($FeAl_2O_3$) for a range of temperatures (293–323 K). Electrical measurements showed an increase in electrical conductivity as the temperatures increased, and this is a traditional behaviour with regard to electrical conductivity in semiconductor materials when increasing the temperature, which in turn leads to an increase in the mobility of the electric charge carriers. Figures 3 and 4 show the linear dependence of electrical conductivity with increasing temperatures and also the hysterical behaviour of electrical conductivity after cooling the samples from 323 to 293 K, which is probably due to the difference in the mechanism of electrical conductivity and also to a decrease in the kinetic energy of the charge carriers.

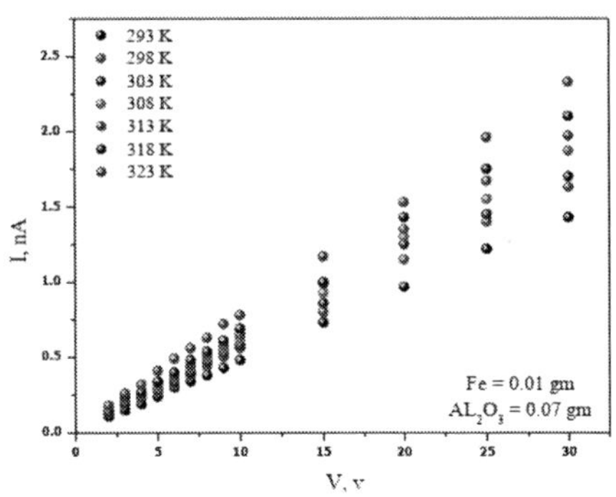

Figure1.The I-V characteristic for sample $FeAl_2O_3$

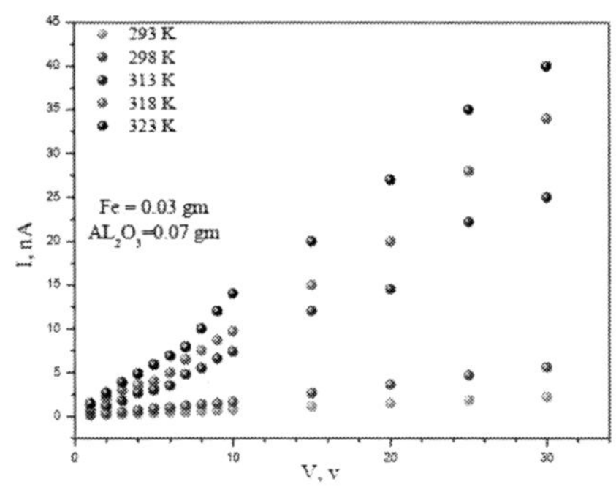

Figure2.The I-V characteristic for sample $FeAl_2O_3$

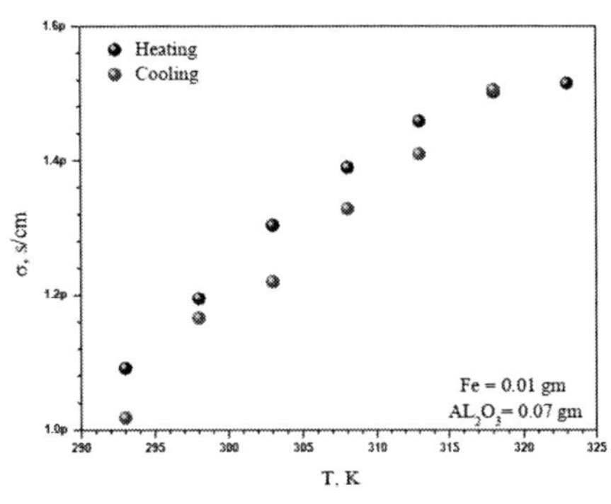

Figure3.Electrical conductivity for sample $FeAL_2O_3$

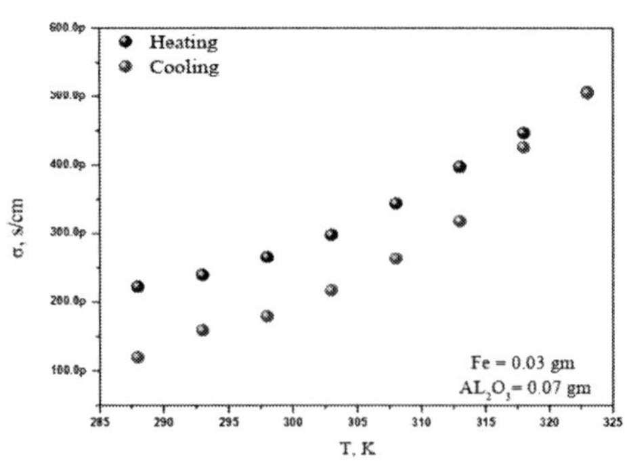

Figure4.Electrical conductivity for sample $FeAL_2O_3$

3.2. Testing conductivity mechanisms for samples $FeAL_2O_3$

3.2.1. Testing the hopping conduction mechanism

The hopping conduction mechanism can be tested by plotting electrical conductivity versus reciprocal temperatures (σ-1000*T-1/4), as shown in Figures 5 and 6, and the plotting of electrical conductivity versus reciprocal temperatures (σ-1000*T-1/3), as shown in Figures 7 and 8, according to the following two equations [24]:

$$\sigma = \exp (Ea/KT1/3) \quad (1)$$
$$\sigma = \exp (Ea/KT1/4) \quad (2)$$

With:
Ea = activation energy (ev)
K = Boltzmann constant (ev/k)
T = temperature (K)

In Figures 5–8, it can be noted that the relationship of conductivity with temperature is non-linear, which indicates the exclusion of this type of conductivity mechanism.

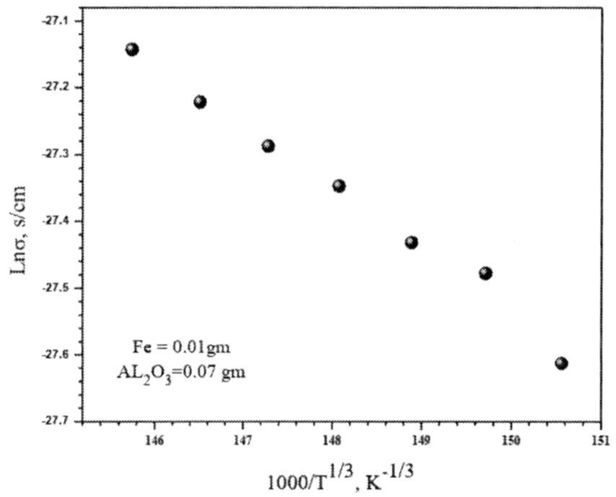

Figure5. The relationship of conductivity with the inverted temperature for composite $FeAL_2O_3$.

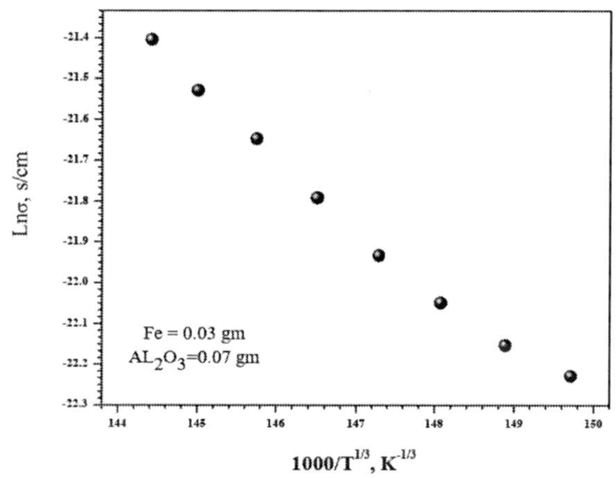

Figure6. The relationship of conductivity with the inverted temperature for composite $FeAL_2O_3$.

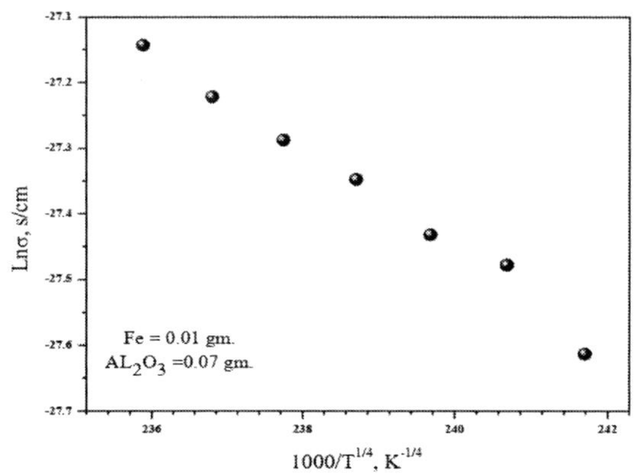

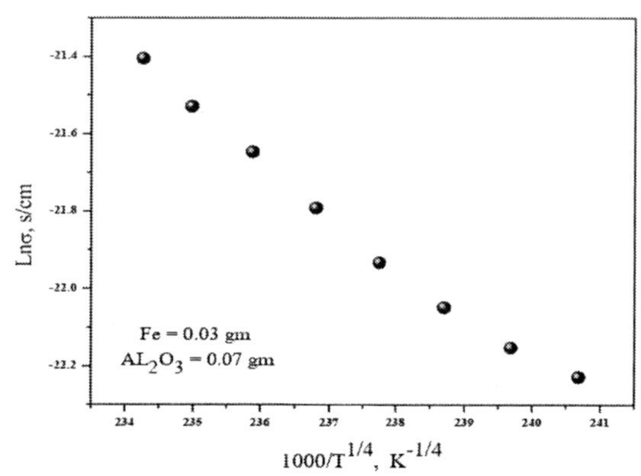

Figure7. The relationship of conductivity with the inverted temperature for composite $FeAL_2O_3$.

Figure8. The relationship of conductivity with the inverted temperature for composite $FeAL_2O_3$.

3.2.2 Testing the Schottky and Poole–Frenkel conduction mechanisms

Schottky emission [25] is one of the causes of non-ohmic behaviour of the current and occurs when voltage is applied on the metal–insulator contact area. This happens when the electrons transfer from the metal to the conduction band in the dielectric. This mechanism is related to the thermal emission of electrons. In addition, the applied field causes a decrease in the voltage barrier between the metal–dielectric joint. This barrier is affected by the surface layer and the presence of surface trapping centre. The Poole–Frenkel effect [26], or glow discharge [27], is based on the same concept of reducing the voltage barrier to the Schottky effect. However, in this case, electrons are released from the capture trap levels to the conduction beam in the compound. In other words, the difference between the two cases is that the Schottky effect causes a decrease in the voltage barrier at the surface while the Poole–Frenkel effect causes a reduction in the voltage barrier inside the compound. At high voltages, Schottky's mechanism is related with Poole–Frenkel's mechanism in the linear relationship of thc I-V feature. Schottky's constant (α_{sh}) is related with Poole–Frenkel's constant (αPF) in the following relationship [28]:

$$\alpha_{sh}=[\frac{q3}{4\pi\epsilon\epsilon_0}]^{1/2} \qquad (3)$$

With:
q = electronic charge (C)
ϵ = dielectric constant
ϵ_0 = dielectric constant in the Space (f/m)

Schottky's constant is related with Poole–Frenkel's constant by the following relationship:

$$\alpha_{P.F} = 2\alpha_{sh} \qquad (4)$$

Figures 9 and 10 show the relationship between the current and the square root of the voltage of the both samples (Fe = 0.01 and 0.03; AL_2O_3 = 0.07). It is clear that the relationship is linear at high fields. The experimental values of the constant α_{exp} for the examined films can be obtained from the slopes in Figures 9 and 10 in the straight line in the high field region. The theoretical values of α_{sh} and $\alpha_{P.F}$ can be determined with the help of Equations 3 and 4. The experimental values α_{exp} in Tables 2 and 3 are close to the theoretical value $\alpha_{P.F}$, which indicates that the Poole–Frenkel effect is a dominant mechanism, so the applied filed is enough to enhance the thermal exited electron inside the composite films.

Table2. Experimental and determined α values from Schottky's and Poole–Frenkel's theories and the dielectric constant for composite sample $FeAL_2O_3$, where Fe = 0.01 gm and AL_2O_3 = 0.07 gm.

T (K)	α_{exp}	α_{sh}	$\alpha_{P.F}$	ϵ
293	0.082	0.045	0.09	0.60
298	0.065	0.037	0.07	0.45
313	0.041	0.027	0.05	0.22
318	0.036	0.018	0.036	0.10
323	0.032	0.016	0.033	0.085

Table3. Experimental and determined α values from Schottky's and Poole–Frenkel's theories and the dielectric constant for composite sample $FeAL_2O_3$, where Fe = 0.03 gm and AL_2O_3 = 0.07 gm.

T (K)	α_{exp}	α_{sh}	$\alpha_{P.F}$	ϵ
293	0.03611	0.0461	0.09219	0.169
298	0.03602	0.04592	0.09184	0.109
303	0.03597	0.04583	0.09165	0.076
308	0.03592	0.04576	0.09152	0.056
313	0.03586	0.04569	0.09139	0.043
318	0.03582	0.04561	0.09122	0.034
323	0.03574	0.04530	0.09060	0.028

Figures 11 and 12 represent the relationship between log Io/T2 versus reciprocal temperature (Richardson plot), where Io represents the extrapolation of current (Fe = 0.01 and 0.03; AL_2O_3 = 0.07). The lowering of the work function (Δϕ) for the thermionic emission due to the Schottky effect can be determined from the following relation [29]:

$$\Delta\phi=\left[\frac{eE}{4\pi\epsilon\epsilon_0}\right]^{1/2} \qquad (5)$$

With:
e = the electronic charge (C)
E = electric field ϵ the dielectric constant (v/m)
These are shown in Figures 13 and 14 for both samples Fe = 0.01 and 0.03 and Al_2O_3 = 0.07.

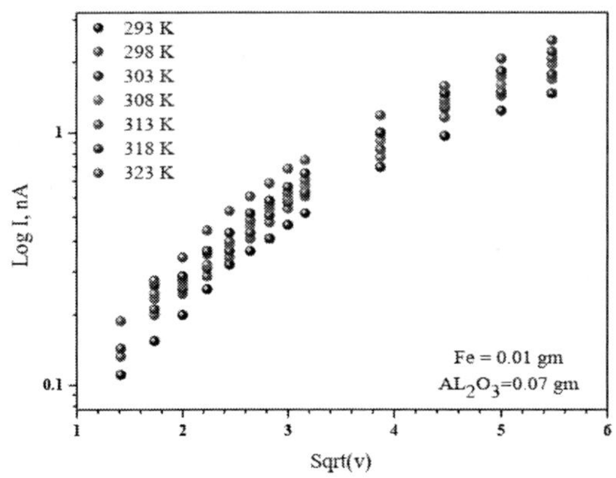

Figure9. Plots of log(I) with square root of V for sample of composite $FeAL_2O_3$ at several temp. (293–323 K).

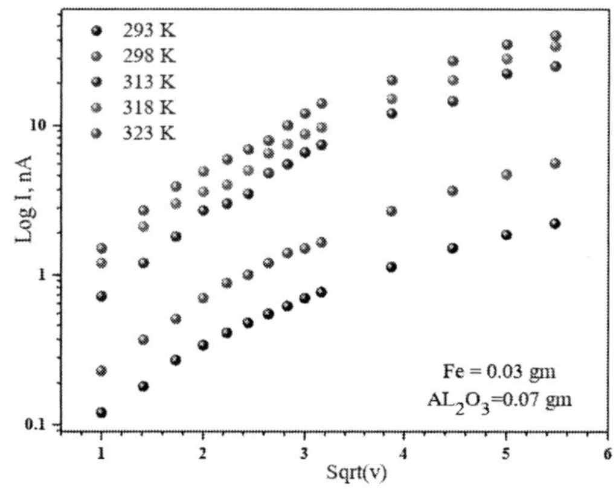

Figure10. Plots of log(I) with square root of V for sample of composite $FeAL_2O_3$ at several temp. (293–323 K).

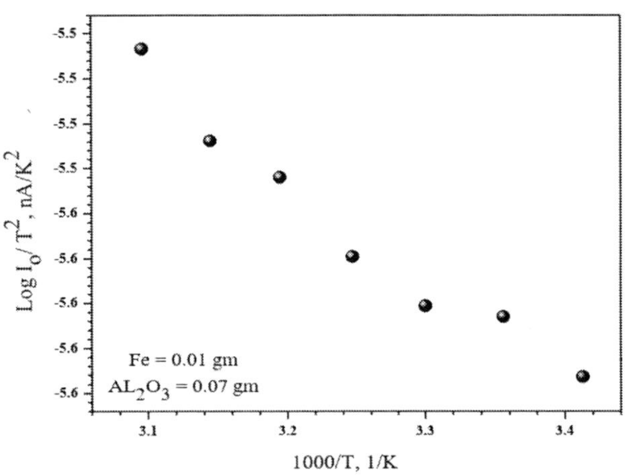

Figure11. Plot of log Io/T2 - 1/T (K-1) for $FeAL_2O_3$ film.

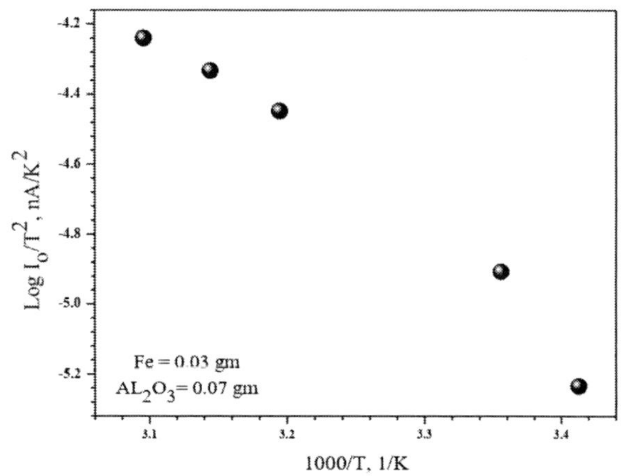

Figure12. Plot of log Io/T2 - 1/T (K-1) for $FeAL_2O_3$ film.

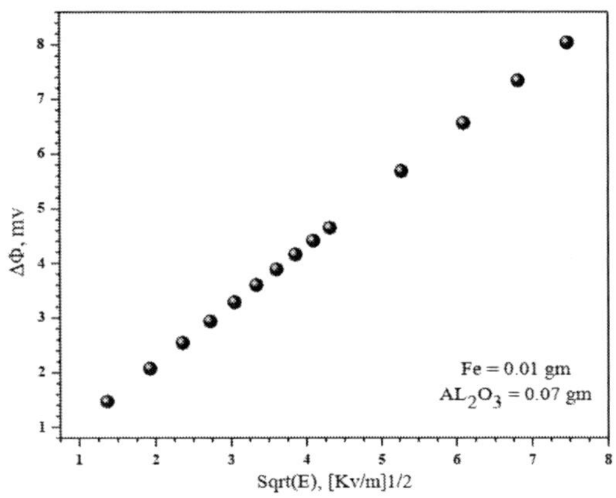

Figure13. The relationship between the lowering potential barrier versus E1/2 for $FeAL_2O_3$ samples at room temperature.

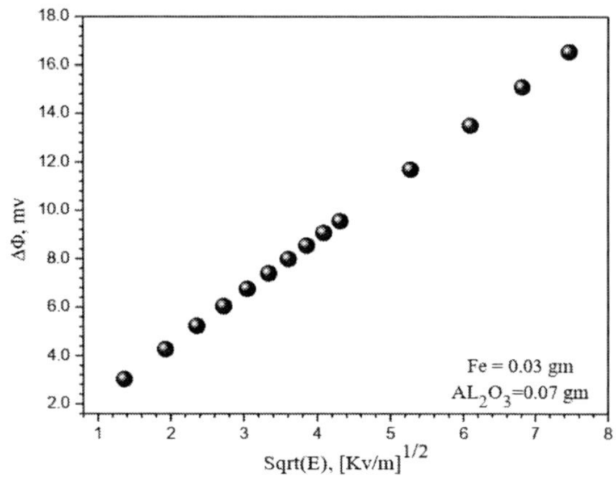

Figure14. The relationship between the lowering potential barrier versus E1/2 for $FeAL_2O_3$ samples at room temperature.

3.3.3 Testing the ion conduction mechanism

The mechanism of ion conduction was tested by drawing the relationship between -L-nσ T1/2 and 1000/T K - 1 according to the following formula [29]:

$$\sigma = \frac{\sigma_0}{\sqrt{T}} \exp\left(-\frac{Ea}{KT}\right) \qquad (6)$$

with:

σ_0 = constant

Figures 15 and 16 show the plots of -L-nσ T1/2 versus 1000/T K - 1, where the ion conduction mechanism can be expected to occur if the data show linear dependence. Furthermore, the current–voltage characteristics do not obey the general ionic equation (hyperbolic sine relationship); the tunnelling mechanism is also not applicable in our study, because it requires very thin films, where the current is independent of temperature, but we have thick films (as shown in Table 4), where the current is dependent on temperature.

Table 4. The thickness and concentrations of both samples.

Composite	Fe	AL_2O_3	Thickness
Fe	0.01 gm	0.07 gm	0.448 mm
AL_2O_3	0.03 gm	0.07 gm	0.538 mm

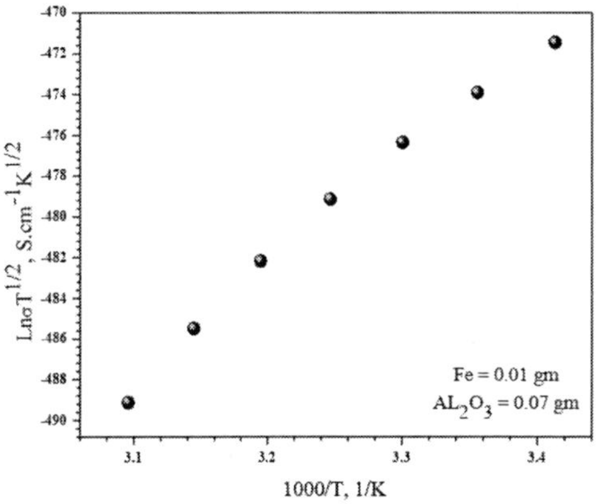

Figure 15.The relationship between Ln (σ $T^{1/2}$) versus 1000/T for $FeAL_2O_3$ for the ionic conduction test.

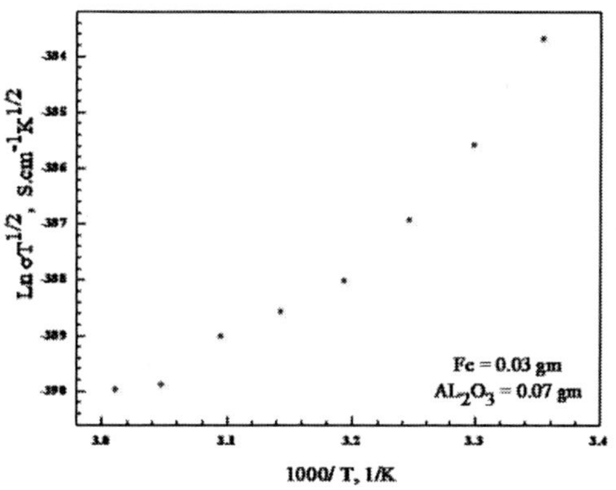

Figure 16.The relationship between Ln (σ $T^{1/2}$) versus 1000/T for $FeAL_2O_3$ for the ionic conduction test.

4. Conclusions

The following conclusions can be drawn from this study of the two samples of iron–aluminium oxide composite ($FeAl_2O_3$), where Fe = 0.01 and 0.03 and Al_2O_3 = 0.07. Electrical properties and determination of the mechanism of electrical conductivity have been investigated in the temperature range 293–323 K; the conduction mechanisms were identified by measuring the current–voltage (I-V) characteristics and current–temperature dependence equations. For sample $FeAl_2O_3$ with Fe = 0.01 gm, Schottky's conduction mechanism is dominant, where its electrical characteristics are explained by comparing α constants and by studying the dependence of conductivity with different work functions. However, for sample $FeAl_2O_3$ with Fe = 0.03 gm, Poole–Frenkel's conduction mechanism is dominant. Other mechanisms (i.e. ion conduction mechanism and hopping conduction mechanism) were also investigated, but they were excluded from our expectations. On the other hand, we found some difference between these two samples in their electrical properties in the values of α constants therefore in conduction mechanism, and finally in some electrical factors (σ, Φ and I), because of the different in thickness and the concentration of metal in each sample.

5.References.

[1] Golden J H, Hawker C J and Ho P S 2001 Designing porous low –K Dielectric *semiconductor international* **24** 79-87

[2] Abdullah O G and Saber D R 2012 optical absorption of PVA films doped with Nickel chloride *applied mechanics and materials* **110-116** 117 -182

[3] XU Y, Chung D D L and Mroz C 2001Thermally conducting Alumium Nitride polymer matrix Composites *Composites Part A: Applied science and manufacturing* **32** 1749- 1757

[4] Mamuny Y p, Davydenko V V, Pissis P and Lebedev E V 2002 Electrical and thermal conductivity of polymer s filied with metal powder *European polymer journal* **38** 1887-1897

[5] Pillai P K C, Narula G K and Tripathi A k 1984 Dielectric properties of poly propylene / poly carbonate poly blend *polymer journal* **16**, 575-578

[6] Stognei O V, Al-Maliki A, Sitnikov A and Makagonov V 2015 Thermoelectric Power of Gradient $Fe_x(Al_2O_3)_{100-x}$ Composite Films *Solid State Phenomena* **233-234** 694-698..

[7] Stognei O V ,Sitnikov AV and Al-Maliky A J 2014 PREPARATION OF Fex (Al2On)100-x CONCENTRATION-GRADIENT COMPOSITE FILMS *Bulletin of the Voronezh State Technical University* **10-1** 7-11

[8] Ajeel Khalid I and Khalid Usama A 2003 Synthesis and study the physical properties of $Sb_2(Te_{1-x}Se_x)_3$ compound *Journal of Basrah Researches/Science* **29** 2
[9] Ajeel Khalid I and Khalid Usama A 2003 Investigation of transport phenomina and determination of carrier concentration and mobility of $Sb_2(Te_{1-x}Se_x)_3$ compound *Journal of Basrah Researches/Science* **29** 2
[10] Ajeel Khalid I and Hussein H F 2003 preparation and studying the dielectric properties of manganese metal phthalocynine (MnPC) Thin films *Iraqi J.Polymers* **7 No.1** 75-82
[11] Chandrala H N , Ramaraj B, Shivakumaraiah, Madhu G M and Siddaramuiah 2013 investigation on the influence of sodium zirconate nanoparticles on the structural characteristics and electrical properties of polyvinyl alcohol nanocomposites films *Journal of alloys and compoundes* **555** 531 -538
[12] Sharma Aa k and Rama C H 1991 optical properties of pure and iron doped cellulose acetate *Journal of materials science letters* **10** 1217-1219
[13] Fakher R K and Ajeel K I International journal 2015 Effect of cobalts chloride on Electrical properties of poly (O-Toluidine) *International Journal Of materials science and Engineering* **3** 4
[14] Ajeel Khalid I and Kareem Q S 2019 Synthesis and characteristics of Polyaniline (PANI) Fillid by Graphene (GR) Nano film *IOP. Con. Journal of Physics: Conference Series* 1234.
[15] Ajeel Khalid I and Kareem Q S 2019 Optical properties for prepared Polyaniline (PANI) Fillid by Graphene (GR) Nanocomposites films *Journal of Basrah Researches/Science* **45** 2
[16] Beles B A, Sheng P, M D Abeles, Coutts B and Arie Y 1975 Structural and electrical properties of granular metal films *Advances in Physics* **24** 407-461
[17] Beloborodov I S, Lopatin A V, Vinokur V M, and Efetov K B 2007 Granular Electronic Systems *Rev. Mod. Phys.* **79** 469-518
[18] Dempsey N et.al 2001 agnetic behavior of Fe:Al2O3 nanocomposite films produced by pulsed laser deposition *Journal of Applied Physics* **90 12** 6268-6274
[19] Batlle X, Labarta A 2002 Finite-size effects in fine particles: magnetic and transport properties *Journal of Physics D: Applied Physics* **35** R15-R42
[20] Hutten A, Sudfeld D, Wojczykowski K, Jutzi Peter and Reiss G 2003 Giant magneto resistance and magnetic aspects in granular structures *Journal of Magnetism and Magnetic Materials* **262** 23-31
[21] Hill R M 1969 Electrical Conduction in Ultra-Thin Metal Films *Proceedings of the Royal Society A* **309** 377-395
[22] Ebade S S 2014 Preparation the copolymers and studying its electrical properties and used electronic application *MSc.thesis submitted to college of pure science University of Basrah/Iraq.*
[23] Abdul Ghafor WAS 2000 Schottky Effect Mechanism in Poly (pyromellitic-4,4-diphenyl sulphone) Films *Journal of Polymer Science: Part B: Polymer Physics* **38** 2507–2514
[24] Abdul Ghafor WAS, W Z Manookien and G A Adam 1997 Iraqi polymer journal **1** 59
[25] Deal B E, Fleming P J and Castro P L 1968 Electrical Properties of Vapor-Deposited Silicon Nitride and Silicon Oxide Films on Silicon *Journal of The Electrochemical Society* **115** 3
[26] Lilly A C and McDowell J R 1968 High-Field Conduction in Films of Mylar and Teflon *Journal of Apply Physics* **39** 141
[27] Frenkel 1938 On Pre-Breakdown Phenomena in Insulators and Electronic Semi-ConductorsJournal of Physical review **54** 467
[28] Abdul Ghafor W A S 1996 Synthesis , characterisation and study of some New Condensated polymers, their Electrical properties and Application as FET *Ph.D.* thesis Basrah university/Iraq.
[29] wenkuo C, Huang C Wen, Chen B Kuan, Li W Bin, Chen P Rong, Ho T Han, Tseng C Guey and Wu T Yi 2013 Enhanced Ionic Conductivity in PAN-PEGME-$LiClO_4$-PC Composite polymer Electrolytes *Int.J.Electrochem.Sci* **8** 3834-3850

IOP Conf. Series: Materials Science and Engineering **987** (2020) 012024 doi:10.1088/1757-899X/987/1/012024

Modelling Low Velocity Impact on Structural Composites

M Yeager, S E Boyd, J M Staniszewski, B A Patterson, D B Knorr Jr and T A Bogetti

U.S. Army Combat Capabilities Development Command Army Research Laboratory, Aberdeen Proving Ground, MD, 21005, USA

E-mail: michael.yeager11.civ@mail.mil

Abstract. Insertion of composite materials into next generation vehicles requires an understanding of their response over the range of relevant loading conditions encountered in the field. This paper focuses on modelling the low velocity, high energy impact on structural composite materials. An experimental procedure, using the four quadrant impact methodology, is presented to characterize composite material response and damage during these events. A model is developed and verified (within 9% for peak force and peak deflection) to predict the composite response to low velocity, high energy events. The model is then exercised to gain a deeper understanding about how mechanical and damage properties influence the structural response of the composite. The model was most sensitive to delamination failure stress, which changed the damage area by a factor of 8 over the range of values examined. Through investigating the effect of boundary conditions, a 67% increase in peak force was found by switching from simply supported to clamped. This work demonstrates a first level analysis and model to predict the composite response to low velocity, high energy loading conditions.

1. Introduction and motivation

A majority of ground vehicles use metal because of its durability, isotropic nature, and ample amounts of data for engineering design [1]. In theory, structural composite materials offer opportunities for improvement over incumbent metal materials [2]. In addition to being lighter weight (i.e., lower density), composites have a large design space which has yet to be fully exploited for vehicle structures [3]. Durability is an issue preventing the widespread use of composite materials in ground vehicle applications [4].

Insertion of composite material into vehicles requires them to be structural and resistant to high impact loads [5]. The response of composite materials to low velocity impacts is an active area of experimentally focused research [6-8]. Modelling the response of composites under the spectrum of loading conditions encountered in the field is necessary for the design of structural composites, but presents a number of challenges [9]. High fidelity composite models also streamline the insertion of new resins into fielded systems as providing model-informed solutions accelerates the design process.

This work focuses on modelling the low velocity, high energy impact on structural composites. An experimental procedure is presented for examining the deformation and damage of thick section composites subjected to low velocity, high energy loading conditions. A finite element based model is developed to accurately predict the composite response during this loading event. The model is verified through comparing the peak deflection, peak load, and damage area with the experimental results. It is then exercised to understand the influence of mechanical properties and damage parameters on composite performance, giving insight into how changing resin dominated properties would influence trends in panel level response. The model is then used to gain information on how changing boundary conditions would affect the panel response, allowing the exploration of selected boundary conditions without having to create a new experimental setup. The objective of this work is to provide a first level analysis to predict the structural response of and delamination in thick section composites as part of a materials-by-design strategy as summarized in Figure 1.

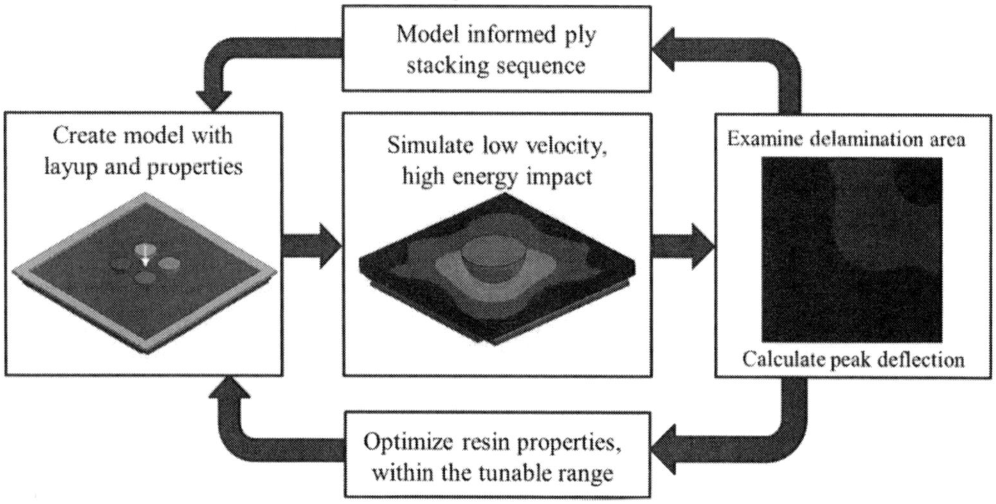

Figure 1. Materials-by-design strategy for the development of structural composites

2. Experimental Methodology

2.1. Impact samples

Composite samples are manufactured for impact testing using an S2-glass fabric infused with a rubber toughened epoxy to a target thickness of 13-mm. A 2-D woven roving fabric of AGY S2-glass fibres [10] with an areal density of 744 g/m2 (24 oz./yd2) was stacked in a quasi-isotropic ([(45/0)5/45]S) lay-up of 22 layers and infused with the rubber toughened, epoxy resin system SC-15 manufactured by Applied Poleramic, Inc. [11]. The infusion was performed using a standard vacuum assisted resin transfer molding (VARTM) procedure with a vacuum pressure of 23-29 mm/Hg and cured in an oven at 60 °C until gelled, then post-cured at 120 °C for at least 4 hours. Four 40.6-cm x 40.6-cm samples were waterjet from a large 89-cm square parent panel to a tolerance of ±0.4-mm. The finished impact samples of S2/SC-15 composites with clamped boundary conditions are listed in Table 1.

Table 1. Thickness, weight, and density of S2/SC-15 impact samples.

S2/SC-15	Clamped
Thickness (cm)	1.509 ± 0.01
Weight (g)	4370 ± 50.9
Density (g/cm^3)	1.798 ± 0.012
Areal Density (kg/m^2)	26.6± 0.18

2.2. Four Quadrant Impact Methodology

The four quadrant impact (FQI) methodology found in [1, 2, 4, 12] was used since no suitable ASTM standard is available for characterizing the durability of thick-section composite laminates under multiple low velocity impacts (LVI). The FQI methodology calls for four successive impacts into a single panel as illustrated in Figure 2, allowing for both a quantitative and qualitative assessment of the durability and delamination resistance of the composite laminate [2]. An impact panel was cut to a

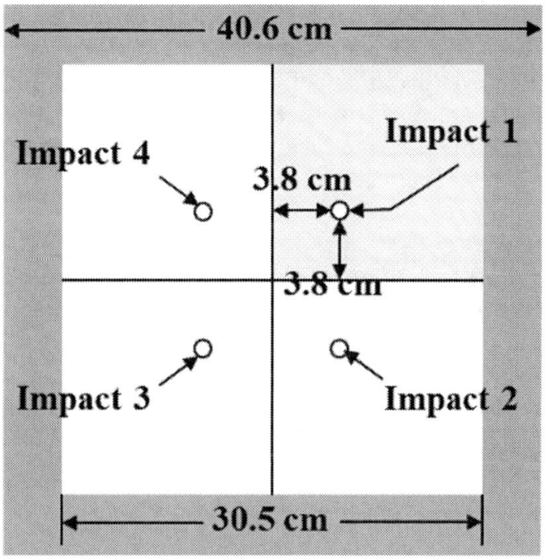

Figure 2. Illustration of impact panel face showing top plate picture frame overlay (grayed portion) and specific location for each of the four successive impact events.

dimension of 40.6-cm by 40.6-cm to ensure a proper fit into the FQI fixture for multiple LVI characterization. The dimensions of the impact samples and FQI fixture table were constrained by the drop tower configuration (clearance between vertical drop rails) [2]. The FQI fixture table allowed for impact samples to be clamped at a prescribed clamping pressure (distributed among 16 bolts with a pneumatic torque wrench) or accommodated a steel insert that converted the fixture to a simply-supported boundary condition for impact as shown in Figure 3.

A Dynatup 8110 drop-tower equipped with a strain-gage based, 222.4 kN load cell TUP and a 5.0 cm diameter hemispherical impactor is used for each impact. Force-time data during each impact event

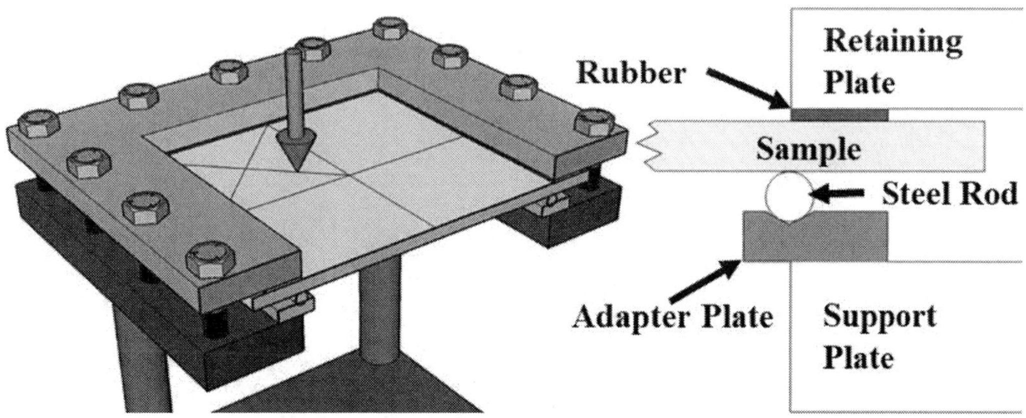

Figure 3. Section of FQI fixture with installed sample (left) and arrow indicating point of impact. Detailed cross-section (right) of sample being supported by simply-supported insert adapter. To achieve a clamped boundary condition for the sample, simply remove the adapter plate, steel rod supports, and rubber clearance strip and apply a uniform clamping torque to all 16 bolts.

was recorded for a 25 ms duration at a sampling rate of 270 kHz, and post-processed according to guidelines outlined in ASTM D7136 [13]. The mass and height of the impactor used in this investigation are 227.4 kg and 96 cm, respectively, yielding impact energy of approximately 2.14 kJ and an impact velocity of 4.34 m/s. The kinematic impactor displacement is calculated from the force-time trace for each impact by integrating twice over Newton's kinematic equations. The kinematic displacement is paired with the measured force to yield a force-deflection profile for each impact event with the implied assumption panel displacements are taken to be equivalent to impactor displacements. In the FQI methodology, impacts are performed in one location or "quadrant" at a time to allow for the non-destructive evaluation of the panel in between impact events.

2.3. Impact Damage Evolution

An important aspect of the FQI methodology is the assessment of damage within the panel between each of the successive impact events using a non-destructive method. A custom-built water tank and gantry system was used to capture ultrasonic c-scans of the panels prior to the first impact and after each of the four successive impact events. Using sectioning and microscopy to examine the cross-sectional views at various locations within the damaged panels, it is confirmed that the c-scans were detecting either one of two types of damage: (1) matrix cracking in the yarns under the impact location or (2) delamination between plies away from the impact location. An example of C-scan images showing the cumulative damage after each of the four impacts is shown in Figure 4.

Figure 4. C-scans of panel damage area for anvil impacts 1-4

3. Numerical model

A finite element model is developed in LS-DYNA to predict the deformation and delamination of the composite during impact. The explicit dynamics solid mechanics model includes the composite panel, supports, top (clamping) plate, and anvils as shown in Figure 5.

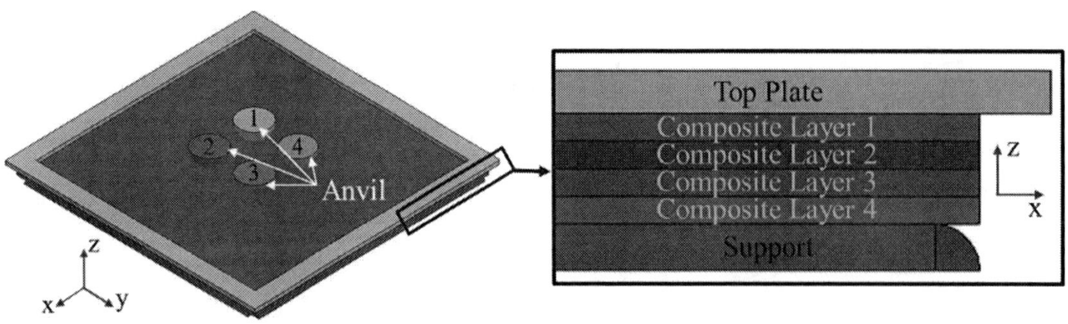

Figure 5. Initial low velocity impact model configuration

IOP Conf. Series: Materials Science and Engineering **987** (2020) 012024 doi:10.1088/1757-899X/987/1/012024

3.1. Model setup and boundary conditions

Initially, the composite panel is between the top plate and supports, with a 1-mm gap between each anvil and the composite plate to ensure no fictitious initial penetrations occur. The anvil impact locations are the same as in the experimental setup (although the model uses four anvils, one at each impact location, for simplicity).

The supports have a fixed location boundary condition in all directions. The first anvil has an initial velocity of 4.48 m/s in the vertical direction (z-direction in Figure 5), while the others begin at rest. The remaining anvils are then given a velocity of 4.48 m/s at 0.03 second intervals (the impact and rebound event for each anvil is completed in less than 0.03 seconds). Multiple boundary conditions are explored for the top plate, with its location being fixed in all directions as the baseline case. The friction coefficient between the panel and tooling is set to 0.3.

To represent delamination damage in the panel, individual layers within the composite panel are held together using a surface to surface tiebreak criterion. The baseline failure stress and distance used in this work are 22 MPa and 1.0 mm, respectively. The nodes initially in contact at the interface move together prior to the failure stress being reached, after which the damage is a linear function of the distance between nodes until it reaches the failure distance and full interfacial failure occurs. Contact between the tooling and composite panel is described using the built-in automatic surface to surface contact algorithm in LS-DYNA.

All tooling (top plate, supports, and anvils) is modelled as rigid bodies. The composite panel is described using an orthotropic composite material model. Although there is an option for including composite damage, the failure parameters are selected such that the delamination is the only failure mode. This is done to eliminate the investment required to characterize unnecessary parameters and ensure the model and experimental failure mechanisms align. The composite properties input into the model are given in Table 2.

Table 2. Composite panel properties input into the model (moduli units are GPa) [9, 14]

E_x	E_y	E_z	ν_{yx}	ν_{zx}	ν_{xy}	G_{xy}	G_{yz}	G_{zx}
27.5	27.5	27.5	0.11	0.18	0.18	7.67	4.28	4.28

3.2. Mesh refinement study

A mesh refinement study is performed to ensure numerical results converge as the number of elements increase. The anvil load-displacement curve from the first impact is used as a characteristic output parameter for the model. Figure 6 shows the force-displacement response of the panel under this loading condition for meshes with a variety of refined element sizes (closer to the anvils on the panel), coarse element sizes (further from the anvils), and a number of elements through the panel thickness. The convergence of results shows the solution is mesh independent with at least 114k elements. The 205k element model is selected for this work moving forward because it represents the converged solution but does not add the extra computational time required for meshes with a larger number of elements. For this mesh, the solid elements in the panel near the anvils are 2.00 mm squares (in the xy-plane), the elements in areas with minimal deformation are 6.00 mm squares (in the xy-plane), and all elements are 1.05 mm thick (z-direction). A similar study is performed for the number of tiebreak layers through the thickness of the panel. It is found that increasing the number of layers beyond three does not influence the results; therefore three tiebreak layers will be used for this work.

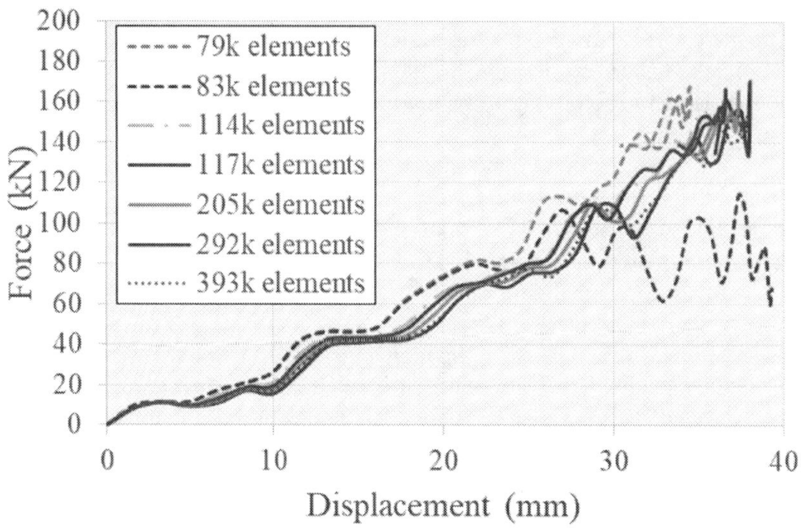

Figure 6. Anvil load-displacement curve for meshes with a range of element sizes

4. Results

4.1. Model verification

Key outputs from the numerical model include peak anvil-panel contact force, peak deflection of the panel, and the delamination area. The top plate is fixed in the x- and y- directions and a 325 kN force (vertical, z-direction) is applied. Such a high force is required to ensure no movement of the top plate during the impact event. The peak force and displacement output from the model are 151 kN and 30.5 mm, respectively, while the experimental results yield values of 150 kN and 28 mm. To explore damage area predictions, only the first anvil hit is used because subsequent impacts have overlapping damage areas. The model acceptably predicts the delamination between the glass layers in the composite, as depicted in Figure 7, with the model results showing the middle tiebreak layer. This was chosen because the tiebreak layers closer to the panel surface exhibited less delamination than the middle layer and the experimental measurement of delamination is cumulative through the thickness. The shape of the damage area matches well, with the only noticeable differences residing under the top plate. The

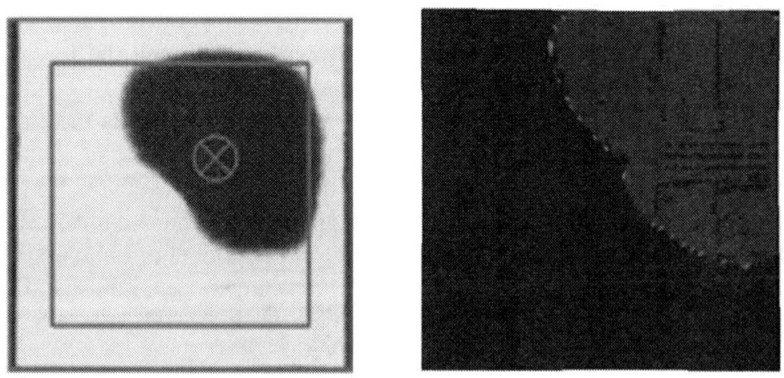
(a) (b)

Figure 7. Damage area after the first anvil impact determined (a) experimentally and (b) with the numerical model

IOP Conf. Series: Materials Science and Engineering **987** (2020) 012024 doi:10.1088/1757-899X/987/1/012024

combination of peak load, peak displacement, and damage area simulation accuracy verify the ability of the model to predict composite mechanical response to low velocity impact.

4.2. Influence of laminate stiffness

A parametric study was conducted to determine which mechanical properties have the most influence on the model, with results tabulated in Table 3. The fiber direction stiffness did not have a significant influence on the peak force or displacement of the panel because there is minimal fiber extension during the event. Panel in-plane shear stiffness (xy-plane) played an important role in the peak anvil force and displacement. This is because in-plane shear is the primary elastic deformation mechanism of the thin panel. New resin formulations which exhibit stiffer in-plane shear responses would be expected to deflect less during this type of an impact event. The influence of out-of-plane (i.e., interlaminar) shear modulus was not significant for this geometry because the panel was thin enough that the out-of-plane shear deformation was minimal. If the thickness was increased substantially, it is expected that the out-of-plane shear stiffness would become more important. The Poisson ratio had a negligible effect on both force and displacement.

Table 3. Composite panel properties input into the model (moduli units are GPa) [9, 14]

	Peak Displacement			Peak Force		
	50%	Baseline	150%	50%	Baseline	150%
E_1, E_2	1.01	1.00	0.98	0.99	1.00	1.02
E_3	1.01	1.00	0.99	0.99	1.00	0.99
G_{12}	1.07	1.00	0.96	0.91	1.00	1.06
G_{23}, G_{31}	1.02	1.00	0.99	1.01	1.00	1.01

4.3. Sensitivity to delamination parameters

With delamination being the primary damage mechanism, it is important to understand how the tiebreak delamination parameters affect model predictions. The parameters governing the tiebreak are the failure stress and complete failure distance. A fixed location boundary condition was applied to the top plate because it eliminates the additional characterization (i.e., clamp pressure or displacement) required for the clamped boundary condition.

The tiebreak failure stress (delamination stress) had a substantial influence on the damage area, as shown in Figure 8. Increasing the failure stress greatly decreased the delamination area, and doubling the failure stress reduced the damage area to a small area under the anvil. The peak deformation only changed about 8% over the range of failure stresses and the peak force saw a negligible difference. This

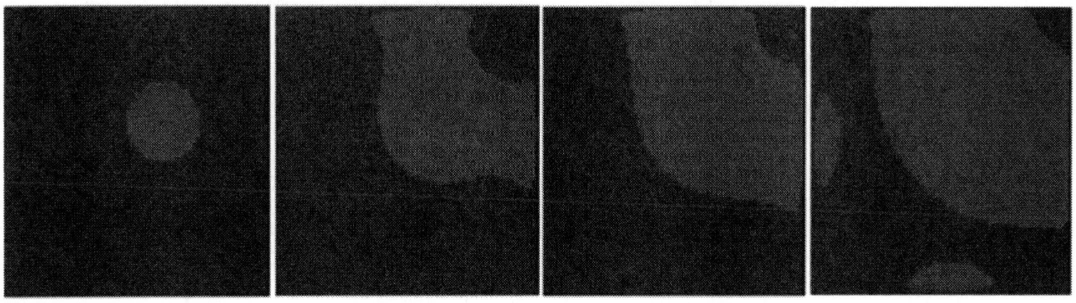

(a) (b) (c) (d)

Figure 8. Damage area predicted for a delamination stress of (a) 58 MPa, (b) 29 MPa, (c) 22 MPa, and (d) 15 MPa is a critical property to optimize when designing new resin formulations for composite structures expected to experience impact events.

The influence of the complete failure distance is important to understand from a modeling standpoint, but it does not clearly align with a specific measureable mechanical property. The critical distance for complete failure was varied from 0.1 mm to 2.0 mm and had a profound influence on the delamination area, described in Figure 9. A smaller failure distance parameter causes delamination to occur at an accelerated rate because there is a larger stress concentration due to the force being distributed over a smaller area, and which results in faster complete delamination. The peak displacement was sensitive (21% change over the range of distances studied) to the failure distance parameter, but the peak force was not.

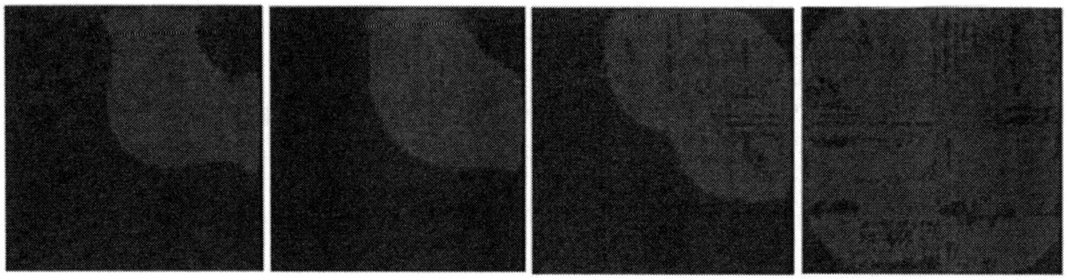

Figure 9. Damage area for complete tiebreak failure at nodal distances of (a) 2.0 mm, (b) 1.0 mm, (c) 0.5 mm, and (d) 0.1 mm

4.4. Influence of boundary conditions

The effect of switching between an applied pressure and fixed location boundary condition on the top plate can be examined by comparing Figure 7-b and Figure 8-c. It is noted that the fixed location boundary condition resulted in the panel experiencing less damage because the added constraint from the pressure boundary condition prevents panel deformation.

Removing the top plate creates an even more simplified boundary condition by eliminating contact between the composite panel and top plate from the low velocity impact event. The model was exercised without a top plate and then compared to the model with a perfect clamp boundary condition (includes top plate and the panel sides cannot slide inwards). The model without a top plate experienced much lower peak forces but higher peak deformations, as seen in Figure 10. This is because removing the top

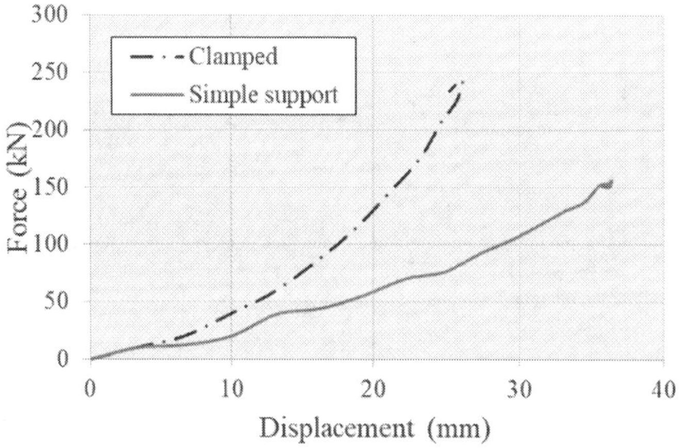

Figure 10. Anvil load-displacement response predicted by model for perfectly clamped and simply supported boundary conditions

plate constraint allows the panel to bend more, which drives down the peak force by spreading the energy absorption over a longer time.

5. Conclusions

An experimental procedure was presented to characterize the response of structural composite materials under low velocity, high energy loading conditions. A model was developed and verified to be capable of predicting the composite deformation and delamination during these events. The model gave insight into how boundary conditions affect composite response, with a clamped boundary condition resulting in a much higher peak load than the simply supported boundary condition. The model was exercised to understand the influence of mechanical and damage parameters on structural performance. It was found that in-plane shear modulus was the most critical elastic property affecting the composite response. The delamination failure stress had a profound influence on the damage area in low velocity, high energy impact events. The trends observed in this study will be useful for providing model-informed guidance for the development of next generation resins for structural composites.

References

[1] Boyd S E, Emerson R and Bogetti T A 2013 Low-Velocity, Multi-Impact Durability Performance of Thick-Section 3WEAVE S2-Glass/SC-15 Composites Toughened with Thermoplastic Polyurethane Inter-Layer Films. Army Research Laboratory.

[2] Boyd S E, Emerson R and Bogetti T A 2012 Multi-Impact Test Method to Assess Damage Tolerance in Thick-Section Composites. In: *SAMPE Technical Conference,*

[3] Dogan A 2019 Single and repeated low-velocity impact response of E-glass fiber-reinforced epoxy and polypropylene composites for different impactor shapes *Journal of Thermoplastic Composite Materials* 089270571988691

[4] Boyd S E, Bogetti T A, Staniszewski J M, Lawrence B D and Walter M S 2018 Enhanced delamination resistance of thick-section glass-epoxy composite laminates using compliant thermoplastic polyurethane interlayers *Composite Structures* **189** 184-91

[5] Sreekantha Reddy T, Rama Subba Reddy P and Madhu V 2019 Low velocity impact studies of E-glass/epoxy composite laminates at different thicknesses and temperatures *Defence Technology* **15** 897-904

[6] Majid D L, Jamal Q M and Manan N H 2018 Low-velocity impact performance of glass fiber, kenaf fiber, and hybrid glass/kenaf fiber reinforced epoxy composite laminates *BioResources* **13** 8839-52

[7] Panettieri E, Fanteria D, Montemurro M and Froustey C 2019 Low-velocity impact tests on carbon/epoxy composite laminates: A benchmark study *Composites Part B: Engineering* **107** 99-21

[8] Patil S, Reddy D M and Reddy M 2018 Low velocity impact analysis on composite structures – A review **1943** 020009

[9] Xiao J R, Gama B A and Gillespie J W 2007 Progressive damage and delamination in plain weave S-2 glass/SC-15 composites under quasi-static punch-shear loading *Composite Structures* **78** 182-96

[10] BGF Industries Danville, NC.

[11] Applied Poleramic Inc. Benicia, CA.

[12] Emerson R, Bogetti T A, Gama B A and Pasupuleti P 2010 A Multi-Hit Impact Method for Assessing the Durability of Thick-Section Composites. In: *SAMPE Technical Conference,* Salt Lake City, UT

[13] 2015 ASTM D7136 / D7136M - 15: Standard Test Method for Measuring the Damage Resistance of a Fiber-Reinforced Polymer Matrix Composite to a Drop-Weight Impact Event.

[14] Staniszewski J 2010 An Improved Design Methodology for Modeling Thick-Section Composite Structures Using a Multi-Scale Approach. In: *Mechanical Engineering*: University of Delaware

Deep Learning-based Phase Prediction of High-Entropy Alloys

Zeyad Yousif Abdoon Al-Shibaany[1, 2], Nadia Alkhafaji[*, 3, 4], Yaser Al-Obaidi[5], Alaa Abdulhasan Atiyah[6]

[1]Cardiff School of Engineering, Cardiff University, Cardiff CF24 3AA, UK.
[2]Biomedical Engineering Department, University of Technology - Iraq, Iraq.
[3]Faculty of Computing, Engineering, and Science, University of South Wales, UK.
[4]Laser and Optoelectronics Engineering Dept., University of Technology - Iraq, Iraq.
[5]Hyundai Engineering Co. Ltd. Iraq Branch, Iraq.
[6]Materials Engineering Department, University of Technology - Iraq, Baghdad, Iraq.

[*]Corresponding author: nadia.a.alkhafaji@gmail.com

Abstract. High-entropy alloys (HEAs) offer a new approach to the design of superior metallic materials, wherein alloys are based on multiple principal elements rather than just one. Deep Neural Networks (DNNs), machine learning tools that are efficiently used for prediction purposes, are transforming fields, from speech recognition to computational medicine. In this study, we extend DNN applications to the field phase prediction of high-entropy alloys. Using the built-in capabilities in TensorFlow and Keras, we train DNNs with different layers and numbers of neurons, achieving a 90% prediction accuracy. The DDN prediction model is examined in detail with different datasets to verify model robustness. Due to the high cost of HEAs and in order to save time, it is important to predict phases in order to design alloy composition. Through this study, we show trained DNNs to be a viable tool for predicting the phases of high-entropy alloys, where 90% phase prediction accuracy was achieved in this work.

Keywords: Deep learning, machine learning, artificial intelligence, prediction, high-entropy alloys.

1. Introduction

Machine learning (ML), a branch of artificial intelligence (AI), is based on the concept that computers (machines) can use data and learn to identify patterns and make predictions. These predictions and pattern-identification capabilities can be done with minimal programming instructions. The high computing power in ML comes from the use of Artificial Neural Networks (ANNs), which are basically computing systems with several interconnected processing elements called neurons that map an array of input variables (features) to one or more outputs (labels). Figure 1 shows schematic of an ANN where the circles refer to the neurons (the processing elements of ANNs) and the arrows refer to the connections between these neurons; the direction of the arrow shows the direction of the data flow through the network.

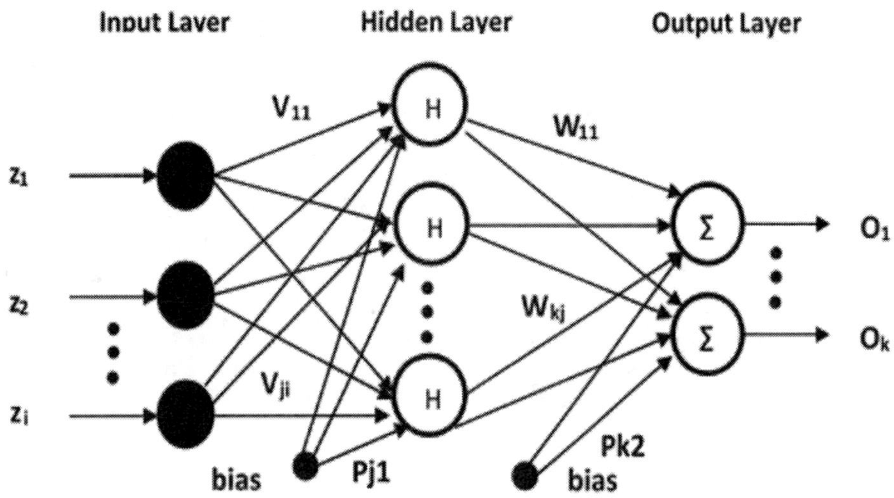

Figure 1: Schematic of Artificial Neural Networks [1]

The parameters of the ANN shown in Figure 1 are listed in Table 1 below.

Table 1. List of the ANN parameters [1].

Parameter	Description
ni	Number nodes in the input layer
nj	Number of nodes in the hidden layer
nk	Number of nodes in the output layer
$\mathbf{V_{ji}}$	Weight between (j) hidden node and (i) input node
$\mathbf{W_{kj}}$	Weight between (k) output node and (j) hidden node
$\mathbf{P_{j1}}$	Weight between (j) hidden node and bias node
$\mathbf{P_{k2}}$	Weight between (k) output node and bias node
h	Hidden layer node
$\mathbf{z_i}$	Input of (i) input node
$\mathbf{O_k}$	Output of (k) output node
H	Activation function

The following steps explain the calculation of the output of ANN:

1. The input data are applied to the input layer. In this layer, the data will not be processed.
2. The data are then passed to the hidden layer, and the output of each neuron at the hidden layer is calculated as follows:

$$hnet_j = H(h_j) \tag{1}$$

Where h_j represents the jth neurons in the hidden layer, and $hnet_j$ represents the jth neurons in the hidden layer.

$$h_j = \sum_{i=1}^{ni} z_i \times V_{ji} + bias \times P_{j1} \tag{2}$$

3. The output of the output layer is then calculated as follows:

$$O_k = \sum_{j=1}^{nj} hnet_j \times W_{kj} + bias \times P_{k2} \tag{3}$$

It is easy to determine the number of neurons at the input layer of any neural network by simply counting the number of independent variables that will be fed into the neural network. The same idea can be applied to the output layer, where the number of neurons is equal to the number of dependent variables. However, choosing the number of neurons at the hidden layer represents a significant challenge that still attracts the interest of many researchers. A binary search technique was used to estimate the number of neurons in the hidden layer where the number chosen was 1, 2, 4, 8, 16, 32, 64, and [2]. However, the binary search method in this work achieved up to 80% accuracy. A comprehensive systematic review of the techniques to count the number of neurons in the hidden layer was conducted by [3]. Based on the literature and for the purposes of this work, Table 2 lists the techniques used to count the number of neurons in the hidden layer in this work where:
Ni: number of neurons in the input layer
No: number of neurons in the output layer
Nh: number of neurons in the hidden layer
Nt: number of training pairs (size of the training dataset)

Table 2. List of techniques for counting the number of neurons in the hidden layer.

No.	Technique	Reference
1	$Nh = \frac{\sqrt{1+8Ni}-1}{2}$	[4]
2	$Nh = Ni - 1$	[5]
3	$Nh = \frac{1}{2} \cdot \frac{Nt}{Ni\ logNt}$	[6]
4	$Nh = \frac{Nt}{Ni}$	[6]
5	$Nh = \sqrt{Ni\ No}$	[7]
6	$Nh = log_2(Ni+1) - No$	[8]
7	$Nh = \frac{4\,Ni^2+3}{Ni^2-8}$	[3]
8	$Nh = \frac{\sqrt{1+8Ni}-1}{2}$	[4]

Deep neural networks are now transforming many research fields, including computational medicine, speech recognition and computer vision [9]. A continuous function can be approximated by a neural network with one hidden layer if this layer has a sufficient number of neurons. However, multi-variable and complex functions are not easy to be approximated with a single hidden layer ANN. This difficulty in approximation can be solved by adding more hidden layers to the ANN (going deeper) [10-13]. Classical ANNs (with one hidden layer of neurons) have been used in the field of computational materials science and engineering in the nineties of the last century [14, 15]; however, deep ANNs have only recently been used in this field [16-18].

High-entropy alloys are defined as alloys with five or more principal elements having equal atom percentage. Each principal element should have a concentration between 5 and 35 atom%. Significant research interest did not develop until after independent papers by Jien-wei Yeh and Brain Cantor. Yeh coined the term 'high-entropy alloy', attributing the high configurational entropy as the mechanism stabilising the solid solution phase. Yeh explored the area of multicomponent alloys independently from 1995 [19], later theorising that a high mixing entropy would play an essential factor in reducing the number of phases in this high order of mixing and resulting in valuable properties [20]. To understand these properties, it is necessary to characterise the chemical ordering and identify order-disorder transitions through efficient simulation and modelling of thermodynamics. One of the main challenges of working on high entropy alloys is the selection of relevant and effective structures of these alloys. The structure represents the basic attributes for alloy or constituent elements of the alloy system. A multicomponent equiatomic alloy will form a single-phase BCC, FCC and multi-phase or more solid solution phases or intermetallic compounds (IM) or an amorphous phase. Designing and implementing an artificial intelligence-based tool to predict those phases will help to implement high entropy alloys design. The main objective of this paper is to utilise deep ANNs that predict body-centred-cubic and face-centred-cubic phases in high-entropy alloys.

1.1. Dataset

The dataset used in this paper to predict the phases of the high-entropy alloys was taken from the work of Miracle and Senkov [21]. The dataset has 18 attributes representing different elements (components of the alloys), and the last attribute is called a phase, which represents the phase of the alloys. A total of 17 elements were chosen as inputs for modelling: Al, Co, Cr, Cu, Fe, Ni, Ti, Mn, Pd, V, Mo, Nb, Si, Ta, Hf, Zr, and W. Figure 2 shows a snapshot of the dataset showing all the variables.

	Al	Co	Cr	Cu	Fe	Ni	Ti	Mn	Pd	V	Mo	Nb	Si	Ta	Hf	Zr	W	phase
0	0.0877	0.1754	0.1754	0.1754	0.1754	0.1754	0.035	0.00	0.0	0.0	0.0	0.0	0.0	0.0	0.0	0.0	0.0	0
1	0.0909	0.0000	0.1818	0.1818	0.1818	0.3636	0.000	0.00	0.0	0.0	0.0	0.0	0.0	0.0	0.0	0.0	0.0	0
2	0.0000	0.2500	0.2500	0.0000	0.2500	0.2500	0.000	0.00	0.0	0.0	0.0	0.0	0.0	0.0	0.0	0.0	0.0	0
3	0.0000	0.2500	0.0000	0.0000	0.2500	0.2500	0.000	0.25	0.0	0.0	0.0	0.0	0.0	0.0	0.0	0.0	0.0	0
4	0.0000	0.2000	0.2000	0.0000	0.2000	0.2000	0.000	0.00	0.2	0.0	0.0	0.0	0.0	0.0	0.0	0.0	0.0	0

Figure 2: Snapshot of the dataset

The values of the 'phase' are 0, 1, and 2, which stand for the FCC, Multiphase and BCC phases, respectively. The dataset was split into training and testing datasets. The training dataset (80% of the overall dataset) was used to train the DNN prediction model, and the rest of the data (20% of the overall dataset) was used as testing data to evaluate the model performance. The dataset was uploaded and examined for any missing values, and it was randomly shuffled to make sure there were no patterns or pre-set data.

1.2. Prediction Model Architecture

The deep neural networks prediction model in this work has an input layer with 17 neurons to handle all the 17 variables values and an output layer with one neuron that represents the 'phase' variable. The main function of this model was to read the values of the 17 elements and make a decision (prediction) about the type of the phase: FCC = 0, Multiphase =1 and BCC = 2. In terms of hidden layers and after trying the different techniques in Table 2, the model was constructed by using two hidden layers with 30 neurons in the first hidden layer and 20 neurons in the second hidden layer. All neurons of the hidden layers have a sigmoid activation function. Figure 3 illustrates the architecture of the DNN prediction model.

In this work, Python programming language was used to develop the artificial intelligence-based prediction models. Python is a general-purpose and high-level programming language, which was created by Guido van Rossum and first released in 1991 [22]. The philosophy of Python emphasises code readability to help programmers write clear, logical code for small and large-scale projects [23]. TensorFlow, which is a machine learning system that operates on a large scale and in heterogeneous environments, is also used in this work to develop the prediction model. TensorFlow supports a variety of applications, but it particularly targets training and inference with deep neural networks [24].

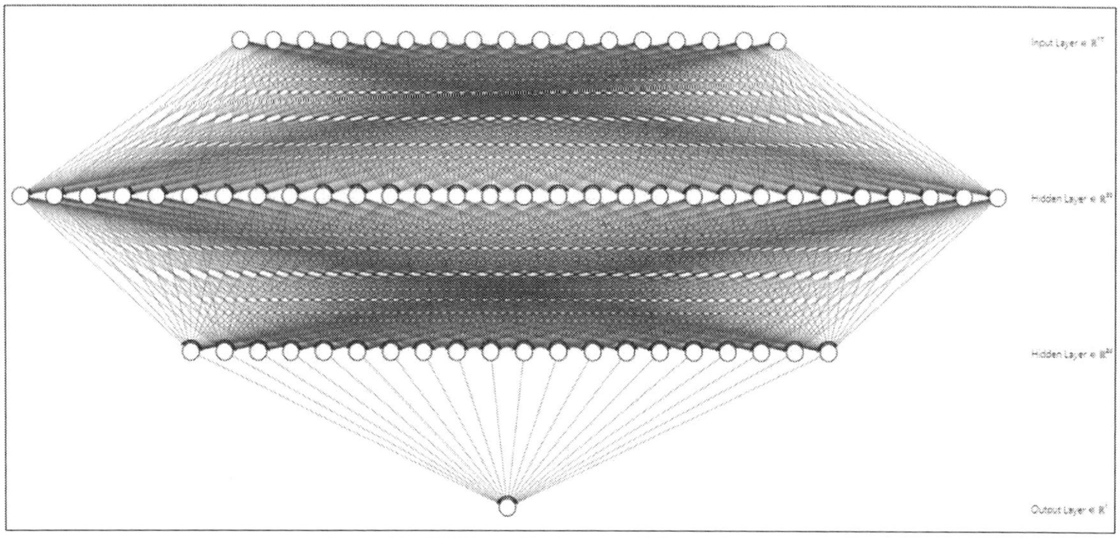

Figure 3: DNN prediction model architecture

2. Results and Discussion

The DNN prediction model was trained for 70k epochs, and the training target was to reduce the error in prediction (loss) to the minimum. Figure 4 shows the performance of the training process and how the loss function declines through the training. The loss function settled after 60k epochs, or 12.33% at the final step of the training process.

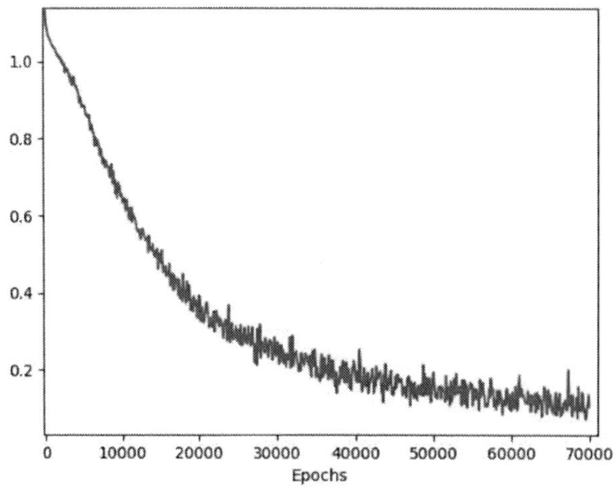

Figure 4: Loss function during the training process

The DNN prediction model was then evaluated by using different datasets (the testing dataset). In order to evaluate the model, the following parameters will be calculated:

- True Positives (TP): these are the correctly predicted positive phases, which means that the value of the actual phase (the one in the testing dataset) is the same as the value of the predicted phase.
- True Negatives (TN): these are the correctly predicted negative phases, which means that the value of the actual phase is not correct, and the value of the predicted phase is not correct as well.
- False Positives (FP): these are the values when the actual phase is not correct, and the predicted phase is correct.
- False Negative (FN): these the values when the actual phase is correct and the predicted phase in not.

Table 3 explains how to find the TP, TN, FP and FN values:

Table 3. Calculation of TP, TN, FP and FN values.

		Predicted Class	
		Class = YES	**Class = NO**
Actual Class	**Class = YES**	True Positive	False Negative
	Class = NO	False Positive	True Negative

Once the TP, TN, FP, and FN values are calculated, the following model measures performance, including accuracy, precision, recall, and F1 scores, which represents the most popular adopted metrics in classification tasks [25]. Accuracy is the most important performance measure of the DNN prediction model, and it is simply a ratio of correctly predicted observation to the total observations and is calculated as follow: *(TP+TN) / (TP+FP+FN+TN)*. Model precision represents the ratio of correctly predicted positive observations to the total predicted positive observations and is calculated as follows: *TP / (TP+FP)*. Recall, which is also called Model Sensitivity, represents the ratio of correctly predicted positive observations to all observations in actual class – yes – and is calculated by using the following formula: *TP / (TP+FN)*. Finally, the F1 score, which is a weighted average of Precision and Recall, is calculated as follows: $2 \times (Recall \times Precision) / (Recall + Precision)$.

The DNN prediction model achieved an overall accuracy of 90%. Table 4 shows the other measures of the DNN prediction model performance:

Table 4. DNN prediction Model Performance.

Phase	Precision	Recall	F1 Score
FCC	0.93	0.86	0.89
Multi-Phase	0.74	0.91	0.82
BCC	1.00	0.93	0.96

In addition to the above measures, a feature importance test was conducted by using random forest methodology. The aim of this test is to figure out the importance of each input variable. Figure 5 shows the importance level of each input (element) in the model. However, further work is ongoing to examine further methodologies for calculating variable importance.

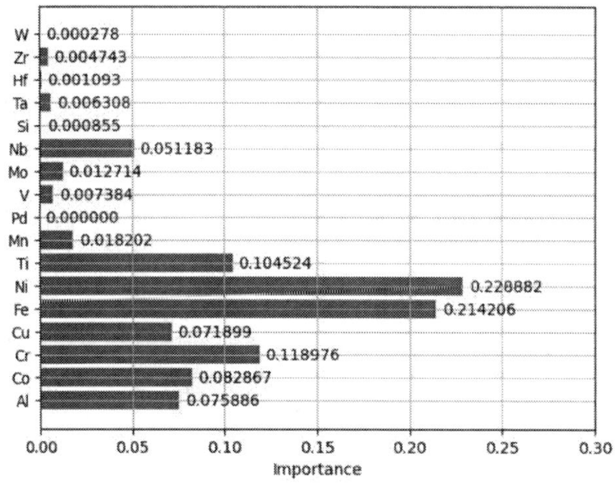

Figure 5: Variable Importance

3. Conclusion

Deep Neural Networks (DNNs) were developed in this work to predict body-centred-cubic, multi-phased and face-centred-cubic phases in high-entropy alloys. Using the built-in capabilities in TensorFlow and Keras, the authors trained DNNs with different layers and numbers of neurons, achieving a 90% prediction accuracy. The DDNs prediction model was examined in detail with different datasets to verify the model robustness. DDNs models used in this work were able to precisely predict the formation of solid solution and intermetallic phases in at least 90% of the cases on datasets. The model performs well on the prediction of the FCC & BCC dual-phase solid-solution as shown in Table 4, possibly due to phase ambiguities. Through this study, trained DNNs are shown to be a viable tool to predict the phases of high-entropy alloys. Further work is ongoing to develop the prediction model and better estimate the variable importance levels.

Authors Contribution

Conceptualisation, Alaa Abdulhasan Atiyah (A.A.A.); methodology, Zeyad Yousif Abdoon Al-Shibaany (Z.Y.A.A.); validation, A.A.A. and Z.Y.A.A.; formal analysis, Nadia Alkhafaji (N.A.), Yaser Al-Obaidi (Y.A.) and Z.Y.A.A.; investigation, Z.Y.A.A., N.A., and Y.A.; writing—original draft preparation, Z.Y.A. and N.A.; writing—review and editing, Z.Y.A. and A.A.A.. All authors have read and agreed to the published version of the manuscript.

References

[1] Z. Y. Al-Shibaany, J. Hedley, and R. Bicker, "Design of an adaptive neural kinematic controller for a National Instrument mobile robot system," presented at the 2012 IEEE International Conference on Control System, Computing and Engineering, 23-25 Nov. 2012, 2012.

[2] C. A. Doukim, J. A. Dargham, and A. Chekima, "Finding the number of hidden neurons for an MLP neural network using coarse to fine search technique," in *10th International Conference on Information Science, Signal Processing and their Applications (ISSPA 2010)*, 2010, pp. 606-609: IEEE.

[3] K. G. Sheela and S. N. Deepa, "Review on methods to fix number of hidden neurons in neural networks," *Mathematical Problems in Engineering,* vol. 2013, 2013.

[4] J.-Y. Li, T. W. Chow, and Y.-L. Yu, "The estimation theory and optimisation algorithm for the number of hidden units in the higher-order feedforward neural network," in *Proceedings of ICNN'95-International Conference on Neural Networks*, 1995, vol. 3, pp. 1229-1233: IEEE.

[5] S. i. Tamura and M. Tateishi, "Capabilities of a four-layered feedforward neural network: four layers versus three," *IEEE Transactions on Neural Networks,* vol. 8, no. 2, pp. 251-255, 1997.

[6] S. Xu and L. Chen, "A novel approach for determining the optimal number of hidden layer neurons for FNN's and its application in data mining," in *5th International Conference on Information Technology and Applications*, 2008, 2008.

[7] K. Shibata and Y. Ikeda, "Effect of number of hidden neurons on learning in large-scale layered neural networks," in *2009 ICCAS-SICE*, 2009, pp. 5008-5013: IEEE.

[8] D. Hunter, H. Yu, M. S. Pukish III, J. Kolbusz, and B. M. Wilamowski, "Selection of proper neural network sizes and architectures—A comparative study," *IEEE Transactions on Industrial Informatics,* vol. 8, no. 2, pp. 228-240, 2012.

[9] I. Goodfellow, Y. Bengio, A. Courville, and Y. Bengio, *Deep learning*. MIT press Cambridge, 2016.

[10] Z. Lu, H. Pu, F. Wang, Z. Hu, and L. Wang, "The expressive power of neural networks: A view from the width," in *Advances in neural information processing systems*, 2017, pp. 6231-6239.

[11] M. Telgarsky, "Benefits of depth in neural networks," *arXiv preprint arXiv:1602.04485,* 2016.

[12] N. Cohen, O. Sharir, and A. Shashua, "On the expressive power of deep learning: A tensor analysis," in *Conference on learning theory*, 2016, pp. 698-728.

[13] R. Eldan and O. Shamir, "The power of depth for feedforward neural networks," in *Conference on learning theory*, 2016, pp. 907-940.

[14] A. Skinner and J. Broughton, "Neural networks in computational materials science: Training algorithms," *Modelling and Simulation in Materials Science and Engineering,* vol. 3, no. 3, p. 371, 1995.

[15] B. HKDH, "Neural networks in materials science," *ISIJ international,* vol. 39, no. 10, pp. 966-979, 1999.

[16] J. Schmidt, M. R. Marques, S. Botti, and M. A. Marques, "Recent advances and applications of machine learning in solid-state materials science," *npj Computational Materials,* vol. 5, no. 1, pp. 1-36, 2019.

[17] A. Agrawal and A. Choudhary, "Deep materials informatics: Applications of deep learning in materials science," *MRS Communications,* vol. 9, no. 3, pp. 779-792, 2019.

[18] M. Picklum and M. Beetz, "MatCALO: Knowledge-enabled machine learning in materials science," *Computational Materials Science,* vol. 163, pp. 50-62, 2019.

[19] C.-Y. Hsu, J.-W. Yeh, S.-K. Chen, and T.-T. Shun, "Wear resistance and high-temperature compression strength of Fcc CuCoNiCrAl 0.5 Fe alloy with boron addition," *Metallurgical and Materials Transactions A,* vol. 35, no. 5, pp. 1465-1469, 2004.

[20] J. W. Yeh *et al.*, "Nanostructured high-entropy alloys with multiple principal elements: novel alloy design concepts and outcomes," *Advanced Engineering Materials,* vol. 6, no. 5, pp. 299-303, 2004.

[21] D. B. Miracle and O. N. Senkov, "A critical review of high entropy alloys and related concepts," *Acta Materialia,* vol. 122, pp. 448-511, 2017/01/01/ 2017.

[22] J. Guttag, *Introduction to computation and programming using Python: With application to understanding data*. MIT Press, 2016.

[23] D. Kuhlman, *A python book: Beginning Python, advanced Python, and python exercises*. Dave Kuhlman Lutz, 2009.

[24] M. Abadi, "TensorFlow: learning functions at scale," presented at the Proceedings of the 21st ACM SIGPLAN International Conference on Functional Programming, Nara, Japan, 2016. Available: https://doi.org/10.1145/2951913.2976746

[25] D. Chicco and G. Jurman, "The advantages of the Matthews correlation coefficient (MCC) over F1 score and accuracy in binary classification evaluation," *BMC Genomics,* vol. 21, no. 1, p. 6, 2020/01/02 2020.

IOP Conf. Series: Materials Science and Engineering **987** (2020) 012026 doi:10.1088/1757-899X/987/1/012026

Preparation and Evaluation of CuO Nanoparticles Using the Sol-Gel Method

Adnan S Jabur[1,*] **and Ali D Ahmed**[1]

[1] Department of Materials Engineering, Engineering College, University of Basrah, Iraq

[*] Correspondence author: adnanjabur@gmail.com

Abstract. Nanomaterials have attractive properties. They are employed in different fields of science, such as materials engineering, medicine, the environment and many technological applications. Transition metal oxides, such as CuO, are strong candidates for many scientific applications due to their unique physical and chemical properties. In this research, CuO nanoparticles were prepared using the nonorganic sol-gel technique under controlled conditions. The precursor of the sol-gel process was cupric chloride with NaOH as the stabilising agent. The aqueous solution was altered with two pH values to study their effect on the final precipitate. The prepared CuO nanoparticles were evaluated by XRD, particle size analyser, SEM and visually.

1. Introduction

After more than two decades of research and development in the field of nanoscience, nanotechnology applications still deliver new and unexpected ways to benefit society. A lot of these applications help improve many technological and industrial sectors. Several metals can be considered as potential primary materials for this purpose, but transition metals are the key elements of the future due to their exceptional characteristics and the ease with which they can be economically synthesised, such as via the sol-gel chemical route. The main reason for selecting CuO among other transition metals oxides is its semiconductivity. Semiconductive materials are widely used in electronic and optoelectronic devices, such as electrochemical cells [1], gas sensors [2], magnetic storage devices [3], field emitters [4], high Tc superconductors [5], nanofluids [6] and catalysts [7]. Considering electrical applications, CuO particles in the nano range have a wide bandgap that nearly equals ZnO. CuO nanoparticles can be mixed with highly purified water (CuO-based nanofluid) to significantly enhance energy transfer, such as in coolants and other thermal applications.

The uniqueness of the nanoscale (10–100 nm) is that the material's aspect ratio or surface-to-volume ratio will be very high, improving many mechanical properties, such as hardness and strength. Furthermore, the material's particles at the nanoscale will no longer be subjected to regular continuum mechanics laws but rather to quantum mechanics laws.

Several nanomaterial synthesis procedures exist, and they are usually classified as top-down or bottom-up processes depending on the process energy and the starting material conditions [8]. Bottom-up wet chemical methods, such as sol-gel, are considered the most efficient and economical. Some experimental results of the sol-gel method showed CuO nanoparticles in the range of 1–10 nm [9], and the smallest CuO nanoparticle obtained by the electrochemical method was nearly 4 nm. This work aims to prepare and evaluate CuO nanoparticles using the nonorganic sol-gel method and examines the effect of pH value on the properties of the prepared particles.

2. Experimental Procedure

The first step of the nonorganic sol-gel method consisted of preparing an aqueous solution of $CuCl_2.2H_2O$ (0.2 M) using deionised water. Then, 4 ml of glacial acetic acid was added to the aqueous solution; it was heated and stirred using a hotplate-stirrer device and then monitored with a digital thermometer and a pH meter, as shown in Figure 1. When the solution temperature reached 100 °C, NaOH pellets were added until the pH value reached 7. During this step, the solution changed gradually to a gel consistency. This chemical reaction was accompanied by a gradual colour change from green to nearly black (Figure 2). The hydrolysis reaction occurred as follows:

$$CuCl_2 + 2NaOH \rightarrow Cu(OH)_2 + 2NaCl$$

The resulting salt was removed by centrifuging and washing the colloid three times with deionised water, and the precipitated $Cu(OH)_2$ particles changed to CuO powder when dried in a furnace at 250 °C. In the second case, the aqueous solution was allowed to exceed the neutral pH value during the addition of NaOH and reached pH 11. The prepared CuO powders were examined by X-ray diffraction (XRD), and a Malvern Zetasizer Nano ZS device and scanning electron microscopy were used for size analysis.

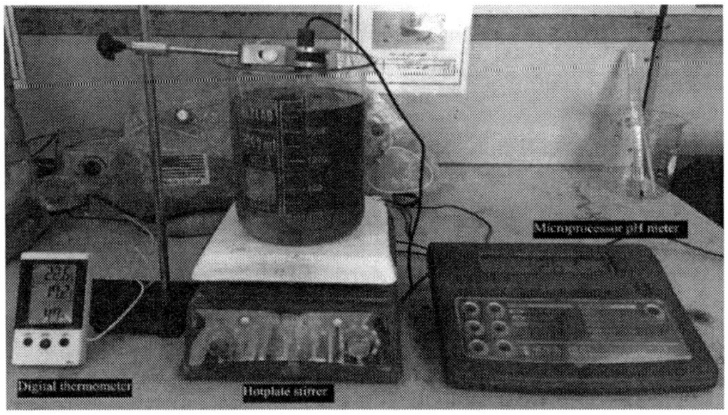

Figure 1 Instruments used in the sol-gel process

Figure 2 Precipitation steps of $Cu(OH)_2$ particles (left to right)

3. Results and Discussion

3.1 X-ray diffraction test

Figures 3 and 4 show the XRD pattern of the prepared CuO particle samples at pH values 7 and 11, respectively. All of the major peaks showed a monoclinic crystal structure of CuO in comparison with the PDF card no. 00-002-1041 (Figure 5). All of the peaks' Miller indices were identified. The average grain size was also calculated from Figure 3 using a Debye-Scherrer equation at approximately 60 nm:

$$d = \frac{K\lambda}{\beta Cos(\theta)}$$

where d is the average particle size, K is a dimensionless shape factor with an approximate value of 0.9, λ is the X-ray wavelength (in this test, it was equal to 0.15406 nm), β is the full width at half maximum

IOP Conf. Series: Materials Science and Engineering **987** (2020) 012026 doi:10.1088/1757-899X/987/1/012026

(FWHM) of the diffraction peak (in radians) and θ is the Bragg's angle or diffraction angle of the peak. Figures 3 and 4 also show that the prepared CuO powders have a good purity.

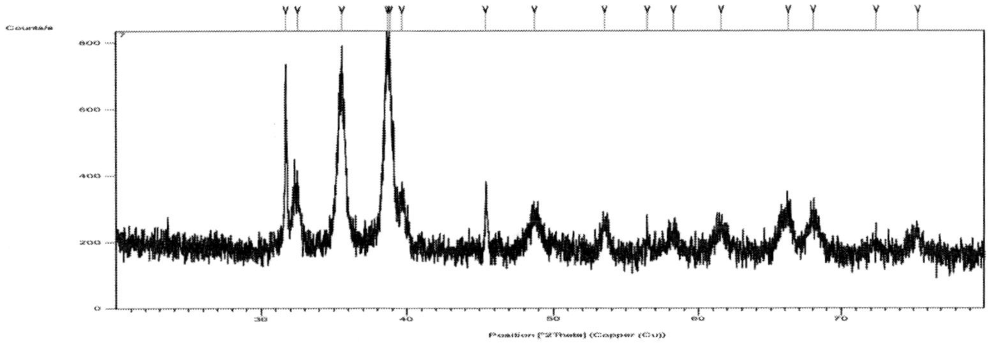

Figure 3 XRD chart of the prepared CuO powder at pH=7

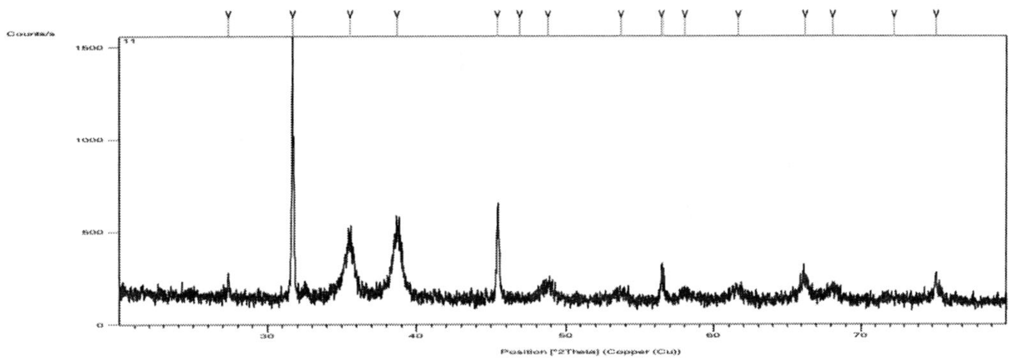

Figure 4 XRD chart of the prepared CuO powder at pH=11

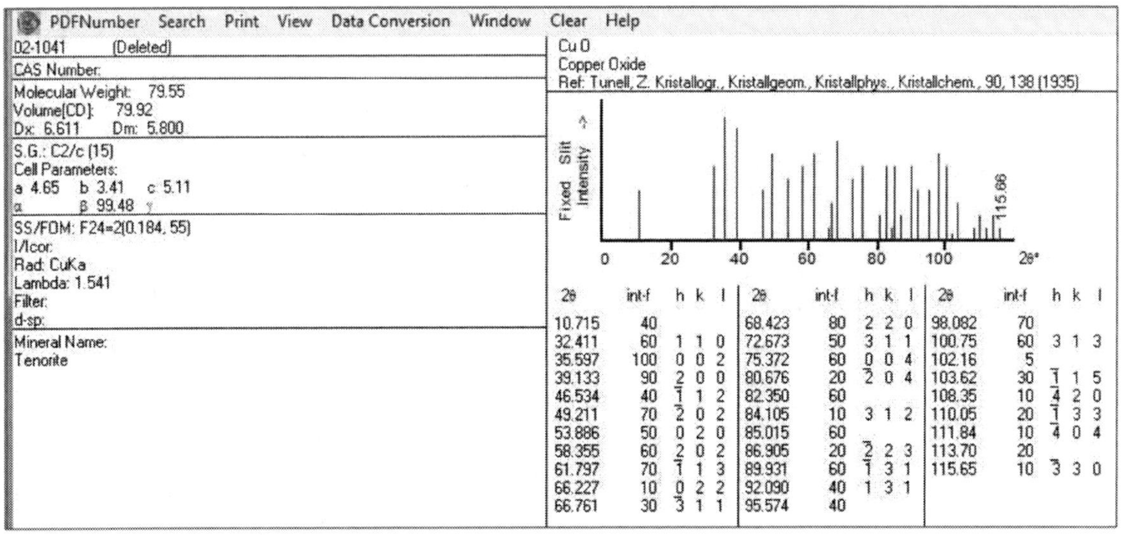

2θ	int-f	h	k	l	2θ	int-f	h	k	l	2θ	int-f	h	k	l
10.715	40				68.423	80	2	2	0	98.082	70			
32.411	60	1	1	0	72.673	50	3	1	1	100.75	60	3	1	3
35.597	100	0	0	2	75.372	60	0	0	4	102.16	5			
39.133	90	2	0	0	80.676	20	$\bar{2}$	0	4	103.62	30	$\bar{1}$	1	5
46.534	40	$\bar{1}$	1	2	82.350	60				108.35	10	$\bar{4}$	2	0
49.211	70	$\bar{2}$	0	2	84.105	10	3	1	2	110.05	20	$\bar{1}$	3	3
53.886	50	0	2	0	85.015	60				111.84	10	$\bar{4}$	0	4
58.355	60	2	0	2	86.905	20	$\bar{2}$	2	3	113.70	20			
61.797	70	$\bar{1}$	1	3	89.931	60	$\bar{1}$	3	1	115.65	10	$\bar{3}$	3	0
66.227	10	0	2	2	92.090	40	1	3	1					
66.761	30	$\bar{3}$	1	1	95.574	40								

Figure 5 PDF card no. 00-002-1041

3.2 Particle size test

Figures 6 and 7 show the size analysis of the prepared CuO particle samples at pH values 7 and 11, respectively. Figure 6 shows that the CuO particles prepared at pH=7 are mostly 60 to 150 nm in size and 30% are 90 nm. In Figure 7, the CuO particles prepared at pH=11 range from 500 to 1,050 nm in size, and 32% are 955 nm. This test shows that increasing the pH value from 7 to 11 increased the average particles' size from 90 to 955 nm.

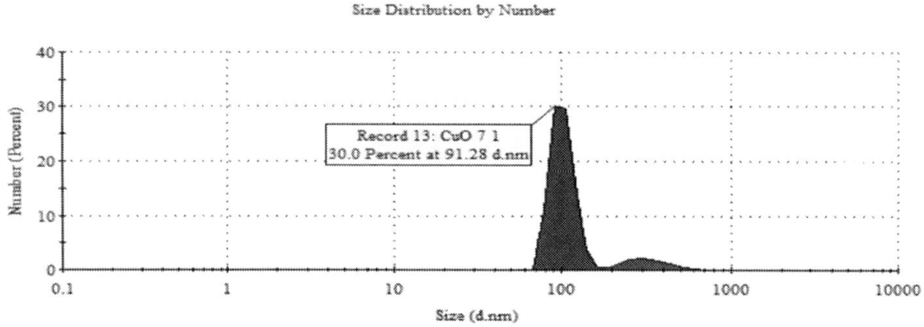

Figure 6 Size distribution of the prepared CuO powder at pH=7

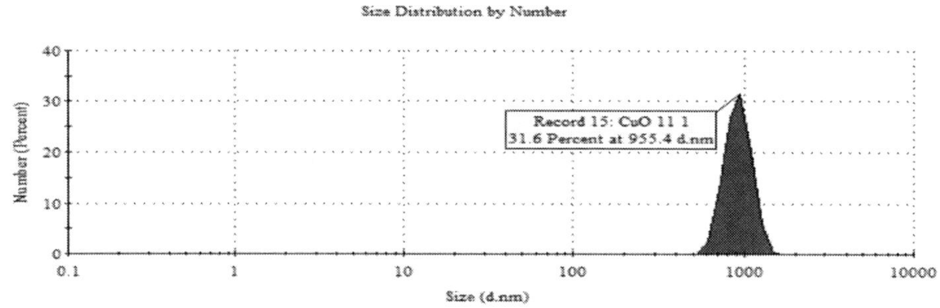

Figure 7 Size distribution of the prepared CuO powder at pH=11

3.3 Scanning electron microscopy test

Figures 8 and 9 include the SEM images of CuO particles prepared at pH values 7 and 11, respectively. Generally, the preparation of these samples caused high agglomeration and bad resolution of the particles on the Al substrate. In Figure 8, the particles were approximately spherical, while the particles prepared at a pH value of 11 were flake-shaped (Figure 9).

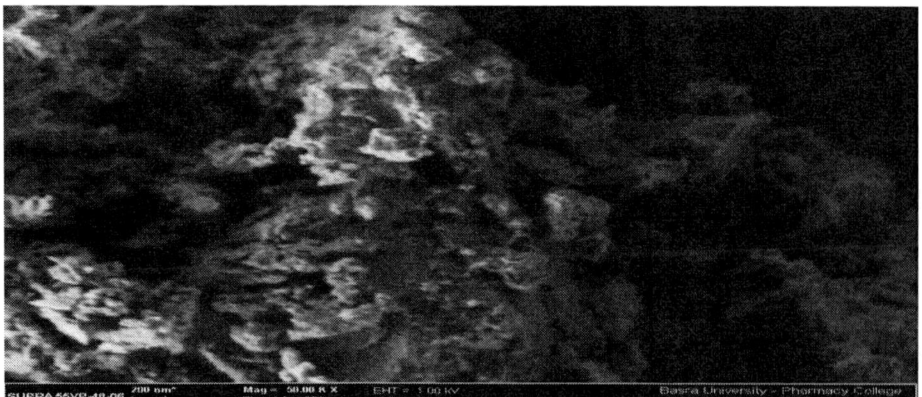

Figure 8 SEM micrograph of CuO particles prepared at pH=7

IOP Conf. Series: Materials Science and Engineering **987** (2020) 012026 doi:10.1088/1757-899X/987/1/012026

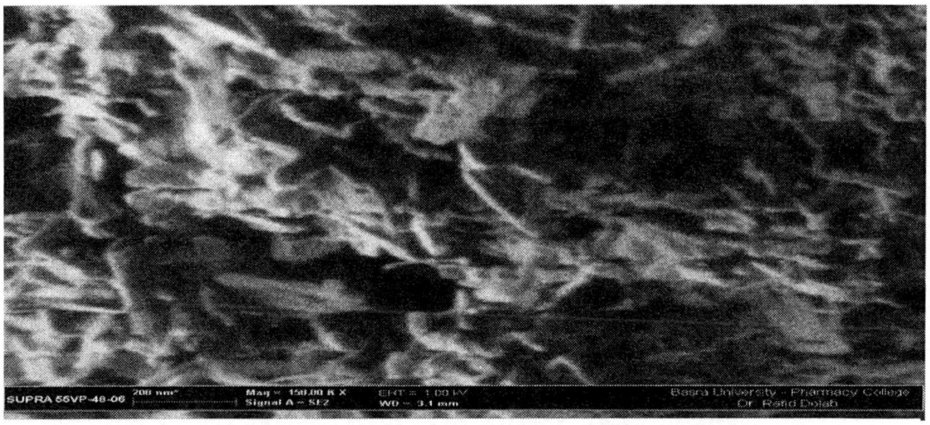

Figure 9 SEM micrograph of CuO particles prepared at pH=11

3.4 Visual test

One gram of each CuO nanoparticle type was mixed homogeneously with deionised water to obtain two nanofluids (Figure 10). Visual examination shows a difference in the nanofluids' colour, as it tends to become darker when the particle size is increased. This is because of the difference in surface energy of particles that react differently with light photons, as explained theoretically by the quantum theory.

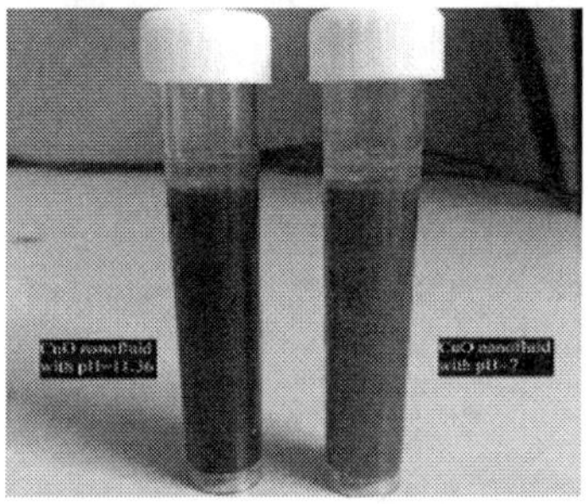

Figure 10 CuO nano-fluids

4. Conclusion

The main conclusions from this work are that the XRD pattern shows a monoclinic structure of CuO nanoparticles with an average grain size of approximately 60 nm (pH=7). Furthermore, the particle size test shows an increase in the average particle size from 90 nm at pH=7 to 955 nm at pH=11. On the other hand, the particles prepared at pH=7 have a spherical shape, while those prepared at pH=11 are flake-shaped. Finally, a visual examination shows a colour change from brown to black as the average particle size increases from 90 to 955 nm.

5. References

[1] Poizot P, Laruelle S, Grugeon S, Dupontl L and Tarascon J M 2000 *Nature* **407** 496 .
[2] Katti V R, Debnath A K, Muthe K P, Kaur M, Dusa A K and Gadkari S C 2003 *Sens. Actuators B Chem.* 245.
[3] Fan H M, Yang L T, Hua W S, Wu X F, Wu Z Y and Xie S S 2004 *Nanotechnology* **15** 37 .
[4] Hsieh C T, Chen J M, Lin H H and Shih H C 2003 *Appl. Phys. Lett.* **83** 3383.
[5] Wu M K, Ashburn J R, Torng C J, Hor P H, Meng R L and Gao L 1987 *Phy. Rev. Lett.* **58** 908.
[6] Chang M H, Liu H Sand Tai C T 2011 *Powder technology* **207** 378.
[7] Kumar N and Kumbhat S, 2016 *Ess. in Nanoscience and Nanotechnology* Jon Wiley & Sons, New Jersey, 1st Edition.
[8] Elseev A A, Lukashin A V, Vertegel A A, Heifets L I, Zhirov A I and Treytyakov Y D 2003 *Mat. Res. Innov* **3** 308.

Structural and Magnetic Characterisations of Pb-Doped MgO Nanoparticles by a Modified Pechini Method

Israa A.Najem[1,4], Shaker J. Edrees[2], Fadhil Abd Rasin[3]
[1] Department of Materials Engineering, University of Technology, Baghdad, Iraq.
[2] Department of Ceramic & Building Materials Engineering, College of Materials Engineering, University of Babylon, Babylon, Iraq.

[3] Department of Materials Engineering, University of Technology, Baghdad, Iraq.
[4] University of Al-Furat Al-Awsat Technical University, Babylon Technical Institute, Babylon, Iraq.
*corresponding author: israadnan2000@gmail.com

Abstract. Doping magnesium oxide nanoparticles (MgO-NPs) forms a good material for magnetisation applications. The Room Temperature Ferromagnetic (RTFM) of Pb^{2+}-doped ions MgO-NPs synthesised by a modified Pechini method are discussed in the present paper. The structural, morphological and magnetic properties of the samples were characterised by X-ray diffraction (XRD), Field Emission scanning electron microscopy (FE-SEM), Energy Dispersive Spectroscopy (EDS) and vibrating sample magnetometer (VSM).
The XRD results showed that the synthesised materials have a single set of peaks in the XRD patterns, corresponding to the cubic phase of MgO-NPs. As the Pb content increased in the host MgO-NPs leading to an increase in the lattice parameter, the interplanar spacing and the crystallite size; however, the intensity decreased.
Small spherical nanoparticles (22.87–29.05nm) were observed in the Pb^{2+}-doped ions MgO samples by (FE-SEM). The purity of the samples was confirmed using EDS spectroscopy.
The pure MgO and doped samples exhibiting RTFM may be attributed to vacancy defects, which caused local magnetisation. The saturation magnetism (Ms) was found to be varied as a function of doping concentration. The maximum (Ms) was found at x=0.015 of $Mg_{0.985}Pb_{0.015}O$ sample. The obtained results suggest that both Pb doping and oxygen vacancies play an important role in the development of room-temperature ferromagnetism.

Keyword: doping, MgO-NPs, PbO, Pechini, RTFM, VSM

1. Introduction

The study and development of nanomaterials and nanostructures are a key ambition in making the industry more innovative and more competitive. Nanomaterials and nanostructures perform a very important supporting function for applications, mainly for providing a notable surface-area-to-volume ratio than conventional structures [1][2][3].
Metal oxide nanoparticles are another field of nanoparticles (NPs), which can be directly used in the application of common chemical reactions, with special electronic and magnetic properties allowing them to be applicable in specific areas [4], such as TiO_2-NPs [5], ZnO-NPs [6], Fe_2O_3-NPs [7] and MgO-NPs [8].
Among these oxides, magnesium oxide nanoparticles with sodium chloride structure have attracted the attention of several researchers due to their exceptionally functional materials that have been widely utilised in superconductors [9], antibacterial materials [10] and magnets [11].

Doping magnesium oxide nanoparticles is an important tool for a wide variety of modifications, with the dopant impurities affecting the properties of metal oxide nanoparticles, such as [12] the optical properties when Zn^{+2}-doped ions affect MgO [13][14], in that the band gap of a MgO lattice, when occupied by transition ions (Ti-, Cr-, Fe-, Ni-, Zn), which can play an important role in plasma display panels [15], the magnetic properties of MgO nanoparticles are modified by doping Mn^{+2} ions [16], and the magnetic field can depend on the thermal conductivity of Fe^{+3} ions doping MgO [17]. The antibacterial activity can markedly be promoted with Ag-doped MgO nanoparticles [18], and the luminescence intensity is greatly enhanced according to the increase of Eu^{3+} concentrations in a MgO lattice [19].
Many researchers note that a bulk rocksalt structure (MgO, HfO, CaO and Al2O3) is nonmagnetic, while their nanoparticle structure confirms room-temp ferromagnetism (RTFM) [20][21]. Hence, the origin of magnetism in MgO-NPs is an issue of some recent interest [22][23] because the rocksalt structure does not include ions with partially filled d or f bands, since the valence band is controlled by O-2p orbitals, and there are no valence d electrons in the rocksalt structure [24][25][26]. Consequently, such magnetisation does not demand the existence of free carriers [27], but it can be correlated with the presence of a slight concentration of intrinsic point defects, isolated cation vacancies, leading to the formation of a high-spin defect site, which can form local magnetism in MgO-NPs [24][25].
Several preparation routes have been involved in preparing a pure and doped MgO-NPs, such as combustion [28], aerogel route [29], hard templating pathway [30], microwave sol-gel method [31], liquid-phase method [32], co-precipitation [33], green synthesis [34], sol-gel method [18], modification of a Pechini sol–gel process [35] and a spinning disk reactor [36].
Among the overall number of the chemical synthesis methods obtained in the preparation of pure and doped MgO-NPs, the Pechini method was a favourite selection due to its main role in the synthesis of magnetic materials, high-temperature superconductors, ceramics and catalysts [37]. The preference for this method includes good stoichiometric control, good homogeneity and good control of particle morphology [38].
The main objective of this work is to detect the influence of Pb^{2+} ions as incorporated into the crystal structure of MgO-NPs nanoparticles, where the samples were prepared by a modified Pechini method. In order to obtain a better understanding of the current work, the XRD, FE-SEM coupled with EDX and VSM tests were investigated to examine changes that take place in the structure with an increase in the fraction of doping.

2. Experimental Method

2.1 Chemicals Used

Magnesium nitrate hexahydrate ($Mg(NO_3)_2{\cdot}6H_2O$) [BDH Chemicals Ltd, England, > 99% purity], lead nitrate ($Pb(NO_3)_2$) [Chemicals PVT, India, > 99% purity], citric acid ($C_6H_8O_7.H_2O$) [CDH, India, >99%], Ethelyen Glygol, EG, ($CH_2(OH).CH_2OH$) [Chemicals PVT. LTD, India, > 99% purity] and distilled water were involved to prepare ($Mg_{1-x}pb_xO$)-NPs samples, where x=(0,0.005,0.01,0.015, 0.02,0.025,0.03), using a modified Pechini method. All raw chemical materials were used without any further purifications.

2.2 Samples Preparation

2.2.1 Synthesis of MgO-NPs Nanoparticles.

The procedure of synthesised MgO-NPs involved the preparation of a stable aqueous solution of the metal salt starting with the magnesium nitrite solution, and the tricarboxylic acid as citric acid solution were dissolved separately in an adequate amount of deionised water (100ml) for each of them, in a molar ratio of 1:1 and stirred for one hour under constant stirring (FALC instrument-F91). Then, citric acid was added into a magnesium nitrite solution to chelate Mg^{2+} ions in the solution; the resulting solution was kept at 90°C for three hours under constant stirring. EG was added with fixed mass ratio, CA/EG, 2:3. The colour of the final mixed solution gradually changed into a yellow shade as a result of water

IOP Conf. Series: Materials Science and Engineering **987** (2020) 012027 doi:10.1088/1757-899X/987/1/012027

evaporation. Finally, a very viscous gel formed with a brown colour. The dehydrated gel was calcinated at a moderate temperature of 420°C for two hours with 3°C /min as a heating rate using a furnace device (PIF 160/15- Made in Turkey) to form a loose powder.
The procedure of the synthesised ($Mg_{1-x}pb_xO$)-NPs samples, where x=(0–0.03), involved the preparation of an extra stable aqueous solution of lead nitrite, dissolved separately in an adequate amount of deionised water under constant stirring for one hour. The citric acid was added into mixed solutions of the magnesium nitrite and lead nitrite to chelate Mg^{2+} and Pb^{2+} in the solution. The same procedure steps were repeated for the preparation of MgO-NPs samples to obtain ($Mg_{1-x}pb_xO$)-NPs samples.

3. Characterisation techniques

All the seven samples of ($Mg_{1-x}pb_xO$)-NPs were analysed by powder XRD, FE-SEM coupled with EDS analysis and VSM. Powder X-ray diffraction patterns were obtained using a Shimadzu-6000 diffractometer with Cu-Kα radiation (λ=1.5406Å). The surface morphologies of the nanoparticles samples were investigated using a field emission scanning electron microscope (TESCAN MIRA3 FEG-SEM) operating at 30kV coupled with an energy-dispersive spectroscopy (EDS) facility for elemental analysis. The magnetic measurements were carried out using a vibrating sample magnetometer (VSM) type cryogenic device at room temperature.

4. Result and Discussion

4.1 *Structural Analysis X-ray diffraction (XRD)*

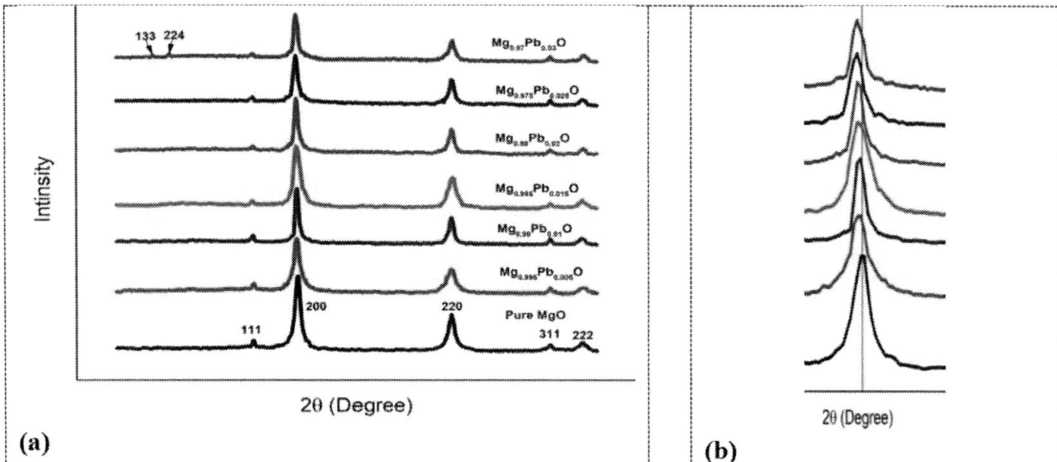

Figure 1. XRD analysis for ($Mg_{1-x}pb_xO$)-NPs samples (a) XRD pattern for ($Mg_{1-x}pb_xO$)-NPs samples calcined at 420°C for 2 hours. (b) XRD shifting for pure and doped ($Mg_{1-x}pb_xO$)-NPs samples.

For characterising crystalline materials, a powerful non-destructive technique was performed to provide information on phases, structures, preferred crystal orientations and other structural parameters, such as crystallinity, width, average grain size and the intensity of the diffraction peaks, which depend on lattice strain. The powder of the prepared ($Mg_{1-x}pb_xO$)-NPs samples using a modified Pechini method were calcined at 420°C for two hours, investigated via XRD technique.
Figure 1a reveals the reflections resulting from the ($Mg_{1-x}pb_xO$)-NPs samples. The pure MgO-NPs and all the doped samples are indexed well to the cubic phase with Fm3m (#225) as a space group.
However, the pure MgO-Nps sample showed five major diffraction peaks at positions 36.6325, 42.6868, 61.9761, 74.3027 and 78.2192. The corresponding lattice planes were (111), (200), (220), (311) and (222), which can be matched with [ICSD PDF code 01-075-0447]. These results are in line with earlier reports[16][18].

IOP Conf. Series: Materials Science and Engineering **987** (2020) 012027 doi:10.1088/1757-899X/987/1/012027

However, the XRD pattern for the $Mg_{0.97}Pb_{0.03}O$ sample appears as an additional weak peak, coinciding with PbO oxide. The strange peaks appear at positions 24.49 and 26.60, corresponding to the lattice planes 133 and 224 respectively, which were matched with [ICSD PDF (no. 45-964)].

The appearance of the strange PbO peaks can be attributed to the hardness of doping Pb^{+2} ions with a high concentration, due to differences in the ionic radius between Pb^{2+} (1.19 Å) ions and Mg^{2+} (0.72 Å) ions. So, the extra amount of Pb^{+2} ions can be segregated as a metal oxide phase (PbO) on the grain and at the grain boundaries of MgO-NPs host lattice [39]. This denotes that at or upward of this doping concentration of x=0.03 and under current experiential conditions, the Pb^{2+} ions cannot be inserted in the lattice sites. However, the additional amount of Pb^{2+} ions existing in an MgO solid solution can undergo oxidation to PbO [40]. Therefore, in the present research, the solubility boundary of Pb^{2+} ions in the MgO host lattice was about 0.005–0.03.

The strongest diffraction peak (200) for the pure structure of MgO-NPs is positioned at (42.6868), while for the doped samples are observed shifting to a lower 2θ peaks position, as cleared in Figure1.b.

Such increasing in shifting can be ascribed to the ionic radius difference between Pb^{2+} ions and Mg^{2+} ions that incorporated in the MgO host lattice.

The shifting of $Mg_{0.97}Pb_{0.03}O$ sample shows disorder shifting due to the incorporation of small fractions of the Pb^{2+} ions in the host lattice and the extra fraction separated as PbO oxide.

The interplanar d-spacing increases from (x=0–0.025) due to increased strain in the lattice of compounds, while it decreases at (x=0.03) doping due to the formation of the PbO phase [14][18]. Table 1 shows the values of d-spacing of pure and doped samples.

It is also evident from Figure 1b that the diffraction peaks shift lightly in this position, indicating a light change in the lattice parameter. Such changing leads to a contraction or expansion in the lattice due to the introduction of a dopant with a different ionic radius in the host lattice. A variation of the lattice constant was proportional to the difference between the ionic radius of the dopant, substituted host lattice atom and the concentration of the dopant [41]. According to Table 1, the lattice parameter for pure MgO-NPs is equal to (4.22832Å), while it is increased for doped samples. The maximum increase in the lattice parameter is (4.22983Å) at (x=0.025) for the $Mg_{0.975}Pb_{0.025}O$ sample.

However, the lattice parameter of the $Mg_{0.97}Pb_{0.03}O$ sample at (x=0.03) is decreased. Such a decrease can be attributed to the formation of the secondary phase of PbO oxide [18][40].

It is also evident from Figure 1b when Pb^{2+} ions were doped in the MgO-NPs host lattice. The intensity of the strongest diffraction peak (2θ=42.643468) along the (200) plane is decreased, and the full width at half maximum (FWHM) is increased. This indicates that the doped samples successfully replaced the Mg^{2+} ions with the Pb^{2+} ions in the MgO-NPs host lattice, and the undoped sample had a better phase purity and crystallinity since the intensity of the XRD peaks can be affected by the composition including occupancy of atomic sites, the scattering ability of constituent atoms, and fractional coordinates[18][42][43].

The average crystallite size (D) was evaluated using Scherrer's formula for the pure and doped MgO-NPs samples along (220) and (200) planes [44].

$$D = \frac{K\lambda}{\beta cos\theta} \dots\dots\dots\dots (1)$$

where D is the crystallite size, λ the X-ray wavelength of the copper anode (1.5406Å), k the shape factor (k = 0.9), β the full width at half maximum (FWHM) of the diffraction peaks, and θ the Bragg diffraction angle [45].

As the diffraction factor of Pb^{2+} ions increased, the average crystallite size (D) along (220) and (200) planes increased for the first six samples due to facilitating the crystalline growth and crystallisation of MgO-NPs, while it decreased for the $Mg_{0.97}Pb_{0.03}O$ sample, as shown in Table 1.

IOP Conf. Series: Materials Science and Engineering **987** (2020) 012027 doi:10.1088/1757-899X/987/1/012027

Table 1. Structural Parameters of Pure and Doped MgO-NPs Calcined at 420°C.

Samples	Space group	Crystalline size along(200) and (220) planes	Lattice parameter Å	d space
MgO-NPs	#225	5.693325	4.22832	2.11949
$Mg_{0.995}Pb_{0.005}O$	#225	5.696727	4.22837	2.11956
$Mg_{0.99}Pb_{0.01}O$	#225	5.91245	4.22848	2.11999
$Mg_{0.985}Pb_{0.015}O$	#225	6.412056	4.22979	2.12026
$Mg_{0.98}Pb_{0.02}O$	#225	10.12461	4.22981	2.12038
$Mg_{0.975}Pb_{0.025}O$	#225	10.24802	4.22983	2.12088
$Mg_{0.97}Pb_{0.03}O$	#225	9.27065	4.22852	2.11952

4.2 Morphology Analysis

The field emission scanning electron microscope (FE-SEM) technique was employed to explore the size and the distribution of all the samples. The images are shown in Figures 2, 3, 4, 5, 6, 7 and 8 – a for synthesised ($Mg_{1-x}pb_xO$)-NPs samples by using the modified Pechini method. The images display the agglomeration of the spherical nanoparticles, which occurs due to the high surface energy of the nanoparticles. The capability to produce agglomeration is very strong, which increases the collision frequency between particles that lead to a higher degree of agglomeration. Chemical bonds may form between particle surfaces as van der Waals forces, which are always present. Additionally, the magnetic dipole forces can potentially participate in internal particle forces[46][47].

The average grain size for all seven samples under investigation is in the (22.87–29.05nm) range, as evident in the histogram in the Figures 2, 3, 4, 5, 6, 7 and 8 – b. Grain size increases with increases in the concentricity of the Pb^{2+} ions, which is in a good convention with the crystallite size as determined from the XRD and statically histogram.

4.2 *Energy-dispersive X-ray spectroscopy EDS*

In order to confirm the composition of the prepared nanocrystals samples, the energy dispersion spectroscopy (EDS) analysis was coupled with the FE-SEM. EDS is a very useful technique for qualitative and quantitative analysis of the relative concentration of an element [48]. EDS spectroscopy, which was used to detach pure and doped MgO-NPs as shown in Figures 2, 3, 4, 5, 6, 7 and 8 – c, is capable of producing elemental distribution maps.

The EDS spectrum specified the presence of magnesium (Mg) and oxygen (O) as the only elements in the MgO-NPs sample that demonstrate high purity of the nanopowder synthesis by the modified Pechini method. It is obvious from chemical mapping that the results for pure MgO-NPs of the Mg/O atomic ratio coincide well with that of the corresponding XRD data and the bulk ratio. A partially higher oxygen atomic ratio compared to that of magnesium is possibly because of moisture absorption from the environment [49].

The EDS analysis of the doped ($Mg_{1-x}pb_xO$)-NPs samples in Figures 2, 3, 4, 5, 6, 7 and 8 – c confirm the presence of Mg, Pb and O as the only elementary components, with the absence of any extra element. The weight percentages are almost equal to their nominal stoichiometry by experimental error.

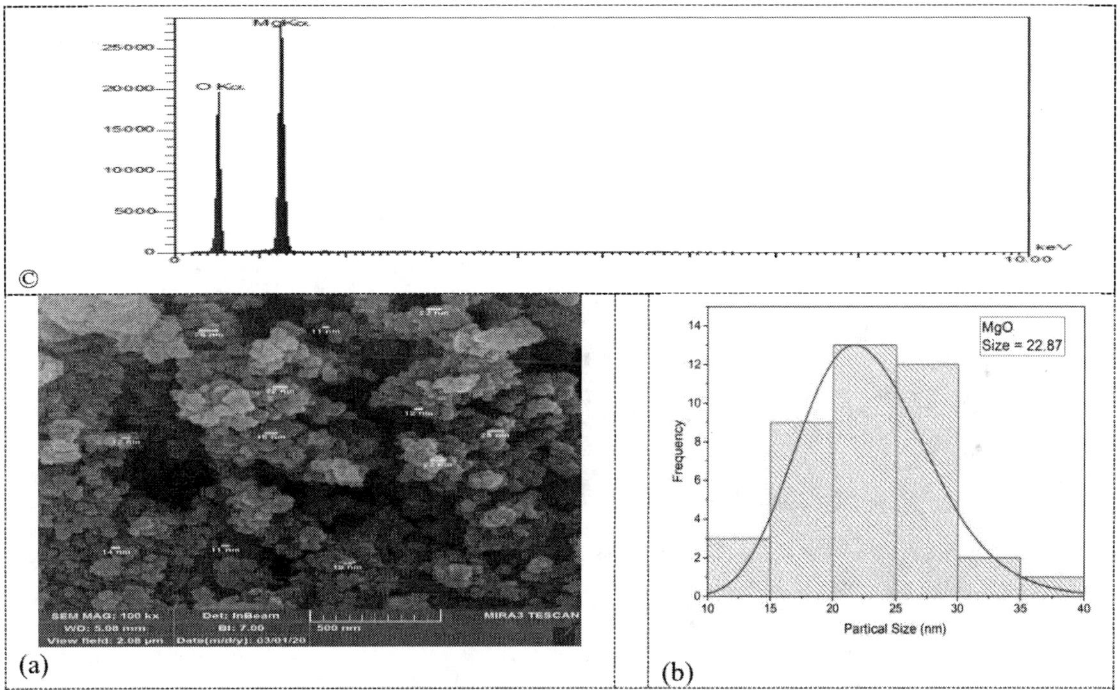

Figure 2. FE-SEM test analysis for MgO-NPs. (a) FE-SEM for MgO-NPs sample. (b) Histogram for MgO-NPs sample. (c) EDS for MgO-NPs sample.

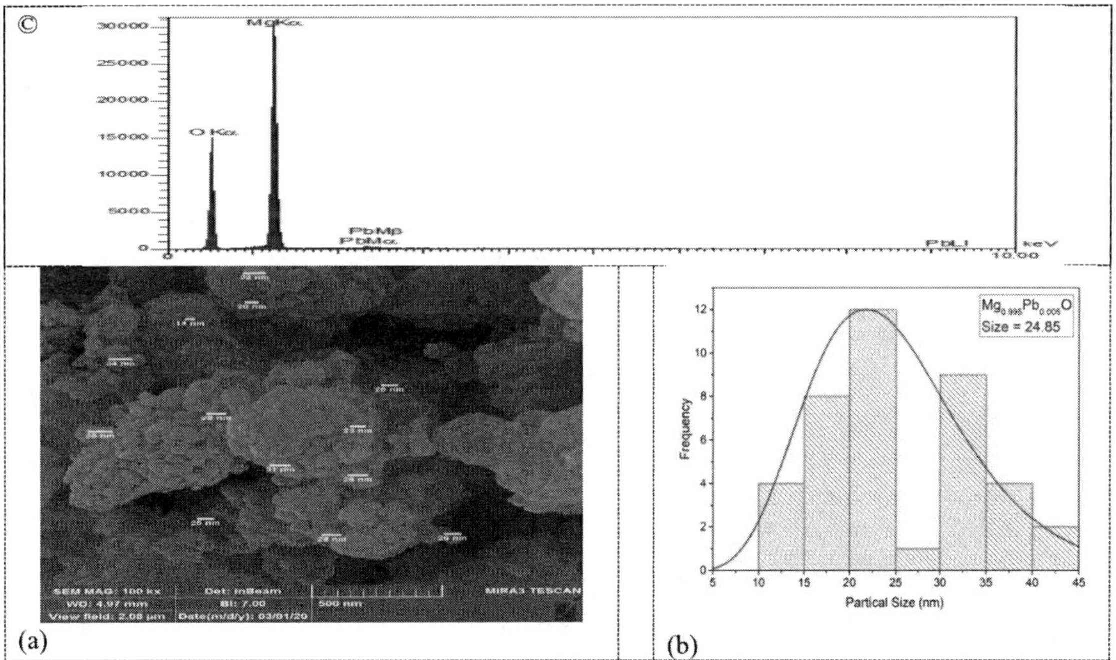

Figure 3. FE-SEM test analysis for $Mg_{0.995}Pb_{0.005}O$ sample. (a) FE-SEM for $Mg_{0.995}Pb_{0.005}O$ sample. (b) Histogram for $Mg_{0.995}Pb_{0.005}O$ sample. (c) EDS for $Mg_{0.995}Pb_{0.005}O$ sample.

IOP Conf. Series: Materials Science and Engineering **987** (2020) 012027 doi:10.1088/1757-899X/987/1/012027

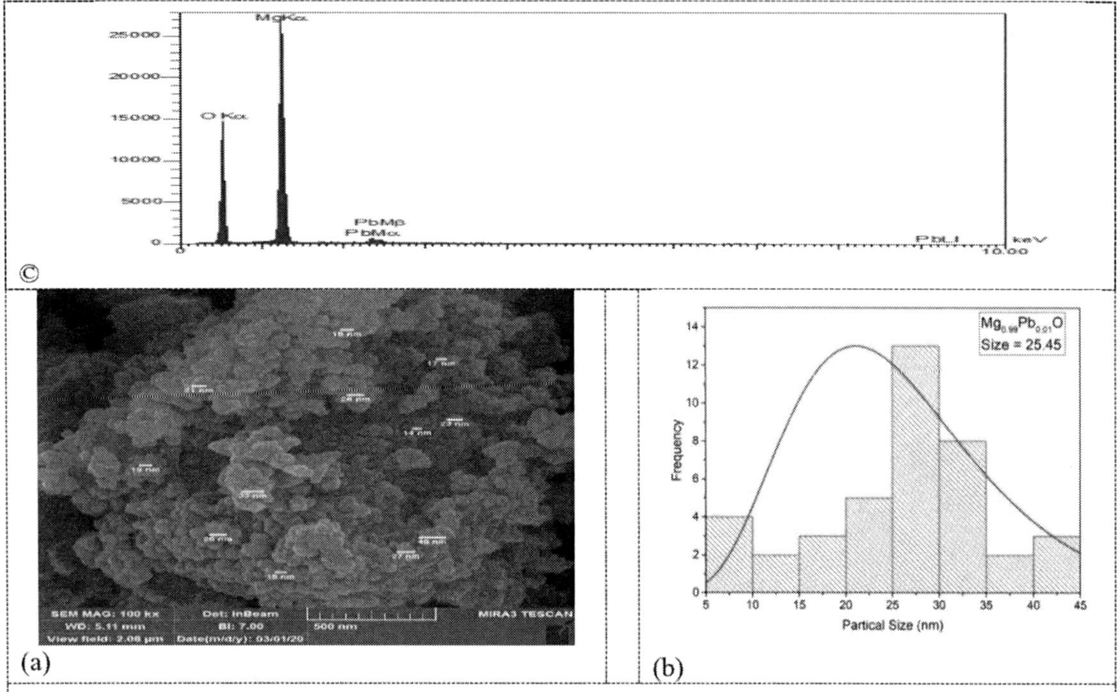

Figure 4. FE-SEM test analysis for $Mg_{0.99}Pb_{0.01}O$ sample. (a) FE-SEM for $Mg_{0.99}Pb_{0.01}O$ sample. (b) Histogram for $Mg_{0.99}Pb_{0.01}O$ sample. (c) EDS for $Mg_{0.99}Pb_{0.01}O$ sample.

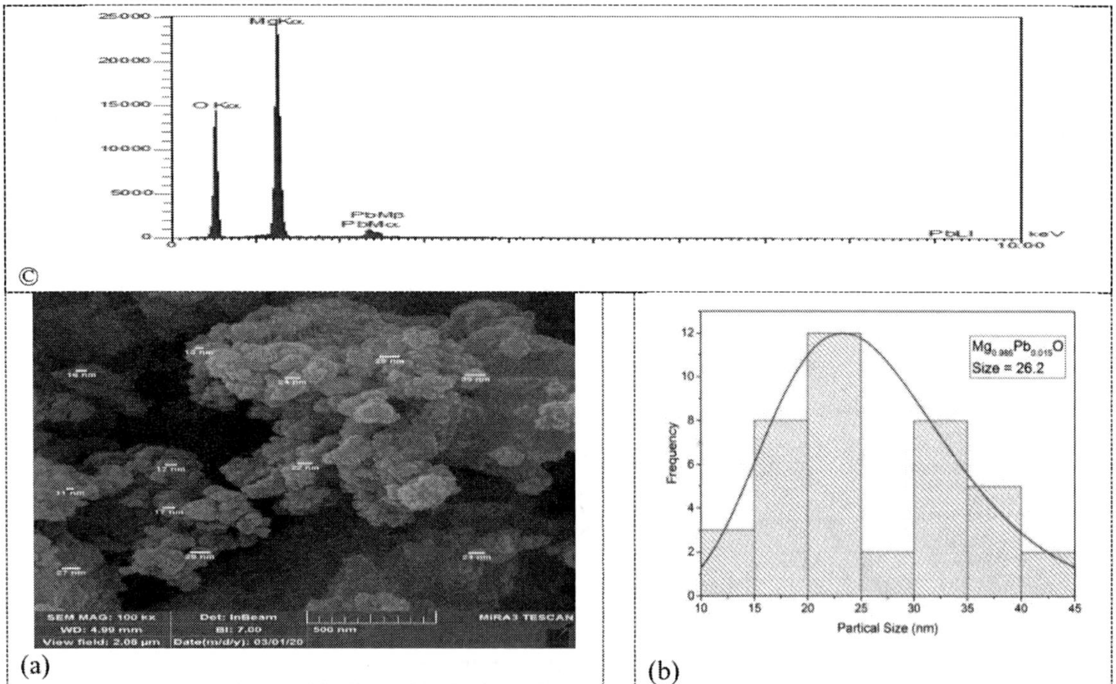

Figure 5. FE-SEM test analysis for $Mg_{0.985}Pb_{0.015}O$ sample. (a) FE-SEM for $Mg_{0.985}Pb_{0.015}O$ sample. (b) Histogram for $Mg_{0.985}Pb_{0.015}O$ sample. (c) EDS for $Mg_{0.985}Pb_{0.015}O$ sample.

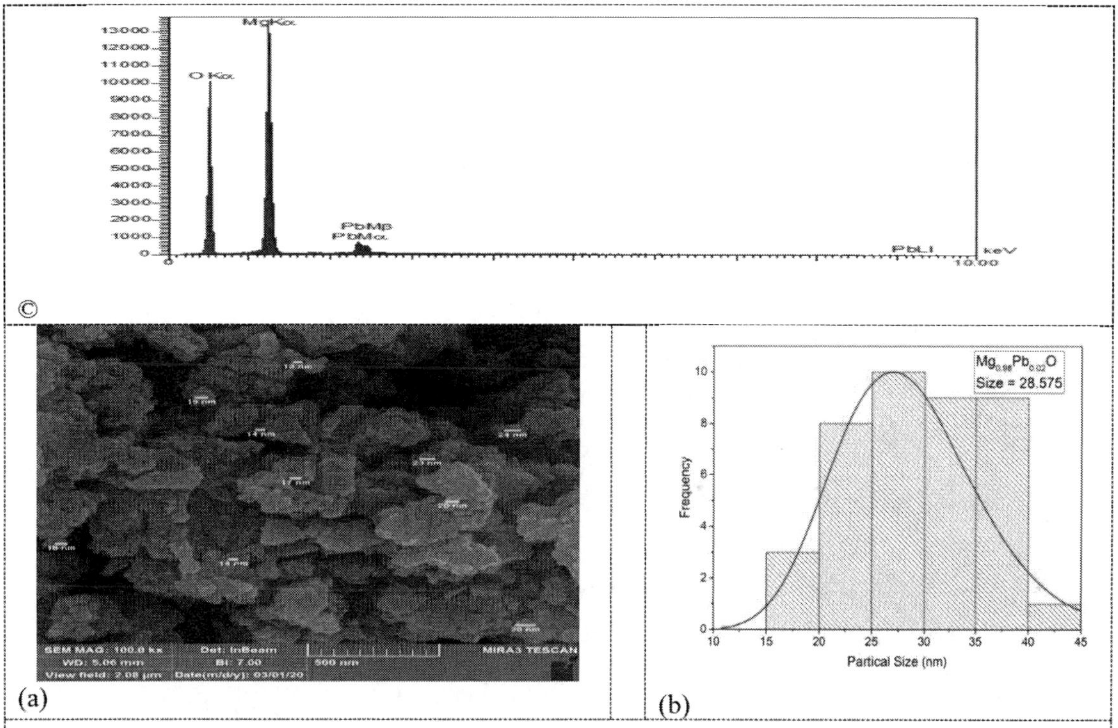

Figure 6. FE-SEM test analysis for $Mg_{0.98}Pb_{0.02}O$ sample. (a) FE-SEM for $Mg_{0.98}Pb_{0.02}O$ sample. (b) Histogram for $Mg_{0.98}Pb_{0.02}O$ sample. (c) EDS for $Mg_{0.98}Pb_{0.02}O$ sample.

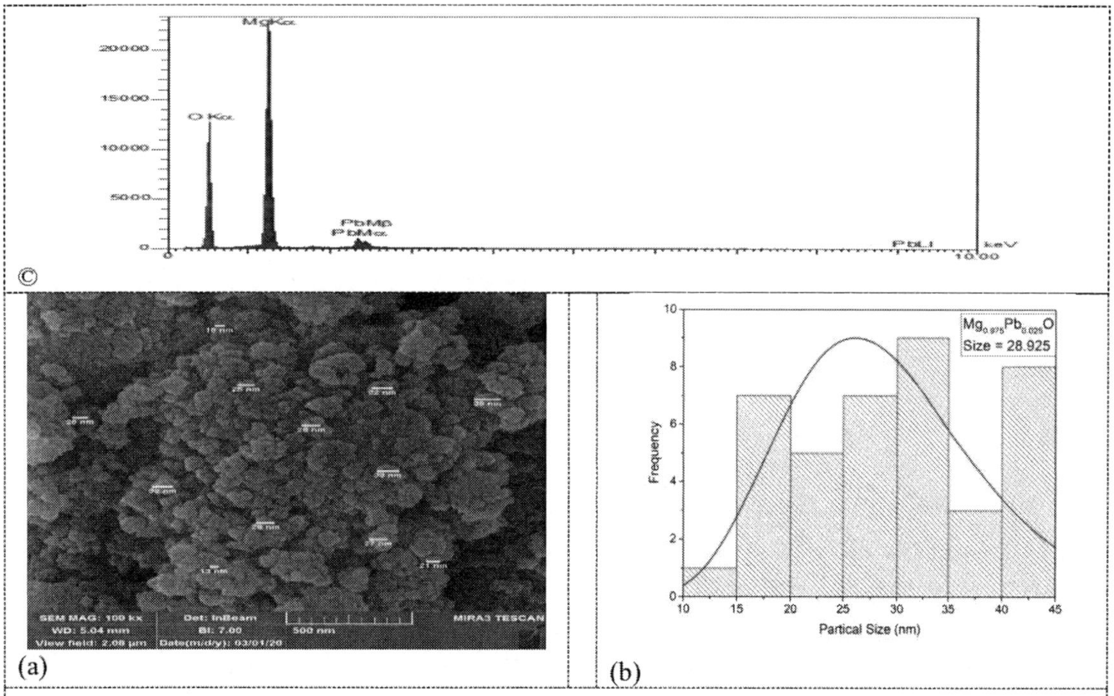

Figure 7. FE-SEM test analysis for $Mg_{0.975}Pb_{0.025}O$ sample. (a) FE-SEM for $Mg_{0.975}Pb_{0.025}O$ sample. (b) Histogram for $Mg_{0.975}Pb_{0.025}O$ sample. (c) EDS for $Mg_{0.975}Pb_{0.025}O$ sample.

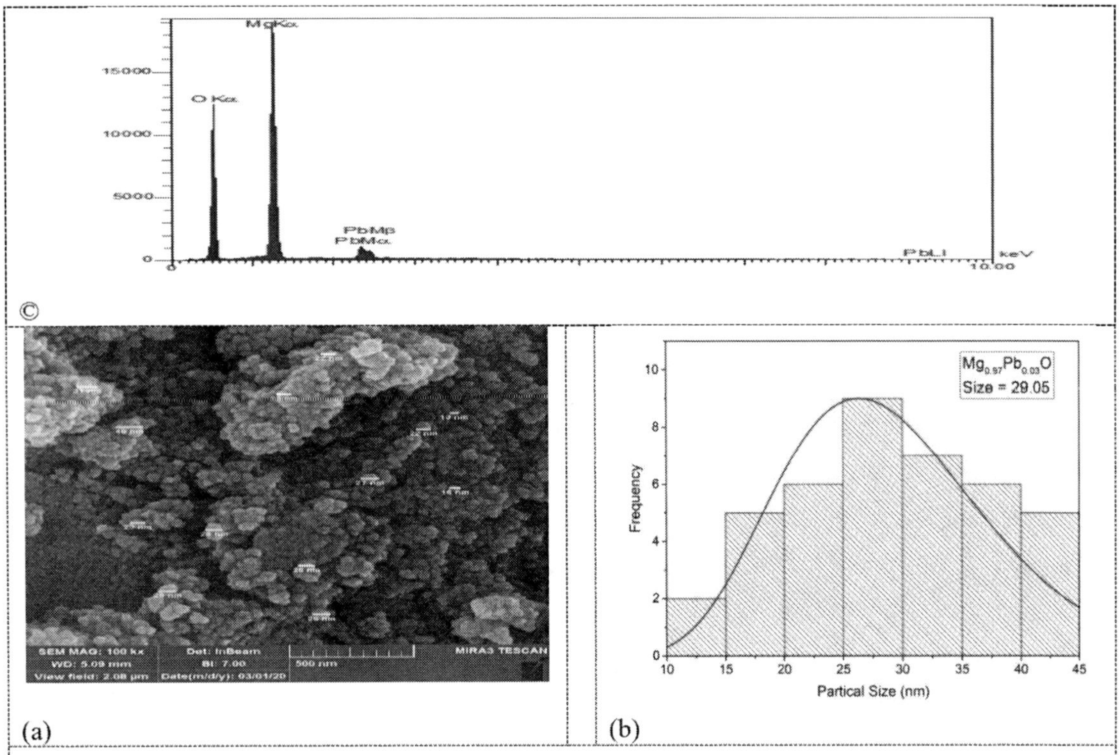

Figure 8. FE-SEM test analysis for $Mg_{0.97}Pb_{0.03}O$ sample. (a) FE-SEM for $Mg_{0.97}Pb_{0.03}O$ sample. (b) Histogram for $Mg_{0.97}Pb_{0.03}O$ sample. (c) EDS for $Mg_{0.97}Pb_{0.03}O$ sample.

4.4 VSM magnetisation

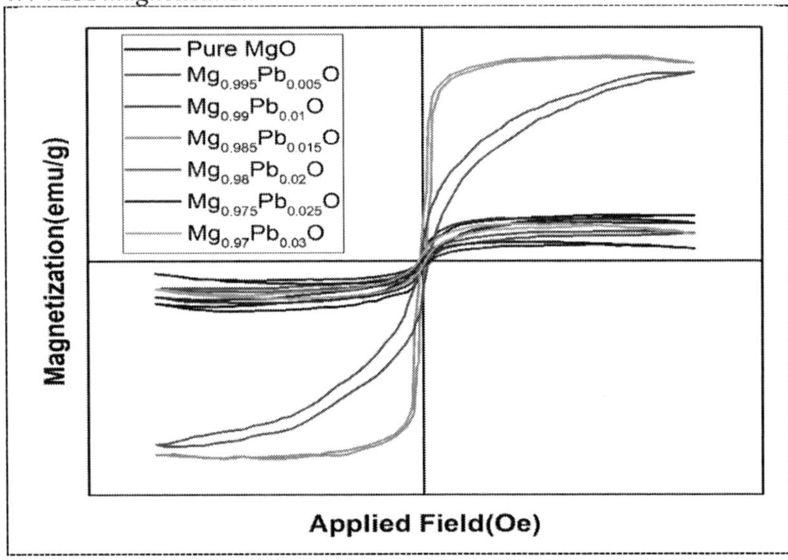

Figure 9. VSM analysis for ($Mg_{1-x}pb_xO$)-NPs samples calcined at 420°C for two hours.

The magnetisation behaviour curves for the prepared nanocrystallites ($Mg_{1-x}pb_xO$) samples were created by modifying the Pechini method and calcined at 420°C for two hours, as plotted in Figure 9. The samples were investigated by vibrating samples magnetometer (VSM). The VSM shows the hysteresis loops for the seven samples examined at room temperature and motivated by defects.

Magnetisation in the pure and doped samples was achieved due to small crystallite size and particle size as evident from XRD and FE-SEM, respectively. The necessary conditions to achieve magnetic order in the lattice of the pure and doped MgO-NPs samples was provided by a certain density of vacancies. Such vacancies or other defects in the lattice were achieved by nearly localised magnetic moments [50]. Such vacancies in the MgO-NPs sample are responsible for magnetisation. In addition, the vacancies were usually labelled as F,V centres, corresponding to a missing oxygen atom or Mg atom in the host lattice of MgO-NPs, respectively [51]. The reason behind the encouraging magnetic moment was the spin polarisation of 2p electrons of O atoms close to the negative charge VMg [20][25]. However, the cations were enclosed by an octahedron of anions with filled valence p bands, and the bonding anion p orbitals with lobes pointing to the cation site would be the generality energetically favoured attraction orbitals to provide the holes. Hence, holes were necessary to charge the recompense by a cation vacancy at the centre of such an octahedron [21][52].

The ferromagnetism in MgO-NPs appeared not only as a result of the VMg but also from the distribution or placement of the vacancies[20]. As Mg^{+2} ions were replaced by Pb^{+2} ions, the saturation magnetisation (Ms) was noted as varying. Such variation was due to the formative localised magnetic moment, which can occur as a result of introducing a cation defect in the MgO-NPs host lattice by doping.

According to Figure 9, the (Ms) decrease for the samples $Mg_{0.995}Pb_{0.005}O$ and $Mg_{0.99}Pb_{0.01}O$ as compared to the pure MgO-NPs sample due to an increase in the crystallite size and particle size, as cleared from XRD and FE-SEM. The magnetisation of $Mg_{0.995}Pb_{0.005}O$ and $Mg_{0.99}Pb_{0.01}O$ are close to a paramagnetic nature. Such magnetisation can occur due to the exchange of one ion with another ion, leading to the reproduction of holes in dopant 2p orbits and the formation of a local magnetic moment on the oxygen sites. Hence, the magnetic moment is a transfer of atoms without d- or f- electrons [25][53]. Consequently, it was convenient to assume that the observed magnetic signal was relative to the vacancies' density [53][54]. Thus, the excessive cation vacancies by doping are expected to change the observed magnetic signal.

The typical magnetisation appears in the $Mg_{0.985}Pb_{0.015}O$ sample, which has a symmetric hysteresis loop, which can be attributed to comparable nanoparticles size and associated with cation inversion [55]. This may happen due to the total replacement of vacancies by cation in a typical situation with an identical localised moment. The $Mg_{0.98}Pb_{0.02}O$ sample has S-shape hysteresis, suggesting superparamagnetic. In the $Mg_{0.975}Pb_{0.025}O$ and $Mg_{0.97}Pb_{0.03}O$ samples, the (Ms) begin to decrease as compared to $Mg_{0.985}Pb_{0.015}O$ due to an increase in the fraction of cations in the lattice by doping. Such an increase leads to the distorting of the lattice structure and a decrease in the localised magnetic moment; moreover, the decrease in the (Ms) of the $Mg_{0.97}Pb_{0.03}O$ sample was due to a separation of a small fraction of Pb^{2+} ions as PbO oxide, as clarified from XRD.

5. Conclusions

Pb^{+2}-doped MgO nanoparticles samples have been successfully synthesised via a modified Pechini method. The calcination temperatures were used 420°C for two hours with 3°C /min as a heating rate to prepare pure and $(Mg_{1-x}pb_xO)$-NPs-doped samples with Pb^{+2} concentrations varying from x=0.05-0.03. The substituted hydroxide precursors have been demonstrated as being effective for the synthesis of Pb^{+2}-doped MgO nanoparticles as they can provide good homogeneity. Powder XRD results confirmed the formation of a single cubic phase with space group Fm3m (#225). As a result of increasing the fraction of doping, the strongest diffraction peak (200) has shifted towards less 2θ values, and the interplanar d-spacing, lattice parameter and crystallinity have been increased. Furthermore, the intensity was reduced, which can be attributed to the ionic radius difference between Pb^{2+}(1.19 Å) and Mg^{2+} (0.72 Å) ions. The morphology surface for pure and doped nanoparticles samples showed a spherical nanoparticle shape. The average grain size increased with increasing the dopant concentration, as confirmed by statically histogram and crystallinity from XRD, which was in the range 22.87–29.05nm for pure and doped samples. The EDS confirmed the presence of Mg, Pb and O as the only elementary components, and the weight percentages were nearly their nominal stoichiometry. The RTFM was motivated by vacancy defects rather than the kind of dopant ions. The saturation magnetisation was noted as being varied, given the dopant incorporated in the host lattice of MgO-NPs. The reason behind such change was the formation of a local magnetic moment, which can vary according to the density of defect and oxygen vacancies. The maximum (Ms) was achieved when x=0.015 of the $Mg_{0.985}Pb_{0.015}O$ sample.

6. References

[1] Wang ZL. Characterization of nanophase materials. vol. 18. 2001. https://doi.org/10.1002/1521-4117(200110)18:3<142::AID-PPSC142>3.0.CO;2-N.

[2] Greim (Chairman) H. Constitution and Procedures of the Commission for the Investigation of Health Hazards of Chemical Compounds in the Work Area. 2004. https://doi.org/10.1002/3527600418.make016.pub2.

[3] Iqbal T, Tufail S, Ghazal S. Synthesis of Silver, Chromium, Manganese, Tin and Iron Nano Particles by Different Techniques. Int J Nanosci Nanotechnol 2017;13:19–52.

[4] Wang S, Gao L. Laser-driven nanomaterials and laser-enabled nanofabrication for industrial applications. Elsevier Inc.; 2019. https://doi.org/10.1016/B978-0-12-815749-7.00007-4.

[5] Wang C, Shi Z hai, Peng L, He W min, Li B liang, Li K zhi. Preparation of carbon foam-loaded nano-TiO2 photocatalyst and its degradation on methyl orange. Surfaces and Interfaces 2017;7:116–24. https://doi.org/10.1016/j.surfin.2017.03.007.

[6] Harun K, Mansor N, Ahmad ZA, Mohamad AA. Electronic Properties of ZnO Nanoparticles Synthesized by Sol-gel Method: A LDA+U Calculation and Experimental Study. Procedia Chem 2016;19:125–32. https://doi.org/10.1016/j.proche.2016.03.125.

[7] Tumanski S. Handbook of magnetic measurements. 2016. https://doi.org/10.1201/b10979.

[8] Tang ZX, Lv BF. MgO nanoparticles as antibacterial agent: Preparation and activity. Brazilian J Chem Eng 2014;31:591–601. https://doi.org/10.1590/0104-6632.20140313s00002813.

[9] Hamid NA, Shamsudin NF, Seec KW. Superconducting properties and mechanical strength of MgO fibres reinforced bulk Bi-2212 superconductor ceramics. Mater Res Innov 2009;13:379–81. https://doi.org/10.1179/143307509X441612.

[10] Tang ZX, Fang XJ, Zhang ZL, Zhou T, Zhang XY, Shi LE. Nanosize MgO as antibacterial agent: Preparation and characteristics. Brazilian J Chem Eng 2012;29:775–81. https://doi.org/10.1590/S0104-66322012000400009.

[11] Dahmardeh A, Davarpanah AM. Short Communication Investigation on Influences of Synthesis Methods on the Magnetic Properties of Trimetallic Nanoparticles of Iron-Cobalt- Manganese Supported by Magnesium Oxide 2015;11:249–56.

[12] Erwin SC, Zu L, Haftel MI, Efros AL, Kennedy TA, Norris DJ. Doping semiconductor nanocrystals. Nature 2005;436:91–4. https://doi.org/10.1038/nature03832.

[13] Sharma U, Jeevanandam P. Synthesis of Zn2+-doped MgO nanoparticles using substituted brucite precursors and studies on their optical properties. J Sol-Gel Sci Technol 2015;75:635–48. https://doi.org/10.1007/s10971-015-3734-0.

[14] Vasanthi V, Kottaisamy M, Anitha K, Ramakrishnan V. Near UV excitable yellow light emitting Zn doped MgO for WLED application. Superlattices Microstruct 2017;106:174–83. https://doi.org/10.1016/j.spmi.2017.03.050.

[15] Lee JW, Ko JH. Defect states of transition metal-doped MgO for secondary electron emission of plasma display panel. J Inf Disp 2014;15:157–61. https://doi.org/10.1080/15980316.2014.955140.

[16] Azzaza S, El-Hilo M, Narayanan S, Judith Vijaya J, Mamouni N, Benyoussef A, et al. Structural, optical and magnetic characterizations of Mn-doped MgO nanoparticles. Mater Chem Phys 2014;143:1500–7. https://doi.org/10.1016/j.matchemphys.2013.12.006.

[17] Fox GT, Wolfmeyer MW, Dillinger JR, Huber DL. Magnetic field dependence of the thermal conductivity of doped MgO. Phys Rev 1968;165:898–901. https://doi.org/10.1103/PhysRev.165.898.

[18] Cai Y, Wu D, Zhu X, Wang W, Tan F, Chen J, et al. Sol-gel preparation of Ag-doped MgO nanoparticles with high efficiency for bacterial inactivation. Ceram Int 2017;43:1066–72. https://doi.org/10.1016/j.ceramint.2016.10.041.

[19] Peng L, Wang Y, Wang Z, Dong Q. Multiplesite structure and photoluminescence properties of Eu3+ doped MgO nanocrystals. Appl Phys A Mater Sci Process 2011;102:387–92. https://doi.org/10.1007/s00339-010-6027-z.

[20] Hassnain Jaffari G, Tahir A, Bah M, Ali A, Bhatti AS, Shah SI. Study of Surface-Active Modes and Defects in Single-Phase Li-Incorporated MgO Nanoparticles. J Phys Chem C

2015;119:28182–9. https://doi.org/10.1021/acs.jpcc.5b10131.

[21] Wang F, Pang Z, Lin L, Fang S, Dai Y, Han S. Magnetism in undoped MgO studied by density functional theory. Phys Rev B - Condens Matter Mater Phys 2009;80:1–7. https://doi.org/10.1103/PhysRevB.80.144424.

[22] Martínez-Boubeta C, Beltrán JI, Balcells L, Konstantinović Z, Valencia S, Schmitz D, et al. Ferromagnetism in transparent thin films of MgO. Phys Rev B - Condens Matter Mater Phys 2010;82:1–7. https://doi.org/10.1103/PhysRevB.82.024405.

[23] Hanish HH, Edrees SJ, Shukur MM. The effect of transition metals incorporation on the structural and magnetic properties of magnesium oxide nanoparticles. Int J Eng Trans A Basics 2020;33:647–56. https://doi.org/10.5829/IJE.2020.33.04A.16.

[24] Xue D, Chai G, Li X, Fan X. Effects of grain size distribution on coercivity and permeability of ferromagnets. J Magn Magn Mater 2008;320:1541–3. https://doi.org/10.1016/j.jmmm.2008.01.004.

[25] Beltrán JI, Muñoz MC, Hafner J. Structural, electronic and magnetic properties of the surfaces of tetragonal and cubic HfO2. New J Phys 2008;10. https://doi.org/10.1088/1367-2630/10/6/063031.

[26] Gallego S, Beltrán JI, Cerdá J, Muñoz MC. Magnetism and half-metallicity at the O surfaces of ceramic oxides. J Phys Condens Matter 2005;17. https://doi.org/10.1088/0953-8984/17/43/L04.

[27] Beltrán JI, Monty C, Balcells L, Martínez-Boubeta C. Possible d0 ferromagnetism in MgO. Solid State Commun 2009;149:1654–7. https://doi.org/10.1016/j.ssc.2009.06.044.

[28] Nassar MY, Mohamed TY, Ahmed IS, Samir I. MgO nanostructure via a sol-gel combustion synthesis method using different fuels: An efficient nano-adsorbent for the removal of some anionic textile dyes. J Mol Liq 2017;225:730–40. https://doi.org/10.1016/j.molliq.2016.10.135.

[29] Jeevanandam P, Klabunde KJ. A study on adsorption of surfactant molecules on magnesium oxide nanocrystals prepared by an aerogel route. Langmuir 2002;18:5309–13. https://doi.org/10.1021/la0200921.

[30] Li WC, Lu AH, Weidenthaler C, Schüth F. Hard-templating pathway to create mesoporous magnesium oxide. Chem Mater 2004;16:5676–81. https://doi.org/10.1021/cm048759n.

[31] Type D. Nucleophilic Chemistry of the Synthesized Magnesium Oxide (Magnesia) Nanoparticles via Microwave @ sol-gel Process for Removal of Sulfurous Pollutant 2020:8–10.

[32] Zaidi B, Belghit S, Ullah MS, Hadjoudja B, Guerraoui A, Gagui S, et al. Investigation of MgO powders synthesized by liquid-phase method. Metallofiz i Noveishie Tekhnologii 2019;41:1121–6. https://doi.org/10.15407/mfint.41.08.1121.

[33] Ganapathi Rao K, Ashok C, Venkateswara Rao K, Shilpa Chakra C. Structural properties of MgO Nanoparticles: Synthesized by Co-Precipitation Technique. Int J Sci Res 2013:43–6.

[34] Mohammad Shafiee MR, Kargar M, Ghashang M. Characterization and low-cost, green synthesis of Zn2+ doped MgO nanoparticles. Green Process Synth 2018;7:248–54. https://doi.org/10.1515/gps-2016-0219.

[35] Ghorbani S, Razavi RS, Loghman-Estarki MR, Alhaji A. Development of MgO–Y2O3 Composite Nanopowder by Pechini Sol–Gel Method: Effect of Synthesis Parameters on Morphology, Particle Size, and Phase Distribution. J Clust Sci 2017;28:1523–39. https://doi.org/10.1007/s10876-017-1162-8.

[36] Tai CY, Tai C Te, Chang MH, Liu HS. Synthesis of magnesium hydroxide and oxide nanoparticles using a spinning disk reactor. Ind Eng Chem Res 2007;46:5536–41. https://doi.org/10.1021/ie060869b.

[37] Pechini MP. (Enno&P and Alway 1967.

[38] Zalapa-Garibay MA, Arizmendi-Moraquecho A, Reyes-López SY. Low temperature synthesis of alpha alumina platelets and acicular mullite in MgO-Al 2 O 3 -SiO 2 system. J Ceram Sci Technol 2019;10:9–18. https://doi.org/10.4416/JCST2018-00043.

[39] Devaraja PB, Avadhani DN, Nagabhushana H, Prashantha SC, Sharma SC, Nagabhushana BM, et al. Luminescence properties of MgO: Fe 3+ nanopowders for WLEDs under NUV excitation prepared via propellant combustion rout e . J Radiat Res Appl Sci 2015;8:362–73. https://doi.org/10.1016/j.jrras.2015.02.001.

[40] Obeid MM, Edrees SJ, Shukur MM. Synthesis and characterization of pure and cobalt doped magnesium oxide nanoparticles: Insight from experimental and theoretical investigation. Superlattices Microstruct 2018;122:124–39. https://doi.org/10.1016/j.spmi.2018.08.015.

[41] Leszczyński M, Litwin-Staszewska E, Suski T, Bąk-Misiuk J, Domagała J. Lattice Constant of Doped Semiconductor. Acta Phys Pol A 1995;88:837–40. https://doi.org/10.12693/aphyspola.88.837.

[42] Rajesh Kumar B, Hymavathi B. X-ray peak profile analysis of solid-state sintered alumina doped zinc oxide ceramics by Williamson–Hall and size-strain plot methods. J Asian Ceram Soc 2017;5:94–103. https://doi.org/10.1016/j.jascer.2017.02.001.

[43] Zhang L, Gonçalves AAS, Jaroniec M. Identification of preferentially exposed crystal facets by X-ray diffraction. RSC Adv 2020;10:5585–9. https://doi.org/10.1039/d0ra00769b.

[44] Ansari A, Ali A, Asif M, Shamsuzzaman. Microwave-assisted MgO NP catalyzed one-pot multicomponent synthesis of polysubstituted steroidal pyridines. New J Chem 2018;42:184–97. https://doi.org/10.1039/c7nj03742b.

[45] Das A, Mandal AC, Roy S, Nambissan PMG. Internal defect structure of calcium doped magnesium oxide nanoparticles studied by positron annihilation spectroscopy. AIP Adv 2018;8. https://doi.org/10.1063/1.5001105.

[46] Theophanides T. Introduction to Infrared Spectroscopy. Infrared Spectrosc - Mater Sci Eng Technol 2012. https://doi.org/10.5772/49106.

[47] Elawam S, Morsi W, Abou-Shady H, Guirguis O. Characterizations of Beta-lead Oxide "Massicot" Nano-particles. Br J Appl Sci Technol 2016;17:1–10. https://doi.org/10.9734/bjast/2016/28143.

[48] Scimeca M, Bischetti S, Lamsira HK, Bonfiglio R, Bonanno E. Energy dispersive X-ray (EDX) microanalysis: A powerful tool in biomedical research and diagnosis. Eur J Histochem 2018;62:89–99. https://doi.org/10.4081/ejh.2018.2841.

[49] Li M, Zhou S, Xu M. Graphene oxide supported magnesium oxide as an efficient cathode catalyst for power generation and wastewater treatment in single chamber microbial fuel cells. Chem Eng J 2017;328:106–16. https://doi.org/10.1016/j.cej.2017.07.031.

[50] Morozov IG, Sathasivam S, Belousova O V., Parkin IP, Kuznetcov M V. Effect of synthesis conditions on room-temperature ferromagnetic properties of Mg-O nanoparticles. J Alloys Compd 2018;765:343–54. https://doi.org/10.1016/j.jallcom.2018.06.211.

[51] Ferrari AM, Pacchioni G. Electronic structure of F and V centers on the MgO surface. J Phys Chem 1995;99:17010–8. https://doi.org/10.1021/j100046a029.

[52] Elfimov IS, Yunoki S, Sawatzky GA. Possible Path to a New Class of Ferromagnetic and Half-Metallic Ferromagnetic Materials. Phys Rev Lett 2002;89:1–4. https://doi.org/10.1103/PhysRevLett.89.216403.

[53] Kumar A, Kumar J, Priya S. Defect and adsorbate induced ferromagnetic spin-order in magnesium oxide nanocrystallites. Appl Phys Lett 2012;100. https://doi.org/10.1063/1.4712058.

[54] Freund MM, Freund F, Batllo F, Field M. Highly Mobile Oxygen Holes in Magnesium Oxide 1989;V:1–4.

[55] Singh JP, Gautam S, Srivastava RC, Asokan K, Kanjilal D, Chae KH. Crystallite size induced crossover from paramagnetism to superparamagnetism in zinc ferrite nanoparticles. Superlattices Microstruct 2015;86:390–4. https://doi.org/10.1016/j.spmi.2015.07.062.

The Effect of Boron Addition on The Microstructure and Corrosion Resistance of Cu-Al-Ni Shape-Memory Alloys Prepared by Powder Technology

ABDULRAHEEM KADHIM ABID ALI[1], NAWAL MOHAMMED DAWOOD[1*]
[1] Department of Metallurgical Engineering, College of Materials Engineering, Babylon University, Babylon, Iraq
[*]Corresponding author: mat.newal.mohammed@uobabylon.edu.iq

Abstract. Due to their promising mechanical characteristics, copper-based shape-memory alloys (SMAs) are crucial for numerous applications. This research inspects the outcomes of various quantities (0.4%, 0.8% and 1.2% wt) of boron (B) additions on the resistance against corrosion and the microstructure of Cu-Al-Ni SMAs using SEM, EDS, XRD and electrochemical tests in a solution of 3.5% NaCl. Numerous measurement baths were used for aged / unaged Cu-Al-Ni-xB SMAs. The electrochemical tests were performed over several periods to confirm the reliability of the corrosion-resistance property, after which the samples were verified using solutions of 3.5% NaCl. The experimental results demonstrated that adding 1.2% wt B followed by an ageing process resulted in an improvement in the corrosion resistance in Cu-Al-Ni SMAs and a decrease in the corrosion rate by 58%.
Keywords: *Cu-Al-Ni alloys; powder metallurgy; electrochemical test; SMA; microstructures.*

1. Introduction

Since the beginning of this century, shape-memory alloys (SMAs) have been considered smart materials with applications that vary from medicine to military, and Cu-based SMAs are extensively in demand as a result of their low cost and useful mechanical and shape-memory properties. Cu-based alloys are an appropriate auxiliary for nickel-titanium (Ni-Ti) alloys in engineering projects, such as buildings, fasteners, bridge-damping elements, oil wells and actuators' requests. Their low cost, simple manufacturing and wide range of temperatures [1] provide additional support. They have a sufficient recovery strain from two per cent to four per cent and a wide variety in temperature transformations (from 100°C to 170°C) [2].

However, when these materials are used in a corrosion-erosion environment, the damage must be considered, particularly in extremely corrosive surroundings, such as oil rigs and coastal fields. Furthermore, SMA systems are widely applicable and simple to use; for example, a portion of the components are an entire system by themselves. Moreover, this property may present galvanic corrosions because SMAs directly interact with other kinds of metals. One of the reasons why SMAs are useful in manufacturing plants is because the Cu-Al-Ni alloy has a strong resistance against corrosion since the formed alumina deposit functions as an inert layer [3].

However, it must be noted that the Cu-Al-Ni alloys become weak post-ageing, which leads to changes in its mechanical behaviour. Once it is exposed to high temperatures, its duration of operation increases [4]. Similarly, more studies have been conducted on Cu-Al-Ni's characteristics in relation to its application ranges than on its extensive demand in industries. There are numerous methods to improve the characteristics of SMAs; one such method is grain fine-tuning and adding minor amounts of the four elements, such as titanium and zirconium, to the SMAs [5, 6].

The Cu-Al-Ni SMAs corrosion resistance increases; for instance, the grains become finer owing to elements, such as manganese and titanium [6, 7]. However, adding cobalt to Cu-Al-Ni can increase both the alloys' mechanical properties and its temperature in the austenitic phases. This means that this combination can be employed at higher temperatures [8].

Although the Cu-Al-Ni alloy has a relatively strong corrosion resistance and mechanical properties, it was less striking in comparison to the extensively utilised NiTi, which has restricted use. However, even though it is used in environments with high corrosion, further studies on the corrosive behaviours of the Cu-Al-Ni SMAs alloy in specific surroundings are required.

Abid Ali et al. [9] measured the effects of iron addition on the corrosive resistance of Cu-Al-Ni SMA using five per cent NaOH and iron (0.4%, 0.8% and 1.2% wt). The outcomes show that the Cu-Al-Ni SMAs have a lower corrosion rate in the austenitic phase compared to the martensitic phase. It was also found that a stronger corrosion resistance was attained by adding 0.4% Fe. Furthermore, Abid and Al-Tai [10] measured the effects of adding silicon (0.4%, 0.8% and 1.2% wt) on the corrosion resistance and dry sliding wear of Cu-Al-Ni SMAs in 3.5% concentration of NaCl. Their results demonstrated that Cu-Al-Ni SMAs have a lower corrosion rate in the austenitic phase compared to the martensitic phase. Gojic et al. [11] verified the microstructures and electro-chemical property of Cu-Al-Ni SMAs on the corrosion resistance of Cu-Al-Ni SMAs in a 0.5 ml NaCl solution, which was examined by linear and open-circuit potential tests. The results showed that the adsorption of chloride ion on the electrode exterior results in a decrease in the EOCP values and the failure of the defensive external oxide scale. Furthermore, Saud et al. [12] examined the effects of tantalum (Ta) on the corrosion resistance, microstructures, damping and shape-memory behaviour of pre-alloyed Cu-Al-Ni SMAs. Polarisation examinations in 3.5% NaCl solutions showed that the corrosion resistance increased with Ta concentrations and created a strong inert layer.

This research aims to evaluate the corrosion resistance and microstructures of Cu-Al-Ni SMAs with boron (B) additions (0.4%, 0.8% and 1.2% wt) in a 3.5% sodium chloride solution. This was done via powder metallurgy.

2. Experimental Procedures

2.1. Sample Preparation

The samples were shaped using the powder technology with the essential powders Cu, Al, Ni and B. Table 1 demonstrates the size of the particles, as well as the purities and compositions of the used powders. The powders were mixed in a planetary ball mill for approximately six hours with a mixing ratio of 1:5 at a weight and rotation speed of 350 rpm. Exactly 0.5 cc ethyl alcohol was used to minimise the temperature that was generated by friction between the balls with walls and powder. After mixing, the wet mixtures were left to dry at room temperature for approximately 30 minutes. After that, they were placed in polyethylene zip locked bags. Later, the mixed powder was pressed in a hydraulic press at 800 MPa using cylindrical stainless-steel die of ten mm diameter. Then, the green specimens were put in a furnace with a vacuum for approximately two hours at 550 °C and were continued to be sintered at 950 °C for about 180 min with the heating rate as ten °C / min for both the sintering stages. The samples were then cooled under the vacuum until they reached room temperature. Next, the sintered samples were heat-treated at 850 °C for 60 minutes and dipped in iced water. Following that, the samples were heated to 300 °C for 1800 min and immersed in water to age the intermetallic compound (γ_2) and steady the martensite phase.

2.2. Preparation of the Samples for Microstructure Observation

The samples were then wet-grinded by 180, 400, 800, 1,000, 1,200, 1,500, 2,000, 2,500 and 3,000 grits SiC papers using a grinding wheel and were later polished with papers and diamonds with a one μm to three μm particle size. All samples were then washed using distilled water and dried in warm air. Later, the samples were treated with an etching solution (2.5 g $FeCl_3$ $6H_2O$ + 48 ml CH_3OH + ten ml HCl) and ferric chloride acid / methanol / hydrochloric acid [13] for ten seconds. They were then used for SEM and EDS tests.

Table 1. Elementals powder specification with compositions.

Properties	Copper	Aluminium	Nickel	Boron
Particle size (μm)	14	15	27	8
Purity (%)	99.95	99.92	99.95	99.94
Prepared	83	13	4	…..
samples	82.6	13	4	0.4
composition	82.6	13	4	0.8
(wt %)	81.8	13	4	1.2

2.3. Material Characterisation

A sample with dimensions of (d = ten mm, t = five mm) was prepared for XRD characteristics using Mini flex2. X-rays with CuKα radiations at 30 mA and 40 Kv were performed. The device's scanning speed was two° / minute with an angle of 2θ, ranging from 30° to 90° with step 0.02° / sec.

2.4. Electro-Chemical Test

The corrosion performance of Cu-Al-Ni SMAs with and without (0.4%, 0.8% and 1.2% wt) the addition of B was achieved at 25 °C in the air in a beaker with 350 ml of 3.5% NaCl (3.5 gm NaCl + 96.5 ml distilled water). The corrosion performance of the alloys was examined using the Tafel potentiostat method WENKING M lap to measure the corrosion current and corrosion potential.

Three electrodes were used for the potentiodynamic polarisation test. A platinum wire was measured for the counter electrodes and reference electrodes for saturated calomel electrodes (SCE) while the sample was the working electrode. The polarisation curves were measured potentiostatically at a scanning rate of 0.35 mVs^{-1}. For all measurements, the electrodes were maintained at the open circuit for at least ten–20 minutes for stabilisation, and then the Eo was obtained. The electrochemical system that was used is shown in Figure 1.

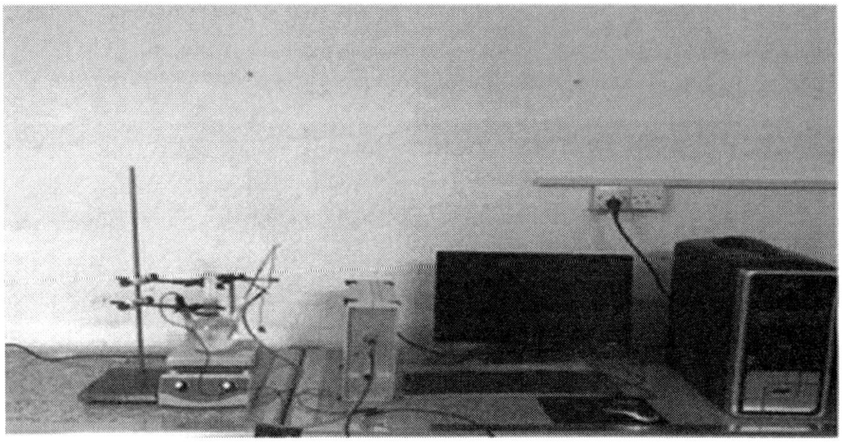

Fig 1. The electrochemical system.

3. Results and Discussions
3.1. Microstructures and X-ray Diffractions for Phase Analysis.

The microstructure and EDS examination of the homogenising Cu-Al-Ni SMA samples with and without alterations after ageing treatments are illustrated in Figures 2–6. The results showed the following two different phases: needle-like for β_1' and plate-like for γ_1'. These two phases were also mentioned in numerous additional revisions [4, 8, 14 & 15]. Like other metallic materials and alloys, the Cu-Al-Ni SMAs grain sizes increased with an increase in the ageing treatments. It was observed that the width of the plates increased, whereas the needles reduced after ageing. Furthermore, the microstructure showed that the grain size of the Cu-Al-Ni-xB SMAs was the finest in the case of increased B content; on the other hand, a different phase, in the formula of precipitates, shaped the matrices. Saud et al. [8] identified the phase as the γ_2 phase as an intermetallic compound. These precipitates tend to isolate the grains' boundaries lengthways, as shown in Figure 1d, and augmented in extent with higher B gratified. This conclusion was also corroborated by an EDX analysis, as shown in Figures 2–6, wherein a higher content of B (1.2% wt) was found in the precipitate. Figure 7 illustrates the XRD results of the aged Cu-Al-Ni SMAs, in which α Cu and intermetallic compounds $AlCu_3$ were found. The other phases that were formed between the elements present in the alloy that play an important role in improving the corrosion resistance of the alloy did not appear in the X-ray diffraction due to their occurrence outside the limits of X-ray detection.

Fig 2. SEM microstructure of (a) Cu-Al-Ni SMAs; (b) Cu-Al-Ni 0.4% B SMAs; (c) Cu-Al-Ni 0.8% B SMAs; and (d) Cu-Al-Ni 12%B SMAs.

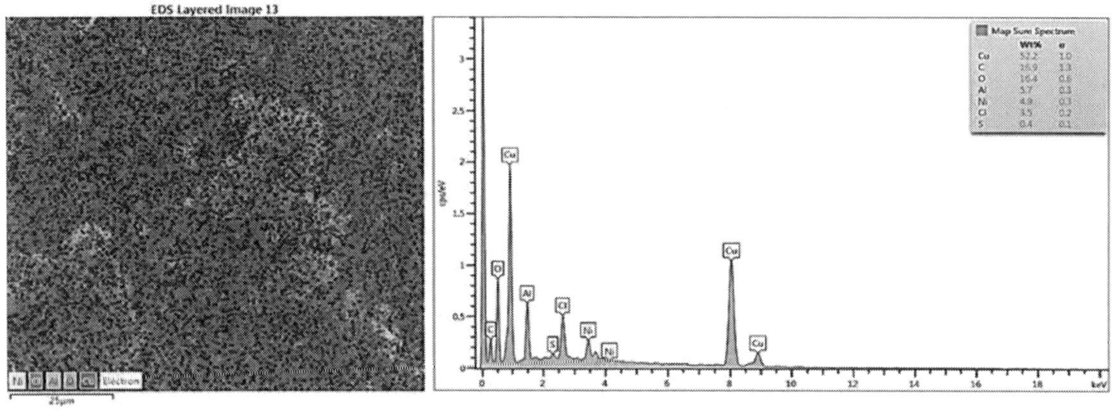

Fig 3. The EDX analysis of Cu-Al-Ni SMAs.

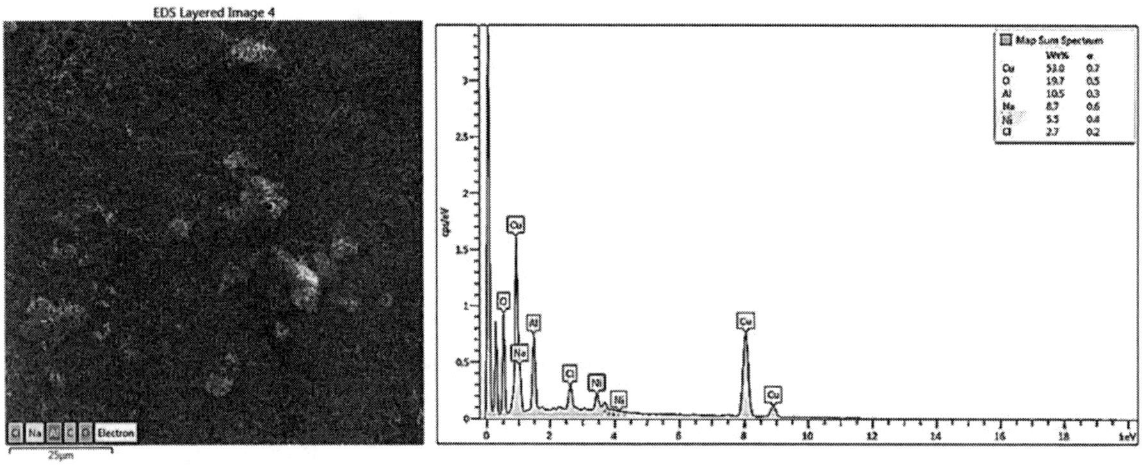

Fig 4. The EDX analysis of Cu-Al-Ni 0.4 B SMAs.

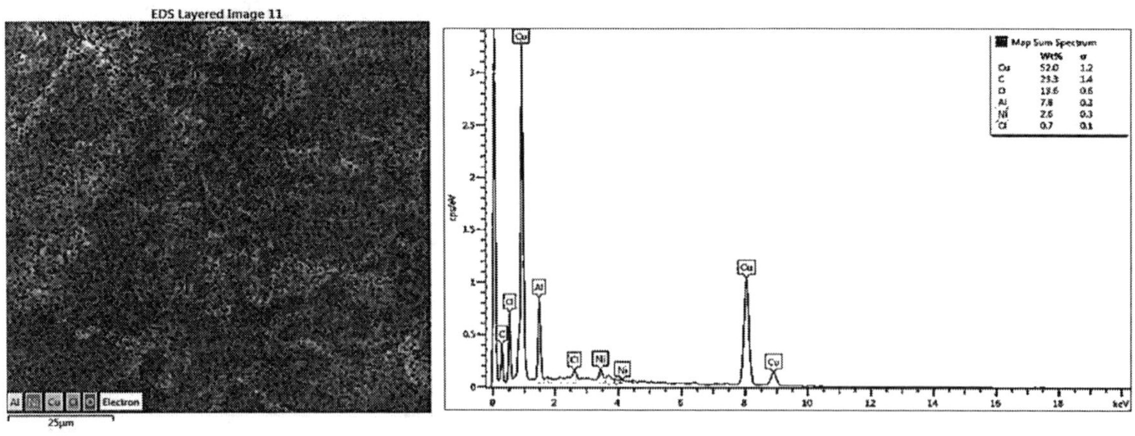

Fig 5. The EDX analysis of Cu-Al-Ni 0.8 B.

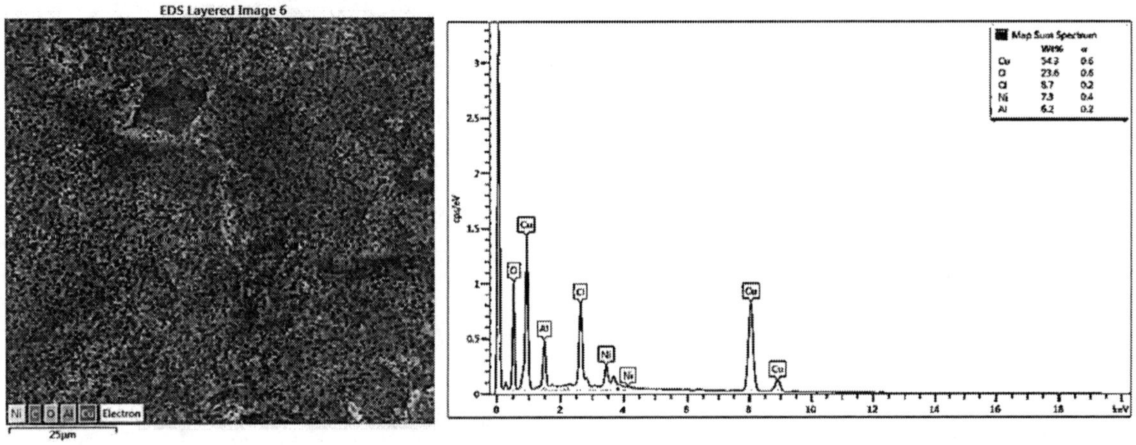

Fig 6. The EDX analysis of Cu-Al-Ni 1.2 B SMAs.

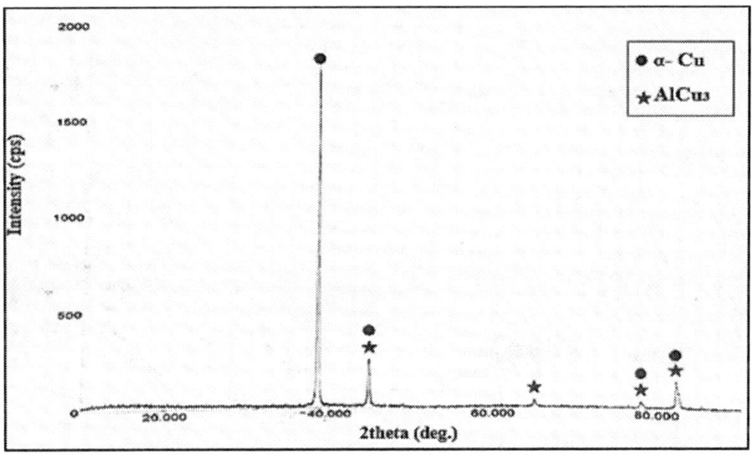

Fig 7. The XRD test results of Cu-Al-Ni SMAs.

3.2 Electrochemical Test Result

Figure 8 shows the polarisation diagrams of Cu-Al-Ni SMAs with and without (0.4%, 0.8% and 1.2% wt) the addition of B. The Cu-Al-Ni SMAs' reduction in anodic polarisation tendency and refinement in the cathodic polarisation performance was also seen [16]. It was observed that the existence of B in the passivation surface improved the resistance of the alloy from chlorine ion in the sodium chloride solution better than in the absence of B [17]. The corrosion rate (mpy), corrosion potential (E_{corr}) and corrosion current density (I_{corr}) of the samples were determined from the polarisation's curves, as shown in Table 2.

Table 2. Polarisation parameters of Cu-Al-Ni SMAs in 3.5% NaCl solution before and after B additions.

Alloys composition	**Corrosion potential E_{corr} (mV)**	**Current density I_{corr} (μA / cm²)**	**Corrosion rate (mpy)**
Cu-Al-Ni	-306	34.6	14.95
Cu-Al-Ni-0.4% B	-596	21.3	9.063
Cu-Al-Ni-0.8% B	-431.4	8.5	3.56
Cu-Al-Ni 1.2% B	-446.8	5.5	2.27

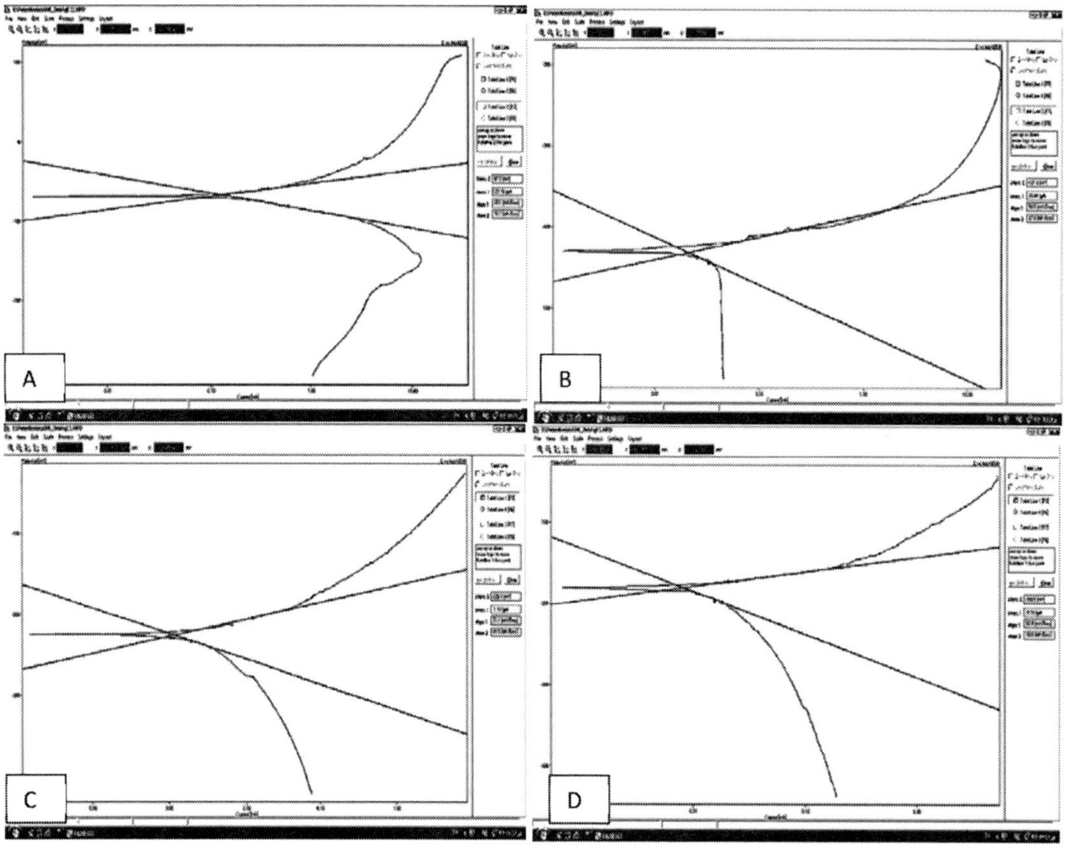

Fig 8. (A) the polarisation of Cu-Al-Ni SMAs; (B) the polarisation of Cu-Al-Ni 0.4%B SMAs; (C) the polarisation of Cu-Al-Ni 0.8%B SMAs; and (D) the polarisation of Cu-Al-Ni 1.2%B S.

I_{corr} is associated with mpy, as shown in the following empirical equation [18, 19]:

$$C.R\ (mpy) = \frac{0.13 \times I_{corr} \times E.W}{A \times \rho} \quad \ldots\ldots \text{Eq1}$$

Where A = area of the surface (cm^2).
ρ = density of the sample (g / cm^2).
I_{corr} = corrosion current density (μA / cm^2).
E.w = equivalent weight.
As shown in Table 2, the corrosion rate of the Cu-Al-Ni SMA is 14.95 mpy. After adding B (0.4%, 0.8% and 1.2% wt), the rate became 9.063, 3.56 and 2.27 mpy, respectively. The creation of corrosion products Al_2O_3 after the corrosion tests in 3.5% NaCl is the cause of the decrease in the corrosion rate from 14.95 mpy in Cu-Al-Ni SMA to 2.27 mpy in Cu-Al-Ni 1.2% B SMA. For the higher effects, the corrosion penetration rate was lower than approximately 20 mpy, which was satisfactory [20, 21].

4. Conclusion

Sintering samples at 550 °C for 120 minutes and 950 °C for 180 minutes (with and without 0.4%, 0.8% and 1.2% wt B) was an efficient method of preparing Cu-Al-Ni alloy structures. The sintered samples were heated at 850 °C for 60 minutes, rapidly immersed in iced water, heated to 300 °C for 1800 minutes and put in water again to age the intermetallic compound (γ_2) and stabilise the martensite phase. The ageing of Cu-Al-Ni SMAs was enough to get γ′ (coarse, martensite plate-like) and ß′ (needle-like shape), whereas the B addition to Cu-Al-Ni SMA tended to refine γ′ (coarse martensite plate-like). The electrochemical test displayed that the corrosion rate decreased from 14.95 mpy to 2.27 mpy when 1.2% wt B was added. Therefore, we conclude that an increase in boron leads to an improvement in the resistance of Cu-Al-Ni SMAs against corrosion.

5. References

[1] C. L'excellent, 2013 "Shape Memory Alloy Handbook" John Wiley and Son, New York, USA.
[2] D. Lagudas, 2006 "Shape Memory Alloy: Engineering Modeling" Springer, Berlin, Germany.
[3] A. Schüssler, H. Exner, 1993, The corrosion of nickel-aluminium bronzes in seawater—I. Protective layers formation and the passivation mechanism. Corros Sci, .34(11):1793–1802.
[4] N. Suresh, U. Ramamurty, 2008, Ageing response and its effect on the functional properties of Cu-Al-Ni shape memory alloys. J Alloys Compds, 449(1–2):113–118.
[5] J.S. Lee, C. M. Wayman, 1986, Grain refinement of a Cu-Al-Ni shape memory alloy by Ti and Zr additions. Trans Japan Inst Met , 27(8):584–591.
[6] S. Saud, E. Hamzah, T. Abubakar, M. Zamri, M. Tanemura, 2014, Influence of Ti additions on the martensitic phase transformation and mechanical properties of Cu-Al-Ni shape memory alloys. J Therm Anal Calorim, 1:111–122,
[7] S. Saud, E. Hamzah, T. Abubakar, HR. Bakhsheshi-Rad, M. Zamri, M. Tanemura, 2014, Effects of Mn additions on the structure, mechanical properties, and corrosion behavior of Cu-Al-Ni shape memory alloys. J Mater Eng Perform , 11:3620–3629.
[8] S. Saud, E. Hamzah, T. Abubakar, HR. Bakhsheshi-Rad, 2015, Thermal aging behavior in Cu-Al-Ni-xCo shape memory alloys. J Therm Anal Calorim , 119(2):1273–1284.

[9] Abdul-Raheem K. AbidAli, and Z. T. K. Al-Tai, 2012, The effect of silicon addition on the dry sliding wear and corrosion behavior of Cu Al Ni shape memory alloy. The Iraqi Journal for Mechanical And Material Engineering, 12(1).

[10] Abdul-Raheem K. AbidAli, and Z. T. K. Al-Tai, 2010 , The effect of iron addition on the dry sliding wear and corrosion behavior of Cu Al Ni shape memory alloy, adopted from Journal of Eng. & Tech, 28(24).

[11] M. Gojic, L. Vrsalovi, S. Kozuh, A. Kneissl, I. Anzel, S. Gudi, B. Kosec, and M. Kliski, 2011, Electrochemical and microstructural study of Cu-Al-Ni shape memory alloy, adopted from The Journal of Alloys and Compounds: 9782–9790.

[12] Safaa N. Saud, E. Hamzah, H.R. Bakhsheshi-Rad, and T.Abubakar. 2017 ,Effect of Ta Additions on the Microstructure, Damping, and Shape Memory Behaviour of Prealloyed Cu-Al-Ni Shape Memory Alloys, adopted from Hindawi Publishing Corporation, Volume1 , Article ID 1789454, 13 p46.

[13] S. Saud, E. Hamzah, H. Bakhsheshi-Rad, and T. Abubaker, 2014,Microstructure and corrosion behavior of Cu-Al- Ni shape memory alloys with Ag nanoparticles, Materials and Corrosion, 66(6).

[14] V. Recartea, RB Pérez-Sáeza, EH Bocanegrac, Nóc ML, and J. San Juan, ,2019, Dependence of the martensitic transformation characteristics on concentration in Cu-Al-Ni shape memory alloys. Mater Sci Eng A 273–275:380–384.

[15] H. Sakamoto, and KI. Shimizu, 1989, Effect of heat treatments on thermally formed martensite phases in mono crystalline Cu-Al-Ni shape memory alloy. ISIJ Int 29(5):395–404,.

[16] B. Chen, C. Liang, D. Fu, and D. Ren, 2005, Corrosion behavior of Cu and the Cu-Zn-Al shape memory alloy in simulated uterine fluid [J]. Contraception,72:221−224;79−89.

[17] E. N Kassab, L. Frotscher, M. Swaminathan, S. Maab, B. Rohweder, M. Gomes, and J G. Eggeler, 2014,Effect of ternary element addition on the corrosion behaviour of NiTi shape memory alloys [J]. Materials and Corrosion, 65:18−22.

[18] Annual Book of ASTM standards, (1988),Wear and erosion, Metal Corrosion, 3(2):G5–87.

[19] Alaa Abdulhasan Atiyah, Abdul-Raheem Kadhum, Abid Ali, Nawal Mohammed Dawood,2014, Characterization of NiTi and NiTiCu porous shape memory alloys prepared by powder metallurgy (Part I), Arabian Journal for Science and Engineering, 40(3):901–913.

[20] W.D. Callister, 2000, "Materials Science and Engineering", 5th ed., United States of America.

[21] Alaa Abdulhasan Atiyah, Abdul-Raheem Kadhum Abid Ali, Nawal Mohammed Dawood, (2019) , Fabrication of porous NiTi shape memory alloy objects by powder metallurgy for biomedical applications, IOP Conf. Series: Materials Science and Engineer 518 03205 doi:10.1088/1757-899X/518/3/032056.

IOP Conf. Series: Materials Science and Engineering **987** (2020) 012029 doi:10.1088/1757-899X/987/1/012029

Characterisation of Copper-Chrome-Reduced Graphene Oxide Electrical Contact Materials Using Powder Metallurgy Technique

Aseel S. Hamza[1*], Abdul Raheem K. Abid Ali[1], Saad Hameed Al Shafaie[1]
[1]College of Materials Engineering, Babylon University, Babylon, Iraq
*Corresponding author: aseeldulamy@yahoo.com

ABSTRACT Copper-chrome alloys are known as contact material alloys with a premium capability for circuit interruption, endurance and erosion resistance and high voltage in a vacuum interrupter. In this research, the alloy fabricated by powder metallurgy mixed Cu, Cr and reduced graphene oxide (rGO). rGO was prepared by electrochemical exfoliation in H_2SO_4 at 11.5v, 0.5pH and dried by exposure to sunlight. rGO was incorporated to characterise the properties. Optical microscopy, X-ray diffraction (XRD), field emission scanning electron microscope (FESEM), energy dispersive analysis (EDS) and Raman spectroscopy were used to determine the identification of phases, morphology, chemical composition and crystal structure orientation of the alloy. Optical microscopy images showed the formation of two distinct areas: a bright area containing Cr-particles and a dark area containing Cu-grains. The XRD patterns exhibited peaks of FCC-Cu, BCC-Cr, Cr_2O_3 and Cr_2C. FESEM revealed the presence of the Cr phase and Cu matrix as well as rGO sheets, while EDS indicated the presence of Cu, Cr, C and O. The Raman peak assessed the degree of rGO disorder in the alloy. The hardest specimen (92 Kg/mm^3) was sintered Cu-20%Cr-1%rGO. Electrical resistance exhibited a minimum value of 12.21 μΩ for the sintered Cu-20%Cr-2%rGO alloy and 6.23 μΩ for aged specimens; compressive strength had the highest value (425.235 MPa) for the sintered Cu-20Cr-1.5%rGO alloy.

1. Introduction

Electrical contact is the interface between the members carrying electrical current in electronic devices that ensure the continuity of the electric circuit and the unit containing the interface [1]. The desired properties of the materials used in electrical contact applications depend on the performance required. The perfect material must have high conductivity, be chemically inert, have a large contact area, high wear resistance, low friction and be low in cost for materials and production [2]. Copper is commonly used in industrial applications because of its excellent electrical and thermal conductivity properties as well as good ductility, toughness, corrosion and oxidation resistance [3]. Copper matrix alloys are characterised by good strength, excellent self-lubricating performance and high electrical conductivity [4]. Over the past 50 years, Cu-Cr-based contact alloys have been established for use in medium voltage tools, such as circuit breakers and contactors in a voltage range of 1 kV–75 kV [5].
Generally, Cu-Cr electrode alloys are manufactured by sintering, and Cr particles improve blocking performance due to its role in refining [6]. However, Cu-Cr system alloys contain only two elements and do not form a single solid solution without being highly entropic [7]. Regarding the drawbacks and expansion of the application of copper and its alloys, it is necessary to develop its properties with heat treatment or a fabrication process of composites to produce new composites with high electrical and thermal conductivity and good mechanical properties [8]. The introduction of secondary particles

as filler, such as graphene in Cu-based alloys, improves their mechanical properties while maintaining good thermal and electrical properties [9]. The graphene–copper alloys exhibit the excellent effect of interfacial bonding and reinforcement [10]. However, many studies focus on the reinforcement of copper composites by a single rGO and determine the effectiveness of the alloying element for the modification of copper–graphene composites [4].

In recent years, an impressive variety of approaches have been developed for the synthesis of Cu-Cr composites. The powder metallurgy technique is a common approach providing a simple pathway for bulk metal-matrix composite production [11]. There are several research papers on Cu-Cr alloys, starting with studies by Mula et al. on Cu-Cr- silicon carbide(SiC) alloys fabricated by a microwave sintering process. The hardest alloy and the best electrical conductivity were ≈2.4 GPa and 60.3% of international annealed copper standard (IACS), respectively, for the alloy of 94%Cu, 6%Cr and 4%SiC sintered at 900 oC. This is perhaps due to the presence of ultra-fine grains, proper bonding between particles and good densification [12]. Gill et al. [13] studied a new Cu-Cr-multiwall carbon nanotube (MWCNT) composite, and the Raman peak assessed the stresses induced and the degree of disorder of MWCNTs in the composite. Transmission electron microscopy (TEM) test indicated oxide and carbide formations, and scanning electron microscopy (SEM) test revealed the existence of MWCNTs. Corrosion resistance was determined by cyclic potentiodynamic polarisation (ASTM F2129-08) in a phosphate buffer saline solution at 37 oC. Imai et al. investigated the properties of a Cu-chromium-vapour-grown carbon fibre (VGCF) alloy manufactured by powder metallurgy and revealed a small decrease in yield strength, while electrical conductivity was higher than that of the Cu-Cr alloy. The Cr concentrated on VGCF using SEM- Energy-dispersive X-ray spectroscopy (EDS) analysis and detected Cr23C6 by TEM analysis [14]. Ozgur et al. studied a Cu-Cr-C alloy manufactured by powder metallurgy. X-ray analysis referred to the formation of Cr-nitride and carbide compounds during sintering and measurements for density increased with an increase of elements [3]. Another study on a Cu-Cr-W-C alloy was conducted by Yanli et al. to examine welding tendencies and the ability to withstand the voltage by adding W and C elements. The force of welding was reduced and arc erosion areas increased [15]. Luo et al. studied a GO/silver-reinforced Cu matrix and demonstrated improvements in thermal and electrical properties compared to pure Cu [16]. Zhang et. al [4] studied Cu-Cr-rGO alloy prepared by wet-mixing, mechanical blending and freeze-drying to obtain the Cu/Cr/rGO mixture powders. Spark plasma sintering for solid-phase sintering assist in faster densification of the mixture powders. A layered rGO network and Cr particles distributed in the copper matrix, compressive yield strength and the hardness of the Cu-Cr-rGO alloys were greater than those of pure copper and single rGO-added alloy and increased with increasing Cr. The rGO/Cr hybrid second phases are believed to be beneficial to the strengthening of the copper matrix.

This study prepared the Cu-Cr-rGO alloy using powder metallurgy. The rGO was prepared by the electrochemical exfoliation of graphite in acidic-based H_2SO_4 electrolyte. The obtained GO solution was ultrasonicated and taken in a petri-dish to dry in the sunlight for a few hours, resulting in the formation of rGO. Sunlight has been used for drying in many studies [17, 18]. Copper, chrome and rGO were mixed, compacted and sintered by powder metallurgy processing; the sintered specimens were aged. The characterisation of alloys was achieved by XRD, EDS, FESEM and Raman analyses and their hardness and electrical resistance properties were studied. Figure 1 displays a diagram of the experimental procedure.

IOP Conf. Series: Materials Science and Engineering **987** (2020) 012029 doi:10.1088/1757-899X/987/1/012029

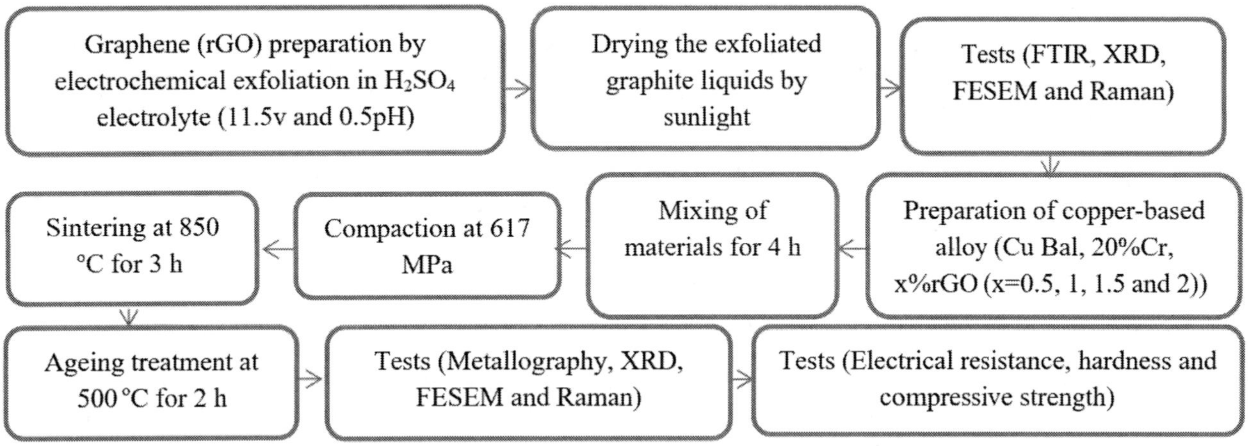

Figure 1 Block diagram of the experimental procedure

2. Materials and Experiment Details

2.1 Materials

In this study, copper, chrome and rGO were used; their properties and proportions are illustrated in Table 1. Cu and Cr were purchased from a trade market. The rGO was synthesised by electrochemical exfoliation in an H_2SO_4 electrolyte at 0.5 pH and 11.5 volts; it was sonicated and then dried in sunlight.

Table 1 Materials used for the alloys in terms of grain size, source and percentage

Element	**Cu**	**Cr**	**rGO**
Particle size	45 μm	63 μm	~100 nm
Source	Korea	Germany	Prepared by electrochemical exfoliation
	Wt. %	**Wt. %**	**Wt. %**
Base alloy	80	20	0
A1	79.5	20	0.5
A2	79.5	20	1
A3	78.5	20	1.5
A4	78	20	2

IOP Conf. Series: Materials Science and Engineering **987** (2020) 012029 doi:10.1088/1757-899X/987/1/012029

2.2 Experiment details

Powder metallurgy processing was developed for the synthesis of copper-based alloys. The elements were mixed for 4 h according to the ratios in Table 1, then they were pressed at 617 MPa to produce disc samples with a diameter of 14 mm and a thickness of 4 mm. Compacted specimens were sintered in a vacuum by heating them to 400 °C then soaking them 30 min; they were then heated to 850 °C and soaked for 3 h. A heating and cooling rate of 7 °C/min was used in sintering, and the sintered specimens were heat-treated by ageing at 450 °C for 2 h, according to previous studies.

3. Characterisation and Testing

3.1 Microstructure analysis

The identification of microstructures was characterised by light optical microscopy (LOM) and FESEM. LOM with a magnification of 20X in an Olympus microscope was used in the metals department of the Materials Engineering College at Babylon University; the microscope was manufactured in Japan. As seen in Figure 2, the disc samples were wet ground, cleaned with water and dried with hot air. After drying, they were stored in polyethene zip-lock bags with silica gel and polished with a fine diamond and a fine cloth then etched with a solution of 5 gram $FeCl_3$ + 10 ml HCl + 100ml H_2O. FESEM investigated the grain size and surface morphology of the powders and sintered specimens and was conducted in the Razi Metallurgy Research Centre in Iran.

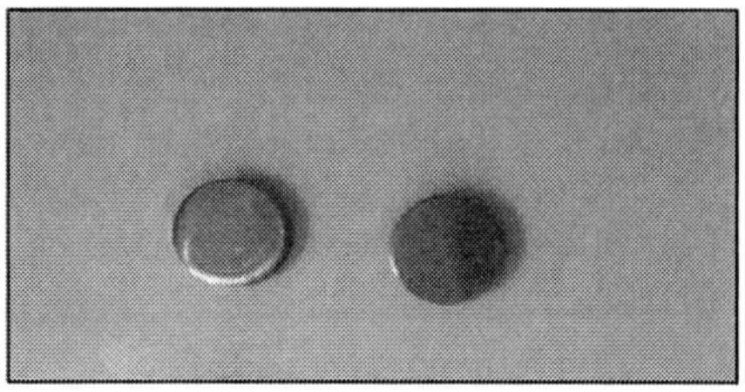

Figure 2 Samples of as-sintered Cu20Cr0.5rGO

3.2 XRD analysis

XRD analysis was performed on the rGO and sintered specimens to confirm the chemical compositions of the materials. This was conducted at the Razi Metallurgy Research Centre in Iran.

3.3 EDS analysis

EDS was conducted at the Razi Metallurgy Research Centre to estimate the chemical composition of the surface of the rGO powder and sintered specimens.

3.4 Raman analysis

The Raman spectrum displays peaks of D and G bands; D was due to structural imperfections created by the attachment of oxygenated groups on the carbon basal plane. Therefore, peak intensity was used as a measure for the degree of disorder. A broad G band, the characteristic of all sp^2-hybridised carbon networks, arose from the first-order scattering. Another peak, usually called the 2D peak, was the overtone of the D peak, which reflected the number of graphene layers. It is attributed to double resonance transitions leading to the production of two opposite momentum phonons. This is different from the D peak, which is active in the presence of defects, while the 2D peak is active in the absence of any defects. Generally, the integrated intensity ratio of ID/IG indicates the degree of oxidation and the size of sp^2 ring clusters in an sp^3/sp^2 hybrid network of carbon atoms [19].

3.5 Hardness analysis

The Vickers macro-hardness test was used to measure hardness with 500 N load for 15 s [11] using the micro-hardness tool model HVS-1000 in the metallurgical laboratory of the Materials Engineering College at Babylon University. Four readings were recorded for each sample. This test shows the variation in hardness, maybe due to the existence of different materials, pores and agglomeration in the surface of the sample.

3.6 Electrical resistance analysis

Electrical resistance was measured for the sintered and aged specimens using an AT512 Precision Ohmmeter in the Polymeric Materials department of the Materials Engineering College at Babylon University. This is shown in Figure 3.

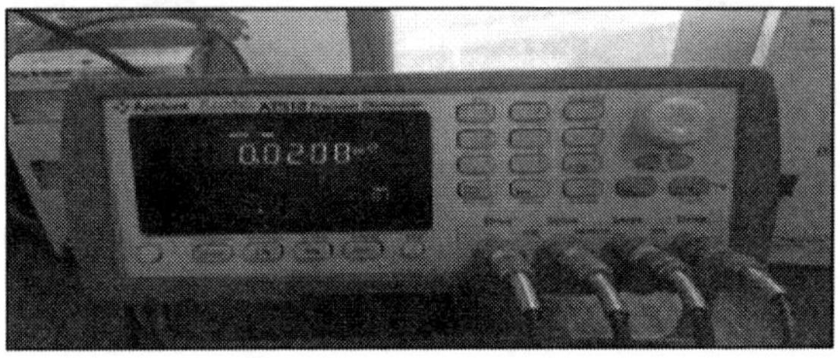

Figure 3 AT512 Precision Ohmmeter instrument

3.7 Compressive strength analysis

Room temperature compressive strength was characterised for sintered alloys using cylindrical specimens with dimensions of 10 mm diameter and 16 mm height. The alloys were 80%Cu20%Cr with different percentages of reduced graphene oxide (0.5, 1, 1.5 and 2%). The test was conducted in the Ceramic Materials department of the Materials Engineering College at Babylon University and is shown in Figure 4.

Figure 4 Hydraulic universal material compression tester (WP 310)

4. Results and Discussion

The rGO was prepared by electrochemical exfoliation and added to the Cu-Cr alloy using the powder metallurgy process. Thermal decomposition can occur to the residual oxygen-containing groups on the rGO nano-sheets and may be removed during the sintering operation. This process is associated with the evolution of gases and causes porosity in the sintered body [16]. One possible solution to this problem is to preheat the green compact at lower temperatures (such as 400 °C) before the final sintering at high temperature. There was a general agreement that the reduction process is never complete up to 800 °C. Crucial questions about the mechanism of binding and growth of the elements also remain open, and the interaction between rGO and the metal remains uncertain. The elements may be non-covalently bound or may have formed metal–C bond. Metal–O–C bonds can also form from the residual oxygen of rGO [20].

4.1 XRD results

XRD analysis exhibited diffraction peaks in the rGO powders and alloys. Diffraction peaks for rGO were absent in the XRD analysis of alloys and were attributed to the total content of the carbon phase below the sensitivity limit [21], as shown in Figure 5. XRD analysis of the exfoliated graphite powders showed peaks at $2\theta \approx 9o$ and $2\theta \approx 24.5$, similar to the peaks of rGO [22]. Figures 6–9 exhibit the XRD patterns of as-sintered specimens. Figure 6 shows peaks of the XRD pattern of the sintered Cu-Cr specimen, and its intensity refers to the ratio of the elements, exhibiting peaks of (111), (200), (220), (311) and (222) and planes at 2θ = 43.25, 50.5, 74 and 90 for FCC-Cu. For BCC-Cr, peaks were exhibited at [110], [200], [211], [220] and [310] and planes at 2θ = 44.5, 64.5, 81.75 and 98 in the agreed XRD plot [12]. The peaks of Cr in the as-sintered Cu-20%Cr-1%rGO and Cu-20%Cr-2%rGO materials in Figures 7 and 9 have slightly less intensity compared to the Cr peaks of the as-sintered Cu-20%Cr and Cu-20%Cr-1.5%rGO specimens in Figures 6 and 8. This is related to the carbide compounds formed due to the coupling of Cr with carbon during sintering treatment and the formation of carbides (Cr_2C) at peaks mentioned previously and confirmed in [24]. The XRD peaks were identical to specific 2θ values for carbides that formed under conditions of 810 °C for 3 h that approach the conditions of the sintering process used in the current study (850 °C for 3 h) [25].

Chromium carbides observed on the as-sintered Cu-20%Cr-1%rGO and Cu-20%Cr-2%rGO specimens at peaks 2θ = 67.25, 75.5, 81.75 and 83°, as shown in Figure 7 and fewer quantities in Figure 9, disappeared for the Cu-20%Cr-1.5%rGO alloy in Figure 8. The defect nature of rGO led to the formation of carbide. Chu et al. obtained a good dispersion of graphene sheets in Cu matrix powder [16]. On the other hand, chromium oxide (Cr_2O_3) was confirmed under XRD spectral analysis at 2θ =24.5, 33.75, 36.25 and 54.75, due to the partial oxidation of the metallic components of samples containing high Cr; this agrees with [12, 3]. The absence of any components between Cu and C was widely agreed upon [26], but there is binding between Cu and residual functional groups of rGO layers, as shown in Figure 18. The diffraction patterns of the composites do not show any effect of graphene, which might be attributed to the small volume fraction and low scattering length of the carbon atom [27].

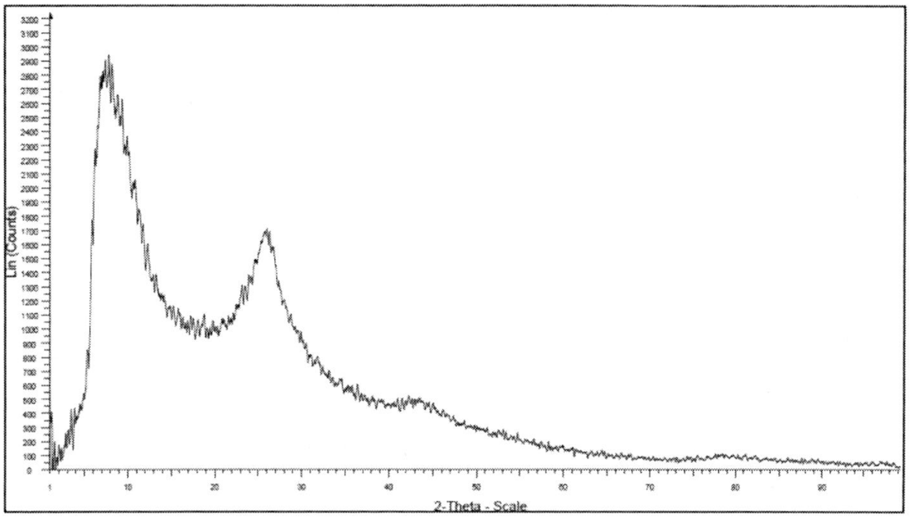

Figure 5 XRD of an as-received rGO powder

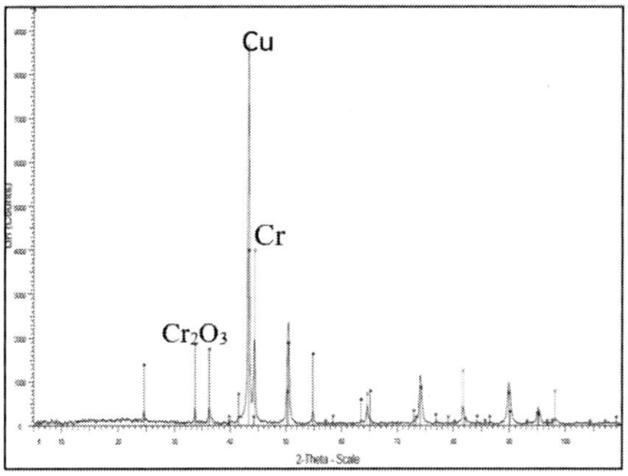

Figure 6 XRD of copper-based alloy (Cu-20%Cr)

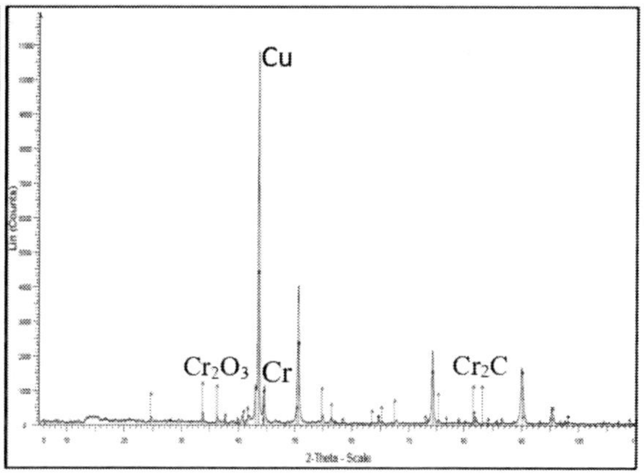

Figure 7 XRD of sintered alloy (Cu-20%Cr-1%rGO)

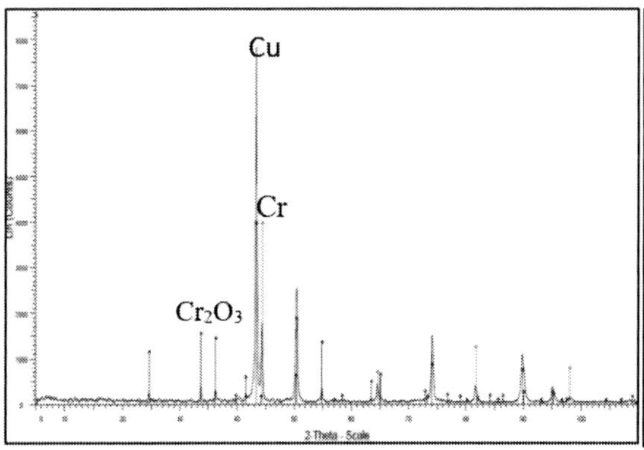

Figure 8 XRD of sintered alloy (Cu-20%Cr-1.5%rGO)

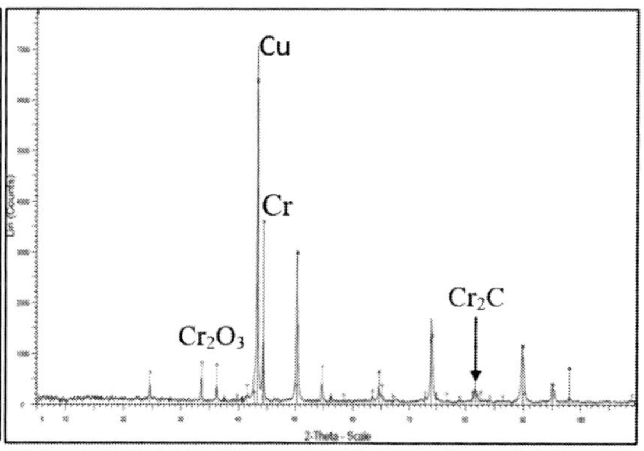

Figure 9 XRD of sintered alloy (Cu-20%Cr-2%rGO)

Copper oxides did not appear in XRD analysis but were observed in Raman analysis. It was expected that residual graphene oxide may be reduced to rGO during the drying process of the graphene liquid and sintering processes at elevated temperatures as well as forming CuO and Cu_2O oxides [21]. The oxides have interfacial interactions with carboxyl or hydroxyl groups by chemisorption, and the bridge formed at oxygen-defect sites and van der Waals interactions between metal oxides and the pristine region of graphene was agreed upon [28].

4.2 Microstructure characterisation results

4.2.1 LOM results

Figure 10 shows the microstructure of as-sintered specimens; the unetched surface is illustrated in Figure 10 and the etched specimen is shown in Figure 11. The microstructure of the alloy shows the formation of two distinct areas with different morphologies: the bright area refers to the number of fine Cr-particles observed beside a few larger Cr-particles, which are elongated, and the dark area contains Cu-grains. The areas formed in this study are confirmed in [29].

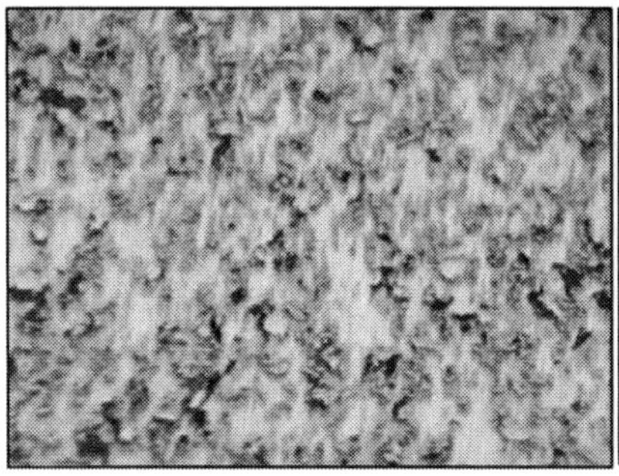

Figure 10 The microstructure of unetched alloy Cu-20%Cr-2%rGO at 20X magnification

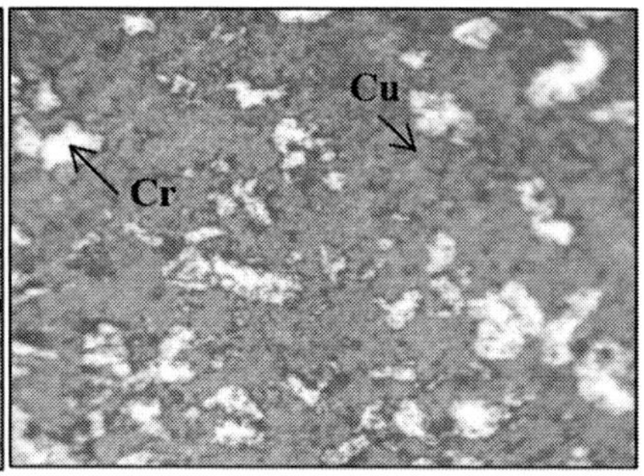

Figure 11 The microstructure of etched alloy Cu-20%Cr-2%rGO at 20X magnification

4.2.2 FESEM results

Figures 12 and 13 demonstrate FESEM for copper particles and layers of rGO, respectively. Figures 14 and 15 show the microstructure of the as-sintered base and Cu-20%Cr-0.5%rGO alloy, respectively, in which the microstructure of the base alloy can be observed in Figure 14. It exhibits a Cu matrix (bold zone) and Cr phase (bright zone) with several Cr particles spread in a line-like shape in interdendritic and grain boundary positions; these were confirmed in [30]. Alteration of the microstructures was achieved in the Cu-Cr-rGO alloys after adding different percentages of rGO. Figure 15 shows important variations in the microstructure, which seems to be surrounding the Cr zone in graphene layers, and the grains become equiaxed after sintering treatment. Gr-sheet distribution into the metal matrix and along the grain boundaries assist the suppression of the grain growth of the copper due to high temperature, which is confirmed by [20].

IOP Conf. Series: Materials Science and Engineering **987** (2020) 012029 doi:10.1088/1757-899X/987/1/012029

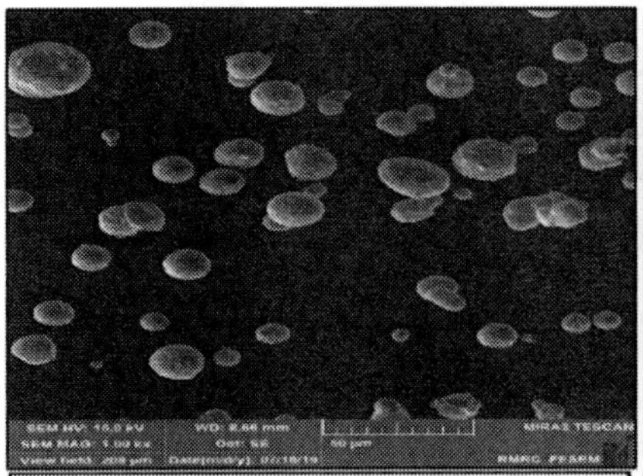

Figure 12 FESEM for grain size of Cu particles

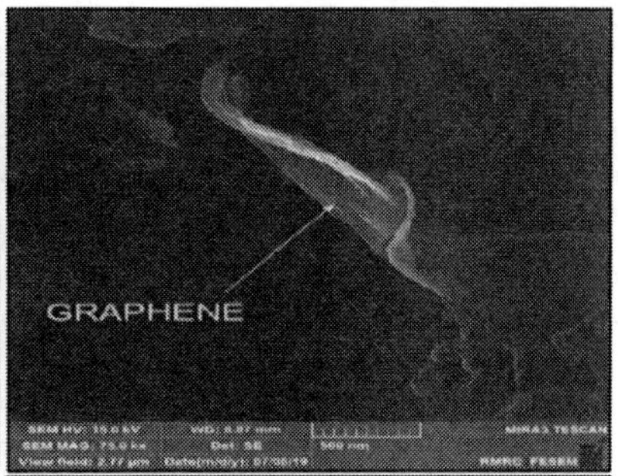

Figure 13 FESEM for rGO layers

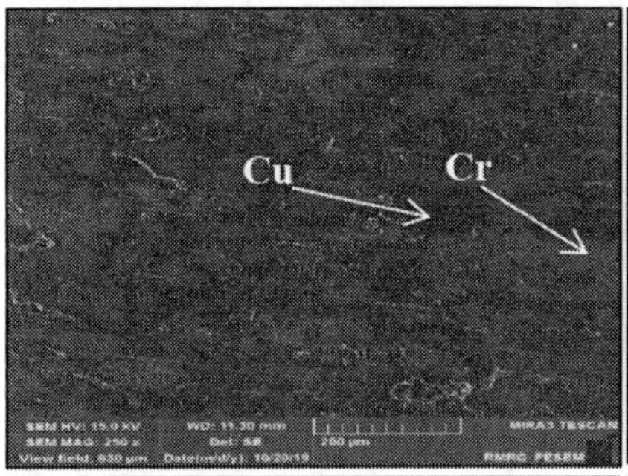

Figure 14 FESEM of as-sintered Cu-20Cr alloys

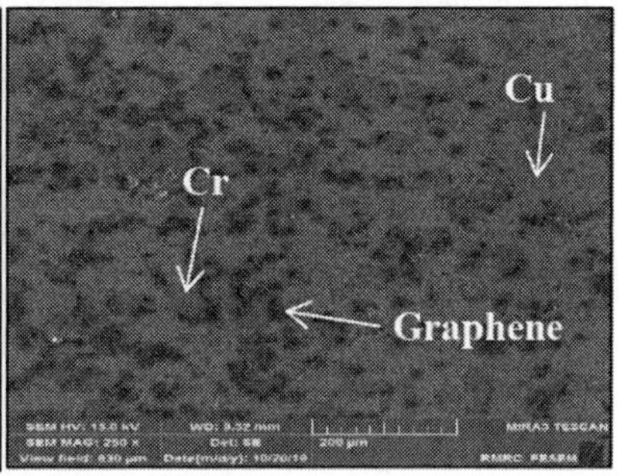

Figure 15 FESEM of as-sintered Cu-20Cr-0.5rGO alloys

Figure 16 shows a semi-coherent interface between the matrix and rGO, as well as an agglomeration of graphene oxide layers that can be seen clearly in Figure 17 due to the high activity of rGO layers. Multilayered rGO sheets were found and fixed in the matrix in thicknesses ranging from 50–125 nm; these work as lubricants under states of adhesive wear. Figure 18 exhibits binding between the matrix Cu and rGO layers after mixing.

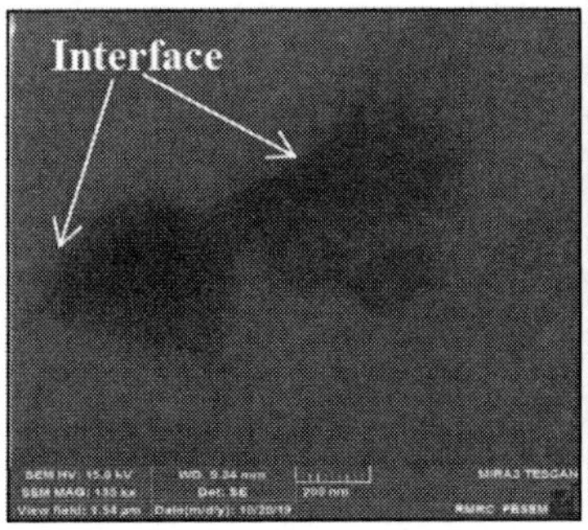

Figure 16 FESEM of Cu-20Cr-0.5rGO alloy

Figure 17 FESEM of Cu-20Cr-1.5rGO alloy

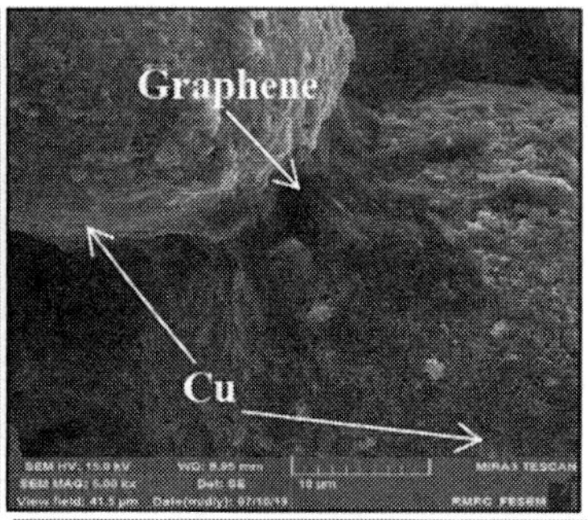

Figure 18 FESEM for binding between Cu and rGO layers.

4.3 EDS results

Figures 19 and 20 present the EDS analysis percentage of Cu, Cr, C and O as well as giving an indication of the formation of interfacial products, which asserts that reaction occurred between the elements during the preparation operation. rGO was induced into the matrix and because of the low solubility of carbon in copper, only mechanical bonding between the two phases occurred [31], but binding did occur with Cu and the functional groups in rGO layers [32], as seen in Figure 18.

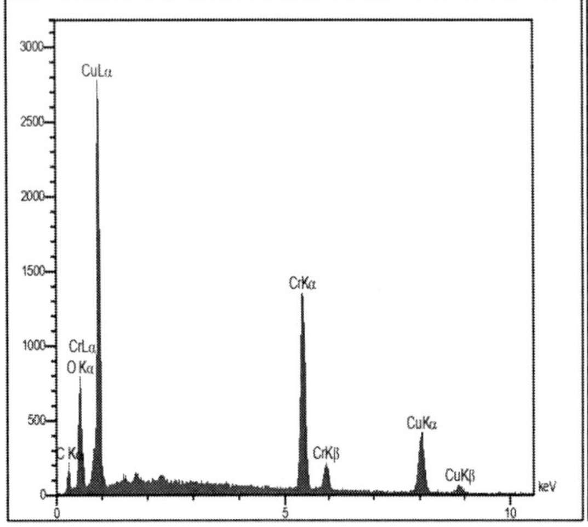

Figure 19 EDS of Cu-Cr-1.5rGO alloy

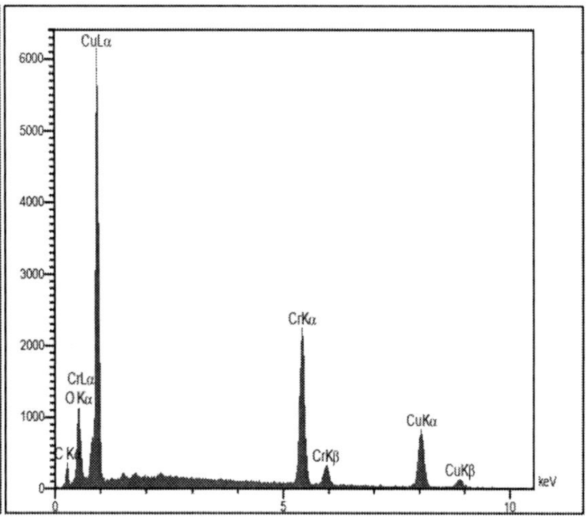

Figure 20 EDS of Cu-Cr-2rGO alloy

4.4 Raman results

Figures 21 and 22 show D, G and 2D peak intensities versus the Raman shift for rGO powder and sintered alloy, respectively. rGO bears too many defects so its Raman spectroscopy is not sensitive [20]. Indeed, the ratio of intensities of D and G peaks change after mixing and sintering processes. The D peak intensity depends on the stress/strain defects at the grain interfaces of the multiphase material [21]. The position of the G peak increases and both the intensity of D and the ID/IG ratio decreases, all of which are expected consequences of greater graphitisation following reduction [33]. The sites for Raman peaks of D, G and 2D were 1,361 cm^{-1}, 1,598 cm^{-1} and 2D≈2,700–2,925 cm^{-1}; as-received rGO shifted to 1,388 cm^{-1}, 1,605 cm^{-1} and 2D≈2,800–2,950 cm^{-1} of as-sintered alloy; this shift was small and referred to the strain formed [21]. The ratio of ID/IG (R) for rGO powder was 0.88, while the as-sintered specimen was 0.74. Prominent peaks of the Raman analysis were observed at 160, 450, 515 and 620 cm^{-1}. This may be corresponding to signals of copper oxides that agreed with results in [34].

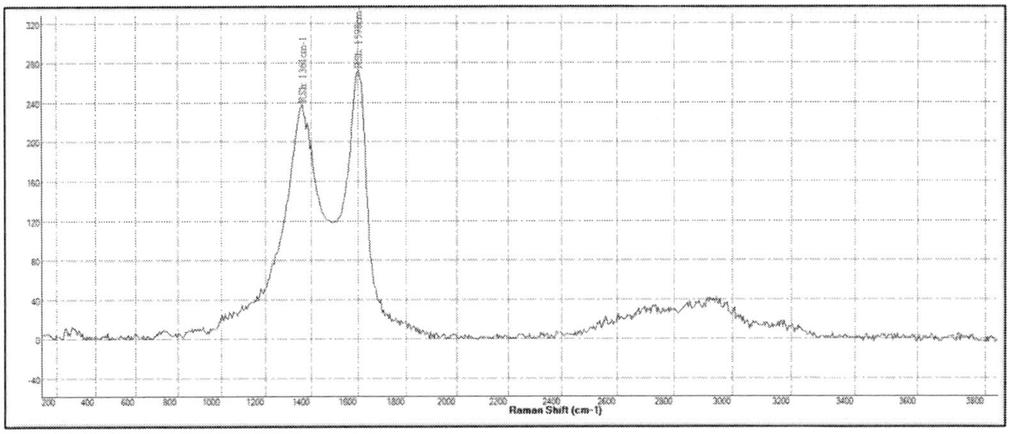

Figure 21 Raman spectrums of the as-received rGO

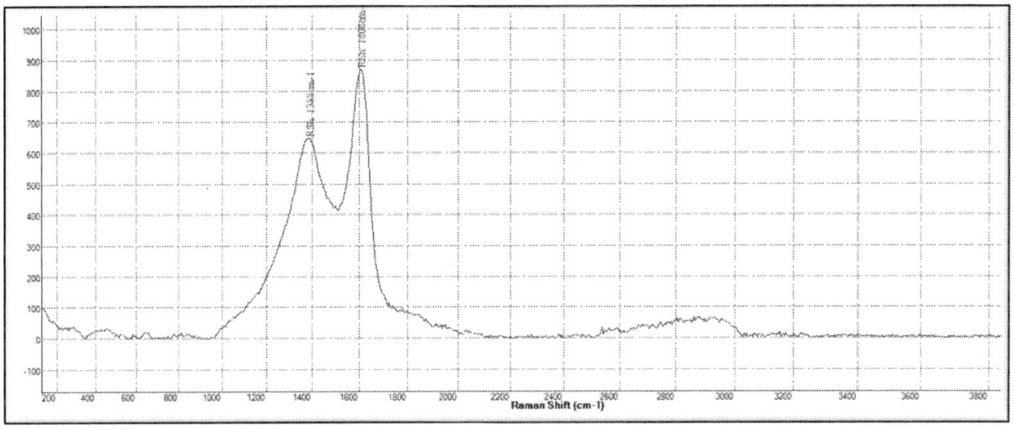

Figure 22 Raman spectrums of the as-sintered Cu-Cr-rGO alloy

4.5 Hardness results

Figure 23 shows the surface micro-hardness values of as-sintered alloys that exhibited fluctuating variations. As noted, the hardness increased with the increase of rGO up to 1%, decreased at 1.5% and then increased at 2%. These results maybe because of the development of microstructures during the sintering process. For the Cu-20%Cr-1%rGO specimen, the increase in hardness may be due to the substantial hardening of Cr/rGO hybrids in the Cu matrix and the formation of carbides as shown in XRD analysis [4]. The Cu-20Cr-1.5%rGO specimen exhibited a decrease in hardness because carbides did not form in the alloy, leading to the segregation of a large number of graphene sheets at grain boundaries, less cohesion of the elements within the matrix, porosity and weak points in the Cu/Cr/rGO binding; it is not clear why carbides did not form [9]. The Cu-20Cr-2%rGO specimen exhibited increases in hardness due to the formation of chromium carbides (Cr_2C) during the sintering process that led to good cohesion and wettability of the elements within the matrix; chromium atoms contained in the Cu matrix also reacted with rGO resulting in strong bonding. In this study, Cr produced oxides and carbides in conjunction with the oxidation/reduction process during sintering; these materials are hard and were detected by XRD and EDS analyses; this agreed with [3].

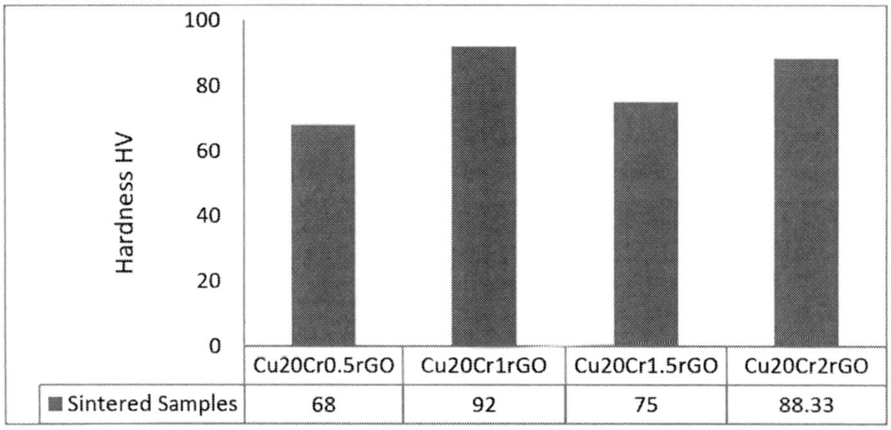

Figure 23 Micro-hardness of as-sintered Cu-Cr-rGO alloy

4.6 Electrical resistance results

Figure 2 shows the electrical resistance values of the sintered and aged alloys at 450 °C for 2 h. As noted, non-uniformity altered in resistance with an increase in rGO, but the values for as-aged samples were less than that estimated for the as-sintered samples. Electrical resistance changed with a fluctuating variation of as-sintered and aged alloys. The sintered Cu-20Cr-0.5%rGO specimen had a higher electrical resistance than that of the base alloy, which may be due to an increase in grain boundaries that decreases the electrical current. However, the electrical resistance decreased for the sintered Cu-20Cr-1%rGO specimen, which may be due to the formation of high oxides and carbides that led to an increase in wettability. Furthermore, the bonding between the matrix and added elements affected the electricity passing through the Cu-Cr matrix [14]. The sintered Cu-20Cr-1.5%rGO specimen observed increases in electrical resistance, probably because carbides were not formed, as deduced by XRD analysis (Figure 8). A large number of graphene sheets in the matrix with less binding and segregated at grain boundaries increased the interfaces and hindered the electrons' movement. Sintered specimens containing Cu-20Cr-2%GO had lower electrical resistance (12.21 μΩ) for as-sintered specimens than for as-aged specimens (6.23 μΩ). This may be due to the reaction between Cr and C in the rGO during the sintering process and the formation of carbides (Cr_2C). Carbides increased the wettability and binding between the elements, which allowed electricity to easily pass through the Cu-Cr matrix. Cr decreased in the matrix and translated to chromium carbides, which have lower resistance [14]. Electrical resistance was measured for the aged samples and showed lower values than those measured in the as-sintered samples, meaning that the ageing treatment reduced the electrical resistance property. This behaviour can be explained by lowering the concentration of elements inside the grains due to their migration to grain boundaries during the ageing treatment [35]. This may also be due to long-range ordering and reduced point defects [36]. Partial reduction for GO may occur and measurements of electrical resistivity have been used to prove the complete reduction of graphene oxide [21].

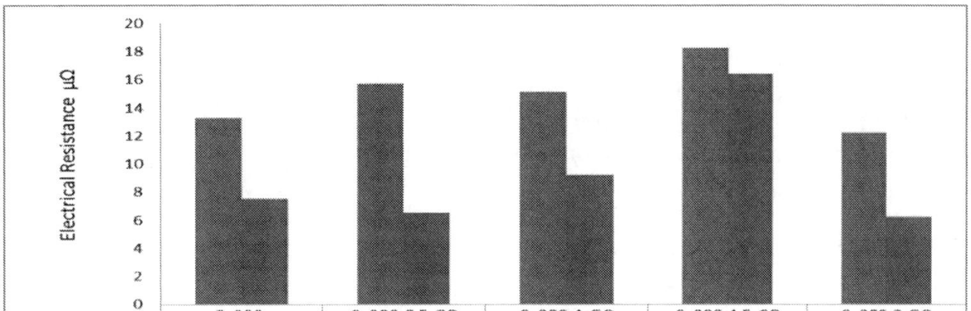

	Cu20Cr	Cu20Cr0.5rGO	Cu20Cr1rGO	Cu20Cr1.5rGO	Cu20Cr2rGO
■ Sintered State	13.3	15.7	15.1	18.23	12.21
■ Aged State	7.5	6.5	9.21	16.4	6.23

Figure 24 Electrical resistance of as-sintered and aged Cu-Cr-rGO specimens

4.7 Compressive strength results

Figure 25 shows the compressive strength values of the sintered alloys. As noted, non-uniformity alteration in compressive strength increased with rGO. The sintered Cu-20Cr-0.5%rGO specimen had lower compressive strength than that for the base alloy, which may be due to the increase of pores and cracks formed during powder metallurgy processing; these form and fracture under compressive loading. The slightly increased compressive strength of the sintered Cu-20Cr-1%rGO specimen maybe because of the formation of oxides and carbides, which lead to increased wettability and bonding between the matrix and added elements; greater compressive strengths means fewer pores and cracks. The Cu-20Cr-1.5%rGO sintered specimen had increased compressive strength and had the highest value; this may be due to the remarkable reinforcement effect of graphene in the composite, which derives from its micro-layered structure that enhances the dispersion of graphene and prevents slippage between graphene flakes. The Cu-20Cr-2%rGO alloy observed a decrease in compressive strength; incorporating a high graphene and chrome content in the Cu matrix may have overstrained the lattice of copper atoms. Furthermore, the segregation and agglomeration of graphene particles could have disintegrated under pressure, leading to the formation of pores, and brittle fracturing as well as presence brittle carbides. There could also have been a mismatch between the constituents that increased the dislocation densities, leading to brittle fractures.

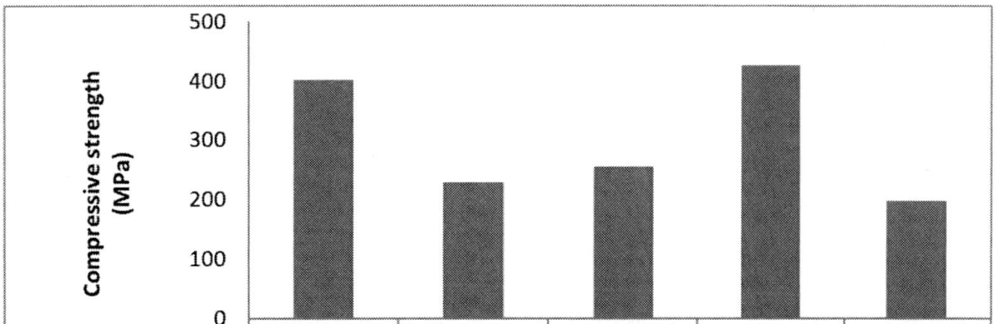

	Cu20Cr	Cu20Cr0.5rGO	Cu20Cr1rGO	Cu20Cr1.5rGO	Cu20Cr2rGO
■ Sintered Samples	401.273	228.6434	254.777	425.235	197.452

Figure 25 Compressive strength of as-sintered Cu20-Cr and Cu-20%Cr-x%rGO specimens (x=0.5, 1, 1.5 and 2%)

5. Conclusion

In summary, rGO was prepared by electrochemical exfoliation and added to Cu-Cr alloys by powder metallurgy to fabricate a Cu-Cr-rGO alloy. These materials play a significant role in the formation of chemical components, which has an important effect on the properties of the composites. The results of XRD, FESEM, EDS, Raman analyses, hardness, electrical resistance and compressive strength tests were obtained for the Cu-Cr-rGO alloys. The remarkable reinforced effect of rGO in the alloys was derived from the structure of the components formed during the sintering process, which enhances the uniform dispersion of graphene layers. The results of the study on the effect of rGO on the properties were:

- XRD, FESEM and Raman analyses approved the formation of rGO by electrochemical exfoliation; FESEM and Raman analyses also demonstrated the dispersion of rGO sheets in the Cu-Cr-based alloy.
- Hardness showed fluctuating values; the hardest material was the as-sintered Cu-20Cr-1%rGO specimen at 92 Kg/mm^3.
- Electrical resistance exhibited a lower value for the Cu-20%Cr-2%rGO as-sintered alloy (12.21 μΩ) than for the as-aged specimen (6.23 μΩ).
- Compressive strength increased with an increase of rGO up to 1.5%. This may be due to the graphene-reinforced composite that enhances the dispersion of graphene and prevents slipping between graphene flakes. Compressive strength decreased at 2%rGO due to copper atoms overstraining the lattice as well as the segregation and agglomeration of graphene particles that disintegrated under pressure. There was also a mismatch between the constituents that increased the dislocation density and led to brittle fractures. The sintered Cu-20Cr-1.5%rGO alloy displayed the highest compressive strength (425.235 MPa).

References

[1] Braunovic , M. , Konchits, V.V. , and Myshkin, N.K.(2006) "Electrical Contacts: Fundamentals, Applications and Technology" Taylor and Francis Group.

[2] Andre, B. (2011) "Nanocomposites for use in sliding electrical contacts" A doctoral dessirtation from the faculty of science and technology, Uppsala University 879.

[3] Ozgün, O. and Ercetin, A. (2017) "Microstructural and mechanical properties of Cr-C reinforced Cu matrix composites produced through powder metallurgy method " , *J. Nat. Sci.*, Vol. **6** No. 2.

[4] Zhang, X. J., Dai, Z. K., Liu, X. R. , Yang, W. C. , He, M. and Yang, Z. R. (2018) "Microstructural Characteristics and Mechanical Behavior of Spark Plasma-Sintered Cu–Cr–rGO Copper Matrix Composites" Germany,, *Acta Metall. Sin.*, Vol. **31**, Issue 7, pp. 761-770.

[5] Gotkhindikar, N. N., Vanjari, S.V. and Nigade, D. B. (2018) "Development of Copper-Chromium electrical contact materials by Powder Forging" India, *I.J.A.R.S.E.* , Vol **7**, special No.03 .

[6] Kimura, W., Kume, Y., Kobashi, M., Kanetake, N. (2014) "Consolidation of Cr-Cu/Cu powder laminated material by compressive torsion processing " in 11th International Conference on Technology of Plasticity, Procedia Engineering 81, pp. 1169 –74.

[7] Gershman, I.S., Gershman, E. I., Mironov, A. E., Rabinovich, G. S. F. and Veldhuis, S. C. (2017)"On Increased Arc Endurance of the Cu-Cr System Materials" , *journal of Entropy*,**19**, pp.386 .

[8] Saboori, A., Moheimani, S. K., Pavese, M., Badini, C. and Fino, P. (2017)"New Nanocomposite Materials with Improved Mechanical Strength and Tailored Coefficient of Thermal Expansion for Electro-Packaging Applications" , *J. Met.* ,7,536.

[9] Manrique, P. H., Lei, X., Xu, R., Zhou, M., Kinloch, I. A. and Young, R. J. (2019) "Copper/graphene composites: a review " Berlin, Germany , *Journal of Mat. Sci.*, **54**, pp. 12236–89.

[10] Sun, L. and Fugetsu, B. (2013) "Massive production of graphene oxide from expanded graphite " Japan, *J. Mater. Sci. Lett.* , Volume **109**, pp. 207-210.

[11] Konakov, V.G., Kurapova, O.Y., Solovyeva, E.N., Lomakin, I.V. and Archakov, I.Y. (2018) "Synthesis, Structure and Mechanical Properties of Bulk Copper-Graphene Composites" Russia, *Adv. Mater. Sci. Eng.*, **57** ,pp.151-157.

[12] Mula, S., Sahani, P., Pratihar, S.K., Mala, S., Kocha, C. C. (2011) "Mechanical properties and electrical conductivity of Cu–Cr and Cu–Cr–4% SiC nanocomposites for thermo-electric applications " *Mater. Sci. Eng. A* 528, pp. 4348–56.

[13] P. Gill and N. Munroe (2012) "Study of Carbon Nanotubes in Cu-Cr Metal Matrix Composites" *J. Mater. Eng. Perform.* , 21, pp.2467–71.

[14] Imai, H., Chen, K. Y., Kondoh, K., Tsai, H.Y. and Umeda, J. (2015) "Effect of Chromium Behavior on Mechanical and Electrical Properties of P/M Copper-Chromium Alloy Dispersed with VGCF" *I.J.M.M.E.* , Vol **9**, No:7.

[15] Y. Chang , W. Zheng, Z. Zhou, Y. Zhai and Y. Wang (2016) "Preparation and Performance of Cu-Cr Contact Materials for Vacuum Switches with Low Contact Pressure" *J. Electron. Mater.*, volume **45**, pp. 5647–54.

[16] Ji, X., Qi, S., Ahmed, R. and Rifat, A. A. (2019) "Graphene-Reinforced Advanced Composite Materials" in Ozkan, C. (ed.) Handbook of Graphene, River Street, Hoboken, USA, Volume **4**, pp. 27–90.

[17] Kumar, P., Subrahmanyam, K. S., Rao, C. N. R. (2011) "Graphene produced by radiation-induced reduction of graphene oxide", *Int. J. Nanosci.*, vol. **10**, issue 4 and 5, pp. 559-566.

[18] Hou, W. C., Chowdhury, I., Goodwin, D. G., Jr., Henderson, W. M., Fairbrother, D. H., Bouchard, D., and Zepp, R. G. (2015) "Photochemical Transformation of Graphene Oxide in Sunlight" *Environ. Sci. Technol.*, **49**, pp.3435−43.

[19] Zhao, J., Liu, L. and Li, F. (2015) "Graphene Oxide: Physics and Applications" Springer .

[20] Dimiev, A. M. and Eigler, S. (2017) "Graphene Oxide Fundamentals and Applications" Wiley , United Kingdom.

[21] Konakov, V. G., Archakov, I. Y. and Kurapova, O. Y. (2019) "Fabrication and Properties of Copper–Graphene Composites" in Ozkan, C. (ed.) Handbook of Graphene , River Street, Hoboken, USA, Volume 4, pp. 285–322.

[22] Thema, F. T. , Moloto, M.J. , Dikio, E.D., Nyangiwe, N.N., kotsedi, L., Maaza, M. and khenfouch, M. (2013) "Synthesis and Characterization of Graphene Thin Films by Chemical Reduction of Exfoliated and Intercalated Graphite Oxide", *J. Chem.* .

[23] Olofinjana, A.O, A. and Tan, K. S. (2010) "Processing of Cu-Cr alloy for combined high strength and high conductivity" *A.J.S.T.D.*, Vol. **26** Issue 2 pp. 11-20.

[24] Schmuecker, S. M., Clouser, D. , Kraus, T. J. and Leonard, B. M. (2017) "Synthesis of Metastable Chromium Carbide Nanomaterials and their Electrocatalytic Activity for the Hydrogen Evolution Reaction" , *The Royal Society of Chemistry*, **46**, pp. 13524–30.

[25] Anderson, K. R. and Groza, J. R. (2001) "Microstructural Size Effects in High-Strength High-Conductivity Cu-Cr-Nb Alloys " *Metallurgical and Materials Transactions-A* , Volume **32**A, pp. 1211–24.

[26] Alam, S.N., Kumar, L. and Sharma, N. (2015) "Development of Cu-Exfoliated Graphite Nanoplatelets (xGnP) Metal Matrix Composite by Powder Metallurgy Route", *Graphene*, **4**, pp. 91-111.

[27] Khobragade, N., Sikdar, K., Kumar, B., Bera, S. and Roy, D. (2019) "Mechanical and electrical properties of copper-graphene nanocomposite fabricated by high pressure torsion " *J. Alloys Compd.*, 776 , pp. 123-132.

[28] Ahmad, M. and Naz, S. (2019) "Graphene-Based Composite Nanostructures: Synthesis, Properties, and Applications" in Ozkan, C. (ed.) Handbook of Graphene, River Street, Hoboken, USA, Volume 4, pp. 203–232.

[29] Hauf, U., Kauffmann, A., Weiss S. K., Feilbach, A., Boening, M., Mueller, F. E. H., Hinrichsen, V. and Heilmaier, M. (2017) "Microstructure Formation and Resistivity Change in Cu-Cr during Rapid Solidification " *J. Met.*, **7**, pp. 478.

[30] Zhang, P., Jie, J., Li, H., Wang, T. and Li, T. (2015) "Microstructure and properties of TiB2 particles reinforced Cu–Cr matrix composite" New York, *J. Mater. Sci.*,**50**, pp. 3320–28.

[31] Ayyappadas, C., Muthuchamy, A., Annamalai, A. Raja and Agrawal D. K. (2017) " An investigation on the effect of sintering mode on various properties of copper-graphene metal matrix composite" *Journal of Advanced Powder Technology* **28**, pp.1760–68.

[32] Wang, L., Yang, Z., Cui, Y., Wei, B., Xu, S., Sheng, J. and Wang, M., Zhu, Y. and Fei, W. (2017) "Graphene-copper composite with micro-layered grains and ultrahigh strength" *Scientific Reports*, **7**, Article number, 41896.

[33] King, A. A. K., Davies, B. R., Noorbehesht, N., Newman, P., Church, T. L., Harris, A. T., Razal, J. M. and Minett, A. I. (2016) "A New Raman Metric for the Characterisation of Graphene oxide and its Derivatives" *Scientific Reports*, **6**, Article number: 19491.

[34] Boukhvalov, D. W., Kurmaev, E. Z., Urbańczyk, E., Dercz, G., Stolarczyk, A., Simka, W., Kukharenko, A. I., Zhidkov, I. S., Slesarev, A. I., Zatsepin, A. F. and Cholakh, S. O. (2018) "Atomic and electronic structure of graphene oxide/Cu interface" *Thin Solid Films* Volume **665**, pp. 99-108.

[35] Islamgaliev, R. K., Nesterov, K. M., Bourgon, J., Valiev, R. Z. and Champion, Y. (2014) "Nanostructured Cu-Cr alloy with high strength and electrical conductivity" Russia , *J. Appl. Phys.* **115**, 194301.

[36] Huuqing, L., Shuisheng, X., Pengyue, W., and Xujun, M. (2007) "Study on improvement of conductivity of Cu-Cr-Zr alloys" China, *Journal of Rare Metals*, Volume **26**, Issue 2, pp. 124-130.

IOP Conf. Series: Materials Science and Engineering **987** (2020) 012030 doi:10.1088/1757-899X/987/1/012030

The Effect of the addition of Ta, Zr and in on the corrosion behaviour of CuAlMn alloys prepared via powder metallurgy

Duaa Amer Ali[a], Ali Hubi Haleem[b], Haydar H.J. Jamal Al-Deen[b],

[a] MSc. Materials Engineering, University of Babylon, Iraq.

[b]Department of Metallurgical Engineering, Materials Engineering Faculty, University of Babylon, Iraq.

alialihhhobi@gmail.com, jaberjd@gmail.com, du3a3030@gmail.com

Abstract. CuAlMn-X (X= Ta, Zr and In) shape memory alloy was fabricated using the powder metallurgy (PM) technique. After sintering at 950 °C, the products were processed using two heat treatments (solution heat treatment at 900 °C and aging at 500 °C). The influence of Tantalum (Ta), Zirconium (Zr) and Indium (In) with three wt.% on the mechanicals properties (hardness), corrosion behaviour and microstructure of CuAlMn shape memory alloys was investigated in this paper. The effect of the addition of Ta, Zr and In alloys was investigated using X-rays diffraction (XRD), open circuit potential, electrochemical tests (Tafel extrapolation method) and Vickers micro-hardness. For the corrosion tests, a 3.5 % NaCl (chloride solution) was used as corrosion solution. The results showed that there is an increase in hardness with addition of the alloy element (Ta, Zr and In) and decrease in the size of martensite plate form, as shown in the microstructures. Furthermore, it was found that the corrosion resistance of Cu-Al-Mn after the addition improved, as shown by the increase in the corrosion potential and decrease in the corrosion current densities. The improvement percentage for CuAlMnTa, CuAlMnIn and CuAlMnZr in the corrosion rate was 97 %, 91 % and 77 %, respectively.

Keywords: *Powder metallurgy, CuAlMn alloys. Cu-based memory alloy, Zirconium addition, Tantalum addition, Indium addition*

1. Introduction

Shape memory alloys (SMAs) are intelligent materials that remember their initial shape when they are heated above their transformation temperature (TT). At a lower temperature than the TT, these properties allow these alloys to distort into new shapes. When the alloys are heated higher than the TT, the crystal structure variant causes the material to go back to the preformed shape [1–2]. SMAs have no diffusion, and the first transformation is characterised by an atomic shearing of the crystal lattice from a high to a low temperature, which represents the structural characteristic called martensite variants [3]. Binary system Cu-Al alloys can be utilised for different applications, such as pumps, valves, bushings and bearings [4]. The high damping capacity of SMAs make them good candidates for seismic damping [5].

The advantages of CuAlMn SMAs are that they are economical, strong and have excellent damping limits within their austenitic to martensitic phase transformation compared to other systems as NiTi. Additionally, grain refining the alloy elements to SMAs can influence the TT (increase or decrease) and shape effects and enhance the stress cracking resistance that is produced during quenching [6]. The addition of manganese (Mn) improved the binding force between atoms, which led to increased diffusion activation energy and a reduced diffusion rate of the atom for re-ordering [7–8].

Powder metallurgy (PM) is a favourable production path, as it provides a net or near shape, is cheap and controls the chemical composition and grain size [9]. The Cu_3Al-essential basis for SMAs has thermos-mechanical characteristics that are related to the shape memory of the SME and super elasticity of the SE. Additional cooling of the Cu_3Al observed these structures form as a body centre cubic (β) → (18R, 9R). The phases that can take place at higher and lower temperatures near the Cu_3Al composition could be determined using the following steps: 1) the disordered β phase (A_2 or body centre cubic structure) is the balanced structure at elevated temperatures, and 2) the balanced phases at low temperatures are the α-Cu, T_3-Cu_3Mn_2Al, γ_2-Cu_9Al_4 and β-Mn phases. The metastabe random body centre cubic (β) phase passes through the following two transition sequences during cooling: beta two structure and DO_3 martensite structure of Cu_3Al or the L_{21} parent phase of the Cu_2AlMn. To summarise the previous phases transformations of the CuAlMn shape memory alloys, the stoichiometric Cu_3Al alloy first $DO_3 \leftrightarrow A_2$ transition is obtained, whereas for concentrations near Cu_2AlMn, the following two second transitions are found: $L_{21} \leftrightarrow B_2$ and $B_2 \leftrightarrow A_2$ [10–15]. The aim of this research is to prepare the CuAlMn SMA using powder metallurgy and investigate the corrosion resistance of CuAlMn SMAs with Ta, Zr and In three % wt. in a 3.5 % NaCl solution.

2. Experimental Section

2.1. **Powders and its Tests**: The purity of the powders was conducted using the laser particle size analyser type Bettersize and a particle size analysis X-ray fluorescent (XRF), as shown in Table 1 and Figure 1.

Table 1: Powder specifications.

Powders	Average particle size (μm)	Purity %	Supplier
Copper	4.827	99.57	India / CDH fined chemical / Central Drug House (P) Ltd.
Aluminium	19.09	99.06	=
Manganese	19.38	99.84	=
Tantalum	11.74	99.86	=
Zirconium	14.23	99.98	=
Indium	18.46	99.2	=

IOP Conf. Series: Materials Science and Engineering **987** (2020) 012030 doi:10.1088/1757-899X/987/1/012030

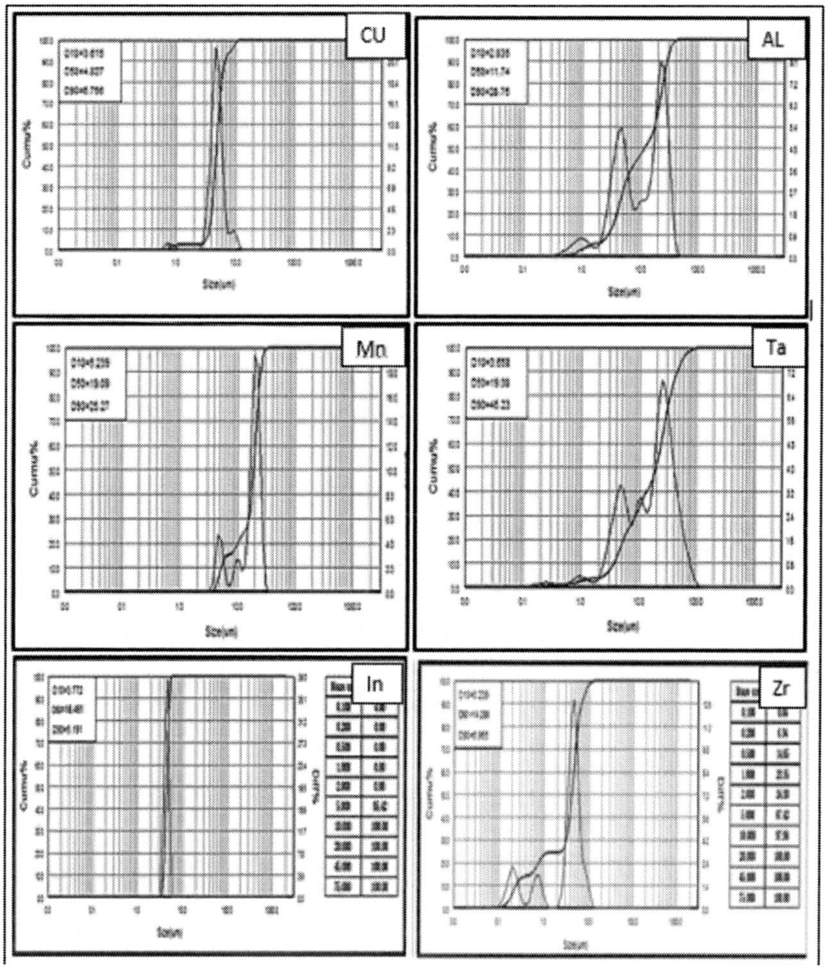

Figure1: Practical size analysis.

2.2. **Preparation of Samples**: The SMAs CuAlMn was prepared using the PM method under an Ar atmosphere to form an ingot. Table 2 shows all samples of CuAlMn and its composites and the percentage of powders in their composition.

Table 2: All samples of CuAlMn and its composites.

Codes	Cu wt.%	Al wt.%	Mn wt.%	Ta wt.%	Zr wt.%	In wt.%
CAM	83	8	9	-	-	-
CAMTa	80	8	9	3	-	-
CAMZr	80	8	9		3	-
CAMIn	80	8	9	-	-	3

The amount of the Cu element varied based on the amount of Ta, Zr and In elements, whereas Al and Mn were kept constant in the alloys. The first-stage powders were mixed in a planetary ball mill for four hours in the presence of ethyl alcohol. Alcohol was added to prevent friction between the mould and powder particles and to reduce the oxidation of the powder particles during the mixing process. During the compacting stage, the mixture was cold compacted in a mould under uni-axial direction using an electric hydraulic press with a compression pressure of nine tons. The dimensions of the mould are shown in Table 3. Cylindrical samples with a diameter of ≈ 12 mm and length of ≈ five mm were produced.

The third stage was sintering, where a heat treatment applied on green was compacted to create bonding between atoms. The sintering furnace consisted of a quartz tube and two ends. One side of the tube was installed into a bottle of water, and the other side was connected to an argon gas cylinder (supply). The other end of tube was connected to a vacuum pump and argon supply. During one sintering process, three specimens were placed into a quartz tube from one end. The samples were placed inside a crucible and submerged into an alumina material crucible. The argon gas pump generated a vacuum atmosphere to prevent oxidation during high temperatures at two bar / min. in quartz tube A. This tube was suspended horizontally into the CARBOLITE tube furnace without touching the wall of the tube furnace. The heating rate of the sintering cycle was set at ten °C / min. Figure 2 explains the sintering procedure and final products.

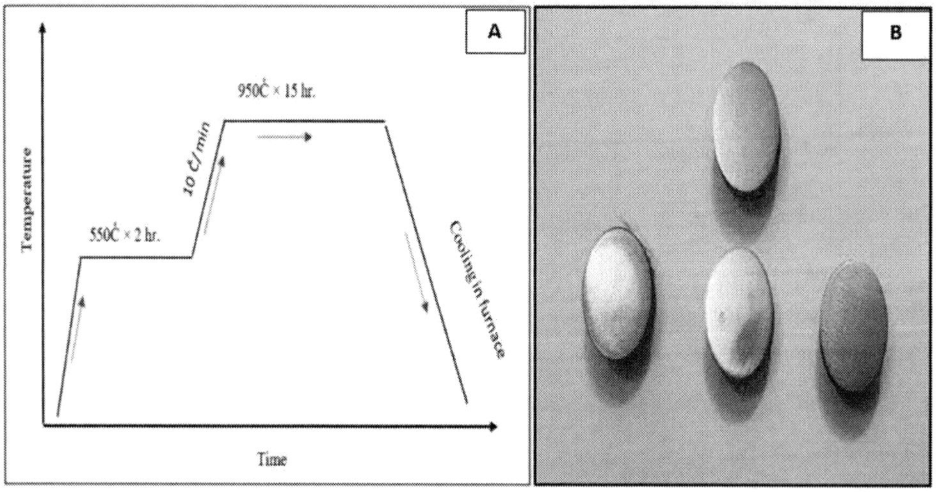

Figure 2: A) sintering heat treatment, B) final production alloys.

2.3. Heat Treatments

Solution heat treatment: the sinter sample is subjected to solution heat treatment, as shown in Figure 3A. The purpose of step quenching is to remove the formation of quench cracks and the pinning of martensitic plates by excess dislocations trapped in quenching from a high temperature. Ageing heat treatment: after the solution heat treatment, the specimens (master alloys with and without additions) were subjected to post-quench isothermal ageing in the austenitic phase. The aging heat treatment cycle is shown in Figure 3B.

IOP Conf. Series: Materials Science and Engineering **987** (2020) 012030 doi:10.1088/1757-899X/987/1/012030

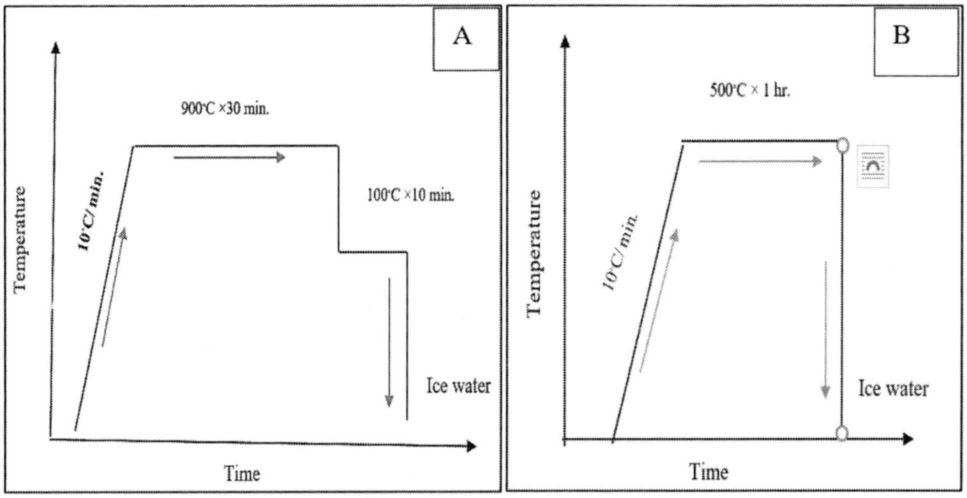

Figure 3: A) solution heat treatment, B) aging heat treatment.

2.4. X-ray diffraction (XRD)

The phases and crystal structures of the alloys were determined using the X-ray diffraction device type shlmadzu XRD-6000 X-Ray diffractometer. The X-Ray diffraction used radiation CuKα with a wavelength of λ = 1.5409 °A 40 kV and 30 mA at room temperature. The diffraction angles (2θ) ranged from 20°–80°.

2.5. Corrosion Test

2.5.1 Open Circuit Potential

Electrochemical investigations were carried out using a three electrode glass cell in 3.5 % NaCl solutions (T = 25 ± one °C). The reference electrode was a saturated calomel electrode, and the counter electrode was a platinum (Pt) electrode.

2.5.2. Electrochemical Tests

Potential-dynamic polarisation was done using a potentiostatic device (winking M lab 200, Germany). The electrochemical cell included specimen alloys that acted as working electrodes. Pt was used as the counter electrode, and the Ag / AgCl electrode was used as a reference electrode. All experiments were conducted at room temperature. The distance between the working and reference electrodes was kept between three to five mm to minimise the ohmic drop across the electrodes. Electrochemical tests were conducted prior to the electrochemical experiments, and each metal sample was grinded and polished to remove the surface oxidation film. The alloys were then degreased ultrasonically for three minutes with ethanol, washed with deionised water and dried. Potentiodynamic polarisation was conducted with a scan rate of + 1.0 mV / s and a start and finish potential of ± 0.25 V vs the open circuit potential (E_{oc}). The corrosion rate (CR) was measured according to Eq. 1 [17–18].

CR =22.85 I_{corr} .... (1)

Where $i$corr is the corrosion's current density.

2.6. Optical Microscope Analysis

The ageing specimens of 12 mm in diameter and four mm in height were ground using SiC paper grits as 180, 400, 800, 1000, 1200 and 2000 then polished using a diamond solution (1 μ). The specimens were etched by ten g $FeCl_3$ + 25 mL HCL + 100 mL distilled water at room temperature. A microscope (1280 XEQMM300TUSB) was used [19].

2.7. Micro-Hardness Measurements

The test was conducted using micro Vickers hardness devices (Digitals Micro Vickers Hardness Tester TH 717) using a load of 500 g for 15 sec with a square-based diamond pyramid.

3. Results and Discussion

3.1. X-ray Diffraction Results

The XRD patterns of CAM SMAs with three different additions of Ta, Zr and In are presented in Figure 4 (A and B) after sintering and aging. It was observed that the formation of phases indicated the formation of both the martensitic and austenite phases. The change in the intensity (I) of the diffraction of the martensite phases peaked based on the limit of the austenite to martensite transformation. Although there was no change observed in the type of martensitic phase with additions, the thickness of the martensite plate decreased, as shown in the OM of the alloys. The results showed that the microstructures of the alloys consist of β and γ` martensites. In these graphs, it is possible to verify the presence of three % Zr, the Cu_5 and Zr phases and Ta_2Al_3 when the addition of Ta was determined [20–21] and [27].

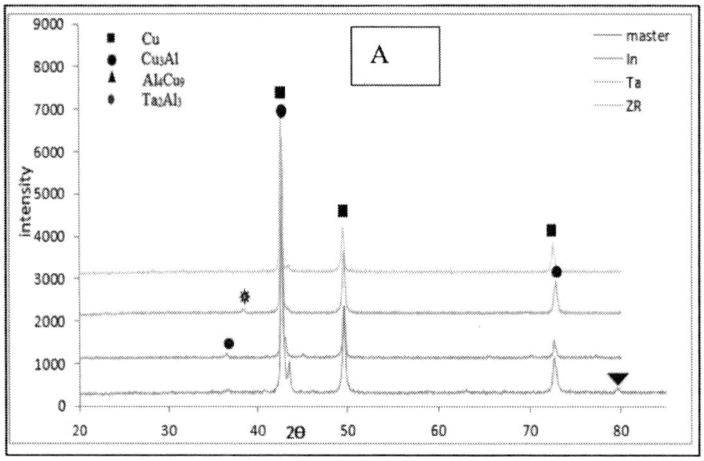

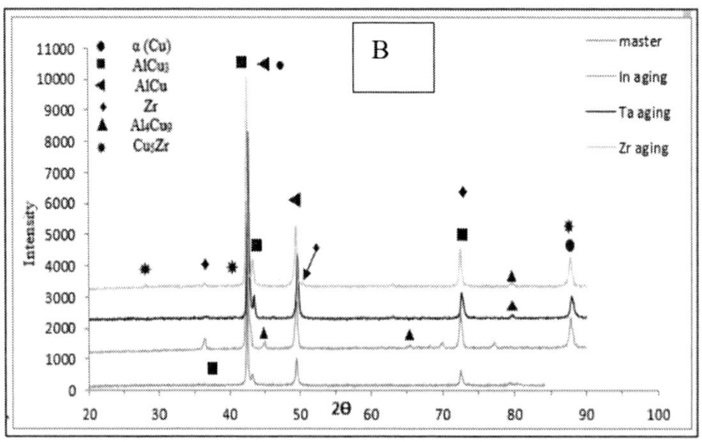

Figure 4 (A and B): XRD after sintering and quenching.

IOP Conf. Series: Materials Science and Engineering **987** (2020) 012030 doi:10.1088/1757-899X/987/1/012030

3.2 Open Circuit Potential

For any closed electrical circuit, anodic and cathodic reactions take place. The sample was placed in a corrosive solution, then a reaction occurred between the surface of the sample and the corrosive medium. This resulted in a potential called the 'open circuit potential'. The forming and breaking of the protective oxide layer were observed due to the potential curve rise and fall that was acquired during the test period. The stability in the potential value is reached (the potential value is fixed) when the anode reaction is equal to the cathodic reaction or the oxidation reaction is equal to the reduction reaction, and therefore, the free energy is equal to zero. The open circuit potential of the CuAlMn (master alloys) with and without additions three % wt. (Ta, Zr and In) is represented in Figure 5. There is a difference between the potential corrosion readings of the CuAlMn alloy and the CuAlMnZr, CuAlMnTa, CuAlMnIn alloys. This means that the alloy shows more nobility when these additions are added. From Figure 5, the CuAlMnIn alloy shows the lowest potential corrosion value. The result of the potential towards more negativity with time means that insufficient oxide layer has formed to prevent the corrosion.

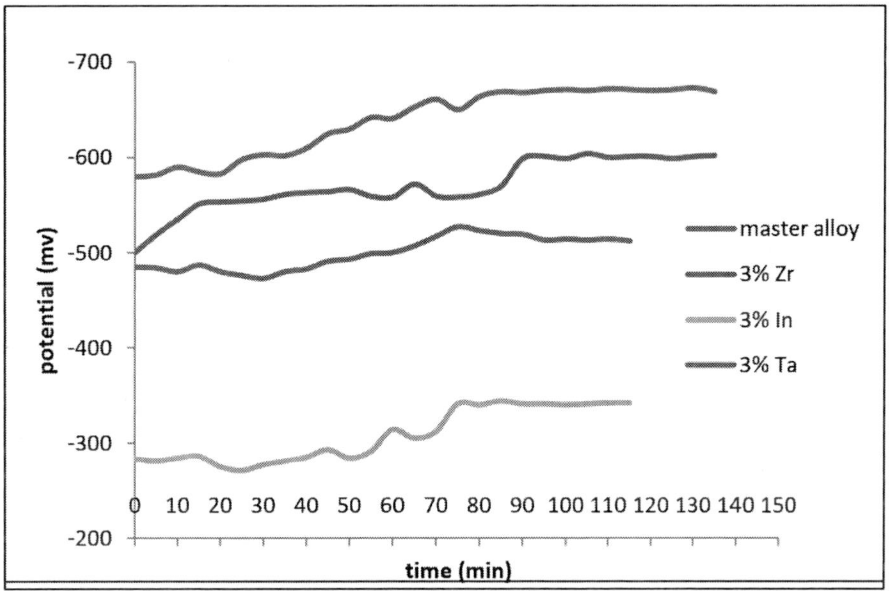

Figure 5: Open circuit potential of samples in 3.5 % NaCl.

3.3. Potential-dynamic polarisation

The potential-dynamic polarisation curve is normally illustrated in a semi logarithm chart comprising cathodic and anodic sections, which represent the effects of the electro-chemical reactions between the sample and corrosion solution. The reaction of the hydrogen evolution is called a cathodic reaction or the reduction of oxygen, and the oxidation reaction is called an anodic reaction or the corrosion process of the alloy. The polarisation curves of the base CuAlMn and CuAlMn-X (x = Ta, Zr and In) SMAs are explained in Fig 6. The corrosion potential (Ecorre) and corrosion current density ($i$corre) samples are presented in Table 5. The addition of Ta, Zr and In to the base CAM improved the C.R value. The curves clearly indicate that the lowest corrosion rate is attributed to CuAlMn-3Ta. The high concentration of the Ta and In ions in the corrosion product film means that it has low solubility in an aggressive chloride-containing solution [22, 23].

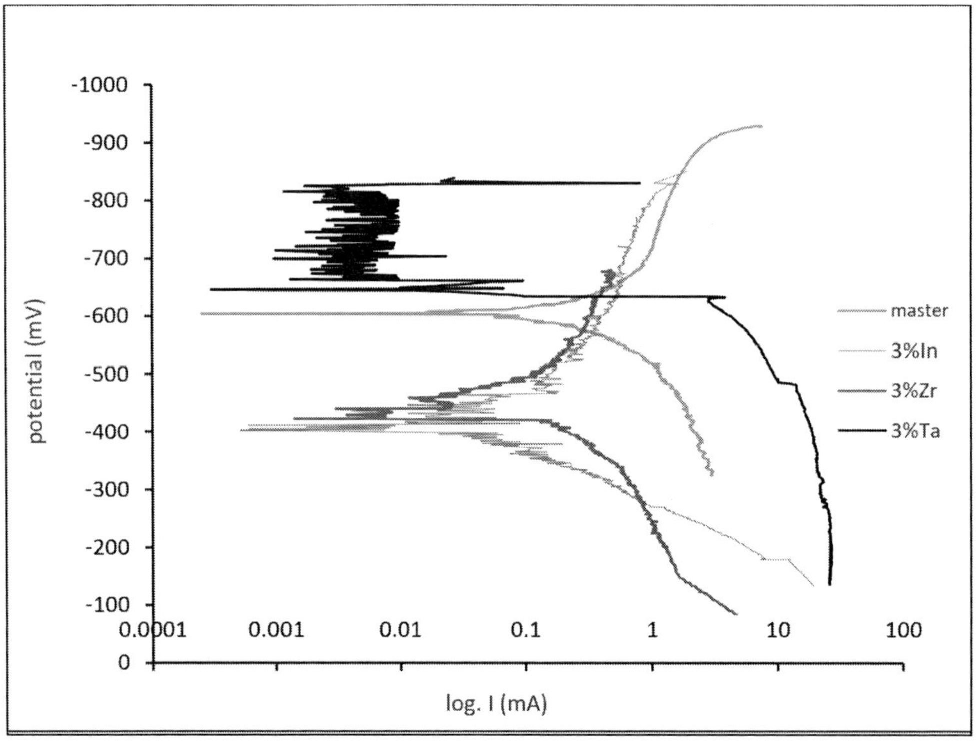

Figure 6: Potential-dynamic polarisation curve for CuAlMn alloy in the aging state in 3.5 % NaCl.

Table 3: Electrochemical parameters of ternary Cu-Al-Mn and quaternary SMAs in the 3.5 wt.% NaCl solution that was obtained from the polarisation test.

Alloys	Current density Icorr (μA / cm2)	Potential *E*corr (mV versus SCE)	Corrosion rate *C .R* (mm / year)	Cathodic slope, βC (mV / decade) versus SCE	Anodic slope, βa (mV / decade) versus SCE
CuAlMn	156.74	-528.5	3.85	-65.9	72.8
CuAlMnTa	13.41	-430.6	0.306	-55.1	72
CuAlMnZr	35.87	-458.2	0.819	-80.5	66.1
CuAlMnIn	3.42	-643.6	0.078	-11.2	18.8

3.4. Optical Microstructures (OM)

The optical microstructures of the CuAlMn samples with quaternary additions is given in Figure 7. At the betatisation at 900 °C, the CuAlMn alloy had a disordered β phase with a body centre cubic structure. After solution annealing at 900 °C and quenching in ice water, the structure consisted of the β_1' martensite and β parent phase at room temperature [27]. The formation of the martensitic is clearly seen in the quenched samples (austenite phase completes the transformation to martensite phase). As seen in Figure 8, all the samples are formed from the complete lath-type martensitic structure with thin martensitic plates or variants. The martensite variants have different orientations in each grain. These martensite variants are known as lath types. A β martensite with thick plates was observed in these alloys and needle-type martensites [24–26]

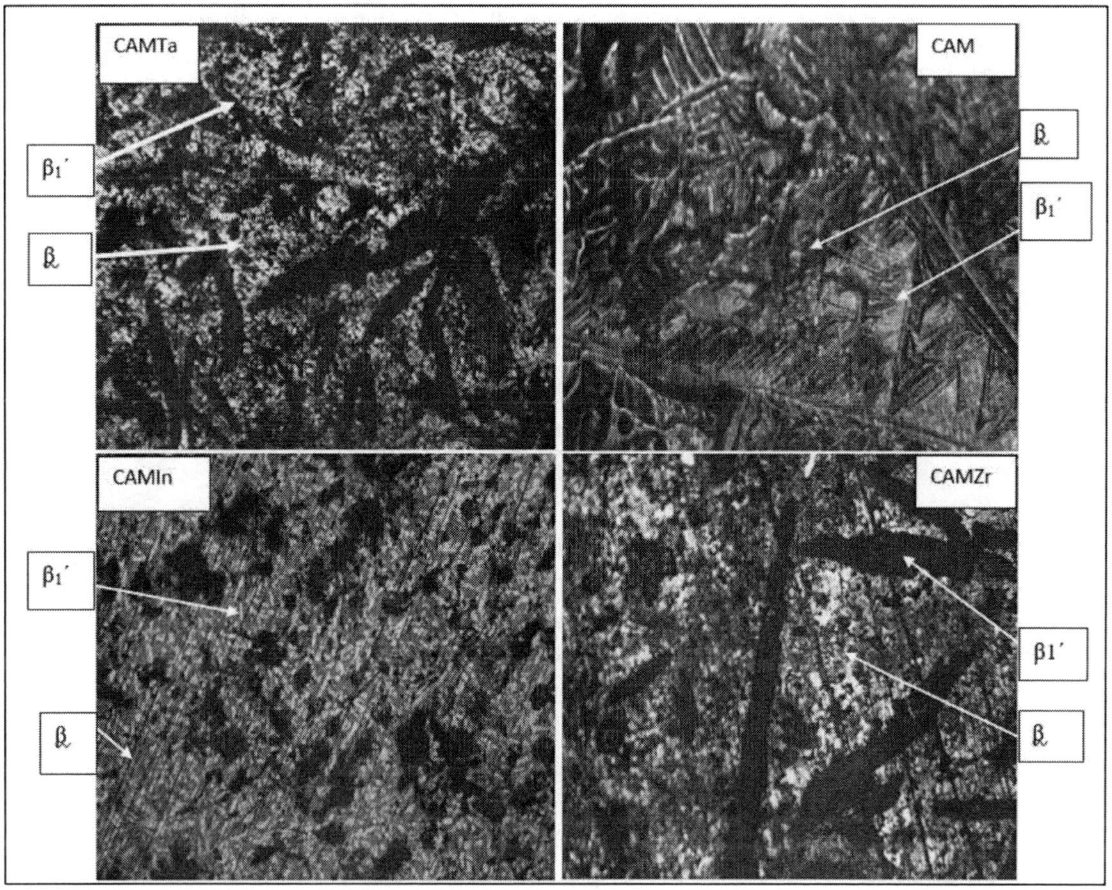

Figure 7: Optical micrographs of CuAlMn SMAs with quaternary alloying addition (100um).

3.5. Micro Hardness Result

Figure 8 shows the micro hardness of Cu8Al9Mn SMA with and without the addition of three wt.% Ta, Zr and In. Each reading represents an average of four readings. This changes into a micro hardness value once Zr, Ta and In are added. It may act as a grain refinement and presence of intermetallic compounds as Cu_5Zr and Ta_2Al_3 are at the grain boundaries of the alloys and dissolution of the elements in the alloy matrix. This result matches with Raied Z. and Safaa N [21, 28].

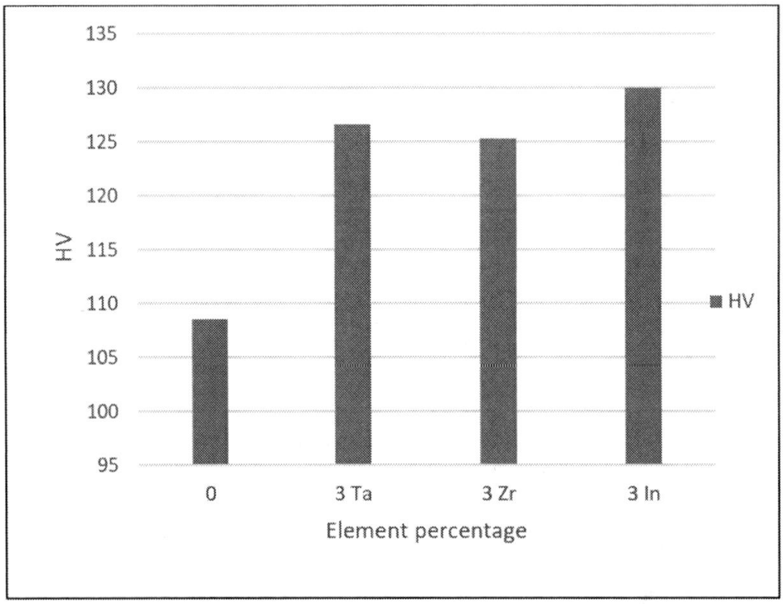

Figure 8: Micro hardness of CuAlMn SMAs.

4. Conclusion

In the corrosion results, we noticed that CuAlMnIn had the lowest corrosion potential, whereas CuAlMnTa had the best corrosion resistance (97.8 %) compared to the master alloy. We also noticed that the appearance of the passive and trance-passive region indicated that the metal gained a protective oxide layer coating. The microstructure results exposed that the addition of a fourth element three wt.% (Ta, Zr and In) can reduce the grain's size of ternary Cu eight % Al nine % Mn SMAs, whereas the In was the best grain refiner. The XRD analysis showed that the types of martensite that were formed were 18R (β'Cu3Al) and a small amount of γ' martensite with a thick plate shape. It was observed that the hardness of the master alloy (Cu eight Al nine Mn) shape memory alloy was improved by adding Ta, Zr and In.

References

[1] Safaa N. Sauda, E. Hamzaha, T. Abubakar and RahelehS 2013 A Review on Influence of Alloying Elements on the Microstructure and Mechanical Properties of Cu-Al-Ni Shape Memory Alloys *Jurnal Teknologi (Sciences & Engineering)* **64:1** 51–56.

[2] Mirko G., Stjepan K., Ivana I., Magdalena S., Tamara H., Borut K., Diana Ć. and Omer B. 2017 Microstructural characterization of Cu82.3Al8.3Mn9.4 shape memory alloy after rolling *Metallurgical and Materials Engineering.*

[3] Canan C., Ayşe T. and İskender Özkul, 2017 Fabrication of CuAlNi shape memory alloy thine film by evaporation Turkish *Journal of Engineering (TUJE)* **1** 2 70-75.

[4] Zhenghua Denga, Haiqing Yina,b,, Cong Zhanga,b, Guofei Zhanga, Tong Zhangc, ZiKui Liud, Haibao Wange and Xuanhui Qua 2019 Sintering mechanism of Cu-9Al alloy prepared from elemental powders *Natural Science: Materials International* 29 425–431.

[5] S. Montecinos A., R. Romero and M. Stipcich 2015 Grain size evolution in Cu-based shape memory alloys J *Mater Sci* 50 3994–4002.

[6] Uwe A., Hennadiy Z., Bernd. and Werner R. 2018 Impact of Alloy Composition and Thermal Stabilization on Martensitic Phase Transformation Structures in CuAlMn Shape Memory Alloys *Materials Research* 21 **2**.

[7] Emine A. and Ilhan A. 2014 Effects of Heat Treatment and Deformation on 2H and 18R Martensites in Cu–9.97%Al–4.62%Mn Alloy, *Arab J Sci Eng* **39**:575–580.

[8] Rupa D., Ashish J., Shahadat H., Abhishek P and V. Sampath 2018 Effect of Alloying Additions on the Properties Affecting Shape Memory Properties of Cu–12.5Al–5Mn Alloy *Research and Technology* 377-390.

[9] Guofei Z., Haiqing Y., Cong Z., Zhenghua D., Ruijie Z., Xue J. and Xuanhui Q. 2020 Effect of Mn on microstructure and properties of Cu-12Al powder metallurgy alloy *Mater. Res. Express* 7 016546.

[10] Sampath, V., and U. S. Mallik 2009 Influence of minor additions of boron and zirconium on shape memory properties and grain refinement of a Cu-Al-Mn shape memory alloy European *Symposium on Martensitic Transformation. EDP Science* 7.

[11] Canan Aksu, Canbay and Tercan Polat 2017 Thermal and structural alternations in CuAlMnNi shape memory alloy by the effect of different pressure applications Physica B **521** 331–338

[12] T. Omori, N. Koeda, Y. Sutou, R. Kainuma, K. Ishida 2007 Superplasticity of Cu-Al-MnNi shape memory alloy Mater. Trans. **11** 2914–2918.

[13] Y. Sutou, N. Koeda, T. Omori, R. Kainuma and K. Ishida 2009 Effects of ageing on bainitic and thermally induced martensitic transformations in ductile Cu-Al-Mn based shape memory alloys Acta Mater. **57** 5748–5758.

[14] Kainuma, S. Takahashi and K. Ishida 1996 Thermoelastic martensite and shape memory effect in ductile Cu-Al-Mn alloys *Metal Mater. Trans.* **27** 2187–2195.

[15] C. Aksu Canbay and Z. Karagoz 2013 The effect of quaternary element on the thermodynamic parameters and structure of CuAlMn shape memory alloys *Physica B* **521** 331-338

[16] Ghulam Gohar, Tareq M. and Asad Shah 2017 Investigation of thermal and mechanical properties of Cu-Al alloys with silver addition prepared by powder metallurgy *Journal of Alloys and Compounds* 11.176

[17] Zhiming S., Ming L. and Andrej A. 2010 Measurement of the corrosion rate of magnesium alloys using Tafel extrapolation *Corrosion Science* **52** 579–588.

[18] Shuiyuan Y., Fan Z., Jialin Wu, Yong Lu, Zhan Shi, Cuiping W., and Xingjun L. 2017 Superelasticity and shape memory effect in Cu–Al–Mn–V shape memory alloys Materials *and Design* **115** 17–25.

[19] C. Pilz , E. Matsumura , A. Paganotti , Cornejo and R. Silva 2020 Microstructure and phase stability of CuAlMnAgZr multicomponent alloys *Materials Chemistry and Physics* **241** 122343.

[20] Raied Z. Alfay, Ayad M. Takhakh and Abdul Raheem K. 2013 Effect of Ta addition on hardness and wear resist of Cu-Al-Ni shape memory alloy fabricated by powder metallurgy *IEEE Business Engineering and Industrial Applications Colloquium (BEIAC)*.

[21] H.-J. Lee, E. Akiyama, H. Habazaki, A. Kawashima, K. Asami and K. Hashimoto 1997 The roles of tantalum and phosphorus in the corrosion behavior of Ni-Ta-P alloys in 12MHCl *Corrosion Science* **39** 2 321–332.

[22] J. Bhattarai, E. Akiyama, H. Habazaki, A. Kawashima, K. Asami, and K.Hashimoto 1998 The passivation behavior of sputter deposited W-Ta alloys in 12 M HCl *Corrosion Science* **40** 4-5 757–779.

[23] Y. Jiao, Y. Wena, N. Lia, J. Hea and J. Tenga 2010 Effect of solution treatment on damping capacity and shape memory effect of a CuAlMn alloy *Journal of Alloys and Compounds* **491** 627–630.

[24] U. Mallik and V. Sampath 2009 Influence of quaternary alloying additions on transformation temperatures and shape memory properties of Cu–Al–Mn shape memory alloy *Journal of Alloys and Compounds* **469** 156–163.

[25] Safaa N., Esah H., Tuty Abu bakar, Mustafa K., and Abdollah B. 2015 Effect of a fourth alloying element on the microstructure and mechanical properties of Cu–Al–Ni shape memory alloys *Materials Research Society 30* 14.

[26] C. Aksu and A. Keskin 2014 Effects of vanadium and cadmium on transformation temperatures of Cu–Al–Mn shape memory alloy *Journal of Thermal Analysis and Calorimetry (JTAC)* **118**, pages1407–1412

[27] S. Kožuh, M. Gojić, I. Ivanić, T. Holjevac Grgurić, B. Kosec, and I. Anžel 2018 The Effect of Heat Treatment on the Microstructure and Mechanical Properties of Cu-Al-Mn Shape Memory Alloy *KUI* **67** 1-2. https://hrcak.srce.hr/file/284196

[28] Safaa N. Saud,1 E. Hamzah, H. R. Bakhsheshi-Rad, and T. Abubakar 2017 Effect of Ta Additions on the Microstructure Damping and Shape Memory behaviors of Pre alloyed Cu-Al-Ni Shape Memory Alloys *Hindawi* **2017** 13.

Assessment of some mechanical properties of PMMA/silica/zirconia nanocomposite as a denture base material

Abdalbseet A Fatalla[1], Mustafa S. Tukmachi[2], Ghasak H Jani[2*]

[1] Professor, Department of Prosthodontics, College of Dentistry, University of Baghdad, Baghdad, Iraq

[2] Lecturer, Department of Prosthodontics, College of Dentistry, University of Baghdad, Baghdad, Iraq

*Corresponding author e-mail: dr.ghasak@yahoo.com

Abstract. In this research, nanocomposites of poly(methyl methacrylate) (PMMA) and a mixture of nano silica (SiO_2) and nano zirconia (ZrO_2) were prepared in different weight percentages of the nano fillers to improve some of the properties of PMMA resin to be used as a denture base material. The nano filles were surface modified with a coupling agent and added to the PMMA in different amounts. Impact strength, transverse strength, hardness and roughness were tested for both control and experimental groups. The results indicate that PMMA/silica/zirconia nanocomposites, prepared with 5% by weight of both types of fillers, had a slight increase in impact strength, transverse strength and hardness with only an insignificant increase in roughness. This kind of composite material may have the potential for application in dental practice.

Keywords: PMMA acrylic resin, nano silica (SiO2), nano zirconia (ZrO2), mechanical properties, nanomaterials.

1. Introduction

Poly(methyl methacrylate) (PMMA) is an excellent organic biocompatible polymeric material used for the construction of denture bases [1]. It was first used for denture production in 1937 and since has been the material of choice for that purpose because of its good physical and mechanical properties [2].

PMMA has various advantages including colour likenesses of the natural gum, good chemical retention with artificial teeth and ease of manipulation and repair [3]. However, PMMA possesses insufficient mechanical strength and surface hardness when used alone. It can also easily be broken during an accident with high impact force or when a patient applies high mastication forces to the denture base [4,5].

In recent years, nanotechnology has been utilised to create materials with improved mechanical and physical properties. Advanced and custom-tailored materials can be produced by integrating nanoparticles into base materials to form composite materials with unique physical and mechanical properties that cannot be attained with the base material alone. This advanced technology is used to fulfil the requirement of the intended application of base materials [6].

Several studies have investigated the properties of composites consisting of different combinations of reinforcing materials such as particles and fibres [7-10], two different types of particles [5], particles and whiskers [1] and two or three types of fibres [11]. The improvement is, in general, better with multiple rather than a single reinforcement type. The idea is that each one of the added reinforcements improves a different material property [12].

Many researchers have focused on studying organic–inorganic hybrid nanocomposite materials and comparing them with macro or microparticle filled composite material; they found that nano filled materials have better enhancements in properties than macro and micro filled materials [13,14]. Nanocomposites have properties of both inorganic (nano filler) as increased rigidity, hardness and durability and organic polymer, which has better processability and flexibility [15,16].

The simplest and most convenient way of fabricating homogeneous hybrid materials is to increase the affinity between organic polymer and inorganic filler phases. Organic–inorganic hybrid materials are generally prepared by incorporating covalent bonds [17,18] or by physical interactions to improve the compatibility between organic and inorganic phases [19,20].

A coupling agent of 3-(methacryloxy)propyl-trimethoxysilane (MPTS) was used to enhance chemical bonding of the added nano filler to the polymer base to form a composite material that has the ability to transmit external forces from the weak polymer base to the much stronger nanoparticles. The nanoparticles should be distributed uniformly within the polymer matrix to prevent the material from weakening, which would reduce the mechanical properties by forming stress concentration areas [21-23].

Silica nano powder is the most used nanomaterial in this field due to its promising and unique properties [24]. Silica nanoparticles can be used for reinforcing denture base polymer due to its rigidity and antimicrobial properties. It was found that PMMA resin incorporated with TiO_2/SiO_2 nanoparticles showed strong antimicrobial activity against cariogenic bacteria [25].

Zirconium oxide (ZrO_2) is a metal oxide that has many benefits such as high mechanical strength, toughness, stiffness, wear resistance, chemical tolerance and good thermal stability, rendering it useful for reinforcing dental materials, such as denture bases [26]. Studies have shown that superior physical and mechanical properties can be obtained when zirconia nanoparticles are incorporated into PMMA polymer [27,28]. Zirconia nano powder is a biocompatible material that has excellent toughness, resistance to physical corrosion, good mechanical strength and abrasion resistance [29].

It was found that the addition of ceramic powder containing a mixture of nano SiO_2/ZrO_2 to PMMA improved optical properties and thermal conductivity. PMMA/SiO_2/ZrO_2 nanocomposites have surprisingly high thermal stability while retaining excellent transparency [30].

There is no report studying the mechanical properties of PMMA/SiO_2/ZrO_2 nanocomposites as a denture base material. The hypothesis of this study is that the mechanical properties (i.e. impact strength, transverse strength, roughness and hardness) of PMMA would be improved after the addition of nano silica/zirconia mixture.

2. Materials and methods

Heat curing poly methyl methacrylate powder (SUPERACRYLP®P PLUS, Czech Republic) and methyl methacrylate liquid monomer (SUPERACRYLP®P PLUS, Czech Republic) were used in this study as denture base material. Silicon dioxide (SiO_2) nanoparticles (20 nm) (Sigma-Aldrich, Germany) and zirconium oxide (ZrO_2) nanoparticles (<100 nm) (Sigma-Aldrich Germany) were used as inorganic fillers for preparing the nanocomposite. Toluene (GCC, UK) was used as a solvent, while 3-(methacryloxy)propyl-trimethoxysilane (Sigma-Aldrich, Germany) was used as a coupling agent to improve the bonding between the organic base and the inorganic filler.

A pilot study was conducted to choose the optimum nano SiO_2/ZrO_2 weight percentages. Specimens were divided into ten groups based on the amount of nano SiO_2 and nano ZrO_2 concentrations, as shown in Table 1.

Table 1. Pilot study specimen grouping.

Group	Percentage of nano SiO_2 by weight	Percentage of nano ZrO_2 by weight	Weight of PMMA powder	Weight of PMMA liquid
Control	0%	0%	44 g	20 ml
A	3%	3%	43.12 g	20 ml
B	5%	3%	42.9 g	20 ml
C	7%	3%	42.68 g	20 ml
D	3%	5%	42.9 g	20 ml
E	5%	5%	42.68 g	20 ml
F	7%	5%	42.46 g	20 ml
G	3%	7%	42.68 g	20 ml
H	5%	7%	42.46 g	20 ml
J	7%	7%	42.24 g	20 ml

Specimens of each pilot study group were tested for transverse strength and hardness to select an optimum group for the main study. According to the results of the pilot study, group E and the control group were selected to be tested for impact strength, transverse strength, roughness and hardness.

2.1. Salinisation of nano fillers

An amount (30 g) of each nano filler was placed in a separate container containing 200 ml of toluene each. Each container was then sonicated using an ultrasonic probe (Soniprep 150, MSE [UK] Ltd, United Kingdom) at ambient temperature for 20 minutes. After that, the mixture was placed in a magnetic stirrer (Lab SH-3, JoanLab, United Kingdom) at room temperature, and 1.5 ml of 3-(methacryloxy)propyl-trimethoxysilane was added drop by drop while stirring the mixture. The mixture was then covered and left for 48 hours. After two days, the toluene was removed by a rotary evaporator (RE-801-AW2, Yamato, Japan) at a speed of 150 rpm and a temperature of 60 °C for half an hour. The nano filler was dried by a vacuum oven (OVA03100, Gallenkamp, United Kingdom) at 60 °C for 20 hours [31].

2.2. Specimens' preparation

Eighty specimens were prepared in the main study (40 for control group and 40 for group E). Ten specimens were tested for each property (impact strength, transverse strength, roughness and hardness). Two metal bar patterns were fabricated to be used for the preparation of stone moulds. The metal pattern for the impact test had dimensions of 80 mm x 10 mm x 4 mm (length x width x thickness, respectively). While the metal pattern used for transverse strength, roughness and hardness tests had dimensions of 65 mm x 10 mm x 2.5 mm (length x width x thickness, respectively). For the control group, specimens were prepared by weighing and mixing PMMA powder and liquid according to the guidelines of the manufacturer. The moulding, packing and polymerising of PMMA were done using conventional flasking technique for dentures. For group E, specimens were prepared by first mixing 5% by weight silane-coated nano silica and 5% by weight silane-coated nano zirconia with PMMA liquid monomer, and the mixture was sonicated by an ultrasonic probe at power of 20 W and frequency of 60 kHz for three minutes to disperse the fillers into the liquid monomer [12]. The suspension of the monomer with the nano filler was instantly added to the PMMA powder and mixed for 30 seconds. Packing and polymerising was done as it was done for the specimens of the control group. Specimens of both groups were then finished, polished and kept in distilled water at 37 °C for two days.

2.3. Impact strength test

Impact strength test was performed at room temperature in accordance with ISO 179-1, 2000 and ANSI/ADA specification no.12, 2002 using Charpy type impact test instrument (Impact Tester N. 43-1, TMI, USA). The unnotched specimen was horizontally supported and struck by a free-swinging pendulum with energy of 2 joules.

Impact strength (I) was calculated in KJ/m2 using the following equation:

$I = E/(bd) \times (10)^3$

Where:

E: the absorbed energy in joules.

b: the specimen width in millimetres.

d: the specimen thickness in millimetres.

2.4. Transverse strength test

Transverse strength test was performed in accordance with ISO 1567, 1999 and ANSI/ADA specification no.12, 2002 using a universal testing machine (WDW-20, Laryee Technology Co. Ltd., China). Specimen was mounted on two parallel supports 50 mm apart for three-point bending test. The specimen was then bent using a rod located at the centre of the two support with a load of 50 kg and a speed of 1 mm/min until fracturing.

Transverse strength (σ) was calculated in MPa using the following equation:

$\sigma = 3Fl/(2bh^2)$

Where:

F: the maximum force applied in Newton.

l: the distance between the supports in millimetres.

b: the specimen width in millimetres.

h: the specimen height in millimetres.

2.5. Surface hardness test

Surface hardness was measured according to ISO 1567, 1999 and ANSI/ADA specification no.12, 2002 using a digital shore-D hardness durometer (HT-6510D, China). The device has a 0.8 mm diameter pointed indenter which tapers to a 1.6 mm cylinder. The indenter is connected to a digital scale. Each specimen was subjected to five indentations at different areas 10 mm apart, and the average of the five readings was calculated.

2.6. Surface roughness test

Surface roughness was measured according with ANSI/ADA specification no.12, 2002 by a profilometer (TR 220, Beijing Time High Technology Ltd., China). The profilometer has a sharp diamond-made surface analyser (stylus). The maximum distance that the stylus can travel is 11 mm. Two measurements were taken on different areas of each specimen, and the average of the two measurements was calculated.

2.7. Atomic force microscopy (AFM)

Surface topography of test specimen was examined by AFM (Solver Nano, NT-MDT SI, Ireland) to visualise the distribution of the filler in the polymer matrix and their impact on surface roughness. The thickness of the specimen should be 2 mm to be put under the scanning probe.

3. Results

The result of particle size distribution of nanoparticles was obtained by using AFM, which showed that the average diameter was 94.33 nm. The result of particle size distribution of nanoparticles of group E (5% wt. nano silica and 5% wt. nano zirconia) is shown in Figure 1 and Table 2.

IOP Conf. Series: Materials Science and Engineering **987** (2020) 012031 doi:10.1088/1757-899X/987/1/012031

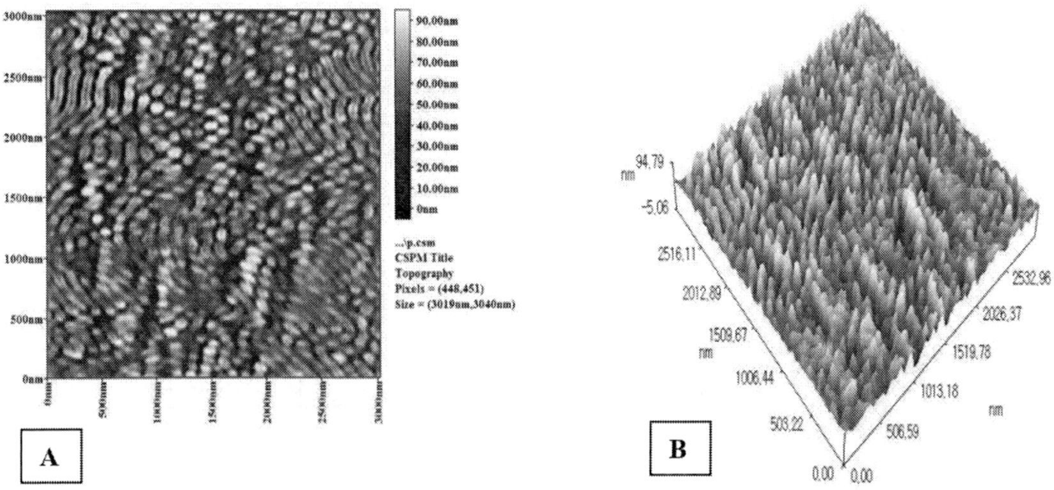

Figure 1. AFM image of group E specimen (PMMA/SiO_2/ZrO_2 nanocomposites). A: two-dimensional; B: three-dimensional.

Table 2. Granularity accumulation distribution of PMMA/SiO_2/ZrO_2 nanocomposites.

Sample: p						**Code:** Sample Code		
Line No.: lineno						**Grain No.:** 387		
Instrument: CSPM						**Date:** 2020-08-16		
Avg. Diameter: 94.33 nm						<=10% Diameter: 75.00 nm		
<=50% Diameter: 90.00 nm						<=90% Diameter: 115.00 nm		
Diameter (nm)	Volume (%)	Cumulation (%)	Diameter (nm)	Volume (%)	Cumulation (%)	Diameter (nm)	Volume (%)	Cumulation (%)
75.00	4.65	4.65	105.00	9.04	75.71	135.00	1.29	98.71
80.00	15.50	20.16	110.00	8.27	83.98	140.00	0.26	98.97
85.00	13.18	33.33	115.00	5.68	89.66	145.00	0.52	99.48
90.00	13.44	46.77	120.00	4.13	93.80	150.00	0.52	100.00
95.00	11.63	58.40	125.00	2.58	96.38			
100.00	8.27	66.67	130.00	1.03	97.42			

Descriptive data (mean values, standard deviation, standard error) of the results of impact strengths, transverse strength, hardness and roughness test for both the control group and group E are presented in Table 3.

Table 3. Descriptive data of the study.

Dependent variables	Groups							
	Control group				Group E (5% wt. nano silica and 5% wt. nano zirconia)			
	No.	mean	SD	SE	No.	mean	SD	SE
Impact strength	10	3.7887	.65270	.20640	10	3.8087	1.10942	.35083
Transverse strength	10	17.1694	2.55035	.80649	10	17.6800	1.47629	.46684
Hardness	10	101.4200	3.35553	1.06111	10	102.2000	1.03280	.32660
Roughness	10	3.2428	.99449	.31448	10	3.3730	97218	.30743

T-test was done to compare the differences between the mean values of control group and group E. P-value of $p < 0.05$ was considered significant as shown in Table 4.

Table 4. Multiple comparisons for study groups.

	Levene's Test for Equality of Variances		t-test for Equality of Means				
	F	Sig.	t	df	Sig. (2-tailed)	Mean Difference	Std. Error Difference
Impact strength	4.988	.038	.049	18	.961	.02000	.40704
Transverse strength	.974	.337	.548	18	.590	.51064	.93186
Hardness	9.316	.007	.703	18	.491	.78000	1.11024
Roughness	.042	.839	.296	18	.770	.13025	.43979

4. Discussion

This study investigated the effect of the addition of nano silica/zirconia mixture on impact strength, transverse strength, hardness and roughness of PMMA denture base material. Surface modification of the nanoparticles was done to enhance chemical bonding between the polymer matrix and the filler phase of the material. The AFM topographic image of PMMA/SiO_2/ZrO_2 nanocomposites shows that nanoparticles were dispersed throughout the polymer matrix in a standardised way. This can be attributed to the use of an ultrasonic probe for the dispersion of the nanoparticles with the liquid monomer. Closer inspection of the image shows that the particle size of the nanocomposite is about 70–140 nm.

4.1. Impact strength test

The impact strength of the PMMA denture base is an important property to investigate as dentures can be subjected to a sudden impact force, such as an external blow to the face or dropping the denture on the floor. This property measures the amount of energy absorbed by the specimen before fracture when subjected to a sudden force [32]. The results of this study show that the impact strength values of the experimental group were higher than the control group; this could be due to high interfacial strength between the nano filler and the matrix as a result of the creation of cross-links or supra-molecular bonds covering or protecting the nano fillers, which in return prevent the spread of cracks. The proliferation of cracks may also be prevented by a strong bonding between the nano fillers and the resin matrix arising from the bonding of the functional groups on the surface of nanoparticles (created by the salinisation) and the polymer chains. Additional improvement of the mechanical properties could be due to nanoparticles filling the spaces between the polymer chains and thus filling the interstitial polymer regions to produce a heterogeneous mixture and to prevent the displacement of the polymer chain segments. In addition, nanoparticles characterised by a large specific surface area thus have the ability to dissipate energy that causes crack propagation and can hence improve impact strength [33]. This result may agree with a previous study in which a mixture of two nano fillers (ZrO_2-Al_2O_3) was added to PMMA, and the impact strength was improved [34].

4.2. Transverse strength test

Transverse strength is one of the mechanical properties that is particularly useful when testing denture base materials. Dentures are always subjected to three-point bending stresses during chewing and mastication, and the transverse strength test is the closest match to the forces inside the mouth. Transverse strength is composed of compression, tension and shear strength [35].

The transverse strength values of this study show that reinforced PMMA composite had higher transverse strength compared to the control group. This may be due to the good distribution of the very fine and various-sized nano fillers within the polymer that enabled them to be embedded between the linear macromolecular chains of the polymer and fill spaces between the chains, thus limiting the segmental motions of the macromolecular chains and improving the transverse strength and rigidity of the resin.

The increase in transverse strength can also be clarified by transformation toughening; as adequate stress arises and cracking starts to propagate, a transformation phase of nano silica and zirconia depletes the energy of crack propagation. The expansion of nano SiO_2 and ZrO_2 crystals also takes place in this phase, which puts the crack under compressive stress, and crack propagation is arrested. A similar finding was reported when zirconia nano filler was added to PMMA, and a significant increase in transverse strength value was obtained [28].

Increased transverse strength may also be attributed to the transition of stress from more flexible polymer to stiffer and more stable filler particles. This result may agree with a study in which nano zirconia was added to heat curing acrylic denture base material, and transverse strength was increased with 5% by wt. nano filler loading and decreased with 7% by wt. nano filler loading [36].

4.3. Surface hardness
Hardness is described as the material's resistance to plastic deformation and is usually calculated under an indentation charge. The current study shows that the increase in hardness of reinforced PMMA was relevant to the amount of filler loading in the polymer matrix and to the dispersion homogeneity of hard and rigid nano zirconia and nano silica within the material. The increase in hardness of the nanocomposite could also be explained by the increase in cross-linking density, which made the polymer more rigid and more resistant to penetration. The slight increase in hardness shown in this study may agree with the result obtained by another study where SiO_2 nanoparticles was added to PMMA in a weight percentage of 5% [37].

4.4. Surface roughness
Surface roughness provokes the adhesion and retention of Candida albicans, which is especially important in denture-induced stomatitis pathogenesis [37]. Therefore, a material should possess a smooth and polished surface so that plaque accumulation is minimised or avoided. Compared with the control group, the surface roughness of modified PMMA was insignificantly increased. This may be attributed to the difference in the particle size of the nano fillers and the acrylic denture base material as well as to the relatively high filler loading amount, which could cause the formation of some filler aggregates on the surface.

5. Conclusion
Within the limits of this study, it can be concluded that the incorporation of a mixture of nano silica and nano zirconia into PMMA denture base material can potentially increase some mechanical properties such as impact strength, transverse strength and hardness with an insignificant effect on surface roughness. To obtain such results, 5% by weight nano silica and 5% by weight nano zirconia was the optimum amount of nano fillers.

References
[1] Zhang XY, Zhang XJ, Huang ZL, Zhu BS, Chen RR. 2014. Hybrid effects of zirconia nanoparticles with aluminum borate whiskers on mechanical properties of denture base resin PMMA. *Dental materials journal*.**33**(1):141-6.

[2] Yu W, Wang X, Tang Q, Guo M, Zhao J. 2014. Reinforcement of denture base PMMA with ZrO2 nanotubes. J Mech Behav Biomed Mater.32:192-7.

[3] Anusavice K, Shen C, Rawls HR. 2012. *Phillips' science of dental materials*. 12th ed. Philadelphia: Saunders.

[4] Çökeliler D, Erkut S, Zemek J, Biederman H, Mutlu M. 2007. Modification of glass fibers to improve reinforcement: A plasma polymerization technique. *Dental Materials*.**23**(3):335-42.

[5] Alhareb AO, Ahmad ZA. 2010. Effect of Al2O3/ZrO2 reinforcement on the mechanical properties of PMMA denture base. *Journal of Reinforced Plastics and Composites*.**30**(1):86-93.

[6] Elmadani AA, Radović I, Tomić NZ, Petrović M, Stojanović DB, Heinemann RJ, et al. 2019. Hybrid denture acrylic composites with nanozirconia and electrospun polystyrene fibers. *PLoS One*.**14**(12 :e0226528):1-18.

[7] Gad MM, Fouda SM, Al-Harbi FA, Näpänkangas R, Raustia A. 2017. PMMA denture base material enhancement: a review of fiber, filler, and nanofiller addition. *International journal of nanomedicine*.**12**:3801-12.

[8] Ismail I. 2017. Development and Performance of Composite from Modified Nano Filler with Plasma Treated Fiber and Heat Cured Acrylic Denture Base Material on Some of Its Properties – In Vitro Study. *International Journal of Science and Research*.**6**(3):371-7.

[9] Muklif O, Ismail I. 2015. Studying the effect of addition a composite of silanized Nano-Al2O3 and plasma treated polypropylene fibers on some physical and mechanical properties of heat cured PMMA denture base material. *Journal of Baghdad College of Dentistry*.**27**(3):22-7.

[10] Gad MM, Al-Thobity AM, Rahoma A, Abualsaud R, Al-Harbi FA, Akhtar S. 2019. Reinforcement of PMMA Denture Base Material with a Mixture of ZrO(2) Nanoparticles and Glass Fibers. *International journal of dentistry*.**2019**:2489393.

[11] Yu SH, Lee Y, Oh S, Cho HW, Oda Y, Bae JM. 2012. Reinforcing effects of different fibers on denture base resin based on the fiber type, concentration, and combination. *Dental materials journal*.**31**(6):1039-46.

[12] Lazouzi G, Vuksanović MM, Tomić NZ, Mitrić M, Petrović M, Radojević V, et al. 2018. Optimized preparation of alumina based fillers for tuning composite properties. *Ceramics International*.**44**(7):7442-9.

[13] Schubert U, Huesing N, Lorenz A. 1995. Hybrid Inorganic-Organic Materials by Sol-Gel Processing of Organofunctional Metal Alkoxides. *Chemistry of Materials*.**7**(11):2010-27.

[14] Mucic RC, Storhoff JJ, Mirkin CA, Letsinger RL. 1998. DNA-Directed Synthesis of Binary Nanoparticle Network Materials. *Journal of the American Chemical Society*.**120**(48):12674-5.

[15] Podsiadlo P, Kaushik AK, Arruda EM, Waas AM, Shim BS, Xu J, et al. 2007. Ultrastrong and Stiff Layered Polymer Nanocomposites. *Science*.**318**(5847):80.

[16] Fan F, Xia Z, Li Q, Li Z, Chen H. 2013. ZrO2/PMMA Nanocomposites: Preparation and Its Dispersion in Polymer Matrix. *Chinese Journal of Chemical Engineering*.**21**(2):113-20.

[17] Wang Q, Liu N, Wang X, Li J, Zhao X, Wang F. 2003. Conductive Hybrids from Water-Borne Conductive Polyaniline and (3-Glycidoxypropyl)trimethoxysilane. *Macromolecules*.**36**(15):5760-4.

[18] Matsuura Y, Miura S, Naito H, Inoue H, Matsukawa K. 2003. Nanostructured polysilane–titania hybrids and their application to porous titania thin films. *Journal of Organometallic Chemistry*.**685**(1):230-4.

[19] Chujo Y, Matsuki H, Kure S, Saegusa T, Yazawa T. 1994. Control of pore size of porous silica by means of pyrolysis of an organic–inorganic polymer hybrid. *Journal of the Chemical Society, Chemical Communications*(5):635-6.

[20] Ogoshi T, Chujo Y. 2004. Synthesis of Photosensitive Organic−Inorganic Polymer Hybrids by Utilizing Caged Photoactivatable Alkoxysilane. *Macromolecules*.**37**(16):5916-22.

[21] Bayou S, Mouzali M, Aloui F, Lecamp L, Lebaudy P. 2013. Simulation of conversion profiles inside a thick dental material photopolymerized in the presence of nanofillers. *Polymer Journal*.**45**(8):863-70.

[22] Conzatti L, Giunco F, Stagnaro P, Patrucco A, Tonin C, Marano C, et al. 2014. Wool fibres functionalised with a silane-based coupling agent for reinforced polypropylene composites. *Composites Part A: Applied Science and Manufacturing*.**61**:51-9.

[23] Fabre-Francke I, Berrebi M, Lavédrine B, Fichet O. 2019. Hybrid PMMA combined with polycarbonate inside interpenetrating polymer network architecture for development of new anti-scratch glass. *Materials Today Communications*.**21**:100582.

[24] Topouzi M, Kontonasaki E, Bikiaris D, Papadopoulou L, Paraskevopoulos KM, Koidis P. 2017. Reinforcement of a PMMA resin for interim fixed prostheses with silica nanoparticles. *Journal of the mechanical behavior of biomedical materials*.**69**:213-22.

[25] Sodagar A, Khalil S, Kassaee MZ, Shahroudi AS, Pourakbari B, Bahador A. 2016. Antimicrobial properties of poly (methyl methacrylate) acrylic resins incorporated with silicon dioxide and titanium dioxide nanoparticles on cariogenic bacteria. *J Orthod Sci*.**5**(1):7-13.

[26] Maji P, Choudhary RB, Majhi M. 2016. Structural, optical and dielectric properties of ZrO2 reinforced polymeric nanocomposite films of polymethylmethacrylate (PMMA). *Optik*.**127**(11):4848-53.

[27] Reyes-Acosta MA, Torres-Huerta AM, Domínguez-Crespo MA, Flores-Vela AI, Dorantes-Rosales HJ, Ramírez-Meneses E. 2015. Influence of ZrO2 nanoparticles and thermal treatment on the properties of PMMA/ZrO2 hybrid coatings. *Journal of Alloys and Compounds*.**643**:S150-S8.

[28] Gad M, ArRejaie AS, Abdel-Halim MS, Rahoma A. 2016. The Reinforcement Effect of Nano-Zirconia on the Transverse Strength of Repaired Acrylic Denture Base. *International journal of dentistry*.**2016**:7094056.

[29] Ergun G, Sahin Z, Ataol AS. 2018. The effects of adding various ratios of zirconium oxide nanoparticles to poly(methyl methacrylate) on physical and mechanical properties. *Journal of oral science*.**60**(2):304-15.

[30] Wang H, Xu P, Zhong W, Shen L, Du Q. 2005. Transparent poly(methyl methacrylate)/silica/zirconia nanocomposites with excellent thermal stabilities. *Polymer Degradation and Stability*.**87**(2):319-27.

[31] Ayad N, Badawi M, Fatah A. 2008. Effect of Reinforcement of High impact acrylic resin with zirconia on some physical and mechanical properties. *Revista de Clínica e Pesquisa Odontológica*.**4**(3):145-51.

[32] Robinson JG, McCabe JF. 1993. Impact strength of acrylic resin denture base materials with surface defects. *Dental Materials*.**9**(6):355-60.

[33] Tukmachi M, Azeez Z, Mohammed D. 2018. Evaluation of Bond Strength of Acrylic Artificial Teeth with Unreinforced and Nano Silica Reinforced Denture Base Material after Chemical Disinfection. *Journal of Research in Medical and Dental Science*.**6**(5):76-82.

[34] Aljafery A, Hussain B. 2015. Effect of addition ZrO2-Al2O3 nanoparticles mixture on some properties and denture base adaptation of heat cured acrylic resin denture base material. *Journal of Baghdad College of Dentistry*.**27**(3):15-21.

[35] Zidan S, Silikas N, Haider J, Alhotan A, Jahantigh J, Yates J. 2020. Evaluation of Equivalent Flexural Strength for Complete Removable Dentures Made of Zirconia-Impregnated PMMA Nanocomposites. *Materials (Basel)*.**13**(11):2580.

[36] Hameed H, Abdul Rahman H. 2015. The effect of addition nano particle ZrO2 on some properties of autoclave processed heat cure acrylic denture base material. *Journal of Baghdad College of Dentistry* **27**(1):32-9.

[37] Alnamel HA, Mudhaffer M. 2014. The effect of Silicon di oxide Nano -Fillers reinforcement on some properties of heat cure polymethyl methacrylate denture base material. *journal of Baghdad College of Dentistry*.**26**(1):32-6.

[38] Zissis AJ, Polyzois GL, Yannikakis SA, Harrison A. 2000. Roughness of denture materials: a comparative study. *The International journal of prosthodontics*.**13**(2):136-40.

IOP Conf. Series: Materials Science and Engineering **987** (2020) 012032 doi:10.1088/1757-899X/987/1/012032

Oxidation of Biocompatible Graphite–Ti Composite after Laser Ablation in Different Atmospheres

B Ludrovcová[1], P Šugár[1], M Sahul[2], J Kováčik[3] and Z Czibor[1]

[1] Institute of Production Technologies, Faculty of Materials Science and Technology, Slovak University of Technology, J. Bottu 25, 917 24 Trnava, Slovakia

[2] Institute of Materials Science, Faculty of Materials Science and Technology, Slovak University of Technology, J. Bottu 25, 917 24 Trnava, Slovakia

[3] Institute of Materials and Machine Mechanics, Slovak Academy of Sciences, Dúbravská cesta 9, 845 13 Bratislava, Slovakia

E-mail: barbora.ludrovcova@stuba.sk

Abstract. The field of biocompatible material surfaces is a widely researched topic. Surface energy, surface topography and surface chemistry are important properties of biocompatible surfaces. These properties contribute to better osseointegration and adhesion of cells to implant surfaces. This article investigates the chemical and phase composition of the surface of a new titanium composite produced by powder metallurgy. Surface oxidation of the graphite–titanium metal matrix composite (TiMMC) after laser beam micromachining (LBMM) is discussed in this paper. Laser micromachining was performed in an argon shielding atmosphere and air. The aim was to determine the influence of the shielding atmosphere and the input parameters of LBMM on the presence of oxygen on the surface. Laser-treated surfaces were examined with scanning electron microscopy (SEM) and energy-dispersive X-ray spectroscopy (EDS). The phase composition was analysed with X-ray diffraction (XRD). Experiments confirmed that an argon shielding atmosphere reduces surface oxidation. The oxidation was also affected by the energy of the laser beam acting on the material. The maximum amount of oxygen detected on the surface after LBMM in air and argon was 38.6 wt. % and 24.2 wt. %, respectively. The presence of TiO, TiO_2 and Ti_2O_3 oxides were detected on the surface after laser ablation in air. In contrast, Ti_2O_3 and TiO oxides were detected after laser ablation in the argon shielding atmosphere.

1. Introduction

Metallic materials are commonly used in orthopaedics and dentistry as hard tissue replacements [1]. Titanium and its alloys, stainless steels and co-based alloys are routinely applied and have an irreplaceable position in the field of implant production [2, 3]. Titanium-based materials in particular possess good biocompatibility, high wear and corrosion resistance and suitable mechanical properties [4].

The success of implantation depends on the interaction of biological tissue with the implant body. After the implant is inserted into living tissue, the interface between the implant and the bone is formed. Interaction between the organic environment and implanted bone replacement is affected by the surface properties of the implant [5]. The osseointegration process can be improved and a reaction of the living tissue can be modified by the surface properties, including surface energy, wettability, electrical charge, topography, the roughness of the implant surface and the surface's chemical composition [6, 7].

Morphological characteristics, such as distance, distribution, shape and height of the protrusions are taken into account when evaluating the topography of the implant surface and the tissue response to it [8]. In terms of topography, important characteristics include the arithmetic average of the roughness profile Ra, area surface roughness Sa, surface flatness and surface area [9]. An extensive review of research in [10] shows that a moderate Sa between 1 and 2 µm elicits the most suitable bone response to the implant surface.

Implant surface modification technologies include chemical modification, grit blasting, laser beam micromachining, bioactive and antibacterial coatings, or a combination of these methods [11–13]. The advantages of applying laser technology are the efficient control of depth and energy density, repeatability, short work cycles and accuracy [14]. Compared to chemical machining, an advantage is non-pollution of the surface of the modified material [9]. LBMM is considered a non-isothermal process that is exhibited by non-uniform energy distribution causing a large thermal gradient. This effect can result in the presence of structural defects and residual stresses [15].

Titanium reacts with light elements, such as O_2, N_2, C from the surrounding atmosphere at high temperatures. Titanium-based oxides and TiN phases presented on the surface of titanium and its alloys are products of diffusion-driven processes that emerge during laser treatment of the surface. Oxygen and nitrogen from the atmosphere or shielding gas create oxides and nitrides of titanium within the process of so-called laser gas alloying. These oxides and nitrides can improve some surface properties, such as hardness, corrosion and wear resistance and biocompatibility [15, 16]. Many authors investigated the reactions and surface characteristics of biocompatible Ti materials during laser irradiation in different atmospheres with different input laser beam machining parameters.

Different surface modifications were evaluated by Wang et al. [9]. Biocompatible Ti6Al4V alloy was processed using three different techniques, namely, grinding, sandblasting and then etching and laser treating by a nanosecond UV laser. In that case, easy cell migration and good proliferation along the grooves produced by laser beam treatment was documented. Additionally, the surface was characterised by the anisotropic wettability and periodic oxygen distribution on the machined surface.

Moura et al. [14] textured Ti6Al4V with an Nd:YAG laser in atmospheric air. The textured surface was grooved and consisted of α-Ti and oxides Ti_6O and Ti_2O.

Rajab et al. [17] developed a hydrophobic surface of Ti6Al4V alloy using picosecond laser etching. This process can reduce the adhesion of bacteria on the surface of the treated alloy. The surface topography consisted of self-organised structures and rounded peaks with an

IOP Conf. Series: Materials Science and Engineering **987** (2020) 012032 doi:10.1088/1757-899X/987/1/012032

average Sa between 0.29 μm and 1.38 μm. The different types of TiO_2 combined with nitrogen – $(TiO_2)_xN_y$ and TiO_xN_y – were formed on the laser-textured surface due to titanium reactivity at high temperatures.

Zeng et al. [15] treated a commercially pure (cp) Ti surface with a continual-wave mode laser at different scanning speeds. The scanning speed rate was reflected in the overall change in the density of energy introduced into the material. The lower scanning speed caused a longer heating time of the material. The O and N reacted with Ti and diffused deeper into the substrate. The resulting surface consisted of TiO_2 and TiN, which appeared in the form of dendrites under the treated surface.

Lavisse et al. [18] investigated the chemical composition of a cp Ti-grade-4 surface after industrial Nd:YAG laser irradiation. The energy density was in a range from 100 to 1,200 $J.cm^{-2}$. A titanium oxynitride TiN_xO_y layer with hexagonal symmetry and a thickness of around 2 μm was formed on the machined surface. They expected the formation of titanium oxynitrocarbides Ti (C, N, O).

Adams et al. [19] treated the surface of cp titanium grade-2 using a nanosecond pulsed-laser with different laser fluencies to grow titanium oxide/oxynitride coatings with various colour appearances according to their thickness. They found that the coatings consisted of three layers, namely, a thin TiO_2 upper coating, TiO middle coating with trace amounts of N and the bottom TiO_xN_{1-x} layer. The dependence of colour on the coating thickness was documented. The thinnest oxide (TiO) coating produced with low fluence was characterised by a gold colour.

Based on previous research, it can be concluded that the topic of laser ablation of pure titanium and its alloys has been studied intensively. However, there is still a lack of information about the laser ablation of graphite–Ti composites produced by powder metallurgy. This article studies the influence of input process parameters in two different atmospheres (air and argon) on chemical and phase composition of the composite ablated surfaces with an emphasis on oxide formation.

2. Experimental Setup

The recently developed powder metallurgy (PM)-processed TiMMC was micro-machined by a laser beam. The nanosecond pulse Yb-doped fibre laser (Lasertec 80Shape from DMG Mori GmbH, Germany) was employed for laser beam machining the graphite–Ti composite. This industrial-grade laser system has a 1,064 nm wavelength, a maximum pulse repetition rate of 100 kHz and 100 W maximal power.

Six square-shaped surfaces with a side length of 6 mm were irradiated by a laser beam. The experimental setup is shown in Figure 1. Within the experiment, cavities were ablated at three different energy densities E_D (0.10, 0.30 and 0.50 $J.mm^{-2}$) and three different lateral pulse distances D_L (0.5, 10 and 20 μm). The material was ablated in two layers with a crosshatching machining strategy, which is depicted in Figure 2.

IOP Conf. Series: Materials Science and Engineering **987** (2020) 012032 doi:10.1088/1757-899X/987/1/012032

Figure 1 Experimental setup and parts of the experiment equipment

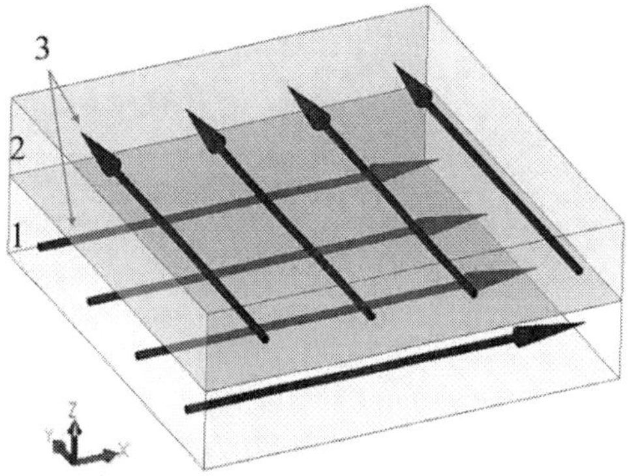

Figure 2 LBM crosshatching strategy scheme: 1 – first layer, 2 – second layer, 3 – laser beam tracks

Constant parameters of the laser beam machining were pulse duration at 120 ns, pulse diameter of 50 μm and transverse pulse distance D_T of 10 μm, which means transverse pulses overlapped O_T by 80%. A schematic representation of the pulse distances and pulse overlap is shown in Figure 3. When machining in a shielding atmosphere, the argon shielding gas flow rate was set at 10 l.min^{-1}. The laser beam machining parameters are listed in Table 1, where f is the pulse frequency and v_s is the laser beam scanning speed.

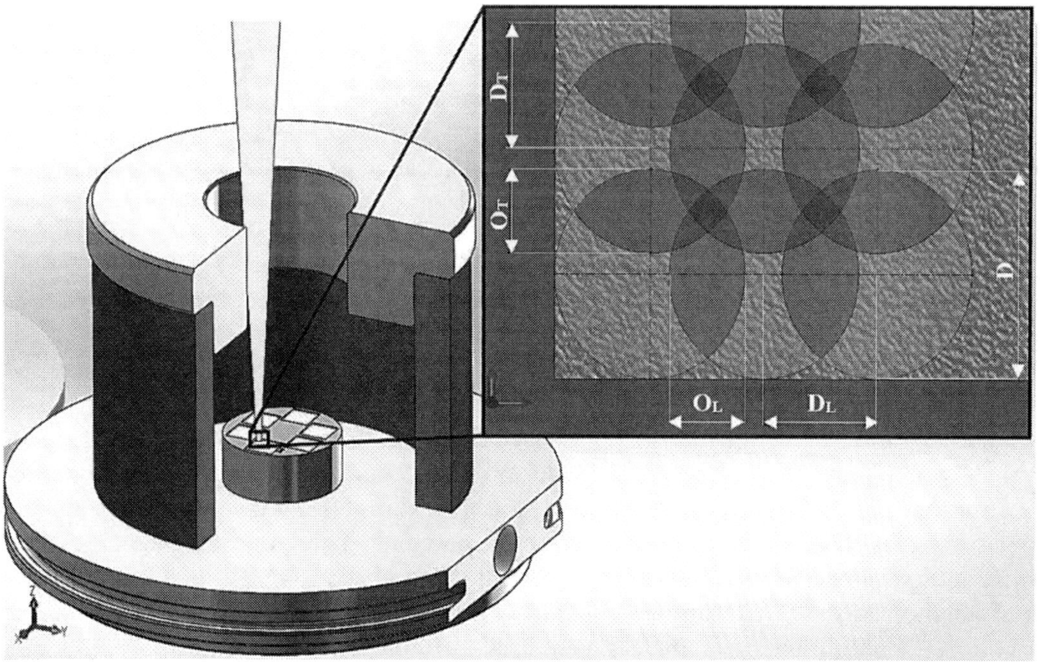

Figure 3 Schematic representation of pulse overlap: D – laser spot diameter, D_L – lateral pulse distance, O_L – lateral pulse overlap, D_T – transverse pulse distance, O_T – transverse pulse overlap

Table 1 Input parameters of laser beam machining the graphite–Ti composite samples

Sample	f [kHz]	v_s [mm.s^{-1}]	D_L [μm]	O_L [%]	E_D [J.mm^{-2}]
A10, A11	20	1000	50	0	**0.10**
A20, A21	20	1000	50	0	**0.30**
A30, A31	20	1000	50	0	**0.50**
B10, B11	100	50	**0.5**	99	0.50
B20, B21	100	1000	**10**	80	0.50
B30, B31	100	2000	**20**	60	0.50

2.1 Materials
The titanium matrix composite produced by powder metallurgy was used as experimental material. Commercially pure Ti powder with 15 vol. % graphite flakes were mixed in a Turbula mixer for 30 min and then compacted [20, 21]. The theoretical density of this mixture was 4.16 g.cm^{-3} when a graphite density of 2.2 g.cm^{-3} and titanium density of 4.506 g.cm^{-3} was considered. Then, the weight percentage was 92.07 wt. % for Ti and 7.93 wt. % for graphite. Cp Ti powder was fabricated using a hydride–dehydride (HDH) process (Kimet Special Metal Precision Casting Co. Ltd., China). A typical feature of this process is the presence of angular-shaped Ti particles below 32 μm in size. Graphite flakes with a purity of 99.9% and an average particle size of 16 μm were used as the reinforcement. The powder size distribution was determined using Fritch Analysette 22 laboratory equipment using wet dispersion. The obtained results were d_{50} = 24.9 μm and d_{90} = 46.3 μm for titanium powder and d_{50} = 5.6 μm and d_{90} = 13.9 μm for graphite flakes. Cold isostatic pressing (at 200 MPa) followed by a hot vacuum press at working temperatures in the range of 450–470 °C and a pressure of 500 MPa were used for compacting. The final sample density was in the range of 4.1–4.15 g.cm^{-3} and the porosity of the finished compact was 2.44 % ± 0.15 % [22].

2.2 Methods
Before laser ablation, the samples were prepared in a standard metallographic procedure that consisted of cutting with the precision saw, hot mounting in conductive resin and grinding up to P1200 (15.3 μm) emery paper. To study the topography of laser-treated surfaces, a JEOL JSM 7600F high-resolution scanning electron microscope (SEM) was utilised. The ablated surfaces were observed in a secondary electron imaging regime (SEI) with the following parameters: U = 15 keV, I = 1.0 nA and WD = 15 mm. The chemical composition was determined with an Oxford Instruments Inca X-Max 50 mm^2 energy-dispersive X-ray spectroscope (EDS) operated at the same parameters. Three measurements of chemical composition were also performed for each sample. The phase composition of the surfaces after laser ablation was studied using a Brucker D8 diffractometer (XRD) with the Cu anode (λ = 1.5406 Å). The following parameters were used during the measurements: U = 40 kV and I = 30 mA. Diffraction patterns were taken in a range from 15° to 105° 2θ with a step size of 0.05°. Finally, the influence of input parameters (laser pulse distance and energy density) and shielding gas (Ar) on the oxygen content presented on the laser-ablated surfaces was evaluated with statistical one-way analysis of variance (one-way ANOVA) followed by Welch's tests using Minitab v.17 software. In this case, the level of significance at 95% (α = 0.05) was chosen.

IOP Conf. Series: Materials Science and Engineering **987** (2020) 012032 doi:10.1088/1757-899X/987/1/012032

3. Results

Table 2 illustrates the chemical composition of graphite–Ti surfaces after laser ablation at different input parameters, namely, energy densities (0.10, 0.30 and 0.50 J.mm^{-2}) and lateral pulse distances (0.5, 10 and 20 μm). Additionally, the influence of air and the protective argon atmosphere was also evaluated. The individual samples were designated in the following way: A1–A3 samples were processed with a different pulse density and B1–B3 samples were treated with different values of lateral pulse distances. The latter digit distinguishes whether the sample was ablated in the air (digit 0) or the shielding Ar (digit 1). Not surprisingly, in most cases, higher oxygen content up to 38.6 ± 0.02 wt. % was documented after the laser ablation without shielding Ar (in the air), which is caused by weak protection of the processed surfaces. Furthermore, the change in energy densities and lateral pulse distances had a strong influence on the oxygen content. As energy density increased from 0.10 to 0.50 J.mm^{-2}, the oxygen content increased from 18.3 ± 0.07 wt. % to 28.1 ± 0.03 wt. % for the samples ablated in the air. Meanwhile, the Ti content decreased from 73.3 ± 0.23 wt. % to 62.4 ± 0.38 wt. % and the C content slightly increased from 8.3 ± 0.23 wt. % to 9.5 ± 0.38 wt. %. When the samples were the same but ablated using shielding Ar, the increase in oxygen content with energy density was less significant (~7 wt. %). As the energy density increased, the oxygen content first increased from 17.2 ± 0.02 wt. % to 24.2 ± 0.47 wt. % and then decreased to 22.5 ± 0.59 wt. %. The Ti and C contents slightly decreased from 72.5 ± 0.21 and 10.3 ± 0.22 wt. % to 68.8 ± 0.72 and 8.8 ± 0.14 wt. %, respectively. Different behaviour was observed in the samples treated with different lateral pulse distances. In general, oxygen content decreased with an increasing lateral pulse distance after processing in the air and Ar atmosphere from 38.6 ± 0.02 wt. % and 23.2 ± 0.41 wt. % to 23.7 ± 0.04 wt. % and 15.4 ± 0.92 wt. %, respectively. Vice versa, the drop in C content was observed for B samples processed in the air (from 15.8 ± 0.29 wt. % to 11.7 ± 0.45 wt. %) and the Ar atmosphere (from 9.5 ± 2.76 wt. % to 7.9 ± 0.24 wt. %). It was possible to adjust the oxygen content by varying the energy density and lateral pulse distances during laser ablation.

Table 2 Chemical composition of graphite–Ti composite as a function of laser input parameters and type of atmosphere

	Air				Ar shielding		
Sample	**Ti ± St. Dev.**	**C ± St. Dev.**	**O_2 ± St. Dev.**	**Sample**	**Ti ± St. Dev.**	**C ± St. Dev.**	**O_2 ± St. Dev.**
A10	73.3±0.23	8.3±0.23	18.3±0.07	A11	72.5±0.21	10.3±0.22	17.2±0.02
A20	70.8±0.11	7.5±0.13	21.8±0.03	A21	65.9±0.64	9.8±0.23	24.2±0.47
A30	62.4±0.38	9.5±0.38	28.1±0.06	A31	68.8±0.72	8.8±0.14	22.5±0.59
B10	45.6±0.29	15.8±0.29	38.6±0.02	B11	67.3±3.13	9.5±2.76	23.2±0.41
B20	64.4±0.13	9.7±0.14	25.9±0.02	B21	68.5±1.99	10.3±0.41	21.2±1.64
B30	64.6±0.41	11.7±0.45	23.7±0.04	B31	76.7±1.03	7.9±0.24	15.4±0.92

Graphically, the amount of oxygen is shown in Figure 4 for A samples and in Figure 5 for B samples.

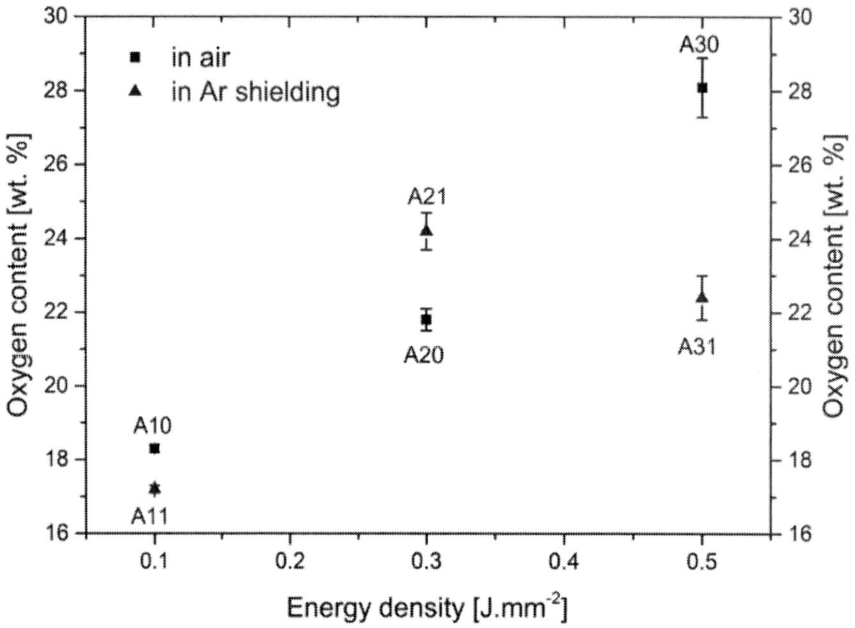

Figure 4 Oxygen content on the surface of the graphite–Ti composite after laser ablation as a function of energy density (0.10, 0.30 and 0.50 J.mm^{-2}) at a constant lateral pulse distance (50 μm)

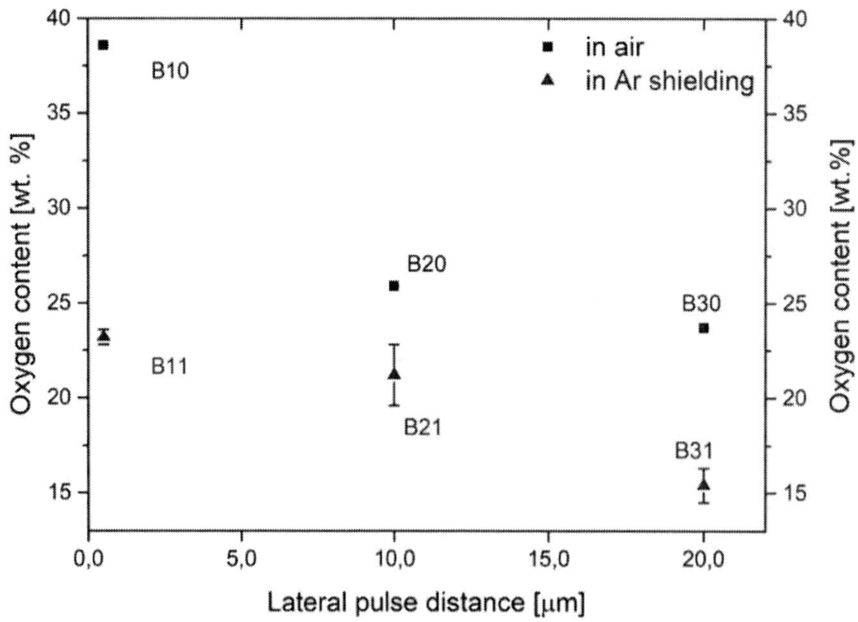

Figure 5 Oxygen content on the surface of the graphite–Ti composite after laser ablation as a function of lateral pulse distance (0.5, 10 and 20 μm) at a constant energy density (0.50 J.mm^{-2})

Figure 6 shows the EDS spectra of laser-ablated surfaces in air and Ar of the A2 and B3 samples accompanied by their surface morphologies. It is clear that the shielding of the samples with Ar gas had no significant impact on their surface morphology.

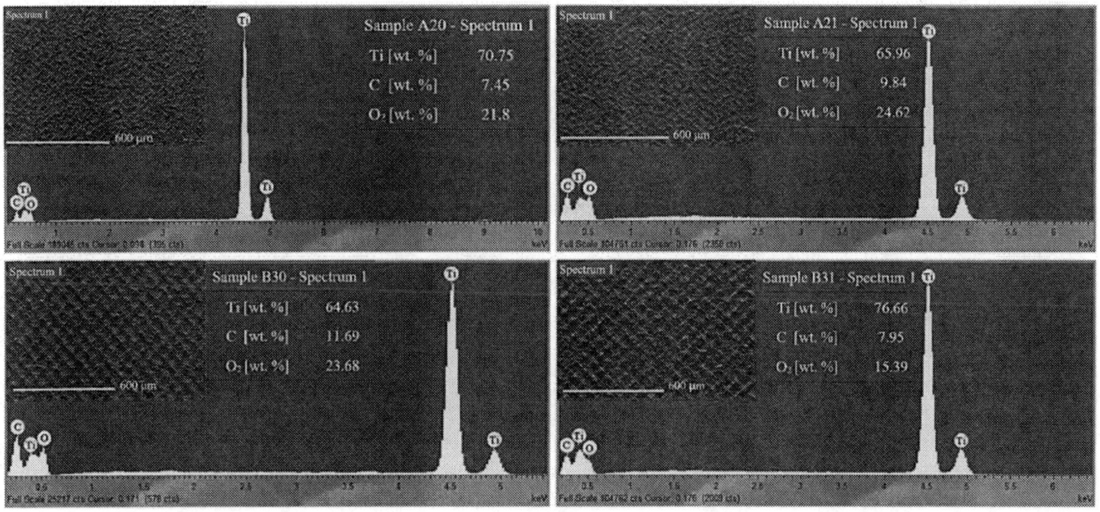

Figure 6 Results of EDS analysis for A-group samples machined at a lateral pulse distance of 50 μm and energy density of 0.30 $J.mm^{-2}$: A20 (machined in air), A21 (machined in Ar) and B-group samples machined at a lateral pulse distance of 20 μm and energy density of 0.50 $J.mm^{-2}$: B30 (machined in air), B31 (machined in Ar)

Figure 7 shows the XRD patterns of the B1 sample before and after laser ablation in air and Ar. The diffraction pattern of the unprocessed sample evicted multiple orientations of (010), (002), (011), (012), (110), (013), (020), (112) and (021) of α-Ti crystal planes (ICSD 98-005-2522). The phase belongs to the hexagonal system (P63/mmc) with the most intense (011) plane at 40.2° 2θ and unit cell parameters of a = b = 2.951 Å, c = 4.686 Å and α = β = 90°, γ = 120°. After laser ablation in shielding Ar, Ti_2O_3 and TiO oxides were detected on the surface of the sample. The TiO oxide (ICDD 00-002-1196) is characterised by a cubic crystal structure with the space group Fm-3m and following unit cell parameters a = b = c = 4.235 Å and α = β = γ = 90°. Compared to TiO oxide, Ti2O3 (ICDD 01-071-1047) belongs to the rhombohedral crystal system (R-3c, a = b = 5.125 Å and c = 13.957 Å, α = β = 90° and γ = 120°). The shielding Ar did not protect the surface of the sample against oxidation. The same types of oxides were also observed after laser ablation in air, without Ar protection. Furthermore, the presence of tetragonal rutile (P42/mnm) TiO_2 oxide (ICDD 01-072-1148) was documented (a = b = 4.594, c = 2.959 Å and α = β = γ = 90°). However, the diffraction patterns of the above-mentioned phases overlap and their precise identification is, therefore, difficult.

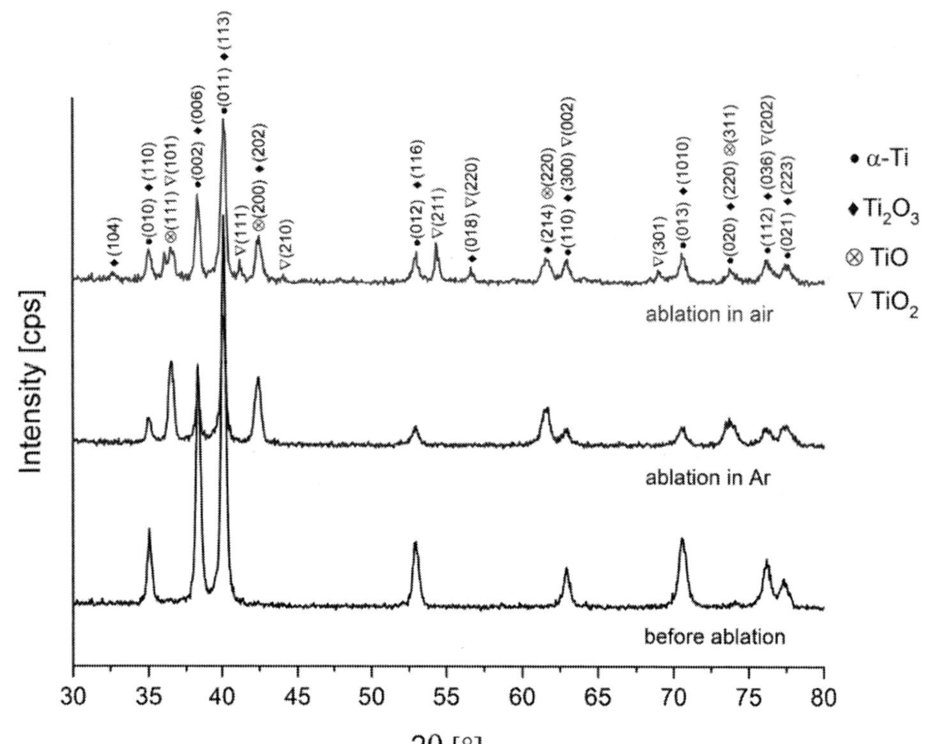

Figure 7 X-ray diffraction patterns of the unmachined sample surface and sample B10 machined in air and B11 machined in Ar shielding atmosphere

The amount of oxygen on the surface of the samples was statistically evaluated by one-way ANOVA – Welch's test. Statistical analysis confirmed that the amount of oxygen was statistically significantly different between samples machined in air and argon.

Table 3 One-way ANOVA for groups of samples machined in air and Ar

Samples	DF	F-value	p-value	R^2
A10 – A11	1	381.73	0.001*	98.96%
A20 – A21	1	53.90	0.018*	93.09%
A30 – A31	1	182.81	0.005*	97.86%
B10 – B11	1	2820.01	0.000*	99.86%
B20 – B21	1	58.14	0.017*	93.56%
B30 – B31	1	163.64	0.006*	97.61%

*$p < 0.05$ means are different for significance level $\alpha = 0.05$

4. Discussion

After the laser ablation of samples in the air environment (A10–A30 and B10–B30), a large amount of oxygen (up to 38.6 ± 0.02 wt. %) was detected on their surfaces. The oxygen content increased along with energy density. This could be explained by the phenomenon of the recoil pressure of laser-induced plasma, which improves the diffusion of atoms in reactive gas (oxygen) to interstitial positions. The influence of laser radiation during ablation causes the dissociation of molecular oxygen (O_2) to atomic oxygen (O). The titanium surface has a better adsorption capacity of atomic oxygen compared to molecular oxygen. It resulted in enhanced diffusion and, therefore, higher oxygen content on the surface of the samples processed in air [23]. In contrast, the oxygen content on the surface of samples B10–B30 decreased with the increase in lateral pulse distance. The heat generated by overlapping laser pulses improves oxygen diffusion into the sample surface efficiently. Oxygen diffusion across the grain boundaries makes them distinct and wider. This effect also supports the formation of oxides on the laser-ablated surfaces. From that point of view, it could be stated that the shorter the laser pulse distance the more heat is generated on the sample surface and the conditions are better for the formation of oxides. So, increasing the lateral pulse distance causes a decrease in oxygen content [24]. Surprisingly, the presence of oxygen up to 24.2 ± 0.47 wt. % was also documented on the surfaces of the samples after laser ablation in the shielding Ar (A11–A31 and B11–B13). This unexpected effect could be easily explained by the insufficient flow of the shielding gas into the molten pool during laser ablation. Additionally, blowing the shielding Ar onto the sample surface had a negative impact on oxygen diffusion and, thus, the growth of oxide layers because of its cooling effect.

Following previous results, the character of the surface morphology was strongly affected by laser ablation parameters, including laser power, pulse frequency and scanning speed [20]. However, the influence of Ar shielding gas on the surface morphology of laser-processed samples was not observed. The absence of diffraction peaks corresponding to carbon that could confirm the presence of graphite after laser ablation of the graphite–Ti composite (sample B1) was confirmed by XRD. This phenomenon is related to the evaporation of graphite during ablation. However, Hu et al. observed a XRD pattern of graphite-Ti composite after laser sintering. They found a narrow peak at 43.3° 2θ that corresponds to (101) diffracting plane of multilayer graphene. This means that graphene existed in the composites even after laser sintering [25]. In contrast, the presence of carbon up to ~ 15 wt. % was observed by EDS analysis after laser ablation in the Ar atmosphere. Regardless of the input parameters and use of shielding gas, carbon was detected in all cases. The amount of carbon ranged from its minimal value of 7.5 ± 0.13wt. % for the A20 sample treated in the air to the maximal content of 15.8 ± 0.29 wt. % for the B11 sample processed in the shielding gas. This suggests that carbon could be present on the surface of laser-treated samples in the form of amorphous graphite undetectable by XRD.

Two types of oxides were detected on the surface of samples after laser ablation in the Ar shielding gas, namely, cubic TiO (ICDD 00-002-1196) and rhombohedral Ti_2O_3 (ICDD 01-071-1047). TiO and Ti_2O_3 oxides have a lower oxygen content, 50 at. % and 60 at. %, respectively. The formation of these oxides took place even after laser ablation in insufficient Ar shielding gas. The level of oxygen present in the environment during laser ablation in the air (without the Ar atmosphere) enabled the formation of oxygen-rich (~67 at. %) tetragonal rutile-type TiO_2 oxide (ICDD 01-072-1148). The diffracting planes of TiO_2 oxide (111), (210), (211) and (301) were clearly distinguished at Bragg angles of 41.2°, 44.0°, 54.3° and 68.9° 2θ, respectively, in the XRD pattern of the sample after laser treatment in the air. In vivo, Hanawa et al. grew a passive film that was mostly based on TiO_2 oxide with a small amount of Ti_2O_3 and TiO oxides. They found that the presence of Ti_2O_3 and TiO oxides depends on the local microenvironment [26]. It was found that the biocompatibility of titanium can be improved by applying a coating covered with bioactive TiO_2 oxide and by using micro- and nanoscale structures [27].

5. Conclusion

In this research, graphite–Ti metal matrix composite produced via low-temperature powder metallurgy was ablated by laser beam in two different atmospheres: air and argon shielding gas. Based on the results of the experiments, the following conclusions can be drawn:

(1) In all cases, the amount of oxygen on the surface decreased when machining in an argon shielding atmosphere.
(2) TiO, TiO_2 and Ti_2O_3 oxides with α-Ti were detected on the machined surface in air.
(3) TiO and Ti_2O_3 with α-Ti were detected on the surface after laser treatment in the argon shielding gas.
(4) Statistical analysis showed significant differences in the amount of oxygen on the surface after machining in air and argon.
(5) Based on the experiment results, it is possible to adjust the oxygen content by both input process parameters and shielding atmosphere. Obviously, it is possible to adjust the oxygen content by varying the energy density and lateral pulse distances.

Acknowledgements

This article was supported by project VEGA 2/0135/20 titled 'Coating of powder metallurgical titanium using electromagnetic radiation under working atmosphere, a study of microstructure and coating properties. Furthermore, the collective authors would like to thank Dr Peter Švec of the Slovak Academy of Sciences at the Institute of Physics for performing X-Ray diffraction analysis.

References

[1] Asri R I M, Harun W S W, Samykano M, Lah N A C, Ghani S A C, Tarlochan F and Raza M R 2017 Corrosion and surface modification on biocompatible metals: A review *Mater. Sci. Eng. C* **77** 1261–74
[2] Kaur M and Singh K 2019 Review on titanium and titanium based alloys as biomaterials for orthopaedic applications *Mater. Sci. Eng. C* **102** 844–62
[3] Muley S V, Vidvans A N, Chaudhari G P and Udainiya S 2016 An assessment of ultra fine grained 316L stainless steel for implant applications *Acta Biomater.* **30** 408–19
[4] Chen Q and Thouas G A 2015 Metallic implant biomaterials *Mater. Sci. Eng. R Reports* **87** 1–57
[5] Su Y, Luo C, Zhang Z, Hermawan H, Zhu D, Huang J, Liang Y, Li G and Ren L 2018 Bioinspired surface functionalization of metallic biomaterials *J. Mech. Behav. Biomed. Mater.* **77** 90–105
[6] Hatamleh M M, Wu X, Alnazzawi A, Watson J and Watts D 2018 Surface characteristics and biocompatibility of cranioplasty titanium implants following different surface treatments *Dent. Mater.* **34** 676–83

[7] Uhlmann E, Schweitzer L, Kieburg H, Spielvogel A and Huth-Herms K 2018 The Effects of Laser Microtexturing of Biomedical Grade 5 Ti-6Al-4V Dental Implants (Abutment) on Biofilm Formation *Procedia CIRP* **68** 184–9

[8] Elias C N, Oshida Y, Lima J H C and Muller C A 2008 Relationship between surface properties (roughness, wettability and morphology) of titanium and dental implant removal torque *J. Mech. Behav. Biomed. Mater.* **1** 234–42

[9] Wang Y, Yu Z, Li K and Hu J 2020 Effects of surface properties of titanium alloys modified by grinding, sandblasting and acidizing and nanosecond laser on cell proliferation and cytoskeleton *Appl. Surf. Sci.* **501** 144279

[10] Wennerberg A and Albrektsson T 2009 Effects of titanium surface topography on bone integration: A systematic review *Clin. Oral Implants Res.* **20** 172–84

[11] Andrukhov O, Huber R, Shi B, Berner S, Rausch-Fan X, Moritz A, Spencer N D and Schedle A 2016 Proliferation, behavior, and differentiation of osteoblasts on surfaces of different microroughness *Dent. Mater.* **32** 1374–84

[12] Rupp F, Liang L, Geis-Gerstorfer J, Scheideler L and Hüttig F 2018 Surface characteristics of dental implants: A review *Dent. Mater.* **34** 40–57

[13] Zafar M S, Farooq I, Awais M, Najeeb S, Khurshid Z and Zohaib S 2019 *Bioactive Surface Coatings for Enhancing Osseointegration of Dental Implants* (Elsevier Ltd.)

[14] Moura C G, Carvalho O, Gonçalves L M V, Cerqueira M F, Nascimento R and Silva F 2019 Laser surface texturing of Ti-6Al-4V by nanosecond laser: Surface characterization, Ti-oxide layer analysis and its electrical insulation performance *Mater. Sci. Eng. C* **104** 109901

[15] Zeng C, Wen H, Zhang B, Sprunger P T and Guo S M 2020 Diffusion of oxygen and nitrogen into titanium under laser irradiation in air *Appl. Surf. Sci.* **505** 144578

[16] Braga F J C, Marques R F C, Filho E de A and Guastaldi A C 2007 Surface modification of Ti dental implants by Nd:YVO 4 laser irradiation *Appl. Surf. Sci.* **253** 9203–8

[17] Rajab F H, Liauw C M, Benson P S, Li L and Whitehead K A 2017 Production of hybrid macro/micro/nano surface structures on Ti6Al4V surfaces by picosecond laser surface texturing and their antifouling characteristics *Colloids Surfaces B Biointerfaces* **160** 688–96

[18] Lavisse L, Jouvard J M, Imhoff L, Heintz O, Korntheuer J, Langlade C, Bourgeois S and Lucas M C M de 2007 Pulsed laser growth and characterization of thin films on titanium substrates *Appl. Surf. Sci.* **253** 8226–30

[19] Adams D P, Murphy R D, Saiz D J, Hirschfeld D A, Rodriguez M A, Kotula P G and Jared B H 2014 Nanosecond pulsed laser irradiation of titanium: Oxide growth and effects on underlying metal *Surf. Coatings Technol.* **248** 38–45

[20] Šugár P, Ludrovcová B, Kováčik J, Sahul M and Šugárová J 2020 Laser-based ablation of titanium-graphite composite for dental application *Materials (Basel).* **13**

[21] Šugár P, Kováčik J, Ludrovcová B, Šugárová J and Sahul M 2019 Laser Beam Machining of Graphite Reinforced Ti Metal Matrix Composite *Dev. Mach. Technol.* **9** 102–12

[22] Balog M, Viskic J, Krizik P, Schauperl Z, Snajdar M, Stanec Z and Catic A 2016 CP Ti fabricated by low temperature extrusion of HDH powder: Application in dentistry *Key Eng. Mater.* **704** 351–9

[23] Kalsoom U i., Bashir S and Ali N 2013 SEM, AFM, EDX and XRD analysis of laser ablated Ti in nonreactive and reactive ambient environments *Surf. Coatings Technol.* **235** 297–302

[24] Jelani M, Bashir S, Rehman M K U, Ahamad R, Faizan-Ul-Haq, Yousaf D, Akram M, Afzal N, Chaudhry M U, Mahmood K, Hayat A and Ahmad S 2013 Effect of laser fluence on surface, structural and mechanical properties of Zr after irradiation in the ambient environment of oxygen *Eur. Phys. J. D* **67** 159

[25] Hu Z, Chen F, Xu J, Ma Z, Guo H, Chen C, Nian Q, Wang X and Zhang M 2018 Fabricating graphene-titanium composites by laser sintering PVA bonding graphene titanium coating: Microstructure and mechanical properties *Compos. Part B Eng.* **134** 133–40

[26] Hanawa T, Asami K and Asaoka K 1998 Repassivation of titanium and surface oxide film regenerated in simulated bioliquid *J. Biomed. Mater. Res.* **40** 530–8

[27] Jäger M, Jennissen H P, Dittrich F, Fischer A and Köhling H L 2017 Antimicrobial and osseointegration properties of nanostructured titanium orthopaedic implans *Materials (Basel).* **10** 1–28

AUTHOR INDEX